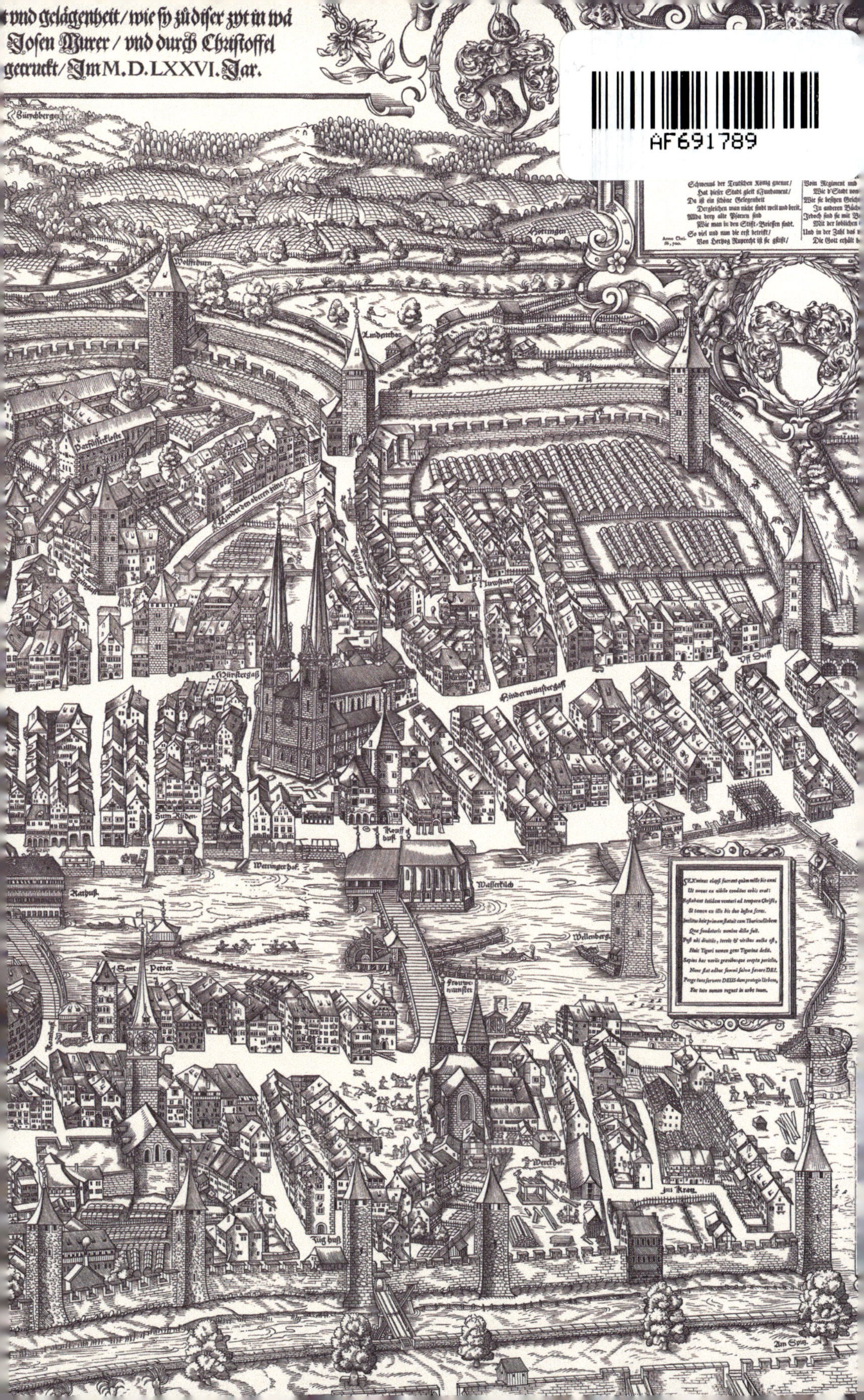
vnd gelägenheit / wie sy zu diser zyt in wä
Josen Murer / vnd durch Christoffel
getruckt / Im M.D.LXXVI. Jar.
Rathuß
Sant Peter
Wasserkilch
Wellenberg

AF691789

Conrad Gessner (1516–1565)

Medieval and Early Modern Philosophy and Science

Editors

C.H. Lüthy (*Radboud University*)

Editorial Consultants

Joël Biard (*University of Tours*)
Jürgen Renn (*Max-Planck-Institute for the History of Science*)
Theo Verbeek (*University of Utrecht*)

VOLUME 38

The titles published in this series are listed at *brill.com/memps*

Conrad Gessner (1516–1565)

Universal Scholar and Natural Scientist of the Renaissance

By

Urs B. Leu

Translated by

Bill C. Ray

BRILL

LEIDEN | BOSTON

This publication has been made possible by the generous support of the Max-Geilinger-Stiftung and the University of Zürich.

Max Geilinger-Stiftung

1962 in Zürich zur Förderung und Verbreitung der Werke des Zürcher Lyrikers und Dramatikers Max Geilinger (1884–1948) errichtet.

Universität Zürich[UZH]

Originally published in German in 2016 as *Conrad Gessner (1516–1565): Universalgelehrter und Naturforscher der Renaissance* by NZZ Libro, Schwabe Verlagsgruppe AG, Basel, Switzerland.

Cover illustration: Rhinoceros, in: Gessner (1551), p. 953, based on Albrecht Dürer's illustration. ZBZ, Abteilung Alte Drucke und Rara: NNN 41.
Flyleaf illustration: Detail from the map of Zürich, known as the Murer-Plan of 1576, showing the Grossmünster church, where the Hohe Schule was located, and Gessner's house. ZBZ, Abteilung Karten und Panoramen: 4 Lb 02:5.

The Library of Congress Cataloging-in-Publication Data is available online at https://catalog.loc.gov
LC record available at https://lccn.loc.gov/2023002130

Typeface for the Latin, Greek, and Cyrillic scripts: "Brill". See and download: brill.com/brill-typeface.

ISSN 2468-6808
ISBN 978-90-04-42126-4 (hardback)
ISBN 978-90-04-54169-6 (e-book)

Copyright 2023 by Koninklijke Brill NV, Leiden, The Netherlands.
Koninklijke Brill NV incorporates the imprints Brill, Brill Nijhoff, Brill Hotei, Brill Schöningh, Brill Fink, Brill mentis, Vandenhoeck & Ruprecht, Böhlau, V&R unipress and Wageningen Academic.
All rights reserved. No part of this publication may be reproduced, translated, stored in a retrieval system, or transmitted in any form or by any means, electronic, mechanical, photocopying, recording or otherwise, without prior written permission from the publisher. Requests for re-use and/or translations must be addressed to Koninklijke Brill NV via brill.com or copyright.com.

This book is printed on acid-free paper and produced in a sustainable manner.

Contents

Preface

The occasion for this biography is the 500th birthday of Zürich's universal scholar, Conrad Gessner: natural scientist, physician, father of the bibliography, author of Switzerland's first printed pharmacological handbook, founder of zoology and discoverer of altitudinal botanical layering. The last Gessner biography that was intended for general reading in German was written by Pastor Johannes Hanhart and published in 1824 in Winterthur.[1] Unfortunately, for various reasons, the polyhistor has become forgotten in his own homeland, and surprisingly, to an even greater degree than in other countries. Since Gessner corresponded and published almost exclusively in Latin, fewer and fewer people are able to read his texts. And, once an author is no longer read, he will fade from memory and become the subject of myth. A goal of the jubilee year is to revitalize the memory of this important and great man in Swiss scientific history, and to stimulate more far-reaching research, starting with his correspondence and critical editions of selected works.

Gessner is not only an interesting personality, or a kind of Leonardo da Vinci of the Swiss Confederation; he is also representative of an epochal threshold, which has been described with the terms humanism, Renaissance and Reformation, and is seen as the transition from the Middle Ages to the modern era. He was a man who belonged to all three of these intellectual-historical movements. As a humanist, he studied and edited antique classics with a special interest in little-known Greek texts. As a devout reformed Protestant, he interpreted nature as the second book – after the Bible – of God's revelation. In this way, compared to his somewhat uncomprehending contemporaries, he legitimized his still largely unknown research and study of nature. As a Renaissance scholar, he actively participated in the study of ancient sources, above all Aristotle, Hippocrates, and Galen; and he promoted the empirical expansion of the medical and natural sciences. He contributed to the development of various fields of expertise, disentangling them from their astrological, superstitious and occult origins to follow the path of inductive methodology. According to his own words, he was aware of living in interesting times, which were especially marked by the rebirth (Renaissance) of the sciences, which is

1 Ley (1929) was published in the journal *Münchener Beiträge zur Geschichte und Literatur der Naturwissenschaften und Medizin* and Fischer (1966) as *Neujahrsblatt der Naturforschenden Gesellschaft in Zürich*, which was primarly sold by the Natural History Society Zürich.

the reason why this association is mentioned in the title of this book.[2] It was a rebirth that led to Europe overtaking China and the Islamic world, with the West subsequently setting the pace for progress in technology and the natural sciences. Up until the fifteenth century, it was well known that the Chinese and Islamic cultures had been more advanced than the European.[3] Two centuries later, the tables had been turned. China had isolated itself during the fifteenth century and fell back on traditional values, while the Islamic world stopped all maritime, economic and technological development after the sea battle of Lepanto in 1571.[4]

This book is primarily based on Gessner's correspondence, introductory dedications and his works, as well as additional sources from the sixteenth century. While the availability of sources is adequate, it is not optimal, which is why this volume features more about the work and less about the life of Zürich's universal scholar. With regard to the spelling of Conrad Gessner's name, this version was chosen because it is the way he most frequently spelled it himself in surviving German-language documents. Occasionally, one encounters "Konrad Gessner," and "Conradus Gesnerus" is clearly the Latin form.[5] Wherever possible, biographical data is provided for other persons mentioned in the text. When available, it appears at the first mention of the individual. At the request of the publisher, we have kept footnotes brief, and they nearly always contain only documentation for individual remarks. References to related literature have been left out, for which I must apologize to various colleagues who may also find that some works are not included in the bibliography.

Special thanks go to Prof. Dr. Susanna Bliggenstorfer, former Director of the Zürich Central Library, and Professor Dr. Lukas Keller, Director of the Zoological Museum of the University of Zürich, who have supported various Gessner anniversary projects in both word and deed. I also owe a great debt of gratitude to PD Dr. Reinhard Bodenmann and Rainer Henrich, lic. theol., for the many helpful clues to the Bullinger and Myconius correspondence they provided. Thus, they are mentioned here and not in each related footnote. I also followed their advice in dating the corresponding letters. Thanks also go to Gertraud Gamper, lic. phil., Dr. Rudolf Gamper and Dr. Katharina Züger, who agreed to tackle the manuscript and make corrections. Finally, I would

2 Gessner (1548), f. 107r. He writes there that the Venetian printer Aldo Manuzio (1449–1515) appeared together with the reborn sciences (*renascentibus studiis*). Gessner, who was born a generation later, was thus a child or contemporary of this renaissance of the sciences. On the rebirth (renaissance) of botany, cf. Gessner's preface in: Bock (1552), f. a_{viii}v.

3 Montgomery/Kumar (2016), pp. 166–252.

4 Menzel (2015), pp. 232–236 and 421–428; Diner (2016).

5 Pyle (2000).

also like to give thanks to the Gessner's Family fund, and especially Urs and Dorothea Gessner, who largely financed the German edition of this biography.

The initiative for the English version came from my dear colleague Arjan van Dijk (Brill), for I alone would not have had the courage to undertake a translation. I also extend heartfelt thanks to the Max Geilinger Foundation in Zürich and to the University of Zürich (especially Rector Prof. Dr. Michael Schaepman), who were convinced of the project from the very first moment and have supported it with a considerable contribution. Naturally, I also wish to thank Bill C. Ray (Juneau, Alaska) for a pleasant intercontinental collaboration in the translation of a challenging text into easily readable English. A big thank you also goes to the editor of the series, Prof. Dr. Christoph Lüthy, who thoroughly proofread the English translation. The English edition has been slightly revised to take into account the literature published up to 2022.

Zürich, October 2022

Abbreviations

AK	Amerbachkorrespondenz, ed. by Alfred Hartmann and Beat R. Jenny, 11 vols., Basel 1942–2010.
BNF	Bibliothèque nationale de France, Paris
CdR	Aimé-Louis Hérminjard (ed.): Correspondance des Réformateurs dans les pays de langue française, 9 vols. Genève and Paris, 1866–1897.
CThB	Correspondance de Théodore de Bèze, recueillie par Hippolyte Aubert, publiée par F. Aubert, H. Meylan and A. Dufour, Genève, 1960ff.
HBBW	Heinrich Bullinger Briefwechsel, ed. by Hans Ulrich Bächtold et al., 20 vols., Zürich 1973–2021.
ÖNB	Österreichische Nationalbibliothek, Vienna
Regest Freuler	Summaries of letters, made by Pastor Freuler, 1862, in: Stadtbibliothek Winterthur, MS. 4° 197a
StAGR	Staatsarchiv Graubünden, Chur
StAZH	Staatsarchiv Zürich
St Gallen KB	Kantonsbibliothek St. Gallen (Vadiana)
VD 16	Verzeichnis der im deutschen Sprachbereich erschienenen Drucke des 16. Jahrhunderts (online: https://opacplus.bib-bvb.de/TouchPoint_touchpoint/start.do?SearchProfile=Altbestand&SearchType=2)
ZBZ	Zentralbibliothek Zürich

Illustrations and Tables

Illustrations

Tables

CHAPTER 1

Early Years

Conrad Gessner was born March 16, 1516[1] in Zürich to the furrier Urs Gessner (†1531) and his wife Agathe Frick (1484–1564).[2] Conrad's grandfather, the master goldsmith Hans Gessner from Nuremberg, became a citizen of Solothurn in 1480, where his two sons Urs and Andreas were born. The latter moved to Zürich, where he was granted citizenship in 1504. Urs followed him in 1511, and for the sum of three Rhenish guilders, also became a citizen. The marriage to Agathe Frick produced five sons and three daughters: Jakob (1508), Küngold (1510), Barbara (1511), Emil (1514), Conrad (1516), Felix (1518), Elisabetha (1520) and Diethelm (1528). While today a flock of eight children might be seen as large family, it was not unusual at the time. Urs's brother Andreas had nine sons and ten daughters with three wives. Among them were Andreas (*1513) and Jakob Gessner (*1527), who made a name for themselves in Zürich as book printers[3] who produced various works by their famous cousin.[4]

By the tender age of about five, Conrad had already left home to be raised by his mother's uncle, Johannes Frick, chaplain of the Grossmünster church. This was due to the meager income of his father, who was barely able to feed his family. One should keep in mind that the economic situation at the time was anything but rosy for the approximately 7,000 inhabitants of the city and the approximately 60,000 in the surrounding countryside. The Valaisan, Thomas Platter the Elder (1499–1582), bemoaned that after extensive travels through Meissen, Silesia, Poland and Hungary, his return to Zürich in 1523 initially meant having to make a living as an unskilled laborer: "So, I returned to Zürich

1 Gessner writes in a letter to Crato von Krafftheim of March 26, 1564, which has survived only in print, that today is his birthday (Gessner [1577], f. 22v). According to his will (Hanhart [1824], p. 285) he was born on Palm Day 1516 (= March 16). It is obvious to consider the 26th of March as a scribal error, which Crato von Krafftheim erroneously repeated in his letter to Gessner of April 27, 1564.

2 The dates of her life are derived from a letter that Conrad Gessner addressed to Theodor Zwinger in Basel on April 7, 1564. In it, he wrote that his mother had died that day at the age of 80. Cf. Gessner (1577), f. 107r.

3 Vischer (1990), pp. 423–462.

4 Another notable descendant of this branch of the family is the well-known goldsmith Abraham Gessner (1552–1613), who died as Schwenckfeldian in Stühlingen in 1613. The children of his son Isaak (*1581) emigrated to Moravia, which offered a new home to various free church groups. Cf. the family registers of the Gessner family in the "Familienarchiv Gessner" in the ZBZ (shelf marks: FA Gessner 1 and 9).

© KONINKLIJKE BRILL NV, LEIDEN, 2023 | DOI:10.1163/9789004541696_002

and stayed with an old woman named Adelheit the hat maker. She usually had five or six girls at her place. They had their own journeyman, who attended to them, but I did not like his character. But I had a good journeyman who was quite skilled, and we had our own small room; I did not care much for the other journeymen. But God knows, I often went hungry and some days there was not a morsel of bread to eat. Then, I would put some water in a pan; ask the women for a little salt; put it in the water and drink it up to quell my hunger. Every week, I had to pay a Zürich schilling for the rent. Then, I would sometimes help the people in the fields, where I got a *batzen* coin per mile, so that I could pay my rent to the women. I also helped to carry wood or did other work, and I was given something to eat. Then I was quite happy and satisfied."[5]

The former Franciscan monk Conrad Pellikan (1478–1556), who began teaching Hebrew in Zürich in 1526, also remembers his early days with the following: "In the monastery, I was a confessor of sincere poverty, but indeed never suffered from deprivation. With regard to money, I was truly a poor man with barely the necessities. Nonetheless, I shared all that I possessed, and my poor housemates, to whom I extended ever more invitations, would share with me as well. First and foremost were two young people, Johannes Fries and Sebastian Faber [Guldinbeck]. I liked their attitude and enthusiasm. Once, when I visited the Greek school, it was a great pleasure to hear them both translate the Gospels from Latin into Greek and explain the grammar they applied. I invited them alternately to take the midday meal with me."[6]

Given the economic importance of Zürich today, it is easy to forget that, in terms of trade and banking, the city was a late bloomer compared to Basel, Geneva and St Gallen. Furthermore, due to purchases of its hinterland, it was deep in debt from the beginning of the fourteenth century into the sixteenth century. The Old Zürich War (1440–1450) also played a role by bringing foreign trade to a virtual standstill, with the result that no impression remains of outstanding merchants from around 1500. Furthermore, at the beginning of the Reformation century, the wool and linen weavers were also limited to producing for the local market. Finally, due to Zwingli's sermons against the activities of Swiss mercenaries, this important source of income was also eliminated, so that financial difficulties were now part of everyday life (ill. 1).[7]

5 Platter (1999), p. 69.

6 Pellikan (1892), p. 106.

7 Lendemann (1996), pp. 131–154.

ILLUSTRATION 1 Portrait of Huldrych Zwingli, painted by Hans Asper (1549).
ZBZ, GRAPHISCHE SAMMLUNG UND FOTOARCHIV: INV. 6

Against this background, one can imagine that Urs Gessner was quite happy that his wife's apparently still robust uncle, the roughly sixty-year-old[8] chaplain Johannes Frick, had taken in the bright young Conrad. Frick had studied

8 Frick died on February 9, 1535 (cf. HBBW 7, p. 241). He was 80 years old. Cf. Gessner (1542), f. α_{ii}r.

in Tübingen and received a bachelor's degree in the liberal arts (*Baccalaureus artium*) in 1484. He was ordained a minister in 1492 and had served as chaplain at the altar of eleven thousand virgins in the central nave of the apostles' chapel in Zürich's Grossmünster church.[9] This provided a steady income – at least until Zwingli's reform plans took effect. These resulted in the secularization of church and monastery properties to be used primarily as schools or in service to the poor.[10] Adopted on September 29, 1523, the *Christenlich ansehen und ordnung* (Christian vision and regulation) required that from now on, for the protection of the "common man", money could no longer be taken for various clerical services such as baptism, giving sacraments or for ringing the church bells at funerals.[11] This new regulation must have also squeezed chaplain Frick's purse, who, in spite of these drastic changes, was still a supporter of the Reformation. He then took over new and less lucrative tasks, for instance, the pastoral care of sick people in 1525, or the supervision of attendance at services by canons and chaplains in 1526.[12] It is very probable that the chaplain's diminished income is the reason why the young Gessner moved to yet another new home in about 1526, namely that of his teacher Oswald Myconius (1488–1552).

It seems as though Gessner spent happy years living with his great-uncle. At that time, the latter still had the wherewithal to send him to the German School, which, unlike the Latin schools, were privately run and required tuition. The school was also relatively independent, receiving school regulation in 1549, but not falling under official oversight until 1586. The students started school there between the ages of five and seven, learning to read, write and calculate.[13] However, Gessner had his great uncle to thank for more than just attendance at the German school; Frick also awakened in him a love for the plant world that stayed with him for the rest of his life. In the preface of his 1542 *Catalogus plantarum*, an index of plant names in four languages, which he dedicated to his later patron Johann Jakob Ammann (1500–1573), he reminisced briefly and wrote: "I think you remember that my great uncle Johannes Frick, a man who was as good as he was dignified, was very well versed in botany. From the house of my father, he took me in as a child, fed me, and would sometimes lead me outside to teach me how to tend his lush little garden. There, he spent pleasant hours with me and told me the names of all the plants so often that, on command, I could immediately bring him any plant that he named from either the

9 Meyer (1986), p. 365 and 551 (see there „Johann Fritz von der Reichenau").

10 Stucki (1996), pp. 196f.

11 *Ein Christenlich ansehen und ordnung* (1523), f. a_{ii}r.

12 HBBW 4, p. 460.

13 Stucki (1996), pp. 246f.

garden or the fields. He often used plants to maintain his health and that of his friends, and managed to live more than 80 years without medical complaints due to the positive effects of the herbs, or (as they say) by the hand of God."[14]

1.1 Gessner's Teachers

After about three years of attendance at the German School, Gessner changed to the Latin School at Grossmünster, ca. 1524–25, which, according to the initial school regulations of 1532, comprised four classes.[15] Likewise in 1525, we find him frequently taking meals with his schoolmaster Oswald Myconius (from Lucerne), who became head of the Latin school at Fraumünster in 1523, and from November 1524 he took up residence in the home previously occupied by the canon Jos Meyer, that stood near the school.[16] Outside of school, the young Gessner was taught by two older housemates of Myconius. Thomas Platter improved his Latin grammar, using the so-called *Donat* as a teaching aid. This was a text based on the late antique grammarian Aelius Donatus (ca. 320–380), which had been used in schools for centuries. Platter recalls in his autobiography: "It also happened, that father Myconius took me into his house. There were many who sat at his table, and I was to practice the *Donat* and declinations with them; including the late Dr Gesnerus. The exercises did me a lot of good."[17]

Gessner's other tutor was the later famous Hebraist Theodor Bibliander. It is unclear which subjects he taught Gessner, but he repeatedly appears in the sources as one of his instructors.[18] Bibliander himself had been introduced to the foundations of Greek and Hebrew in 1525 by the philologist Jakob Wiesendanger (1499–1525), also named Ceporinus,[19] who had recently been given a position in Zürich. As his letters later revealed, he worked as an assistant to Myconius at the Fraumünster school and remained closely associated with his host for the rest of his life.[20] Gessner also maintained close relations to Myconius, who took his responsibilities more seriously than his predecessor

14 Gessner (1542), f. a_{ii}r/v. Here, as elsewhere, slightly modernized translation based on Hanhart (1824), p. 3f.
15 Ernst (1879), pp. 88f.
16 Spillmann (1962), p. 435.
17 Platter (1999), p. 75.
18 Gessner (1545), f. 179v; Simmler (1566), f. 4r.
19 Riedweg (2000).
20 Egli (1901), pp. 5f.

Wolfgang Knöwell.[21] Thomas Platter commented: "There was a school director named master Wolfgang Knöwell from Baar near Zug, with a graduate degree of the University of Paris, and who was called *Gran Diabell* (big devil) while in Paris. He was a tall, upright man who had little regard for school but had an eye out for the pretty girls, whom he could barely resist, etc." Aside from the fact that this Knöwell had a weakness for attractive women, it seems that Zürich's women themselves were not shy about their charms. The aforementioned Hebraist Conrad Pellikan noted in his autobiography: "Zwingli diligently oversaw the necessary expenses, so on the whole it never occurred to me in the slightest to take a wife, especially since I find the dressiness and the indecent behavior of Zürich's women and maidens offputting."[22]

Gessner not only had Myconius to thank, but many years later he fondly remembered many other teachers, for example, Thomas Platter. In 1558, Gessner let Platter's son Felix know that he could refuse him no request due to the many great good deeds his father had done for him.[23] Also in this circle of former teachers was the humanist Peter Hasenfratz from Frauenfeld, whose Latin name was Petrus Dasypodius (ca. 1495–1559), and who was a teacher at the Latin school of the Fraumünster church.[24] He produced one of the first, if not the first, Latin-German/German-Latin dictionary.[25]

Two other names repeatedly appear when discussing Gessner's teachers and those known to have influenced him in his youth: Johann Jakob Ammann and Rudolf Collin (1499–1578). Both taught at the *Hohen Schule* or *Schola Tigurina*[26] that was founded by Zwingli in 1525 (ill. 2). This was primarily a reformed institution of higher learning that served, above all, to educate prospective reformed pastors. The two were hired to succeed Ceporinus in teaching Latin and Greek language and literature. Having to share a canon's benefice,

21 The schoolmaster's annual salary was rather modest, increased by 24 guilders for Myconius in 1524, and finally amounted to 80 guilders in 1529, 100 guilders in 1543 (consisting of 20 Mütt Kernen, 15 Malter Haber, 15 buckets of wine and 60 guilders of money), 135 guilders in 1571 (consisting of 42 Mütt Kernen, 10 Malter Haber, 25 buckets of wine and 58 guilders of money). Cf. Spillmann (1962), p. 441.

22 Pellikan (1892), p. 105.

23 Gessner (1577), f. 97r.

24 Büeler (1920), pp. 4–9.

25 In the preface to the reader to Josua Maler's *Die Teütsch spraach* of 1561 (f. *7r), which Gessner wrote, we read: „Petrus Dasypodius, vir pius et eruditus, praeceptor olim meus, … Hic enim primum nostra memoria Dictionarium Latinogermanicum, et viceversa Germanicolatinum, cum appendice vocabulorum in locos et classes apte digestorum, magno omnium applausu edidit."

26 A modern history of this important educational institution is still lacking. Welcome beginnings can be found in: Bächtold (1999); Goeing (2017).

however, they were financially disadvantaged, which led to Collin working on the side as a ropemaker. He wrote about this in his autobiography: "Because my income was so small, I was busy with both academics and ropes for three whole years. And ropemaking went well for me. I actually had three shops: one in Zürich, which was my own; the second was in Wesen, which belonged to Jos Hartmann, also a ropemaker; and the third was in Walenstadt, which was managed in my name by the baker Hans Wildhaber. Later, as my income grew, I gave up ropemaking and devoted myself to scholarship, where I sufficiently applied myself (I hope) to my professional activities with all my might."[27]

In 1529, another canon's benefice became available and Amman and Collin were each able to enjoy receiving income from their own prebends. In that year, the improved finances also made it possible for Conrad, now 13, to move into Ammann's home,[28] where he lived at no cost for three years. Later, Gessner even expressed his thanks twice in print for the support during difficult times: first, in the preface of the *Catalogus plantarum* in 1542, and also in the short autobiography which appeared in his publication *Bibliotheca universalis* in 1545.[29] In the latter, Gessner wrote that during these three gratis years, Ammann was especially responsible for introducing him to the classical languages.[30] In his personal copy of the *Bibliotheca universalis*, Gessner later deleted "gratis" (*gratuito*) by hand. For what reason or how he might have returned the favor is not known.[31] In any case, his gratitude regarding Ammann seems sincere. At another time, he gave him a mushroom as a gift and memorialized him in a four-line poem, published in the preface of Gessner's *Epistolae medicinales*.[32]

27 Collin (1859), p. 218.

28 Ammann lived since 1531 in the so called „Grünen Zweig" (Green Branch) at the „Oberen Kirchgasse." Cf. Vögelin (1878), p. 343.

29 Gessner (1545), f. 179v/180r.

30 Gessner's first biographer Josias Simmler wrote (1566), f. 4r/v: „Eo tempore audivit ipsum Ammianum et Rodolphum Collinum Graecae linguae doctorem, qui tum Quintiliani oratorias institutiones et Plutarchi vitas interpretabantur, et se in dialecticis et oratoriis praeceptis, et in Graecae linguae studio, quo plurimum afficiebatur, diligenter exercuit." Ammann's personal copy of Quintilian has numerous handwritten annotations and is kept in the ZBZ (call number: Gal St 224). According to the entry on the flyleaf, Ammann received the volume as a gift from Heinrich Glarean when he studied in Paris and lived there in Glarean's student burse.

31 Students were often admitted to supplement the income. Cf. Ernst (1879), p. 62. A letter from Bullinger to Johannes Travers of the Grisons, dated September 14, 1543, indicates that such a student paid in Zürich 25 or 26 guilders annually. Cf. HBBW 13, p. 255.

32 Gessner (1577), f. β_3v (the two hexameters have no date): „Mitto tibi fungum, non qualem Claudius edit,// Pars Strobili, pars est mellis arundinei.// Suavior hic multo, multoque salubrior idem,// Quam quos e putri terra vapore parit."

ILLUSTRATION 2 Detail from the Murer-Plan of 1576 with the Grossmünster church, where the *Hohe Schule* was located, and Gessner's house, where he lived from 1552 at the latest (reprint 1859/60).

ZBZ, ABTEILUNG KARTEN UND PANORAMEN: 4 LB 02:5

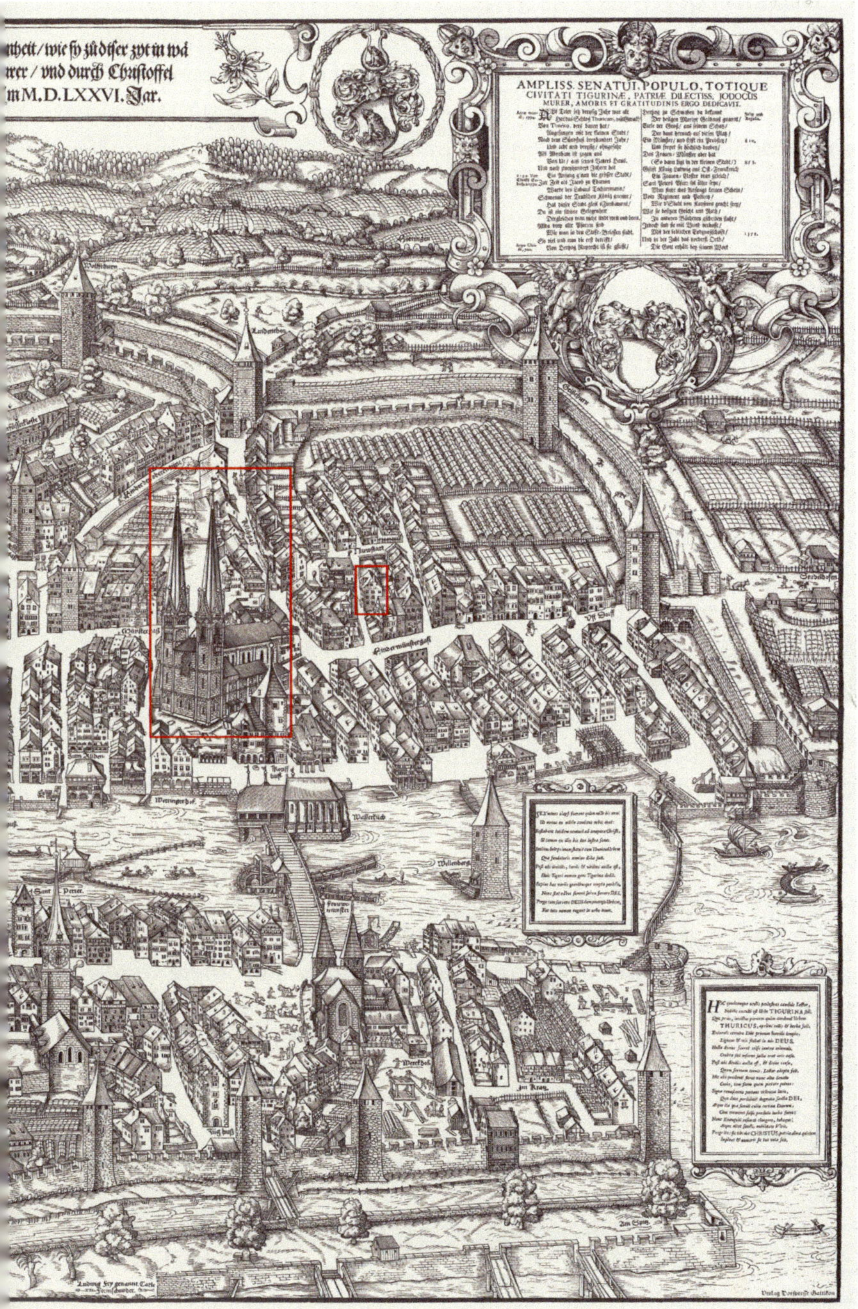
AMPLISS. SENATUI, POPULO, TOTIQUE
CIVITATI TIGURINÆ, PATRIÆ DILECTISS. IODOCUS
MURER, AMORIS ET GRATITUDINIS ERGO DEDICAVIT.

1.2 Humanism in Zürich

Conrad learned about various animals from his father, Urs Gessner, who processed their pelts as a furrier. His great uncle introduced him to the world of plants and remedies, and the above-mentioned teachers provided him with a solid philological foundation. This generally outlines the three most important subject areas that would preoccupy Gessner for the rest of his life. During Gessner's attendance at the *Hohen Schule*, probably from 1529–1531, it was primarily the lessons from Ammann and Collin – which more or less covered the Faculty of Arts – that advanced his language skills. It is not clear to what extent Gessner took part in theological classes by the other two professors, Zwingli and Pellikan, or how often he took in public interpretations of the Bible in German by Myconius. It can nonetheless be assumed, that he did absorb some of it, since he remained devoted to the Zürich Reformation throughout his life.[33]

With the founding of the *Hohe Schule*, Zwingli not only expanded Zürich's school system, but he also introduced new elements into its curriculum. Studying and interpreting the Bible required solid knowledge of the so-called *tres linguae sacrae* (Hebrew, Greek and Latin), which is why he strongly promoted Semitic and classical studies. This was also in keeping with the spirit of Renaissance humanism. He shared his humanistic enthusiasm for ancient authors with students such as Johann Jakob Ammann and the later Anabaptist leader Conrad Grebel (*c.*1498–1526) by conducting private readings of the Greek philosopher Plato.[34] He also studied works written by the Florentine neoplatonists Giovanni Pico della Mirandola (1463–1494) and Giovanni Francesco Pico della Mirandola (1469–1533),[35] all authors that would later be read by Gessner.[36] A glance at Zwingli's private library shows that his fascination for antiquity even included texts from the natural sciences, since there were also relevant works by Aristotle,[37] Pliny[38] and Strabo[39] in his study.[40] However, such versatile reading material was not only to be found in the reformer's home.

33 Leu (1990).

34 Letter from Conrad Grebel to Oswald Myconius, November 4, 1521 (StAZH E I 1 [Teil 2], no. 35).

35 Backus (1990/91); Serrai (1990), pp. 52–57; Schindler (1996). Cf. Zwinglis own copies: Leu and Weidmann (2019), pp. 35–37.

36 Leu et al. (2008), pp. 197f. and 200.

37 The copy belongs today to the ZBZ, call number: V E 21.

38 The copy belongs today to the ZBZ, call number: V P 200 & a.

39 The copy belongs today to the ZBZ, call number: V H 11.

40 Leu and Weidmann (2019).

It also shaped the school's academic life. The German theologian Andreas Karlstadt (*c.*1486–1541), who Zwingli brought to Froschauer as a proofreader in December 1530, put it this way: "I might add, among the historians, they read Plutarch and among the Greek comedic poets, Aristophanes. They interpret the rules of rhetoric by Fabius and cover dialectic – not just sipping from the puddles but drinking directly from the well, and not with just half a mouthful, but with full gulps whenever possible. Pliny's wonderful hidden treasures get so much attention, they compete for these treasures and make them their own. Cato, Virgil's instructive *Georgica*, Varro, Columella, Vegetius, Palladios and others who write about agriculture are researched and worked on daily. This is a benefit to both the profession and public life, through which they can more thoroughly perceive the wisdom and goodness of God by his works."[41]

This humanistic approach to education, which was initiated to a great degree by Zwingli, but also by Erasmus of Rotterdam (1464/69–1536),[42] found expression not at least through theater performances of works by various antique authors such as Aristophanes, Homer and Terence.[43] The first and only comedy performed in ancient Greek during the sixteenth century in Zürich appeared on stage on January 1, 1531. It was the play *Ploutos* (wealth) from the pen of the Greek poet Aristophanes (ca. 450/444–380 BC) with the following story line: Chremylos, a poor farmer from Attica, goes on a pilgrimage to Delphi with his servant Karion to ask Apollo what he must do to at least improve the financial situation of his son. Apollo answers that he must invite to his home the first person he encounters after leaving the temple. The person he meets is blind and dressed in dirty rags, but he turns out to be Ploutos, the god of wealth. Chremylos promises Ploutos that he will do what he can to heal his blindness. He discusses the situation with his neighbor Blepsidemos, and they decide to take Ploutos to the Asklepios temple. Suddenly, Penia the

41 Mörikofer (1869), pp. 340f.

42 Christ-von Wedel and Leu (2007).

43 Cf. Brunnschweiler (1989), pp. 267f. On February 28, 1544, the first book of Homer's Iliad was performed in the form of a Latin comedy at the home of Rudolf Gwalther (1519–1586). The autograph of Johann Rudolf Gwalther is kept in the ZBZ (Ms C 86a, no. 2). Whether the tragedies of Euripides translated into Latin by Rudolf Collin were also performed eludes us (cf. ZBZ, MS. C 93). According to the words of the schoolmaster Georg Binder, such performances seem to have been much more frequent than has been handed down or is known. Cf. Binder (1535), f. $a_{ii}v/a_{iii}r$: „Ich hab nun etwan vil iaren hie Zürich mit minen knaben vil der Latinischen und Griechischen comoedien Terentii und Aristophanis gespylt / damit die jugennd geuept erlernte der red (die sust an iro selbs todt) ein waesen und laeben gaeben mit der action und ußspraechen / nit nun das die gedaechtnuß gesterckt und etliche guote sprüch behalten wurdent. Dann sust in heiliger gschrifft erfunden das vil bas dienet zuo besserung deß laebens und guotem byspyl."

goddess of poverty appears and warns the men not to restore Ploutos's sight, so that he can recognize people again and distribute wealth to them, for poverty is what keeps people alive. Without financial hardship, no one would work, and everyone would indulge themselves in idleness. However, Chremylos knows what it means to be poor, and he chases Penia away with insults and reproaches. Ploutos's treatment is successful; he can see once again, and he distributes wealth among those who are good and just, including Chremylos. Because of all the resulting social upheaval, Zeus sends his messenger Hermes to report that the house of Chremylos is to be destroyed, because apparently no one feels the need to make sacrifices to Zeus any more. Hermes himself is also starving and asks Chremylos for a job. In the end, Ploutos is celebrated and brought to the treasury located atop the Acropolis. Consequently, this is a piece about the advantages and disadvantages of wealth and the question of how it should be properly distributed. At the time, this topic seemed to be taken from real life in Zürich.

Rudolf Collin wrote a prologue for the play that was read aloud before the performance by Georg Binder († 1545), schoolmaster of the Latin school at the Grossmünster church. In it, the reference to Zürich being the new Athens expresses the self-regard held by Zürich's educational elite: "Athens is silent; Greece is silent; but Zürich still stands, and this is where the muse of Attica dwells."[44] The point of the play was described as follows: "The god of wealth makes only good and modest people rich. Whoever has not yet received such a gift should strive diligently for modesty, and they will soon see greater prosperity."[45] This sentence has a double meaning: because it corresponds completely to the work ethic of the Reformation – that success shines on those who are good and work hard – but also because those who lead a modest lifestyle are thankful for the things that they have and are internally "richer" and more satisfied. Both aspects appealed to Zürich's students and instructors, who frequently experienced poverty. Therefore, it is not surprising that because of the moral and pedagogic aspects of the play, which endured for centuries, it became the most popular of Aristophanes's comedies in the Middle Ages and the Renaissance. It was the most transcribed play in Byzantium, and was often translated by humanists into either Latin or the vernacular during the fifteenth and sixteenth centuries.[46] So, the small circle of people in Zürich were

44 Hug (1874), p. 23.
45 Hug (1874), p. 23.
46 Holzberg (2010), p. 209; Hertel (1969).

in good company, when they rehearsed and performed *Ploutos* in the so-called "Ofenzimmer" of the Grossmünster church. Zwingli wrote the music for it.[47]

Zürich's central library stores three copies of Aristophane's work that belonged to those involved in this production. They were: Conrad Gessner,[48] Rudolf Collin[49] and another actor whose handwriting could not be identified.[50] There are notes in all three citing which actors played which roles (ill. 3). When assembled,[51] the following list emerges:

Prologue narrator: Georg Binder († 1545)
Karion: Johannes Fries (1505–1565)
Choir director: Jos Haas
Sycophanta: Conrad Grebel
Chremylos: Sebastian Guldinbeck († 1565)[52]
Chremylos's wife: Christoph Clauser († 1552)
Blepsidemos: Leonhard Hospinian (1508/10–1564)
Old woman: Nikolaus Zehnder (1503?–1553)
Priest: Leonhard Hospinian
Ploutos: Georg Binder
Penia: Conrad Gessner (1516–1565)
Neanias (youth): Gerold Meyer von Knonau (1509–1531)
Hermes: Conrad Gessner
Fair-minded man: [no entry]
Second fair-minded man: [according to Gessner's private copy, this role was cut]

Here we see that the 15-year-old Gessner took the small role of Hermes as well as the larger role of Penia (poverty), who appears in the second act. The part was tailor made for him, since some of Penia's lines were virtually autobiographical, such as the statement: "I'm poverty, who has been your housemate for years!"[53]

47 Hug (1874), p. 35.
48 ZBZ, call number: SV 187.
49 ZBZ, call number: 20.133.
50 ZBZ, call number: Zw 270.
51 Cf. also Collin's autograph preface with the incomplete distribution of roles, which he had noted down on a sheet of paper and sent to Rudolf Gwalther on April 11, 1562. The document is kept in the ZBZ, MS. D 75, no. 28, f. 144 and 145.
52 Also called Schmid (Faber). Bullinger praises his knowledge of Greek when listing the scholars, cf. HBBW 8, p. 277.
53 Aristophanes (1910), p. 337.

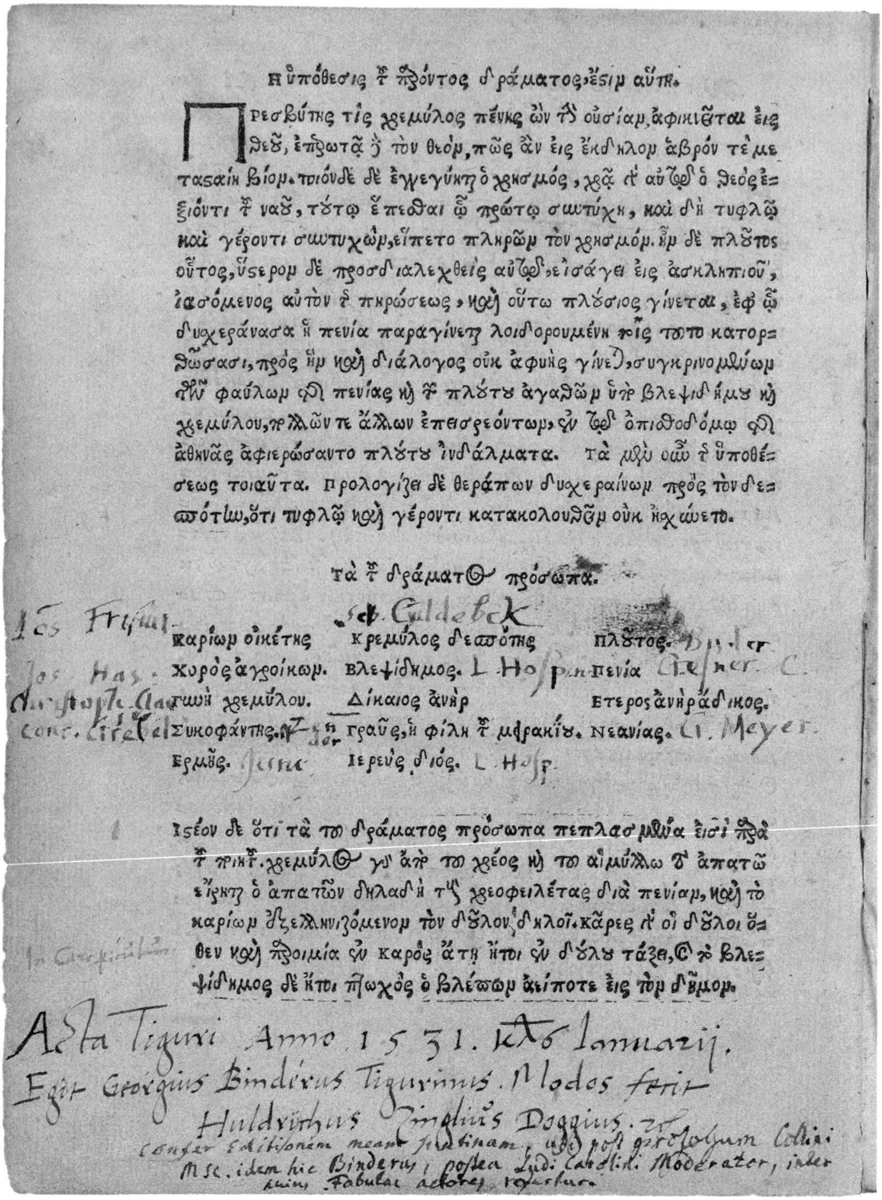

ILLUSTRATION 3 Cast of the comedy *Ploutos* by Aristophanes, which was performed in Zürich on January 1, 1531. Gessner played the part of *Penia* (poverty). Aristophanes, *Comoediae*, Paris 1528, f.a$_{ii}$v.

ZBZ, ABTEILUNG ALTE DRUCKE UND RARA: ZWINGLI 270

Gessner's participation in this production testifies to his outstanding skill with the Greek language, which he seems to have mastered at such a young age. The other actors, aside from the already mentioned Georg Binder, were the later schoolmasters Johannes Fries, Sebastian Guldinbeck and Leonhard Hospinian (Wirth); as well as Christoph Clauser, who had held the office of municipal physician since 1531; Nikolaus Zehnder, the first deacon at St Peter's church; as well as Zwingli's stepson Gerold Meyer von Knonau, who died that same year in the Second War of Kappel. Conrad Grebel must have been a lesser-known nobleman who got married in 1526 and was mentioned in Bernhard Wyss's chronicle. Jos Haas was hired as canon in Embrach in 1525 and worked as a pastor in Ottenbach after the Battle of Kappel.[54]

Johannes Rütiner (1501–1556/57), a merchant and councilman from St Gallen, was informed about the play by Georg Binder. Rütiner then noted in his *Diarium*: "In Zürich there are some young people, namely Georg Binder's students, who performed Aristophanes's play *Ploutos* and the sixth book of Homer's *Odyssey* in Greek in the presence of Zwingli. The pious man shed tears of joy. In the same year, they also did Ovid's book *Fasti*. Likewise, the group performed the poem *Cyclops*, as well as *Galathea* by Antonius Thylesius[55] from Cosenza."[56] In addition to this passage by Rütiner, the schoolmaster Binder himself also suggests that theater life in Zürich during the Reformation was quite active and varied: "For many years here in Zürich, I and my boys have presented numerous Latin and Greek comedies by Terence and Aristophanes. In this way, the youths practice and learn to give life and character to speech (which is otherwise lifeless) with the help of action and diction, and are not just enhancing their memory and learning a few good speeches. Otherwise they read the Scriptures, which serves as a role model and is much more useful for the improvement of one's life."[57]

The fact that Gessner was by far the youngest member in this troupe of actors, but was still completely accepted, shows an acknowledgement of his extraordinary intellectual skills, and that he was well integrated in the local world of scholars. It is not surprising that on March 1, 1531 he was included among those awarded a stipend by Zwingli with support and encouragement from the Students' Office (*Studentenamt*) that was established in 1525. The

54 Egli (1904), p. 13.

55 Works written by Thylesius have also been read and possessed by Bullinger, Gessner and Gwalther. See: Leu and Weidmann (2004), Leu et al. (2008), Leu and Weidmann (2020).

56 Rütiner (1996), vol. 1/1, p. 287. Rütiner lists the event under August 1534, which cannot be correct, because Zwingli was then three years dead. It seems much more that he was only then informed about it by Binder.

57 Binder (1535), f. $a_{ii}v/a_{iii}r$.

application he submitted to Zwingli on October 27, 1530 read: "Dear Zwingli, I wish very much for you to be informed about the marvelous situation I find myself in. I have it all, and I have nothing; everything is gone, but there is nothing missing, just as Gnatho[58] explained it according to the comedian. I have completely given myself over to my studies with equal enthusiasm for both Greek and Latin literature. How easy it would be, however, for all my hopes to be dashed. My great-uncle Johannes Frick has one foot in the grave; and once he dies, my teacher, Johann Jakob Ammann, will not wish to continue putting me up; my father cannot take me in, since he is burdened by a horde of children that he can barely feed; and none of my other relatives are able to care for me. Then, what will happen to poor Conrad? So please, I'm begging you gracious Zwingli, to employ your philanthropy that helps so many people, and like Sebastian Guldinbeck, Johannes Fries, and Benedikt Finsler, include me among those receiving a stipend. If you grant my wish, you can be sure that I will never forget the goodness shown to me; that I am grateful; and that I will exert the maximum effort in applying myself to scholarship. Farewell! Written in the house of Johann Jakob Ammann, October 27, 1530."[59] In the end, Gessner received between 10 to 12 guilders per year,[60] which at the time was equivalent to about three months income for a country pastor.[61]

The year 1531, that had started so promisingly with the Aristophanes production and acceptance into the circle of stipend recipients, ended with the Second War of Kappel as a catastrophe for Zürich. More than ten percent of the men who were fit for combat were killed, including Zwingli and Urs Gessner.[62] The 55,500 pounds (or 27,750 guilders) cost of the war strained the city's treasury; and it was questionable whether Reformation efforts could carry on.[63] Myconius wrote to the Memmingen reformer, Simprecht Schenk (ca. 1485–1559), on November 29, 1531 that he could not rule out either his own death or exile. The Evangelical preachings were threatened with demise,

58 Terence, *Eunuchus*, II,2,12.

59 Quoted according to Fischer (1966), p. 11. Latin text in: Zwingli (1929), p. 325f. Fischer dated the letter to 1530.

60 In the two lists of scholarship holders handed down by Bullinger, one time 10 guilders are mentioned (ZBZ, MS. F 95, p. 3), the other time 12 guilders (ZBZ, MS. Car C 44, p. 917).

61 Schoolmaster Georg Binder earned comparatively 80 guilders, and Zwingli, in his various functions, about 350 guilders per year. Cf. Spillmann (1962), p. 64; Corrodi-Sulzer (1923), p. 178.

62 Zwingli died on October 11 and Gessner's father in the last battle of the Second Kappeler War, the so-called Battle of Gubel, on October 24, 1531. Cf. Simmler (1566), f. 4v; Meyer (1976), pp. 179–184.

63 Meyer (1976), p. 288; Meyer (1981), p. 44.

without even a glimmer of hope. The best were dead, the populace was terrified, and only the wicked rejoiced.[64]

The death of his spiritual father, Zwingli, shook the young Gessner to the core, and according to his own words, even more so than the death of his actual father and other relatives.[65] On August 10, 1532, he complained to Theodor Bibliander and Heinrich Bullinger, Zwingli's successor in office, that no one had yet had the courage to give such an exceptional man a fitting memorial: "Honored teachers, whenever I think to myself about the great lack of gratitude in our times, my soul is wounded with the most bitter pain. As Pliny said, loyalty is so rare in friendship and the dead are soon forgotten. None of us here utters a single word of thanks to that most courageous hero, the father of the fatherland and light of the world, Huldrych Zwingli. … And yes, I am right to call the man a hero; he who no violence, no threats, no envy, and no persecution could cause to veer from the course of truth and justice. For this, he should not only be praised in word, but it would be most appropriate to recreate his likeness in gold, ivory, silver and Parian marble. This praiseworthy hero of virtue is celebrated abroad, but here, where he deserves it the most, his name lies in the dirt. He is given no honor, no mention and no thanks for his good deeds. This is what hurts me, tortures me, and gnaws at my heart. Just hearing his name, I am choked with tears and my soul is wrapped in pain."[66]

If the pain and sadness were not enough, the young Gessner was now quite alone. His uncle Andreas Gessner was severely wounded in the Kappel War,[67] and his host Johann Jakob Ammann could no longer take him in, because, as canon in the Grossmünster cathedral chapter, he had to pay a certain amount to support the Evangelical clergymen who had been expelled by the Catholics.[68] Furthermore, Myconius was so afflicted by the resurgence of the Old Believers,[69] that he fled to Basel.[70] Stripped of all support, the gifted young man stood abandoned and without prospects, facing a grim future. What was to become of him?

---

64 ZBZ, MS. F 81, p. 304.

65 Cf. Gessner to Myconius, April 17, 1536, Kantonsbibliothek St Gallen (Vadiana), MS. 33 (VBS 4), 24.

66 Cf. the Latin original in: HBBW 2, pp. 173f.

67 Hanhart (1824), p. 12.

68 Letter form Myconius to Capito, June 15, 1532: ZBZ, MS. F 81, 311.

69 Certain representatives of the Old Believer party were prepared to go to extremes. Cf. Platter (1999), p. 108.

70 Fischer (1966), p. 12.

CHAPTER 2

The Traveling Student

At a dead-end, Gessner turned to his former teacher, Oswald Myconius (ill. 4). He described the fate of the gifted youth from Zürich to his colleague in Strasbourg, the reformer and Hebraist Wolfgang Capito (1478–1541), with whom he had been corresponding since at least 1519. Capito then suggested that Myconius send Gessner to him, so that such a talent would not go to waste. Myconius wrote a letter of recommendation to Capito on June 15, 1532 in which he extolled the intellectual abilities and character of the young man. He ended the letter with the hint that one should not be alarmed by Gessner's physical size, as he was just 16 years old and compliant.[1] Gessner delivered the letter himself and resided in Strasbourg from June to November 1532. In his autobiography, which was published 13 years later in the *Bibliotheca universalis*,[2] he made a brief and succinct mention: "Because I could find no patrons for my studies in my home town, I moved to Strasbourg, where for several months, I was in useful scholarly service to the esteemed and venerably-remembered Capito."[3] This short, dry account suggests that he did not have the fondest memories of his stay with Capito. Although he had been able to improve his Hebraic skills,[4] and it was surely a challenge to teach ancient Greek to the much older book printer Wendelin Rihel (1490–1555),[5] he found little purpose in his other activities. He still yearned in his thoughts for his cherished Zwingli, for whom he wrote poems of mourning in Greek.[6] Finally, he poured his heart out to Zwingli's successor, Heinrich Bullinger. At the end of October 1532 he wrote to him: "My present situation causes me such distressing insecurity that I have no idea what to do. I come to you seeking refuge and begging for help; here there is neither advice nor hope. If you do not accept my plea, then I am finished. For what kind of hope remains for the wretched one, who, with nothing to show

1 Letter from Myconius to Capito, June 15, 1532: ZBZ, MS. F 81, 311.

2 Gessner (1545), f. 180r.

3 Stotz (1965), P. 213.

4 Simmler (1566), f. 4v. Hebrew lectures had been offered in Strasbourg since 1525. Cf. Schindling (1977), pp. 262f.

5 Gessner (1577), f. 138r.

6 The handwritten Greek poems are preserved in: ZBZ, MS. C 50a, no. 65. The accompanying letter to Bullinger and Bibliander is dated August 10, 1532. The latter is published in: HBBW 2, pp. 173f. Cf. edition and translation in: Vogel (2019).

© KONINKLIJKE BRILL NV, LEIDEN, 2023 | DOI:10.1163/9789004541696_003

ILLUSTRATION 4 Portrait of Oswald Myconius. Copper engraving by Johann Jakob Schönauer (1695–1767), possibly after a late-seventeenth-century painting. ZBZ, GRAPHISCHE SAMMLUNG UND FOTOARCHIV: MYCONIUS, OSWALD I,1

for it, loses the irretrievable time of his youth. Yes, my youth is running away from me; the most advantageous time to learn is leaving me behind … When so much must be done for the master, you can guess how little time remains for one's own studies. Not an hour, not even a moment is to be found in the day that I can call my own, in which I can freely apply myself to scholarship. … So, in my misery, I'm asking that you give me some sort of counsel out of human kindness. If it were possible for a stipend to be given that is not smaller than the other one, I would return to Zürich as soon as allowed. If my request is denied, then the muses will abandon me. I hope to hear from you soon in order to know what will become of me."[7]

There is another of Gessner's letters that dates from this dark time in Strasbourg that he sent to Leo Jud (1482–1542), the second most important ecclesiastic in Zürich. Jud worked as the successor to Zwingli in Einsiedeln, and as pastor at the St Peter's church in Zürich from 1522, and had fought alongside Zwingli for the cause of the Reformation. Worried about his education and future, Gessner also applied to him for a travel grant. He began with the great men of antiquity: the Greek poets Homer and Aesop, then to Lycurgus the Spartan lawmaker, and further to the philosopher Aristotle, all of whom had benefitted from educational travels. He had twice been promised a raise in his stipend but never received it. Those in Zürich, including Binder, Ammann, Pellikan, Andreas Gessner, Collin and others had recommended that he spend time in foreign lands. Hence, he found it baffling that they would all recommend it, but, in his case, no one would fund it. From the end of the letter, one may surmise that the cathedral chapter's administrator had denied support, because canon law did not provide for youths to be given grants to continue their education in other countries. To Gessner, this was completely incomprehensible.[8]

Finally, Capito released the young man to return to Zürich and gave him a letter written on November 17, 1532 requesting that Bullinger and Jud assist him with a stipend to study abroad.[9] On December 14, Gessner informed Myconius in Basel that, as before, the schoolmasters had not given him the freedom to continue his studies in the way he had hoped for, because he had already received a stipend for nearly two years. Myconius had been his teacher since he was a small boy, and Gessner was asking him to also be his advisor in the future.[10] About three weeks later, on January 7, 1533, he thanked Myconius for his letter (unfortunately lost) in which the Basel teacher suggested to him that true freedom does not play out through external things, but within a person's inner life. Gessner agreed with this view as follows: "Then, I will diligently carry on my studies, whether I can achieve greater freedom with the help of the muses. Indeed, true freedom only resides in the soul, adorned with knowledge

7 Cf. Hanhart (1824), p. 17f.; letter published in: HBBW 2, pp. 252f.

8 Letter from Gessner to Leo Jud from the house of Capito from 1532: ZBZ, MS. A 51, no. 53, f. 96r–97r.

9 HBBW 2, p. 261.

10 Letter from Gessner to Myconius, December 14, 1532 (dating according to lic. theol. Rainer Henrich): St Gallen KB (Vadiana), MS. 33 (VBS 4), 97.

and filled with true piety."[11] Gessner also makes reference to the tense situation in Zürich, where things have not yet settled down since the Kappel defeat and the Catholics' claim to power.[12]

A short time later, Gessner's wish to study abroad was unexpectedly granted and his quarterly stipend of 10 pounds (= 5 guilders) was raised to 14 pounds (= 7 guilders),[13] which was actually still too little for foreign study. On February 18, 1533, Gessner and his eleven-year-older companion, the later schoolmaster and philologist Johannes Fries (1505–1565), left the city on the Limmat River for university studies in France. It is unclear whether it might have been the undated poem, penned in Greek in 15 hexameters, that Gessner and Fries sent to Bullinger, that had managed to soften the hearts of the head of the church and its high officials. Until now, Gessner's handwritten poetic plea for a foreign travel grant, with its reference to Calliope, the muse of poetry, has gone unnoticed in Zürich's Central Library. The English translation reads (ill. 5):

> To the honored patron and protector Heinrich Bullinger.
> You, blessed one, heard from your mother Calliope that there is
> nothing better for people than to be reasonable.
> Zeus gives only small portions of prudence to widely scattered people
> if they remain inexperienced.
> When among men, however, the understanding heart yearns for
> the study of many cities of many peoples, Zeus has always awarded
> him wisdom, and he will learn what he wishes to know about
> the good and the bad on this earth. But he demands a man with stamina,
> who is willing to try anything; is happy to make an effort; who lets
> himself be taught and will question the people with knowledge he encounters. For not without
> effort the far-seeing son of Kronos (Zeus) will give completion
> to the words; otherwise dark clouds will gather around the
> thoughts of the ignorant, and the darkness of the cloud will keep him
> from reaching the blooming meadow of perfection,
> and he will remain a burden upon the earth:
> Like wild animals, he runs helter-skelter from helpful wisdom
> as if it were a great evil.

---

11 Hanhart (1824), p. 22. This letter from January 7, 1553, is preserved in: St Gallen KB (Vadiana), MS. 32 (VBS 3), 147.

12 Cf. on the difficult situation in Zürich and Bullinger's skilful political approach: Campi (2004), pp. 116–126.

13 *Studentenamtsrechnung*, StAZH, G II 39.1.

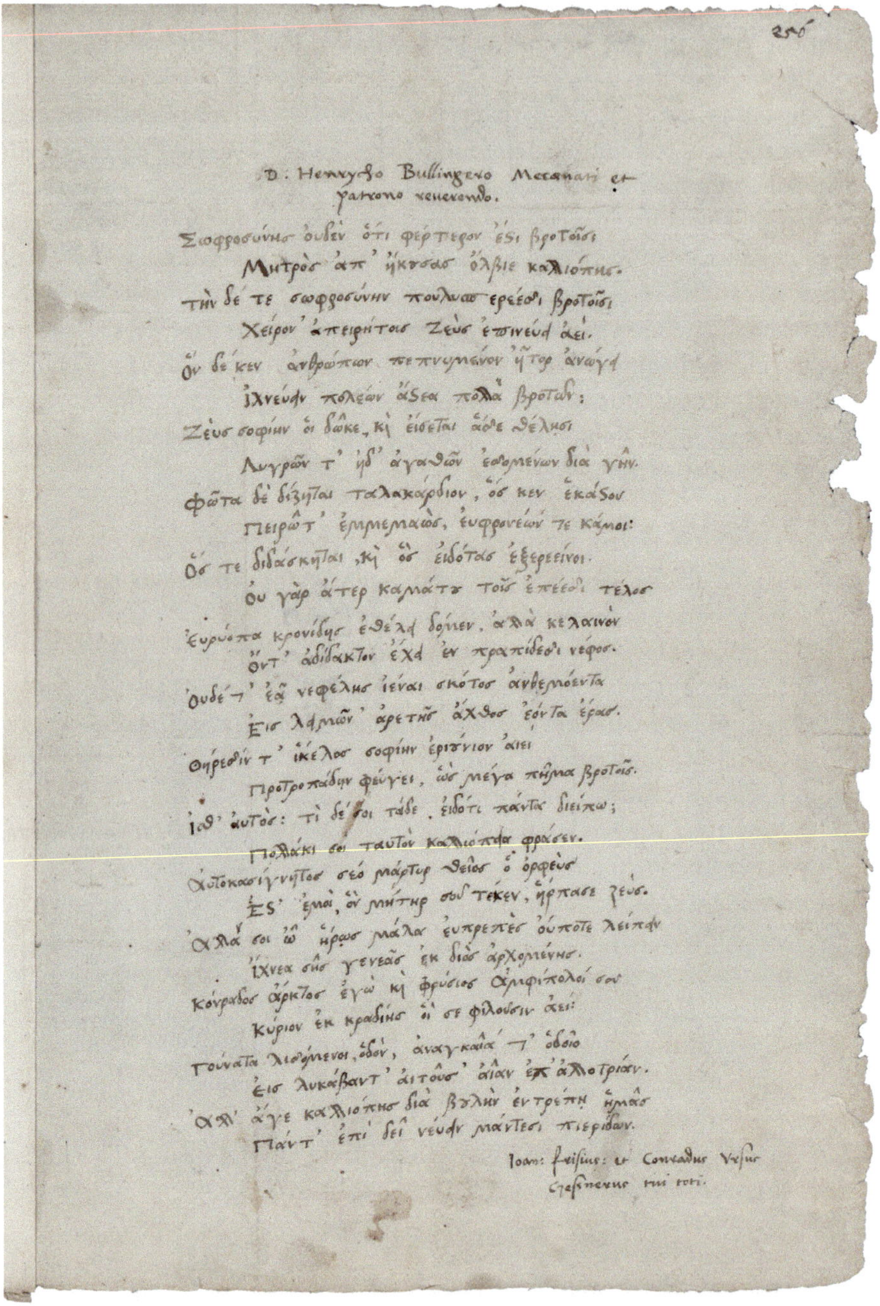

ILLUSTRATION 5 A Greek poem by Gessner and Fries that was sent to Bullinger.
ZBZ, HANDSCHRIFTENABTEILUNG: MS. F 62, F. 256R

But you are aware of all this yourself;
what shall I explain to one who already knows?
Calliope has often made this all clear to you.
As a witness, I have your natural brother, the godly Orpheus,
whom your mother bore and Zeus carried away.
Indeed, it befits you well, my patron, that you have never left behind
the traces of your dynasty that Zeus brought forth.
Your servants, myself, Conrad Urs,[14] and Fries,
who love you from the heart, are asking your lordship,
your knees embraced, for permission
to travel in a foreign land and for what is necessary for the journey for one year. Well then,
consider us according to the wish of Calliope, it is up to the priest of the Muses to grant everything. Johannes Fries and Conrad Urs Gessner, at your service.[15]

Gessner was striving for a well-rounded education, while Fries, who had recently lost his wife and child, required a master's degree in order to become a schoolmaster (*Ludi magister*) in Zürich.[16] At the time, it was generally advisable not to make such a journey by yourself, so as to avoid being an easy target for criminal activity. Their first extended stop along the way was in Basel with Myconius, who had succeeded Johannes Oecolampadius (1482–1531) as head of the Basel church. A week after leaving Zürich, the two let Bullinger know that they were at a standstill, because snow, rain and strong winds were hindering travel. Additionally, they were waiting for the arrival of French and other colleagues who were to travel on with them. It was not until after they had already departed their home town that it occurred to them how helpful letters of recommendation from the Zürich city council would be in making their journey more secure. They asked Bullinger to have such documents sent to them.[17]

2.1 Paris and Bourges

They subsequently headed for Paris, which at the time was a major city with 200,000 inhabitants. Due to encountering such throngs of people, however,

14 Gessner uses his father's name Urs as a middle name.
15 ZBZ, MS. F 62, p. 256.
16 Bührer (2002), p. 161.
17 HBBW 2, p. 74.

they could not find appropriate lodgings. Furthermore, they were facing yearly costs of over 40 guilders (= 80 pounds) which was far beyond their budget,[18] so they continued on to Bourges, where Louis XI had established a university in 1463. From 1527, it had been under the auspices of Margarete of Angoulême (1492–1549), who was a sister of the French king, Francis I (1494–1547), a supporter of the Reformation who had made it safe for Protestants. The university had seen its best days in the second half of the sixteenth century, and when Gessner and Fries attended, it was in a poor condition. Fries wrote to Bullinger on May 14, 1533 that aside from law, all the other important disciplines had become completely torpid.[19] Gessner shared this impression and passed it along in letters sent to Bullinger and Myconius on April 14.[20] Not only did he mention that it was primarily due to Andreas Alciatus (1492–1550) that civil law had been uplifted and was flourishing,[21] but he also complained about the high costs of living. He let the church leader in Zürich know that Fries was still undecided as to whether he would stay or not, and because of their financial predicament, he asked for the rest of the stipend to be sent. Additionally, he accepted the offer to teach the children of Melchior Volmar (1497–1561), a philologist from Rottweil.[22] Within this circle, Gessner made the acquaintance of the Swabish nobleman, Count Froben Christoph von Zimmern (1519–1566)[23] as well as the later reformer from Geneva, Theodor Beza (1519–1605), who studied Hebrew and Greek under Volmar in Bourges. In a letter to Gessner from August 22, 1549, Beza recalled the period during which they had studied together and dedicated a polemical work to him against the Catholic theologian Johannes Cochlaeus (1479–1552).[24] In the letters mentioned above, both Fries and Gessner expressed the urgent need for financial support. This did not go unheard in Zürich. Bullinger contacted Myconius in Basel and Berchtold Haller (1492–1536) in Bern about arranging for money to be sent to the two.[25]

Fries returned to Paris (ill. 6). He probably left Bourges in summer or early fall, since rector Nicolas Cop began the new academic year on November 1. At this point, Cop gave a speech that would go down in the university's history, which, of course, Fries could not yet have known. He spoke publicly in favor of the reform concept of the justification of man by faith, which resulted in

18 HBBW 2, pp. 105 and 126f.
19 HBBW 2, p. 126.
20 Letter from Gessner to Myconius, April 14, 1533: St Gallen KB (Vadiana), MS. 33 (VBS 4), 22.
21 Dotzauer (1971), pp. 21–27.
22 HBBW 3, pp. 105f.
23 Jenny (1959), p. 72.
24 Aubert (1960), pp. 49–55.
25 HBBW 3, no. 251 and 258.

a violent reaction from the faculty of theology.[26] For his part, Gessner stayed in Bourges for a year[27] and seems to have made one or the other interesting acquaintance. This is evidenced, for example, by the handwriting in a commentary on Homer's Odyssey by the Greek grammarian and lexicographer Didymos Chalkenteros (65 BC–10 AD) that was printed in Paris in 1530. In it, the lawyer and later rector of the University of Basel, Johannes Sphyractes (1508–1578), wrote a dedication to his highly esteemed friend and brother Conrad Gessner and refers to the time they spent together in Bourges.[28]

In his autobiography of 1545, Gessner reports matter-of-factly about his time in Bourges: "From there (Strasbourg) I returned to my home town and, with the support of a grant, travelled to France with Johannes Fries, who was like a brother to me. I spent a year in Bourges as an educator; the stipend was not enough to cover the costs necessary for independent study. It seemed to me, that by teaching others, I could make a great deal of progress, since there was little time left to devote myself to reading Latin or Greek authors."[29]

Bullinger asked Gessner about their well-being on May 12, 1534 and expressed his astonishment that they had not reported anything about their studies or professors. He assured them of further financial support from the book printer in Zürich, Christoph Froschauer, who could pass along the money at the Frankfurt book fair through a printer they knew.[30] That May, Gessner was already in Paris and never received the letter, and sent a complaint to Bullinger on August 26 that he and Fries had heard nothing more out of Zürich for almost a year. They had also written to Ammann asking for money with assurances that they were hard-working and thrifty.[31]

While Fries in the winter of 1535/36 was rigorously learning French in the city on the Seine and attaining a master's degree in the faculty of arts,[32] Gessner was going through a period of disorientation. He was without a spiritual mentor; someone who could give guidance to an 18-year-old. For sure he still had in mind Johann Jakob Ammann's recommendation, who had suggested

26 Tuilier (1994), pp. 323f.

27 Simmler (1566), f. 4v.

28 The dedication reads literally: "D. Conrado Gessnero Tigurino charissimo suo amico ac fratri, Jo. Sphyractes Basileiens. D. D. 1534. Bituritigbus." The volume is kept in the ZBZ and bears the call number: 20.384. By unknown means, another volume from the possession of Sphyractes, which he owned during his time in Bourges, has entered the holdings of the ZBZ (call number: 22.13). It is an epigraphic work on Roman antiquity published by Giacomo Mazzocchi under the title *Epigrammata antiquae urbis* (Rome 1521).

29 Stotz (1967), p. 213.

30 HBBW 4, pp. 172f.

31 HBBW 4, pp. 290f.

32 Bührer (2002), p. 166.

ILLUSTRATION 6 A map of Paris, in: Georg Braun, *Beschreibung und Contrafactur der vornembsten Stätt der Welt*, Cologne 1582, plan no. 8.
ZBZ, ABTEILUNG ALTE DRUCKE UND RARA: T 45

PARIS pour vray est la maison royalle,
Du dieu Phœbus en splendeur radiale
C'est Cyrrhea pleine de bons espritz,
Tresuigoureux, faisans divers escriptz,
C'est Chrysea en metaulx habondante
Grece de pris en livres florissante
Cum Privilegio

him before they left Zürich for France, that he should study medicine,[33] for which he dedicated himself to reading the doctors of antiquity. Indeed, a decade later, Gessner described his situation at that time with critical words: "Drawn by its fame and its university, in the following year I went to Paris in order to fully apply my ambitions to academics. However, I do not know how – perhaps because of the pressing poverty or through wilful negligence – but when I had just turned 18, I avoided philosophy and the more difficult disciplines, and I followed the most exciting aspects of study, which had been my tendency since childhood. When there is no one to admonish or give counsel, it is an age when one indulges himself, and I pointlessly raced through the various authors: Greek and Latin writers, poets, doctors, philologists, and also some dialecticians and rhetoricians. Although at times I skipped over a lot and read through only a few books in order to satisfy my spirit with sheer variety and diversity in the reading, which for many is an obstacle to progress. Therefore, it is important for me to write about this and set an example for others. They must consider how damaging it is to leave adolescents to their own discretions without a phoenix to teach and admonish them – an Achilles, master of words and mighty deeds. Those who send young people off to study somewhere should be sharply reproached if, in order to save money, they do not also provide an instructor. The extent of the damage is greater that one can describe in just a few words."[34]

Perhaps if Gessner had studied in Paris a decade earlier, he would have been included in the group of students associated with the Swiss humanist Heinrich Glarean (1488–1563), as Conrad Grebel (1498–1526) from Zürich was.[35] However, by 1522, Glarean had moved to Basel, and in 1529 to Freiburg (Breisgau). Not only was the young Gessner in need of supervision, but the situation was further exacerbated by a change in the religious mood with the so-called *Affaire des Placards*. On the night of October 17, 1534, in Paris and other French cities, Protestants hung up placards protesting the Catholic mass. This triggered a wave of persecutions that led to the first death sentence being carried out by November 13. The affair was also a turning point in the history of Parisian universities: "The repression that followed the *Affaire des Placards* led the kingdom and the university to a political system that ultimately had to permit a successful Counter-Reformation."[36] While Fries continued to hold

33 Gessner (1542), f. a_{ii}r.

34 Stotz (1967), p. 213.

35 Büchi (1928), pp. 381f.

36 Tilier (1994), p. 335.

out during this tense climate, others left the country, including John Calvin (1509–1564) and Conrad Gessner.

Gessner left Paris in early December 1534 and travelled with Caspar Mellinger and a certain Rauchwolff[37] to Strasbourg, where he stayed at the house of the reformer Martin Bucer, who was not present at the time. The first letter he sent from there was to his trusted patron Oswald Myconius. He wrote to him on December 21, 1534 that Fries had stayed in Paris because of a swollen foot and the desire to improve his French language skills. As for himself, he would and could no longer stand to watch the tyranny against Protestants, even though he had been in no personal danger. The day before his departure, a distinguished man from Flanders had been beaten to death; although it was said that he was actually a German and that whoever killed him would receive an indulgence. Several hundred people had been arrested, including scholars – some on mere suspicion. Their hands were chopped off, their tongues cut out, they were buried alive, or administered something new called the "little fire," which amounted to being roasted to death. He passed along greetings from his travel companion and housemate, Mellinger.[38] This was actually Caspar Petri, the son of the printer Johannes Petri in Basel. He took Greek lessons in the summer of 1537 from the later Zürcher theologian Rudolf Gwalther (1519–1586),[39] matriculated at the university of Basel in 1537/38, became a medical doctor in 1545, and assumed the chair for theoretical medicine until 1551. Afterward, he was the house physician of Duke Georg of Württemberg.

Six days later, Gessner sent a letter to Bullinger in which he again described horrific scenes of persecution against Protestants. He reported further: "Before all of this upheaval, godfearing people hungrily read your writings, bought them enthusiastically and spoke of you with the greatest respect and admiration. Now, however, nearly all private homes have been searched, and such pious texts must be burned or otherwise thrown in the Seine. I myself and a Spanish scholar friend of mine had many such texts. When they were discovered by our landlord, he would not tolerate them unless we received absolution from a priest. We were able to present the matter to the archbishop of Paris's brother, who at the time was his deputy and a supporter of pure Evangelical

37 Gessner (1577), f. 60v. Until now, this Rauchwolff was identified with Leonhard Rauwolf (1540–1596), but he could not possibly have studied in France in 1541.

38 Letter from Gessner to Myconius, December 21, 1534: St Gallen KB (Vadiana), MS. 32 (VBS 3), 272.

39 This hitherto unknown information can be gleaned from a note that Gwalther placed on the title page of an anthology of printed material kept in the ZBZ under the call number: 5.311. The entry reads, "25 Aug: 1537 Greca rudimenta prolegere coepi D. Gasparo Meling Bassiliensi [sic!]." Cf. Leu and Weidmann (2020), pp. 19 and 179.

doctrine. Thus, we were fortunate to escape the gathering storm. In the meantime, this man has also been arrested. They have not yet tried to attack his brother, the archbishop, despite being well aware of his faithfulness to the Gospels. So, the assailants fill the air with clamor and shouting. Nonetheless, there are several thousand followers of the pure and true teachings who stick together and breathe quiet courage."[40]

Shortly after this, he gave Bullinger an overview of his expenditures. He admitted that Fries had been spending less than he did, for Gessner seems to have been buying books, which compared to today's prices were relatively expensive: "I bought many books, and indeed the best ones, which every day I hope will arrive here. I was gone for nearly two years and had to pay 25 crowns per year for my lodgings. Traveling was also expensive; first to Paris, then to Bourges, and then the return through Paris to Strasbourg. While I could actually list it all out, in short, I do not wish to overload you with the cost of every piece of clothing, book and other necessities. I can indeed provide you with an accounting, just as the people of Athens received from Pericles. If you calculate that one bushel of grain costs one crown, then my yearly stipend is worth about 25 crowns. In the last two years, I have already spent 65 crowns, or about 15 more than my allowance."[41]

From this, it appears that the money from Zürich was just enough for his accommodations. Whether meals were included is unclear, but he obviously must have lived very frugally. Gessner added that Peter Collin had paid the missing funds, who left him his stipend in the amount of 22 crowns, and he requested that Bullinger compensate Collin.[42]

---

40 Hanhart (1824), pp. 38f.; edited in: HBBW 4, pp. 457–461.

41 Hanhart (1824), p. 40; edited in: HBBW 4, pp. 457–461.

42 Gessner's stipend was 28 guilders annually and 6 Mütt of kernels or the money for it. At the beginning of 1533, one Mütt of grain was equivalent to 3.4 pounds or 1.7 guilders. 6 Mütt therefore had an equivalent value of 10.2 guilders. Added to the 28 guilders, one arrives at an annual stipend of 38.2 guilders. This amount corresponds approximately with the annual 25 crowns mentioned by Gessner according to the following equations: 1 guilder = 15 batzen, 38.2 guilders = 573 batzen; 1 crown = 25 batzen, 25 crowns = 625 batzen. The sometimes strongly fluctuating grain prices must be taken into account. Cf. HBBW 4, p. 459.

CHAPTER 3

Marriage and Medical Studies

In Strasbourg, Gessner broadened his Hebrew language skills and hoped to find work as a teacher. Martin Bucer wrote glowingly to Bullinger about the young man, that he was born to a higher calling, and with assurances that he would arrange for him to study rhetoric and mathematics under the theologian and Aristotle expert, Simon Grynaeus (1493–1541) in Tübingen.[1] Nonetheless, the school administration ordered Gessner back to Zürich, despite a February 4 request from Myconius to the head of Zürich's church not to force his return home.[2] Gessner had now returned to Basel in order to avoid breaking off his studies abroad.[3] Nonetheless, the opinions of Bucer and Myconius were disregarded and ignored in Zürich.

In either Strasbourg or Basel, the 19-year-old fell hopelessly in love with a pretty young woman named Barbara Singysen. She must have completely turned his head, because it seems, within only a few weeks, he had married her.[4] He was obviously not completely comfortable with the matter, because he chose not to travel directly from Basel to Zürich, but spent instead some time in the mineral baths at Baden. There, he wrote on March 24, 1535 to Heinrich Nüscheler, the student office administrator, that he needed more money. At this point, it was also quite clear from his writings how enraptured he must have been with his sweetheart: "I had hoped to easily cover everything here at the baths with the money I received from you. But, what can I say? Fate, which rules over everything and stirs things up, dashed my hopes – as it is wont to do. This is only a figure of speech, since there is no fate if we believe in God and providence, and unlike heathens, we do not speak of fate blindly spinning her fickle wheel. Instead, the past and future are arranged in a predestined order that leads the willing onward and draws away the unwilling. But I should stop speaking obscurely and trust you and your friendly hearts, and lay out my intentions under the condition that you keep it to yourselves until I have returned to Zürich. I have taken a wife. She is not rich, but young, modest and chaste (the most beautiful of gifts). Should I also add that she is delightfully beautiful, but it is not for the sake of beauty but for the sake of

1 HBBW 5, pp. 75f.
2 Simmler (1566), f. 5r.
3 HBBW 5, p. 84.
4 The announcement date of the marriage in Zürich was May 4, 1535. Cf. Fischer (1966), p. 16.

© KONINKLIJKE BRILL NV, LEIDEN, 2023 | DOI:10.1163/9789004541696_004

character that one should choose a wife. I followed the advice of the wise man who was questioned by a youth who was desired by two women. Whom should I marry, the rich or the poor one? The wise man answered: take the one most suited to you – choose the one who behooves you. I will never leave her as long as I live, even if everyone is against my marriage; not just because I love her with all my heart, but because I have also given my word of faithfulness to her. Furthermore, she will not hinder my academic activities, but rather encourage them. Please send four guilders through this carter from Strasbourg, and if necessary, make a plea on my behalf. If you act as a father to me, you will surely find me ever the obedient son. Farewell. If you think it would be better to do so, then broach the issue with Bullinger and show him this letter. But show only him, for he is my singular and greatest patron. I will be home for the Easter festival celebration. Once again, farewell."[5]

Nothing more is known about Barbara Singysen. She likely came from a family of modest means in Liestal near Basel, possibly with the surname Bantli, since Gessner's widow was named Barbara Bantli in relevant documents. In Zürich's death and marriage records, there is no mention of her early death or of a second marriage by Gessner, so one would assume that Barbara Singysen and Barbara Bantli are one and the same person.[6]

Gessner's first biographer, Josias Simmler (1530–1576), dryly remarked that he did not know who might have advised him to go through with this untimely marriage,[7] which mildly suggests that those around Gessner were anything but supportive. Clearly, people had invested in the young man with certain hopes and expectations, which were now shattered. Once back in Zürich, Gessner became aware that, in Basel, Myconius was enraged about the spontaneous marriage, so he wrote a letter to him on June 30 to try to smooth things over. He admitted to his mentor that his actions had been ill-considered and were the result of youthful recklessness. The marriage, however, would not hinder his studies, but would bring with it a new drive and urgency. He begged Myconius not to shun him. In closing, he complained about how he was being treated in Zürich because of this issue: "I have been shunted off to teach grammar and beginner classes at the lowest grade, and my stipend has not been increased in the least. To the great disadvantage of my studies, I teach at the school for three hours every day, and I would gladly do anything to be free from this work. I would much rather prove by deeds what I am capable of accomplishing, than praise and promote myself. I could accomplish a lot in the natural sciences,

5 Hanhart (1824), pp. 44–46.
6 Guyer (1973), p. 82.
7 Simmler (1566), f. 5r.

pharmacology and logic as well as the three scholarly languages, since I have also been working on Hebrew for some time."[8]

For the following nine months, all the sources fall silent about Gessner's life. In his autobiography, however, he did look back on this gloomy period of teaching at the lowest grade level: "When I returned to Strasbourg from France, I was called back to my home town. Shortly after arriving, due to my untimely marriage, I was shoved in a corner of the public school for a while, where I taught the boys elementary grammar for a good portion of the day. I had only a minimal stipend, so I had no choice but to go hungry. In my free time, I read books by physicians."[9] On April 17, 1536, he again sought contact with Myconius, whose silence weighed heavily upon him. In this letter, he once more expressed his interest in medicine by offering to present Myconius with a test of his medical knowledge. Being involved in this science "seemed most appropriate in my younger years. Theology will include me among her students when I reach a more mature age. Since I already know that piety is appropriate at any age, then indeed, the study of theology will not be carried out successfully without scientific knowledge in many other areas. It will be my goal one day, but first I will acquire all the necessary tools."[10]

A short time later, Gessner apparently traveled to visit Myconius in Basel. In any case, the latter wrote to Bullinger on June 2 that he had recently sent Gessner off without writing a letter to him.[11] Obviously, Gessner had once again won over his mentor in Basel, since on June 12, 1536 he sent him a letter and thanked him for recommending him to Johann Jakob Ammann, who was waiting for an opportunity to propose another promotion of Gessner to Bullinger. If the plan worked, then, to an extent, Myconius would have saved Gessner's life, for what was a life without science? He asked Myconius not to think that he had lost interest in the classical studies just because he had developed a preference for medicine.[12]

Four days later, Myconius picked up his pen and wrote to his Zürich colleague Bullinger to ask him to improve Gessner's situation (ill. 7). Every scholar who knew the young man spoke ill of those in Zürich who would let such a talent be wasted. In light of the dirth of brilliant people, it pained Myconius, above all, that Gessner was not being supported, in particular because of what a great asset he could potentially be to Zürich.[13] On July 3, Gessner complained

8 Hanhart (1824), pp. 48f.

9 Stotz (1967), p. 213.

10 Hanhart (1824), p. 50. Manuscript version in: St Gallen KB (Vadiana), MS. 33, (VBS 4), 24.

11 StAZH, E II 336, f. 153 r/v (new: 172 r/v).

12 St Gallen KB (Vadiana), MS. 33 (VBS 4), 34.

13 HBBW 6, p. 336.

ILLUSTRATION 7 Portrait of Heinrich Bullinger by an unknown sixteenth-century artist. ZBZ, GRAPHISCHE SAMMLUNG UND FOTOARCHIV: INV. 483

again to Myconius about his difficult situation. Ammann had still not spoken to Bullinger and he felt that an increased stipend would not be approved. Gessner felt that the head of Zürich's church was not well disposed towards him and often yearned to have Zwingli back. He then planned to inform the canons that he would have to leave Zürich in order to avoid going into debt. In France, he had already spent more than his stipend allowed. Most of the books purchased there had been lost in a shipwreck. Furthermore, he had given 12 guilders to his brother and spent another 10 guilders on a pledge. Most regrettable, however, was that his studies had languished. He had no time to study and no money for books.[14]

Regardless how bad his situation was, Gessner did not give up, and he devoted every free minute to reading a variety of texts, especially by medical authors. Along with the same July 3 letter to Myconius, he also sent back a medical work by a certain Epiphanius. Gessner's judgment of the book contained his earliest praise for the Greek physician Galen (ca. 130–200), whom he held in high regard. While Epiphanius seems to owe almost everything good he brings to the Arab physicians, Gessner misses the mention of Galen in his work. Whereby: "Galen is unanimously considered by all scholars to be the only one who could revive the sadly deteriorated state of pharmacology, and the only one who deserves to be read through and through in secondary school. When I speak of Galen, then I also mean Hippocrates, who would be of little use to us without Galen."[15] Like Aristotle, Galen remained an important groundbreaking figure throughout Gessner's life. In his 1545 *Bibliotheca universalis* he placed Galen, Aristotle and Plato together on the same level. What the latter two meant to philosophy and the sciences, Galen was to medicine.[16]

For a year and a half, it seems that no one in Zürich was looking out for Gessner's future. He was to be taught a lesson and was left to brood about it. It is not known exactly when Antoine Saunier,[17] a pastor from Geneva enquired in Zürich for an unmarried teacher with good Greek and Latin skills to lead the Geneva School, and proposed Gessner. Pellikan reported to Fries on July 13, 1536, however, that this too had been rejected. He then asked if Fries might imagine himself committing to such a position for a good salary, but Fries did not act on it.[18] There is no evidence whether Gessner being married was the reason for officials in Zürich not to accommodate Saunier's request, or because it was part of their disciplinary action. However, one has the impression that it was probably the latter.

3.1 Basel

Sometime in the following months, the tide turned for Gessner. Prior to November 11, 1536, he was back with Fries in Basel again,[19] where he wrote to Bullinger to request that arrangements be made for the grain ration allotted

---

14 St Gallen KB (Vadiana), MS. 33 (VBS 4), 37.
15 Hanhart (1824), p. 54.
16 Gessner (1545), f. 169v.
17 Bodenmann (2016), pp. 118 and 243–245.
18 ZBZ, MS. S 41, no. 20. Edited in: CdR, vol. 4, pp. 78–81.
19 Bührer (2002), pp. 166f.

by the student office ([*Studentenamt*] six bushels of grain per year = about 35 cubic ft.) to be passed along to the person responsible for shipping. Gessner wanted all six bushels, and Fries was content with five.[20] On December 30, Gessner wrote again to Bullinger to thank him for his benevolence and to swear that he would give his all, so that no one would regret their charitable acts on his behalf. He also asked that the church leader should care that the Greek commentaries on Aristotle be purchased by the Grossmünster church library (Stiftsbibliothek), which was also the library for the *Hohe Schule*. He continued: "For a thorough understanding of philosophy based on the sources, the Greek commentaries on Aristotle are indispensible. However, copies are extremely expensive, and if one of us were to purchase one or the other, it would deplete his entire stipend. Thus, you will surely be doing a good deed in support of our studies by contributing these titles to the library for common use. Fries and certainly all the other students would make the same request to you. There is no greater treasure with which you could grace our library. If you support this, the others will also agree, and when the purchase is approved, Fries and I will make sure that they are obtained at the best price."[21]

It is possible that Bullinger responded to Gessner's proposal and ordered two corresponding works, namely, translations of the *Eudemian Ethics* (*Magna moralia*, Paris 1535) and the *Nicomachean Ethics* (Paris 1541) by Aristotle and translated by the Byzantine humanist Johannes Argyropoulos (1415–1487) with commentaries by him and other authors. Both titles ended up in the 1540s in the *Stiftsbibliothek* and were bound together in a volume that is currently preserved by the ZBZ.[22] A search in what remains of the *Stiftsbibliothek* at the Grossmünster for other titles that may have been purchased at Gessner's urging has been unsuccessful. This is not surprising, since the library's yearly budget was only ten guilders.[23] If one looks through Gessner's private library, a relevant volume may be found of a commentary on Aristotle by the late antique Aristotle scholar, Alexander of Aphrodisias (ca. 200 AD), which was published in Paris in 1536. It may be that Gessner had purchased it himself[24] because it was not available in the library. As a result of relatively high book prices, the holdings of private and public book collections complemented each other in Zürich during the early modern period.

---

20 HBBW 6, p. 468. Cf. also the letter from Bibliander to Oporinus of December 20, 1536, in which he sends his regards to Gessner (ZBZ, MS. S 41, no. 206).

21 HBBW 6, p. 499.

22 ZBZ, call number: V E 24. Cf. Germann (1994), p. 322.

23 Leu (2004), p. 82.

24 Leu et al. (2008), p. 46.

In Basel's university register, it is noted that Gessner enrolled at the university in the spring 1537 for the usual fee of six shillings (about 1/7 guilder).[25] According to his autobiography, he received the same stipend as before in Bourges and Paris, which was insufficient for the study of medicine. This meant he had to keep an eye open for other sources of income.[26] This ten-month stay in the city on the Rhine was the longest period spent there during his lifetime. In addition to studying medicine, Gessner also devoted himself to collecting plants[27] and doing work on the side in the burgeoning printing trade, which brought him welcomed earnings. A degree of insight into the prosperity of certain printers can be gained from the following quote by Thomas Platter, who founded a printing business in Basel in 1535 with Johannes Oporinus (1507–1568), Balthasar Ruch and Robert Winter (ca. 1500–1554):[28] "However, because I saw how good Hervagius and other printers had it – with very large profits for little work – I thought that I would like to be a printer too. Likewise, D. Oporinus, who had done much proofreading in printing shops, thought the same. There was also a good typesetter in the 'Zum Sessel' printing house, Balthasar Ruch, he was peacockish and also wished to be rich; he was a good employee of mine and Oporinus. Indeed, we had good plans, but no money. Oporinus's brother-in-law Ruprecht Winter had a spouse who wanted to be a printer's wife. She saw how nicely printers' wives were dressed, and how some flaunted it because they had so many (too many) nice possessions. She urged her husband Ruoprecht to become a printer together with the brother-in-law Oporinus. So, that was our group of four: Oporinus, Ruoprecht, Balthasaar and myself. We bought the tools from Andreas Cratander, who, together with his son Polycarpus, had become a bookkeeper, because his wife no longer wanted to have to deal with all the mess (so she said). We gave him 800 guilders for the tools, to be paid over a certain period."[29] The proofreader Peter Schmid from Zürich is a good example of how not only printers, but also their employees, sometimes possessed considerable property. On June 26, 1561, he sold his house to Bernhard Sprüngli († 1568) for the price of 1,014 guilders.[30]

---

25 His enrollment took place in 1537 under the rectorate of Andreas Bodenstein, called Karlstadt (1486–1541), who was close to the Anabaptists. Cf. Wackernagel (1956), p. 16.

26 Gessner (1545), f. 180r.

27 Buess (1948), p. 6.

28 Steinmann (1967), p. 11.

29 Platter (1999), p. 118.

30 Cf. the handwritten annotation by Wolfgang Haller (1525–1601) on this date in: *Lassbüchlin sampt der Schrybtafel, Mässen und Jarmarckten, uffs Jar M.D.LXI. Jar*, Zürich, Christoph Froschauer the Younger, 1560, ZBZ, shelmark: MS. D 270_3.

3.2 First Publications

Gessner's first publications were released during these months. In the summer of 1537, an anonymous medical student published five comprehensive tables that summarized the content of the *Compendiaria in artem medendi introductio* by the Tübingen medical professor Leonhart Fuchs. At the foot of the title page one finds an epigram in Greek and Latin that is attributed to a certain Thrasyboulos Gessneros, which is a Greek version of Conrad Gessner (ill. 8).[31] It is possible that he was also the anonymous student mentioned in the subtitle.[32] Apart from this, he worked for a considerable time for the book printer Johannes Walder († 1541), originally from Zürich. He was well-known to Gessner's teacher Thomas Platter in Zürich, who asked him to be the godfather of his son Felix, born on October 28, 1536.[33] Gessner's work involved revising a Greek-Latin dictionary by the Italian scholar Guarino Favorino (1450–1537). He greatly expanded the vocabulary, but inexplicably, and to Gessner's great chagrin, Walder used only a small portion of the studiously compiled additions.[34] Furthermore, the printer had the gall to refrain from mentioning Gessner in either the title page or the foreword, when the work was published in September 1537. Nonetheless, in 1548, Gessner was magnanimous enough to underscore and compliment the quality of Walder's printings.[35] The dictionary was printed at least fourteen times during the sixteenth century[36] and was reworked by Gessner three or four times. In doing so, he not only added to the content but also corrected countless absurd typesetting mistakes that resulted from the ignorance of the proofreaders in the printing house.[37] In the third edition of September 1543, there is a famous foreword by Gessner with the title *About the Use and Excellence of the Greek Language*. It is well known that humanists and reformers held the three holy languages (*tres linguae sacrae*) in high regard, since the Bible and works by the church fathers were written in Hebrew, Greek and Latin.[38] Gessner and various humanist colleagues of his showed their great enthusiasm for the Greek language and Greek manuscripts. From this foreword, it was clear that Gessner held the ancient Greek language in higher esteem than the Latin commonly used by scholars. He compared the

31 "Kon" = Old High German "bold," "rath" = "council" (= Greek: Thrasy-boulos).

32 Hieronymus (2005), vol. 1, pp. 229–231.

33 Platter (1999), p. 121.

34 Stotz (1967), p. 213.

35 Gessner (1545), f. 87r; cf. Wolkenhauser (2002), p. 319 (misquoted here as a preface to Book VII, but is Book VIII).

36 Wellisch (1984), pp. 31–34.

37 Gessner (1562), f. f. A_2r/v.

38 Leu (1990), pp. 275–287.

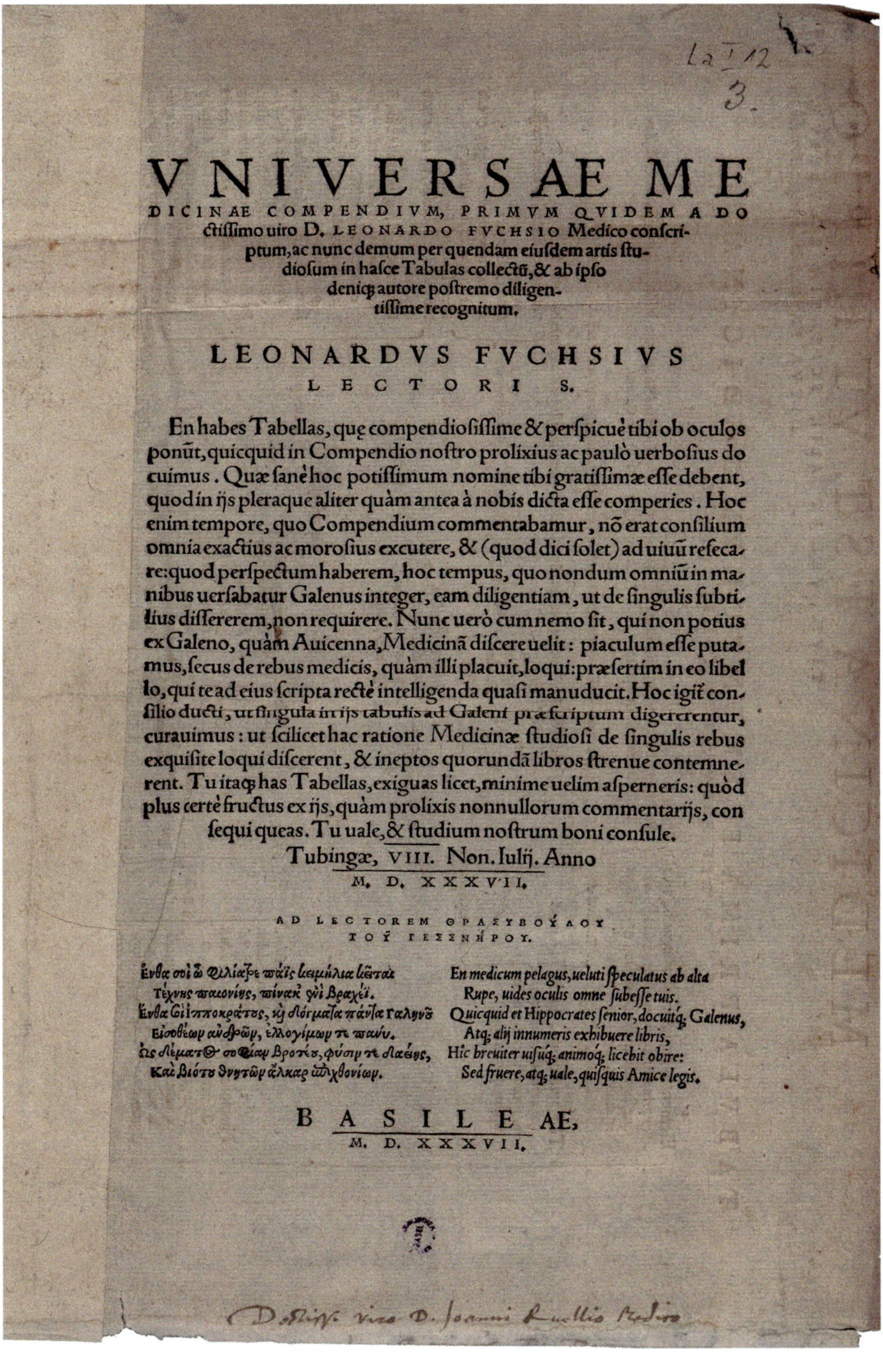

ILLUSTRATION 8 The Greek and Latin *distichon* on the title page is by Conrad Gessner. *Universae medicinae compendium* ..., by Leonhart Fuchs, Tübingen 1537.
UNIVERSITY LIBRARY BASEL: LA I 12.3

former to the sun and the latter to the moon. Every educated person should have a command of Greek, since most scientific terms originated from this language, as did, for example, the elemental tools of rhetoric and poetry, the foundations of mathematics, and medical nomenclature, not to mention, theology. God had the New Testament written in this language. It is through Greek that the individual receives new dimensions of human and divine wisdom. With regard to the area where ancient Greek had been spoken, Gessner knew that it had been much larger when compared to his own time: It had been spoken in Greece, Asia Minor, Ionia, parts of Italy and even among the Gallic druids.

Gessner's firm knowledge of Greek, which was not very common during the era of the humanists,[39] did not go unnoticed, as was shown during the course of the year 1537.[40] In Basel, an opportunity opened for his former student colleague Johannes Fries, who became the head of the newly created *pedagogium*, the first class of the Faculty of Arts. At the same time, he was also chosen to replace Benedikt Finsler as rector of the Fraumünster School in Zürich. Fries informed Bullinger on March 7, 1537 that, due to his commitments in Basel, he would not be able to fill the post in Zürich. He added that Gessner could use such a position much more than himself.[41] Regardless, the gentlemen in Zürich were insistent, and after they had turned away offers for him from Geneva and Lausanne, they ordered Fries back to the city on the Limmat River in the fall of 1537. They had their own plans for him.[42]

In consultation with Bullinger, a solution was also found for the now 21-year-old Gessner. Myconius had recommended him to the theologians Peter Kunz (1480–1545) and Caspar Megander (1495–1545) in Bern. In 1537, the latter played an important role in the search for someone to be in charge of the newly founded Latin school (*collège*) in Lausanne. He was also responsible for finding

39 Even the reformer Heinrich Bullinger said of himself that his knowledge of foreign languages was not the best. Cf. Bullinger (1904), p. 128.

40 In general, Gessner must have been an excellent classical philologist, because even in later years he was asked for advice in matters of classical languages, for example by the St Gallen theologian Josua Kessler, who sent him a Greek funeral poem on the recently deceased Vadian for review. Gessner corrected the five distichs, but they were not published until 1620. Kessler also contributed a dedicatory poem to the *Novum dictionariolum puerorum Latinogermanicum* (1556, fol. $*_3$v) by Johannes Fries. Fries thanked him in a letter and mentioned that Kessler's poem had been corrected in certain points before printing in consultation with colleagues in Zürich, which probably refers to Gessner, who had studied with Fries in Paris. Cf. Müller (2017); Letter from Fries to Kessler in Kantonsbibliothek St Gallen (Vadiana), VadSlg MS. 37:115 (I owe this reference to Dr. Clemens Müller, St Gallen).

41 HBBW 7, pp. 87f.

42 Bührer (2002), pp. 169f.

teachers of Greek and Hebrew in the likewise newly established *Académie*.[43] This was after the Savoy canton of Waadt had been conquered by Bern in 1536 and had converted to the reformed faith.[44] On August 18, the Bern city council hired Gessner as a professor of Greek in Lausanne,[45] and on September 19, he left Basel.[46] Formerly an episcopal see, Lausanne had between five thousand and six thousand residents at the beginning of the sixteenth century.[47] Looking back in 1543, Gessner gave thanks for his position to the councilmen and mayors at the time, Hans Jakob von Wattenwyl (1506–1560) and Hans Franz Nägeli (1497–1579) in his preface for an edition of *Sententiae* (proverbs and adages) by the Macedonian Johannes Stobaeus (fifth century AD). It cannot be substantiated whether contact to his student colleague in France might have played a role, namely, the Bern patrician Johann Steiger (1519–1581),[48] to whom he dedicated *Appendix quadrupedum viviparorum* (*Inventoy of Four-footed, Live-bearing Animals*), published in 1554.[49]

---

43 Crousaz (2012), p. 539.

44 Locher (1979), pp. 554–559.

45 HBBW 7, pp. 225f.

46 HBBW 7, pp. 241–243. In this letter, Gessner approached Bullinger about an inheritance dispute and asked the antistes for help.

47 Biaudet (1982), p. 171.

48 Hanhart (1824), p. 34, refers to the family closeness to the Bern mayors (*Schultheiss*) Nägeli, whose daughter Steiger married. However, he seems to have overlooked the fact that this second marriage was not consummated until 1567, two years after Gessner's death. Cf. Johann Steiger, in: Neues Schweitzerisches Museum 1 (1793), p. 901.

49 Published at the end of Gessner (1554), there p. 1.

CHAPTER 4

Three Happy Years in Lausanne

In his 1520 work, *To the Christian Nobility of the Christian Nation*, Martin Luther had already spoken out for the establishment of Protestant educational institutions. In Zürich, Zwingli had also fostered this idea and brought it to fruition in the aforementioned *Hohen Schule*, founded in 1525. Zürich's example was followed by Bern in 1528, Lausanne in 1537 and Geneva in 1559. While the Latin school in Zürich was meant to last five years, the one in Lausanne lasted for seven. The label "Latin" school should not obscure the fact that the attending 7–17 year-old students were also learning ancient Greek. The so-called *Collège* was followed by a three or four-year program at the *Académie* where the intended emphasis was on theology, Greek, Hebrew and above all the education of reformed pastors.[1] The Swiss model of the reformed *Hohen Schule* lived on further in other countries, for instance, in Herborn and Pinczow (northeast of Krakow, Poland).[2]

In Zürich, five professors were hired, in Bern there were three (from 1574 four), and four each in Lausanne and Geneva. Starting in 1541, Gessner assumed the ancillary role of *Professio physica*, a position exclusive to the city on the Limmat. It was originally planned that the two pastors from Lausanne would run the Theology Department, namely Pierre Caroli[3] (1480–1545) covering the Old Testament and Pierre Viret (1511–1572) the New Testament. However, because Caroli accused Calvin and Viret of anti-Trinitarianism, the Bern government stripped him of all offices on June 3, 1537. As a result, the teaching of both the Old and New Testament fell completely on Viret's shoulders.[4] For the Hebrew professorship, they had considered Evander (Benedikt Kienisen), a stipend holder from Zürich.[5] In the end, however, they chose the fairly unknown Jean Adoutout, who replaced Hymbert Pecolet († 1548) in 1538,[6] whose house near the cathedral in Lausanne was next to Gessner's (ill. 9).[7] Others in Gessner's circle of friends in Lausanne were the doctor and theologian Beat Comte († 1578), who took over Caroli's position as pastor, but in 1545

1 Im Hof (1978), pp. 246f.; Fatio (1987), p. 49.
2 Büsser (1999), p. 20; Im Hof (1978), p. 258.
3 Cf. Bodenmann (2016).
4 Crousaz (2012), pp. 72 and 534.
5 CdR 4, p. 167.
6 Crousaz (2012), pp. 72, 528 and 540f.
7 Olivier (1951), p. 407; CdR 4, pp. 333–335.

© KONINKLIJKE BRILL NV, LEIDEN, 2023 | DOI:10.1163/9789004541696_005

also had to give it up due to theological differences.[8] During Gessner's stay in the western part of Switzerland, he also had a long friendship with Jean Ribit († 1564), who married Agnes Rosin from Zürich in 1538.[9] For a time, he was possibly a boarder at Gessner's home in Lausanne[10] and in 1541 became Gessner's successor as professor of Greek. In 1546 Gessner published a collection of ancient Greek and Byzantine texts by Antonius Melissa (eleventh century) and Maximus the Confessor (580–662) as well as the early Christian apologists Theophilus of Antioch († 181) and Tatian († 170). Ribit strongly urged him to add a Latin translation to the Greek text,[11] whereby he translated two large sections into Latin himself.[12] The letters between Gessner and Ribit are extant until 1554 and cover various topics from botany to biblical interpretations and even include illustrations of fish, sent by Ribit to Gessner in Zürich.[13]

Gessner had good memories of the three years spent with the aforementioned friends and looked back fondly in his 1545 autobiography: "As soon as my year in Basel was over, I suddenly had the opportunity to teach Greek in Lausanne on Lake Geneva – and with a large stipend from the high council of Bern. So I taught there for three years and lived well with the heartfelt friendship of pious and scholarly men. Among others were Pierre Viret, Béat Comte, the professor of Hebrew Himbert, and Jean Ribit, who became my successor."[14] The warm-hearted closeness with Ribit and Viret is also expressed in corresponding passages in Gessner's *Bibliotheca universalis* of 1545, where he characterizes Ribit as a very scholarly gentleman and his best friend. Here, he describes Viret as well-versed in theology and other sciences, as an expert in the *tres liguae sacrae*, a responsible head of the Vaud church, a gifted teacher, a person of integrity and a living role model. He wished him a long life in order to remain an asset to the church, both vocally and in writing, for a long time to come.[15]

The substantial grant that he mentioned was for 200 guilders, two bushels of grain, and two buckets[16] of wine per year, which at the time amounted

---

8 Crousaz (2012), p. 535.
9 Crousaz (2012), p. 542.
10 CdR 5, pp. 310–312.
11 Antonius Melissa and Maximus the Confessor (1546), f. $\dagger_3$v.
12 Gessner (1562), f. A_7r.
13 The correspondence is preserved in: Bibliothèque nationale de France (Paris), Département des manuscripts, call number: Latin 8641, f. 9v–70r. I owe this reference to PD Dr. Reinhard Bodenmann (Brugg).
14 Stotz (1967), p. 214.
15 Gessner (1545), f. 450v and 554r.
16 1 bucket = ca. 100 liters.

ILLUSTRATION 9 A city map of Lausanne, in: Matthäus Merian, *Topographia Helvetiae, Rhaetiae et Valesiae*, Frankfurt/M. 1642. Gessner lived across from the cathedral apse.

ZBZ, ALTE DRUCKE UND RARA: T 137

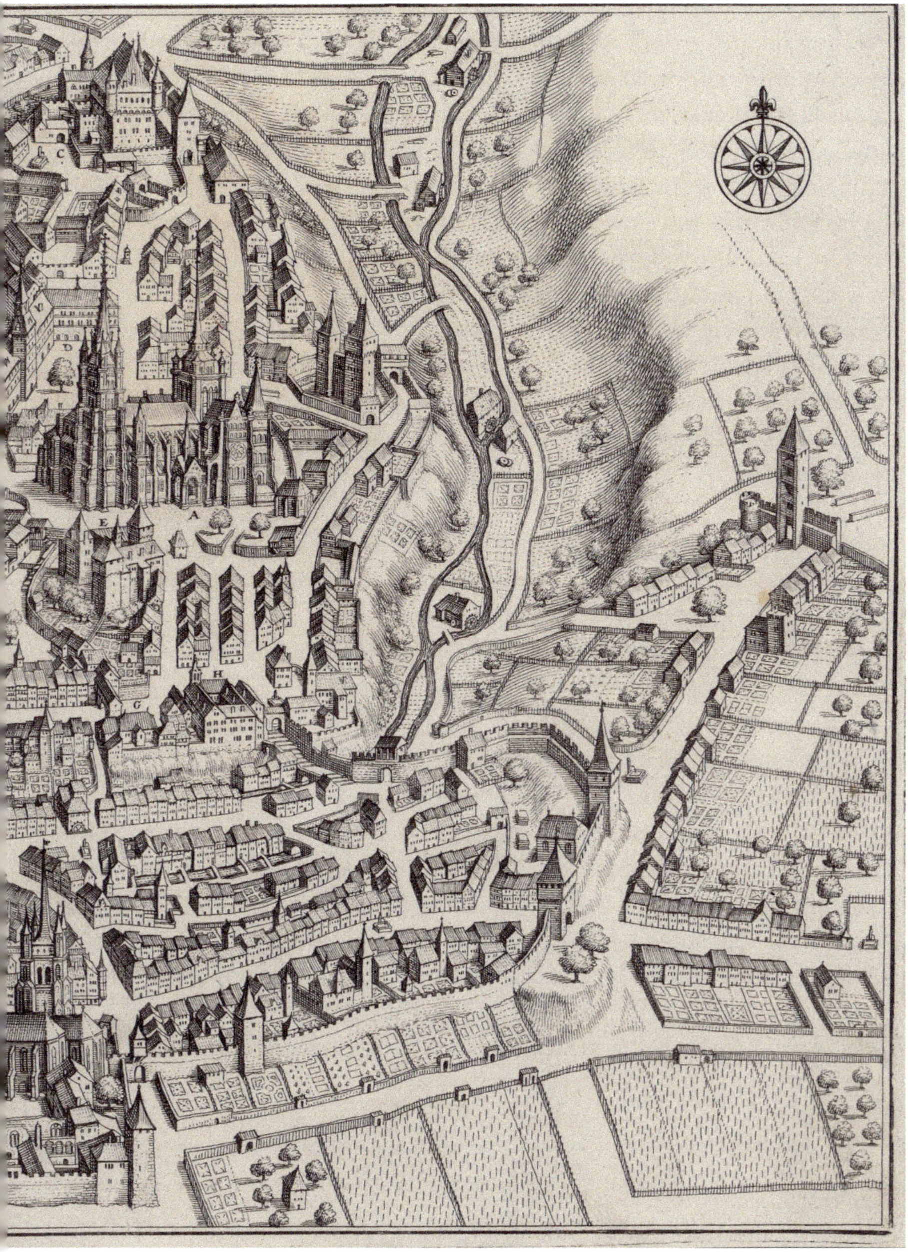

to a very good income for a 21 year-old.[17] The oldest document with a view to revealing details – the *Leges scholae Lausannensis* of 1547[18] – contains no trustworthy information about the content of Gessner's teaching. Only one letter to his three-year-younger colleague in Zürich, Rudolf Gwalther, written on June 24, 1539, gives minimal insights into what Gessner was undertaking with his roughly dozen students.[19] He wrote that he had read through various works by Aristotle, and now he was reading the *Theriaca* by Nicander (2nd century BC). Additionally, he had attended a lecture on Genesis by Pécolet, and a presentation about the Gospel of Matthew by Viret. At that time, they were also beginning with the prophet Isaiah in Hebrew. He mentioned in a letter to Calvin on September 28, 1544, that he was also focusing on the French language,[20] which he had learned in Bourges and Paris, but of which he did not yet have a good command.[21] The suggestion that he was reading Nicander's *Theriaca* along with the students shows that he was very free with his teaching subjects. Indeed, what did this educational poem about animal toxins have to do with giving prospective pastors an education in Greek? It is unclear whether the Orphic poems, handwritten in Greek, that Johannes Fries sent to Lausanne were also used in his classes.[22]

Gessner's presence in Lausanne was reason enough for his younger colleague in Zürich, Rudolf Gwalther, to travel there and spend at least seven months in western Switzerland.[23] Since the fall of 1538, Gwalther had studied at the University of Basel and lived at the home of Myconius. On July 13 1539 he wrote letters to Bullinger and also Myconius, and from the latter, he asked for forgiveness for his sudden and unexpected departure. Due to pressure from his traveling companion, there had not been an opportunity to say goodbye to Myconius and his wife. He was now living with Béat Comte, because too much German was spoken at Gessner's house, and he very much wanted to learn French.[24] Further, he let Bullinger know that he could study more freely there than in Basel, and that he was working on a Latin translation[25] of the *Onomastikon* by Julius Pollux (2nd century). This was a Greek dictionary of

17 CdR 6, p. 341f.
18 Junod / Meylan (1947). About the Greek professorship, ibid., p. 29.
19 Crousaz (2012), pp. 255–329.
20 CdR 5, pp. 334f.
21 Calvin (1873), p. 745; cf. Crousaz (2012), pp. 235f.
22 On the parchment binding there is the following dedication by the hand of Johannes Fries: "A Monßieur Conrad Gesner … Losanne." ZBZ, MS. C 296.
23 Gwalther's last letter to Bullinger from French speaking part of Switzerland is dated January 24, 1540. HBBW 10, pp. 39–43.
24 CdR 5, p. 365.
25 Gessner (1553), f. 295v.

synonyms, organized by subject areas, that was published in 1541 by Robert Winter in Basel.[26] Additionally, he was reading the New Testament in French, attending theological lectures and studying Cicero. He asked Bullinger for permission to spend a year in Lausanne,[27] which was approved a short time later.[28] In another letter from Gwalther to Bullinger, written on August 16, 1539, he mentioned that Gessner and his wife were spending time at the hot springs in Leukerbad.[29] Fourteen years later, in a work he published about thermal baths, Gessner briefly mentioned this visit, which had lasted nearly twenty days. He wrote that the hot springs there had warmed him and strengthened his muscles, but that they also dried them out.[30]

In another later publication, which was about the ascent of Mt Pilatus near Lucerne, Gessner mentioned that he had also explored the Savoy Alps during his time in Lausanne. With reference to his description of Lake Pilatus, he wrote: "I also remember seeing a similar lake in the Savoy while on one of the highest mountains close to the Cluses fortress. However, there was no superstition associated with it. It was nearly round, very small, but so deep that it was believed to extend down through the entire mountain. The following account was considered proof: Once, an ox fell into the lake, and some time later, its head and horns were found at the foot of the mountain by a wellspring near Cluses."[31] Gessner used these excursions in western Switzerland and the Valais canton to collect plants for botanical study. Because the botanical information from antiquity was confined to the Mediterranean area, his study of alpine plants was still unexplored territory.[32]

4.1 Works from Western Switzerland

During his time in Lausanne, the first scientific publications were released under Gessner's name. In his autobiography of 1545, he mentions five of them. The first was *Historia plantarum et vires ex Dioscoride, Paulo Aegineta, Theophrasto, Plinio, & recentioribus Graecis*, which was published in 1541 by Robert Winter in Basel. It is an alphabetized index of plants and their medicinal qualities as

26 Hieronymus (1992), pp. 112–114.
27 HBBW 9, pp. 176–179.
28 HBBW 9, pp. 213–218.
29 HBBW 9, pp. 199f.
30 Gessner (1553), p. 295v.
31 Hanhart (1824), p. 182. Cf. the original text in: Gessner (1555), p. 53.
32 Gessner refers in various manuscripts and printed works to plants he saw in Lausanne or Savoy. For example, in Gessner (1555), pp. 22, 23 and 31.

described by the Greek physician Pedanius Dioscorides (1st century), Paul of Aegina (7th century) and earlier Greek authors, as well as the ancient natural scientists Theophrast (371–287 BC) and Pliny the Elder (23–79 AD). Along with the Latin plant names, Gessner also included the Greek and sometimes Arabic names. The book's dedication to his colleague from Basel student days, the patrician Heinrich Billing († 1541), was written by Gessner in Lausanne on August 9 of (probably) 1541.[33] In it, he mentions that the idea for this medicinal handbook came from Zürich's municipal physician, Christoph Clauser, whom Gessner had visited a few months before.[34] The work apparently met a certain level of demand, as it was also printed in Paris and Venice in the same year. In July 1562, Gessner made handwritten edits in his own personal copy from the Venice printing, possibly with an eye toward a new edition.[35]

The next work that was mentioned in his autobiography was titled *Apparatus et delectus simplicium medicamentorum*, which was published in Lyon by Jean and François Frellon in 1542. The dedication to Christoph Clauser, dated January 13, 1541, was written in Lyon after Gessner had already left Lausanne. While the aforementioned title was intended for those studying medicine, Gessner saw the *Apparatus* as a useful work for apothecaries.[36] Within, he not only describes matter derived from plants that can be used for the creation of medicines, but also materials from animal and mineral sources. In addition, he also included compilations from the works of Galen and Paul of Aegina that mention ingredients used in remedies.

Gessner titled the third work from the Lausanne period his *Compendium ex Actuarii Zachariae libris de differentiis urinarum, iudiciis et praevidentiis*, which was printed in 1541 by Christoph Froschauer (ca. 1490–1564) in Zürich. This was a condensed Latin version of perhaps the most important uroscopic work about the varieties of urine (*De urinis*) written by the Byzantine physician Johannes Zacharias Actuarius (1275–1328). It had first been published in Latin in 1519, but Gessner referred back to the Greek manuscripts, in which he corrected various faulty passages that were likely due to the transcribers' poor Greek skills.[37] The inspection of urine was one of the most important diagnostic medical procedures in the early modern era.[38] Even in Gessner's

33 HBBW 9, p. 278.
34 Gessner (1541), f. α_3r.
35 On the title page of Gessner's copy, which is kept in the ZBZ (call number: D 459_2), there is a corresponding handwritten note by him.
36 Gessner (1545), f. 181r.
37 Christoffel (1953), pp. 119f.
38 In the course of the sixteenth century, there was a certain turning away from urine-analysis because it was recognized that it could only be used as a diagnostic tool to a limited extent. Cf. Stolberg (2009), pp. 167–212.

correspondence, his patients' urine is mentioned repeatedly.[39] It is no wonder that the doctor in Jost Ammann's 1568 book of occupations is depicted holding a flask of urine (ill. 10). The accompanying text reads:

> I am a doctor of medicine
> Who can clearly see from urine
> A person's burdensome disease
> With God's mercy, I can ease
> With a prescription or a syrup
> To cause his illness to stop
> To make a person well again
> The Arabs invented this medicine.[40]

Gessner added a short text to the Actuarius compendium called *Sylvula experimentorum Galeni*, in which empirically proven information is compiled about the application and effectiveness of various remedies found in the works of Galen. This included the treatment of suppurating abscesses with urine.[41] It was noted that not only human urine had such uses but also that of animals. This was especially evident in a letter that Gessner wrote on April 3, 1565 to his friend and colleague from Augsburg, Adolph Occo III (1524–1606). Here, he mentions that a friend had recommended drinking goat urine to treat stones in the urinary tract, but that he preferred using a solution made from beanstalk lye.[42]

The fourth publication, which appeared in March of 1540 in Basel, had been written during his year in Lausanne and, according to the foreword, was again dedicated to medical school graduates. It is attached to the Latin translation of *De medicamentorum compositione* by Johannes Zacharias Actuarius, provided by the French court physician Jean Ruel (1479–1537) and printed by Robert Winter in Basel. Gessner wrote also the *Succiduorum medicaminum tabula* for it. This listed alternative remedies, in case those described by Actuarius were not available. They were compiled from Galen, Dioscorides, Paul of Aegina and the works of the Byzantine doctor Aëtius of Amida (502–575).

The fifth and last of the publications mentioned by Gessner that originated in western Switzerland was the *Catalogus plantarum*, which was also published in Zürich by Froschauer in 1542. It was a 300-page index of plant names in Greek, Latin, German and French. Before the introduction of Carl Linnaeus's

39 Gessner (1577), f. 125r–129v.
40 Ammann (1960), p. 11.
41 Gessner (1541a), f. 159v.
42 Gessner (1577), f. 73v.

Der Doctor.

Ich bin ein Doctor der Artzney/
An dem Harn kan ich sehen frey
Was kranckheit ein Menschn thut beladn
Dem kan ich helffen mit Gotts gnadn
Durch ein Syrup oder Recept
Das seiner kranckheit widerstrebt/
Daß der Mensch wider werd gesund/
Arabo die Artzney erfund.

D iij Der

ILLUSTRATION 10 Doctor with a flask of urine, in: Jost Ammann, *Stände und Handwerker mit Versen von Hans Sachs*, Frankfurt/M. 1568 (facsimile, Munich 1884), f. D_{iii}r.
ZBZ, ABTEILUNG ALTE DRUCKE UND RARA: IV B 395E

binary Latin nomenclature for plants and animals, these names were not standardized, and each language had its own name for a specific organism. For example, in Gessner's catalog, the camomile was called: Anthemis, Ἄνθεμις, Camillenbluomen and Camille; and the lily of the valley was listed as Ephemerum, Ἐφήμερον, Meyenryslin and Muguet. When there was uncertainty as to which plant was cited in a prescription or textbook, then that which was common and unproblematic in the vernacular might have especially negative consequences for international medical and pharmacological intercourse. Because the Greek and Latin names were derived mostly from authors in antiquity who lived in the Mediterranean basin, Gessner was not always able to find the corresponding German or French word. He was occasionally unsuccessful in pulling together the multilingual designations, an example being the so-called *Schmärwurtz* (white turnip), which is listed in three different contexts.[43] From the interesting autobiographical foreword of the *Catalogus*, which he probably wrote along with the concluding epilogue in 1542, he mentions that for four years he read all available botanical authors, and collected many hitherto unknown species of plants from the fields and mountains. *De natura stirpium* by Jean Ruel was useful to Gessner's book[44] because the plant names were sometimes given in other languages. Due to spending three years working on this book in French-speaking Lausanne, it was somewhat difficult to determine the German names for some plants. For this, the *Kreutterbuch* by Hieronymus Bock (1498–1554) was of great assistance. A decade later, he looked back on this publication with a certain amount of pride, because some of his observations about specific plants had never been made or published by anyone before him.[45] The *Catalogus plantarum* was soon imitated. In 1546, the English doctor William Turner (1508–1568), who had visited Gessner around 1540,[46] presented a work entitled *The names of herbes in Greek, Latin, Englishe, Duche and Frenche*.[47] In 1557, Anton Schneeberger (1530–1581) followed up with a catalog of Latin-Polish plant names.[48]

When one takes an overview of these five publications and compares their development, one cannot avoid the impression that they are thematically organized compilations taken from the fruits of Gessner's reading, so-called

43 Fischer et al. (1967), p. 184.
44 Gessner's copy is preserved in the Biblioteca Angelica in Rome, call number: TT.14.21. Cf. Leu et al. (2008), pp. 217f.
45 Gessner (1551), f. α_2r.
46 Gessner (1555), p. 34.
47 Wellisch (1978), p. 94.
48 Cf. p. 106.

loci-collections that he formulated and published.[49] As previously noted, Gessner spent the spring or early summer of 1540 in Zürich. It is likely that the school master and those awarding stipendia encouraged him to return to studying medicine, a suggestion to which Gessner was very receptive.[50] There was also a decision made in favor of one of the most illustrious and important European institutions for the study of medicine – Montpellier University.

---

49 Cf. p. 89.

50 Stotz (1967), p. 214. Hanhart (1824), p. 79, claims that Christoph Clauser advised him to resume his medical studies, but this cannot be substantiated.

CHAPTER 5

Montpellier

Gessner left Lausanne at the end of August or early September 1540.[1] Pierre Viret was unable to immediately fill the vacancy, and on October 22 he turned to John Calvin in Geneva to help find a suitable candidate. Ultimately, Gessner's former student, Jean Ribit, was chosen as a worthy replacement and confirmed by the Bern city council on January 29, 1541.[2] In the meantime, Gessner was off to Montpellier, which, along with Salerno, had for centuries been among the most renowned centers for medicine in the Western world. In both locations, Greek as well as Arab medicine were taught, cultivated, and often instructed by Jewish doctors.[3] The faculty of medicine had an exemplary reputation, and during Gessner's attendance it admitted between 30 and 50 students each year. By comparison, the law faculty only took in a scant dozen.[4] In 1537, none other than the French writer François Rabelais (1494–1553) graduated there as a medical doctor, and twelve years after Gessner's stay, the famous Basel physician Felix Platter (1536–1614) also attended for five years. Among others, Gessner studied under Guillaume Rondelet (1507–1566), who made a lasting name for himself in the fields of anatomy and ichthyology (ill. 11). Rondelet conducted numerous dissections of human corpses and founded the anatomical theater at the university in 1556. Roughly a third of Platter's diary is dedicated to his university days and contains interesting information about contemporary student life in this southern French city.[5] Foreign guests made up about a quarter of the student body during Gessner's and Platter's time there, with most coming from the German empire, Switzerland or Italy. More than 60 percent stayed for less than two years.[6]

It was not Gessner's intention to study there for long since he hoped instead to be taken into the medical practice of a famous physician. In the end this did not come to pass. In his autobiography, he recalls: "As I departed Lausanne and arrived there, I found no scholarly physician who would take me into their home. I had actually expected to gain more scientific knowledge from domestic contact than from public lectures. So I did not stay long, and after a bit more

1 AK XI/2, p. 988.
2 Crousaz (2012), pp. 219 and 542.
3 Goody (2010), pp. 43–61.
4 Julia/Revel (1989), p. 39.
5 Platter (1976).
6 Julia (1995), pp. 91f. and 95.

© KONINKLIJKE BRILL NV, LEIDEN, 2023 | DOI:10.1163/9789004541696_006

ILLUSTRATION 11 A 1545 portrait of Guillaume Rondelet by an unknown artist.
MEDICAL FACULTY OF THE UNIVERSITY OF MONTPELLIER

study of anatomy and botany, I returned to Germany [Basel]."[7] It is therefore no surprise that Gessner's name is not to be found among the 40 new admissions of 1540.[8]

The dedication to the Spanish doctor Petrus Iacobus in the foreword of Gessner's aforementioned *Compendium ex Actuarii Zachariae libris de differentiis*

7 Stotz (1967), p. 214.

8 Gouron (1957), pp. 89–91.

urinarum (Zürich 1541) relates their scientific excursions together to inland areas and sea coasts. He used the opportunity to strongly remind Iacobus of his earlier promise to send him seeds from Spanish plants.[9] Twenty-five years later, among others, he included an illustration of a piece of coral in his book about minerals and fossils. Unfortunately, however, it could no longer be determined which of the illustrated fragments had come from the coast of southern France.[10] In a letter to the Nuremberg physician and botanist Joachim Camerarius the Younger (1534–1598) dated August 27, 1565, he mentions having found a piece of coral in Montpellier and had thought it was a plant:[11] "With the letter, I also received a piece of sea life about which you wrote that in Italy, had been classified in different ways: corallium imperfectum, adarcen, antipathes etc. I have clearly stated my opinion in a book I am sending to you about forms of stone and petrifications. There, you will find an illustration of it that I had previously cut. For I had found the plant myself on the beach near Montpellier 25 years ago and kept it with others that had not been named with certainty."[12]

In Montpellier, Gessner did not secure a position with Rondelet or any other medical luminary. However, the two did become acquainted and exchanged correspondence for many years. Two decades later, Gessner still considered the Frenchman a medical eminence, which is why he wrote on October 24, 1561 to ask him for advice concerning the varicose veins that periodically caused his wife great suffering. The letter was sent to Johannes Bauhin the Younger (1541–1612), who lived in Rondelet's home while studying in Montpellier that year. It was intended for Bauhin to ask Rondelet for advice in the matter.[13] A few weeks later, on November 9, 1561, in another show of respect, Gessner wished to know from Bauhin what Rondelet had recently published and what he thought of Gessner's essay about oxymel (a mix of honey and vinegar).[14] Rondelet's lecture on Dioscorides, which Bauhin had written down for himself in 1563, was also of special interest to Gessner, and for this reason Bauhin sent him his notes from which Gessner had a copy made for himself in 1564.[15] It was not until the end of 1565, however, that Bauhin would retrieve his notes. With

9 Gessner (1541a), f. A_3r.
10 Gessner (1565), f. 132r/v.
11 Corals were described as plants until the eighteenth century. Cf. Anton von Schouppé (1991), pp. 27f. and 33.
12 Rath (1950), p. 166.
13 Bauhin (1591), p. 103.
14 Bauhin (1591), p. 105f.
15 Bauhin (1591), pp. 117, 138, 141 and 145.

no other option for getting the letters to France, they were sent to Bauhin's father in Basel with the request that they be passed along to his son.[16]

One possible reason that Gessner did not directly reach out to Rondelet, but went through Bauhin instead, was the assumption that since his marriage to a younger woman, Rondelet would no longer have much time for social contacts.[17] Conversely, Rondelet inquired of Gessner on June 30, 1562 as to why he had stopped writing to him for the previous three years.[18] It is also possible that unrest caused by the French religious wars had contributed to communication difficulties between the two scholars.

Although Gessner did not find the mentoring he had hoped for in Montpellier, he nonetheless recommended the medical faculty to others. On May 22, 1555 he created an academic plan for his two medical pupils, Caspar Wolf (1532–1601) and Georg Keller (1533–1603), in which he envisioned the former at Montpellier and the latter at Padua.[19] On June 21, 1555, Wolf enrolled at Montpellier.[20] On October 6, 1560, Gessner recommended to the emperor's Protestant physician, Johannes Crato von Krafftheim (1519–1585),[21] that his son Gottfried take up the study of anatomy and botany at Montpellier. The young man should study diligently there through the winter, then move to Paris to attend medical sessions and lectures in Greek about Aristotle's physics. And should the opportunities arise, it would be advisable for him to travel to the University of Padua, and also to Florence for a practice-oriented specialization.[22] Ultimately, Crato von Krafftheim decided to send his son to Paris as well as to Leuven and Italy.[23]

It is not known exactly when Gessner left Montpellier. However, in his autobiography, he says that he did not waste much time there,[24] and subsequently

---

16 Bauhin (1591), p. 162.

17 Gessner (1577), f. 1r.

18 Letter from Rondelet to Gessner, June 30, 1562, copy preserved in the Stadtbibliothek Winterthur, MS. 4° 196, pp. 99f.

19 ZBZ, MS. S 85, no. 11.

20 Gouron (1957), p. 134.

21 On Crato's commitment to Protestantism at court, see Louthan (1994). On his correspondence with Gessner, cf. Huth and Walter (2019).

22 Gessner (1577), f. 6v.

23 Gessner (1577), f. 8r.

24 Gessner (1545), f. 180v, and Simmler (1566), f. 6v, write that he did not stay long (non diu) in Montpellier. If it had been three to four months, as has been stated in the Gessner biographies so far, both would probably have used a different choice of words. In addition, if he had really arrived in Basel at the end of January 1541 and had taken the doctoral exam already one month later, he would have had only one month to "listen to lectures of famous physicians for quite a long time, among others with Alban Torinus and Sebastian Sinckeler, and to hold disputations," as can be read in Simmler (1566), f. 6v. But how could

returned to Basel. It is possible that by October of 1540 he was already back in the city on the Rhine to prepare for his doctorate. The return trip from Languedoc had taken two weeks.[25]

Simmler describe three to four months in Montpellier as "not long" (non diu), whereas one month in Basel as "quite a long time" (aliquanto tempore)? Gessner must therefore have spent more than a month at the University of Basel between his departure from Montpellier and his doctorate, and must have traveled from Basel to Lyon at short notice.

25 Platter (1976), p. 145.

CHAPTER 6

Lyon

In January 1541, Gessner interrupted his studies in Basel to attend a book fair in Lyon. There were many books on offer and an especially rich assortment of local publications. In 1494 King Charles VIII (1470–1498) had established four yearly book fairs[1] with the following schedule: "Four fairs are to be held in Lyon, each lasting fifteen continuous working days without interruption. The first one starts on the first Monday after *Quasimodo* (the Sunday following Easter Sunday); the second on the fourth day of August; the third on the third day of November; and the fourth on the first Monday after the Feast of the Kings."[2] Gessner's previously-mentioned booklet *Sylvula experimentorum Galeni*, published in 1541, was dedicated to the Lyon physician Claude Millet and makes mention of the pleasant time spent with him during Gessner's stay in "this year" (*hoc anno*).[3] This year can only mean 1541 since a visit during the previous year while returning from Montpellier – as earlier biographers have continually claimed – can be dismissed. Unfortunately, Millet never acknowledged the dedication, leaving Gessner still vexed 23 years later, as he mentioned in a letter to Bauhin.[4]

The dedication in the foreword in his *Apparatus et delectus simplicium medicamentorum* (Lyon 1542) is dated as follows: "*Ludguni in peregrinatione, 1541, Idibus Ianuarii*" (in Lyon, on travel, in 1541, the Ides of January). This clearly means that he visited the three kings fair on January 13, 1541. The fairs held in Lyon were among the most important such events in Europe: "Many foreign merchants have established their homes here, the center of all commercial activity. Geneva lost its importance with the end of Louis XI's reign. Antwerp is just beginning to develop. In the first half of the sixteenth century, Lyon is at its peak and is the city of major European fairs."[5]

The city had sixty thousand residents at the time, and was located in a geographically advantageous site for transport at the mouth of the Saône tributary of the Rhone River. This made it not only a significant economic and trading center, but like Basel, Paris or Venice, it was also one of the important humanistic book and printing cities of Europe. Roughly two thousand ships per year

1 Boucher (2001), pp. 46f.
2 Brésard (1914), p. 21.
3 Gessner (1541a), p. 159.
4 Bauhin (1591), p. 129.
5 Brésard (1914), p. 75.

© KONINKLIJKE BRILL NV, LEIDEN, 2023 | DOI:10.1163/9789004541696_007

docked there to exchange goods.[6] Lyon was the first printing center in which books were published in the French language.[7] In fact, more French publications were produced there than in Paris. Moreover, the Lyon printers gained a fine reputation for high-quality medical and legal publications.[8] Next to Paris, Lyon remained France's most important printing center, where between ten thousand and fifteen thousand titles were printed during the sixteenth century. Starting in the 1560s, the religious wars brought a certain thematic and personnel shift, but this resulted in only a relatively small decline in production.[9] By comparison, Basel experienced a massive drop in quantity and quality starting in 1580.[10]

During Gessner's stay in Lyon, there were about sixty active printing houses and thirty booksellers.[11] Many were located in the Rue Mercière or adjacent side streets (ill. 12).[12] Gessner's previously-mentioned *Apparatus et delectus simplicium medicamentorum* was published by Jean and François Frellon, whose business was located in the Rue Mercière in a house with the coat of arms of Cologne.[13] It is likely that Gessner submitted his manuscript to them during his January 1541 visit. In the *Apparatus*' dedication to Christoph Clauser, Gessner expressed a wish to dedicate the book to another person as well: the Grisons physician Albert Belfort of Chiavenna,[14] whose hospitality he had enjoyed in Lyon for several days.[15] In the above-mentioned dedication to Claude Millet, he also remembered a dining acquaintance, a little known Italian doctor named Bartolomaeus.[16]

We cannot know of everyone whom Gessner encountered and became acquainted with in Lyon, but there must have been some individuals, especially printers and booksellers, with whom he maintained contact throughout his life. In his opinion the Lyon book printers were among the best during the Reformation century. On March 17, 1561, he strongly encouraged Crato von Krafftheim to finish his book about the art of distillation (*De destillationibus*). Gessner wished to help him find a good book printer in Basel, or preferably in Lyon. There, it would be easier to secure a royal printing privilege that would

6 Rossiaud (2012), p. 53.
7 Brésard (1914), p. 187.
8 Pettegree (2010), p. 48.
9 Martin (1982), p. 442.
10 Leu (2013).
11 Gascon (1971), p. 628.
12 Cf. Moledina (2012); Krumenacker (2009), p. 30; Rossiaud (2012), p. 17.
13 Rue Mercière à L'Ecu de Cologne (house with Cologne's coat of arms). Cf. Baudrier, vol. 5 (1964/65), p. 157.
14 Bernhard (2019), pp. 341f. and 345.
15 Gessner (1542a), f. a_4v.
16 Gessner (1541a), p. 159.

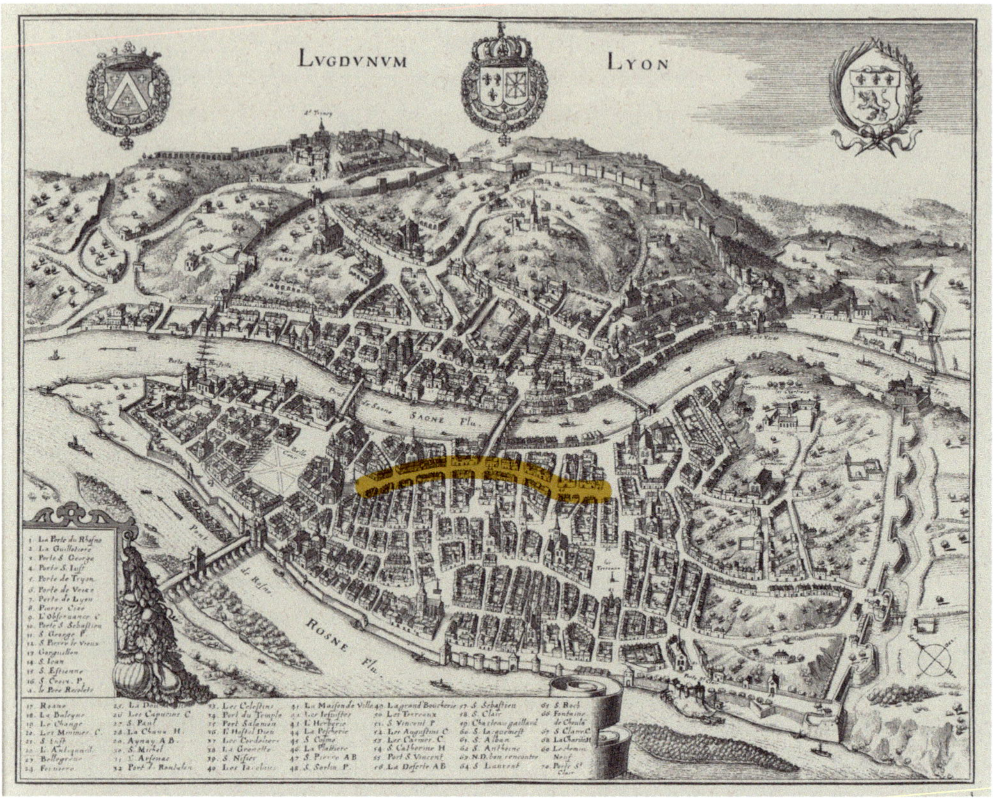

ILLUSTRATION 12 City map of Lyon, in: Martin Zeiller, *Topographie Galliae*, Frankfurt/M. 1656. Running through the book printing quarter is the Rue Mercière. Among others, the printing house of Etienne Dolet (1509–1546) and the book shop of Guillaume Rouillé were located here. At the end (left) was the Rue Thomassin, where the printer Sebastian Gryphius (1493–1556) was located.
ZBZ, ABTEILUNG ALTE DRUCKE UND RARA: T 148

forbid unauthorized printing by others. Furthermore, nearly all Lyon printers would do meticulous work.[17] A few months later, he wrote to Johannes Hospinian (1515–1575) in Basel to inform him that in case of the book *Analytika* by the physician and Aristotelian Jacob Schegk (1511–1587), whom he held in high regard, were not printed, then he would try to find a printer in France, specifically in Lyon.[18] The Lyon printer Guillaume Rouillé (1518–1589) received the highest praise from Gessner. He wrote to Bauhin on November 14, 1563 that he would prefer that his own great botanical work (*Historia plantarum*) be printed by Rouillé rather than Froschauer in Zürich, but unfortunately, he

17 Gessner (1577), f. 8r.
18 Gessner (1577), f. 102v.

had already promised it to the latter.[19] Rouillé had a preference for botanical books. About a third of the 800 works he printed contained medical or natural science content. Among these was the most important work by Jacques Daléchamps (1523–1588), the two-volume *Historia generalis plantarum* (1587), published in folio format with hundreds of illustrations.[20] There was also contact between publishers in Zürich and Lyon in the 1580s. It is highly probable that Christoph Froschauer the Younger (1532–1585) arranged for an emblem book by the French poetess Georgette de Montenay (1540–1581), which contained many copperplate engravings, to be printed in Lyon in 1584.[21]

In Gessner's correspondence, we find that he was interested in relevant new French publications. He frequently made enquiries about books to Johannes Bauhin, who began work in Lyon as a physician in 1563. For example, he asked whether Bauhin could send him the book about fractured skulls by the Parisian surgeon Ambroise Paré (1510–1590),[22] or whether the work by the Arab doctor Averroes (1126–1198) had been published in Lyon yet and what it might cost.[23] Gessner's Lyon bookseller was named Antoine Gryphius (1527–1599), who did not want to take over the famous printing house of his father Sebastian (1493–1556) and concentrated instead on book sales.[24] He helped the polyhistor from Zürich with the procurement of French books, but also with Italian titles, since the enquiries Gessner sent to Italy often went unanswered.[25] The shipping of books seems to have functioned quite well; at least it does not appear that there were complaints from Gessner. It is worth noting that the two shipments to Zürich in the winter months of 1563/64, that were mentioned in correspondence, contained a total of only six titles, which did not make for particularly heavy packages.[26] One title was *In librum de insomniis Hippocratis commentarius* (Lyon 1561) by Julius Caesar Scaliger, which Gryphius sent to Gessner in November 1563 – and that he had already received as a gift from Robert Constantin (1530–1605).[27] Apparently Gryphius also sent books that he thought would be of interest to Gessner, or he sent orders from colleagues for him to pass along.

---

19 Bauhin, (1591), p. 126.

20 Zemon-Davis (1966), pp. 81 and 93. He had Gessner to thank for various illustrations. Cf. Christ (1917), pp. 146f.

21 Leemann-van Elck (1940), pp. 155f.

22 Bauhin (1591), p. 132.

23 Bauhin (1591), p. 138.

24 Moledina (2012), p. 55.

25 Bauhin (1591), p. 130f.

26 Letter from Gryphius to Gessner dated November 17, 1563 (StAGR, D V/37 C 36.06.34). Another book dispatch to which Gessner replied on March 24, 1564: Bauhin (1591), p. 148.

27 Leu et al. (2008), pp. 221f.

According to a letter from Gessner to Bauhin on February 11, 1564, there were no problems worth mentioning involved in shipping books. However, sending possibly heavier shipments to Lyon presented challenges. He wrote that he had not yet sent metals to Bauhin's relative, the Flemish pharmacist Valérand Dourez.[28] There were few couriers to be found who were willing to transport heavy loads to Lyon. It was easier to send such goods to Antwerp or Frankfurt than to Lyon.[29] Sending letters also had its pitfalls, and Lyon was no exception. A letter sent from Zürich to the city on the Rhone took from four to ten weeks to arrive. Frequently, they never reached the addressee.[30] Along with books, the residents of Zürich also bought *simplicia* (ingredients for remedies) from Lyon. In a postscript to his letter of October 28, 1563, Gessner wrote to Bauhin that his relative, the merchant Anton Gessner, was in Lyon to buy *simplicia*. It would be nice if Bauhin and Dourez would help him find them at a good price.[31]

Another person with whom Gessner corresponded for many years was the physician and botanist Jacques Daléchamps, who had studied at Montpellier from 1545 to 1547 and began practicing medicine in Lyon in 1552. It seems as though Gessner never met him personally, but based on the eight surviving letters from the period 1555–1562 and comments from other written sources, the two had a good relationship. The oldest letter Daléchamps wrote to Gessner is dated "December 1" but without the year. It must have been written shortly after the publication of Gessner's bird book in 1555. Daléchamps wrote that he recognized many of the birds in the book for which he also had illustrations, and he would gladly send some of the bird images to Gessner. Along with the letter, he included a small book about the plague, which today is a rarity.[32] The remaining correspondence is a scientific exchange with regard to plants, animals, medical remedies and books. On January 8, 1562, Gessner found himself compelled to intervene in Daléchamps's re-conversion to Catholicism. His condemnation was expressed in a long, heartfelt but thoughtful letter that was also an expressive personal declaration of Gessner's own faith,[33] which is why a longer excerpt from it is quoted here: "I was stricken and deeply pained when I learned that you, a man of such erudition and meaningful influence, had fallen away from our faith. Perhaps you were driven to this by the blind

28 Legré (1904), pp. 89–105.

29 Bauhin (1591), p. 143.

30 Bauhin (1591), p. 126; letter from Daléchamps, September 6, 1561 (BN Paris, MS. lat. 13063, pp. 257f.).

31 Bauhin (1591), p. 131.

32 Letter from Daléchamps to Gessner, December 1, [1555], StAGR, shelf mark: D V/37 C 36.06.48.

33 Gessner (1746), pp. 133–150.

ambition and inappropriate behavior by some of the innovators [Protestants] that you have complained about. But you should have remained so true to the pure word of God that no violence and injustice could have turned you away. Excuse me my dear Daléchamps that I speak so frankly to you, because of my affection for you, and all that you have done for me. It is as if we have had the deepest bond since childhood; as if I were born your brother. You have been too anxious about your earthly progress, that you have not taken these words to heart: 'Seek first the kingdom of God and his righteousness, then all else will come unto you.'[34] You have seen more splendor, greater influence, profit and honor in the enemies of our religion – and on our side, mistakes and missteps that I will not deny. However, they are truly not so great that you should convert to an un-Christian church and entrust your sacred soul to it … could you not decide to set aside your scholarly activities for a while and devote yourself entirely to the Holy Scriptures? First, lift up your heart and humbly beseech God, the source of all mercy, to open the eyes of your mind so that you may understand what you read;[35] to give you the strength of will to love and faithfully confess what you have recognized as the true good and return it all to the Lord and his glory. Oh, if you will choose to do this, you will quickly learn to know and better love, with more certainty and more thoroughly that which the truly pious are meant to know and love. Now leave worldliness behind, disavow the childish games of vanity and profit; lift yourself to higher thoughts, listen to a better teacher who calls out to us: 'I am the way, the truth and life itself.'[36] Who shows us a better way than Christ? Can it be those of you who have designated a multitude of saints and their intercessors to whom you dedicate your good works in the hope of salvation; those who have invented a purgatory and offer it for sale? Who shows us another truth? You, who would add to the Holy Scriptures and words of Jesus Christ the judgments of church fathers, councils and the Pope, as if the truth were insufficient or not perfect enough. Those of you who condemn the articles of faith known as the 'apostolic profession of faith', which were universally accepted in the first Christian church; who drag those to their death as heretics who reject additional words as necessary for salvation. Forgive me, my brother, for writing without eloquence or a clear order that which instills my innermost affection for you and my fervent wish to bring you back from the wrong and destructive path. I will say it again, set aside the worldliness in which you are so skillful, stop thinking only about that which is useful or brings honor. Pause for a while from the fascinating sciences in which you excel over others, and halt your study of natural history – in

34 Gospel of Matthew 6,33.

35 Acts 8,30.

36 Gospel of John 14,6.

which I also delight – and devote yourself exclusively to the study of God's pure teachings. Turn to your inner self, and realize that you are a person, a mortal person, and according to the natural order, you are not far from your goal of life [death]. Turn to reading the Holy Scriptures with all of your spiritual strength; to pious meditations and devotional prayer; implore the merciful Father in Heaven to give you the strength and will to know thyself. Shed tears, just like St Peter did when he was unfaithful to his master, and he who shows mercy to all who call on him will do so for you as well. As I write this, there are also hot tears in my own eyes, falling because of my love for you and my sadness about the situation in France. I implore you to save yourself in the holy sanctuary of the Bible's Old and New Testaments. Observe the pure truth, free from all external influences, regardless of how much we value them. Whether good or bad, give up all your extra human accoutrements and advantages: luxuries, wealth, honor, comforts and even life on earth itself. Let yourself be moved by the simple, unvarnished truth. I will not refer you to Luther, Calvin or Viret, although I consider them to be instruments of God and most of their words to be the holy truth. I realize that they are men and capable of error. I recommend to you the Holy Scripture, the oldest written document, the purest source of the truth, the most secure and simple guide to salvation. It is the only path for us to take in order to become true religious scholars. Let yourself be taught by the patriarchs, the prophets, the apostles, and especially by Jesus Christ himself. Become his student and hear his words."[37]

It would seem that Daléchamps experienced the letter as overburdening, so he broke off contact to the scholar from Zürich. On October 28, 1563, Gessner asked Bauhin if he had any news of the colleague from Lyon,[38] since Daléchamps had not answered him. Daléchamps had not actually distanced himself from all Protestants, otherwise he would not have given so much of his material to Bauhin.[39] Bauhin was indeed far removed from Catholicism, but he was not an orthodox reformed protestant. While he did have certain sympathies for the antitrinitarians, this was something that Daléchamps was not likely to have known.[40] Therefore, it is telling that Gessner frequently wrote to Bauhin with news about the fate of Bernardino Ochino (1487–1564), an antitrinitarian who had been exiled from Zürich in 1563.[41] He had probably received questions from Bauhin, who had an interest in his like-minded comrade.

37 Hanhart (1824), pp. 171–174.

38 Bauhin (1591), p. 130.

39 Bauhin (1591), p. 157.

40 Krumenacker (2009), pp. 236–239.

41 Bauhin (1591), pp. 133, 138 and 143. Recent research has largely rehabilitated Ochino theologically, cf. Campi (2022).

CHAPTER 7

Doctorate in Basel

When the book fair in Lyon concluded at the end of January 1541, Gessner returned to Basel, which, nearly 20 years later, he referred to as his second hometown (ill. 13).[1] He visited the city on the Rhine frequently and likely found accommodations with his colleague Wilhelm Gratarolus (1516–1568).[2] Gratarolus found work in Basel in 1552 and also enjoyed Gessner's reciprocal hospitality.[3] On a Saturday, either the 19th or 26th of February, or at the latest on March 5, 1541 Gessner acquired a doctorate of medicine from the University of Basel.[4] His first biographer, Josias Simmler, wrote that his studies were primarily under Alban Torinus (1489–1550) from Winterthur and Sebastian Sinckeler († 1547), a physician from southern Germany.[5] The latter might be referred to as Gessner's doctoral advisor[6] to whom he hand-wrote a dedication in a copy of his *Historia plantarum* (1541).[7] He also added a dedication epistle to him in the commentary by the Byzantine scholar Michael of Ephesus (twelfth[h] century) on *De iuventute, senectute, vita et morte* (*On Youth, Old Age, Life and Death*), by Aristotle. Gessner had translated it and two other examinations of Aristotle's works by Michael of Ephesus, *De longitudine ac brevitate vitae* (*On the Length and Brevity of Life*) and *De divinatione per somnum* (*On Dream Interpretation*) from Greek into Latin. Gessner's dedication forewords to the three works that were published by Bartholomäus Westheimer (1499–1567) in Basel date from August 1541.[8]

Written to Sinckeler, Gessner notes in the first one that the translation was meant for students since to his knowledge there had not yet been a Latin version of the text. He dedicated *On Youth, Old Age, Life and Death* to him, because Aristotle addressed the question as to where the feelings of the soul are located, whether in the heart or some other part of the body since this had also been the subject of Gessner's doctoral exam. Aristotle believed that the

1 Gessner (1560a), p. 280.
2 Cf. Gessner's thanks to Gratarolus for his hospitality in the dedicatory epistle of Gessner (1556), pp. 245–247; cf. Jüttner (1969), pp. 134–150.
3 Letter from Philippus Gallicius to Bullinger from 1552 or later, cf. StAZH: E II 365, 103b.
4 AK XI/2, p. 988.
5 Simmler (1566), f. 6v.
6 AK XI/2, p. 989.
7 The copy is today preserved in the University Library Basel, call number: Lo X 1:1.
8 Interestingly, Conrad Pellikan also dealt with these Aristotelian writings in 1541. Cf. Pellikan (1892), p. 124.

© KONINKLIJKE BRILL NV, LEIDEN, 2023 | DOI:10.1163/9789004541696_008

ILLUSTRATION 13 City map of Basel, in: Mattäus Merian, *Topographia Helvetiae, Rhaetiae et Valesiae*, Frankfurt/M. 1642.
ZBZ, ABTEILUNG ALTE DRUCKE UND RARA: T 137

heart was the location and source of human feelings, whereas Gessner and contemporary doctors attributed this function to the brain. Gessner's disputation theses, which are difficult to understand for today's readers and have only been handed down in a roundabout way, read as follows:

> "An cerebrum sit, principium sensus et motus, an cor?" (Is the brain the principle of perception and movement or the heart?)
>
> "An qui crescunt plurimum habeant calidi innati?" (Does what grows have a particularly large amount of natural heat?)
>
> "An qualitates formae sint elementorum?" (Are the qualities the forms of the elements?)[9]

9 Gessner (1541b), p. 3f. The presumed disputation theses of Gessner are reproduced in Hanhart (1824), p. 86f.

Gessner dedicated the second text in the collection by Michael of Ephesus to the Glarus councilman and historian, Aegidius Tschudi (1505–1572), who by 1558 was one of his patients.[10] He expressed thanks for the gracious hospitality during his excursion to the Glarus Alps in the summer of 1541.[11] The third tract, on dream interpretation, was dedicated to his colleague Melchior Wirz (1519–1558), an acquaintance since childhood with whom he had taught mathematics at the *Hohen Schule* in 1541/42.[12] Gessner wrote that the title of the booklet might well cause offence to many people because the Old Testament forbade the observation of dreams.[13] He then explained that he did not care much for superstitions. He was convinced, however, that Christians would find dreams useful if they were to read the works of ancient authors and differentiate between superstition and the facts of natural science. Most dreams were clearly without significance and paying them too much heed would be profane. However, since they are generated by human nature, and humans were created by God, then one should not dismiss them. In fact, a doctor could determine a patient's physical condition based on the kind of dreams they have had. According to the humoral pathology, that is, the four-humor theory taught in that era, the four bodily fluids – yellow bile, black bile, blood and phlegm – determine a person's health, feelings and inclinations. Furthermore, these elements also influence dreams. The medical observation of dreams yields conclusions about the mental state of the patient as well as the interrelationship of the four humors in his/her body, and therefore the condition of their health. Gessner continued that, due to space issues, he left out the categories of dreams as enumerated by the church fathers Gregory of Nazianzus (around 326–390) and Augustine (354–430).[14] He warned that dreams that give predictions of the future should only be accepted with caution since they are both ambiguous and dubious. One should not give weight to such analyses, except for those by divine interpreters such as the patriarch Joseph in the Old Testament.[15] In this instance and in others, the Bible is generally a reliable guide for discerning whether or not one is on the wrong path.[16]

Therefore, Gessner did not accept dream interpretation as a way to look into the future, but as a diagnostic tool associated with humoral pathology, which only disappeared from medicine in the nineteenth century. It seems that over time his restrained attitude was set aside, and he did occasionally engage in some non-medical dream interpretations. He was not the only one; the Zürich

10 Milt (1950).
11 Gessner (1541b), pp. 41f.
12 HBBW 6, p. 402.
13 Cf. Deut. 13,2–6.
14 Cf. Näf (2004), p. 159.
15 Cf. Gen. 41.
16 Gessner (1541b), pp. 78–80.

theologian and later antistes Rudolf Gwalther even became somewhat famous for it. On February 8, 1547 Georg Fröhlich from Augsburg asked Bullinger if either Gwalther or Gessner could interpret two dreams for him. He assured the head of the church in Zürich that it had nothing to do with superstition, but he nonetheless had the feeling of some sort of pending disaster.[17] Two related notes were found in Gessner's estate that were published by his successor Caspar Wolf in 1586 in which two dreams from 1561 and 1563 are mentioned that autobiographically hint at his demise.[18] Josias Simmler also related that Gessner had dreamed in 1564 of being bitten by a snake, which he interpreted as a prediction of his own death from the plague.[19]

Gessner's lifelong pride in achieving a doctorate from the University of Basel is exemplified in the dedication he wrote for Basel's Galen publication of January 23, 1562, which was addressed to the university rector at the time, Basilius Amerbach (1533–1591), and to the professorial staff.[20] He referred to himself as a pupil, even as a schoolboy and son of the university.[21] He took the opportunity to praise Basel's educational system and its scholars, especially Thomas Platter, Oswald Myconius, the cosmographer and Hebraist Sebastian Münster (1488–1552) as well as the jurist Bonifacius Amerbach (1495–1562). He finished with a great compliment to the printing house of Hieronymus Froben (1501–1563) and Nicolaus Episcopius (1501–1564). Out of gratitude for his eloquent dedication, in spring of 1563, the University of Basel gave him a gift of 5 silver coins.[22] At the time, it was customary to honor such dedications in this way, which likely provided some motivation for one or the other of Gessner's texts.

---

17 The letter has not yet been published and is kept in the StAZH, E II 346, 224.

18 Gessner (1586), second part: *Scholia et Annotationes*, pp. 190f. Cf. Baumann (1953), p. 147.

19 Simmler (1566), f. 18r.

20 AK XI/2, pp. 974–994.

21 AK XI/2, pp. 980.

22 Letter from Simon Sulzer to Gessner, April 30, 1563 (StAGR, D V/37 C 36.06.26).

CHAPTER 8

Return to Zürich

Myconius wrote to Bullinger on March 8, 1541 that Gessner was returning home with the highest of academic degrees. Although Gessner was relatively young, in Myconius's judgment, the title of medical doctor was deserved. He was suited to this profession more for his natural inclinations than his scientific qualifications. Because he was in need of a practice, however, Bullinger should arrange for him to gain some experience.[1] Little is known of Gessner's medical activities until 1554, when he succeeded Christoph Clauser as municipal physician. Even his being named *Poliater*, or second municipal physician, in 1552 is recorded in only one surviving eighteenth-century text.[2] During the 1540s, there are only three places in the correspondence with Bullinger where his medical activities are mentioned. On June 12, 1543, Myconius wrote to Bullinger that he had just learned that Clauser, Zürich's municipal physician, was suffering from the death of his wife, and that his paralysis was especially regrettable. His successor would probably be Gessner, otherwise Myconius would recommend the carrier of the letter, Philipp von Hertenstein, who was about to receive his doctorate.[3] Although there is not a word about Gessner's work as a physician, Myconius would certainly not have suggested him as a possible successor to Clauser had he not already been a practicing doctor in Zürich. In December 1547, Ambrosius Blarer from Konstanz contacted Bullinger to recommend that a young man who was interested in medicine spend a period of time with Gessner.[4] This would indicate that Gessner had a fairly good reputation, although, on December 11, 1549, Bullinger wrote to the Memmingen physician Ulrich Wolfhart that Gessner paid more attention to his publications than to the ailments of his patients.[5] Bullinger worried because there were no outstanding doctors to be found, whether in Lucerne, Zug, Schwyz, Uri, Glarus, Thurgau or Aargau. Whenever Zürich did have a good doctor, Bullinger wrote, people from all around streamed into the city. But even when there was a good and skillful doctor in Zürich, one was not enough for so many residents, which is why he sought to persuade Wolfhart to take a position

1 HBBW 11, p. 98.
2 Johannes Gessner: *Oratio de variis Physicae et Matheseos in Gymnasiis et Scholis Turicensibus fatis, …*, die 12. Sept. 1772, p. 649 (ZBZ, MS. L 430).
3 HBBW 13, p. 150.
4 Schiess (1910), p. 674f.
5 StAZH E II 335, 2120.

© KONINKLIJKE BRILL NV, LEIDEN, 2023 | DOI:10.1163/9789004541696_009

in Zürich.[6] Whether Bullinger's judgment was right or wrong remains to be seen, but it is certain that Gessner did not go down in scientific history as a medical luminary. His fame was based much more on his innovative zoological, botanical and bibliographic works.

8.1 Travels in the Alps

While in Lausanne, Gessner had explored the Alps of Savoy and Valais, and a few months after arriving in Zürich in the summer of 1541 he set out for the Glarus mountains. So as not to arrive empty-handed at the home of his friend Jakob Vogel,[7] a civil servant in Glarus, he wrote a booklet with the title *De lacte et lactariis* (*About Milk and Milk Products*). In the dedication, he noted: "There is much more that I am drawn to than just the spectacle of the mountains' massiveness. They are very high, but aside from that, I hear that they are very lush with vegetation. This, along with your inviting friendship, has overcome me with desire to visit. However, to avoid coming with no gift for such a valued friend, and as circumstances allowed, I chose to write down everything known from the ancients about milk and milk products that you might somehow appreciate. This seems somewhat appropriate to your people, of whom a large part live from milk. They convert it to various sorts of edibles, including famous gratable cheeses that are seasoned with aromatic herbs [Schabziger cheese]. This is well-received by all the foreigners to whom it is customarily marketed."[8]

The "collection of material" about milk, as Gessner calls it,[9] is mostly composed of extensive quotes by ancient authors, above all Galen. The first part of the dedication to Jakob Vogel, printed in 1541, became famous as an expression of the Renaissance's new view of nature. Gessner's trek fits in with a line of humanistic mountain ascents that had begun already in the early fourteenth century by various scholars who had an interest in nature and which include Jean Buridan (Mont Ventoux) in 1334, Francesco Petrarch (Mont Ventoux) in 1336, Pietro Bembo (Mt Aetna) in 1494 and Leonardo da Vinci (Monboso) in 1511.[10]

---

6 HBBW 10, p. 95.
7 Lateinisch: Iacobus Avienus. Cf. Winteler (1956).
8 Gessner (1996), p. 11.
9 Gessner (1996), p. 95.
10 Grupp (2008), p. 28.

Gessner combines an appreciation for mountainous terrain with praise for the Creator, and he expresses a wish to study the world of mountains on a regular basis: "For as long as God lets me live, most scholarly Avienus [Vogel], I have decided to climb several mountains every year – or at least one, when the flowers are in full bloom. In part, this is to gain knowledge, but also for worthwhile physical exercise and to satisfy my urges. Can you imagine how great the pleasure is and what a joy it is for an enthusiast to behold and marvel at the infinite enormity of the mountains, and to lift one's head up among the clouds, so to speak? I cannot express how the spirit will be moved at such wondrous heights and how amazing it is to observe the ultimate master builder. On the other hand, those who are lazy by nature marvel at nothing, remain passively at home, never tread the world stage and stay hidden in a corner like a dormouse in winter. They do not realize that mankind is placed on this earth in order to draw conclusions from the wonders of the world about something greater, even about the existence of God himself. They suffer from such laziness that, like swine, they only stare at the ground and never lift their faces to see the heavens or look up to see the stars above. Each wants to roll in the mud and lie in it, completely possessed by an addiction to profit and ignoble tendencies. Those who seek wisdom will go forth to observe the spectacles of this earthly paradise with physical and spiritual eyes, not least of which are the steep and soaring mountaintops; the dauntingly deep vales; the rough flanks, striving to reach the sky; the steep boulders and the shaded forests."[11] Gessner was informed in a letter from his cousin Otto Werdmüller (1513–1552) that Myconius had liked the booklet about milk and its products.[12]

8.2 Publications on Classics

During his first years in Zürich as a freshly-minted physician, Gessner published several philological works, but only one medical piece. This was a list of laxatives and vomit-inducing substances, which was included in a related work by the Italian doctor Antonio Musa Brassavola (1500–1570), entitled *Enumeratio medicamentorum purgantium* and published in Basel in 1543. Gessner's publishing efforts were primarily focused on the translation and publication of little-known Greek texts. For example, he translated from Greek into Latin a short *Compendium de syllogismis* by an unknown author. In essence, it was an

11 Gessner (1996), p. 5. Cf. Boscani Leoni (2019).

12 Gessner referred to this letter from Myconius to Werdmüller in his letter to Myconius of November 21, 1541. Cf. Kantonsbibliothek St Gallen (Vadiana), MS. 34 (VBS 5), 76.

approximately 30-page, student-oriented introduction to Aristotelian logic with a foreword addressed to Otto Werdmüller.[13] It is possible that the booklet was also used by Gessner and Werdmüller in their classes at the *Hohen Schule* in Zürich, which will be covered in our following chapter. This brief *Compendium* was published by Robert Winter in Basel in 1542 as an appendix to two works about Aristotle: one from the pen of the late ancient neoplatonist Porphyrius (234–305), and the other by the Byzantine scholar Michael Psellos (1017–1078).[14]

That same year, Gessner published three more Latin translations of Greek texts through Froschauer in Zürich. In March 1542, he dedicated them to his three friends from Lausanne, Pierre Viret, Béat Comte and Jean Ribit. The first was titled *Moralis interpretatio errorum Ulyssis Homerici* (*Moral Interpretation of the Wanderings of Homer's Odysseus*). Since antiquity, Homeric epics had been analyzed on many levels.[15] In this treatment, selected travels by Odysseus were psychologically interpreted as if they had never actually taken place. Instead, they might have been the expression of the internal needs and challenges of a person who had suffered a great deal and was attempting to cope. The Greek text had already been published by the German humanist Vincentius Opsopoeus (ca. 1485–1539) in Hagenau.[16] Gessner's Latin translation was intended for a larger audience.

The second text, with a dedication to Béat Comte, also examined an incident in the life of Odysseus. The author was the aforementioned neoplatonist Porphyrius, and his piece was titled *Commentatio Porphyrii Philosophi de Nympharum antro in XIII. libro Odysseae Homericae* (*Commentary by Porphyrius About the Cave of the Nymphs in the 13th Book of the* Odyssey). Again, Gessner translated the text into Latin, which included an allegorical interpretation found in the *Odyssey* (8, 102–112) about the grotto on Ithaca.[17] Porphyrius, a student of the famous Neoplatonist Plotinus (205–270), wrote a 15-book work opposing Christians. Gessner refers to this in his foreword and lets Comte know that he did not take Porphyrius's enemity against Christians too seriously, because he had described Christ as a wise and pious man, as the Florentine Renaissance

13 The preface can be found in paraphrased form in: Hieronymus (1992), p. 230.

14 The title of the book is: *Porphyrii institutiones quinque vocum, ad Chrysaorium. … His coniunximus in dialecticae studiosorum gratiam, Michaelis Pselli peripatetici, Blemmidae, Georgii Pachymerii, in eosdem Porphyrii et Aristotelis libros epitomen, atque Paraphrasin: una cum compendio de syllogismis, autoris incerti. …*, Basel, [Robert Winter], 1542. Wellisch (1984), p. 40, is wrong when he claims that Gessner translated Porphyrius' work into Latin.

15 Buffière (1973).

16 VD 16 X 72.

17 Alt (1998).

philosopher and Neoplatonist Marsilio Ficino (1433–1499) had pointed out in a commentary on Plotinus. And Plotinus could also not have been overly adverse to Christianity, since he was a student of Ammonios Sakkas, a Christian († ca. 242),[18] and also friends with Origenes (185–254), a church father. Ultimately, positive comments made about Christ by heathens would honor God, including those from the philosopher Numenius of Apamea (2nd century), who had been quoted by Porphyrius and who had considered Moses a prophet.[19]

The third part of Gessner's Homeric collection was by the Neoplatonist philosopher Proclus (412–485) with the title *Ex commentariis Procli Lycii, Philosophi Platonici in libros Platonis de Republicae apologiae quaedam pro Homero, et fabularum aliquot enarrationes* (*From Proclus Lycaeus's Commentaries on Plato's Work Concerning the State, Several Defenses of Homer, and a Variety of his Narratives*). According to Gessner, the text that he translated into Latin had been poorly edited in a Basel publication.[20] He dedicated it to Jean Ribit and made the remarkable claim that the Platonists shared some of the core principles of Christianity. Gessner left it hanging as to whether these Christian truths had been imported from Egypt, gained from reading Old Testament Scripture, received from divine communication, or perhaps discovered in some other manner. As presented, the booklet was to make clear why the poets of antiquity were considered to be theologians and why their poetry was seen as theology and an expression of primeval wisdom (*prima philosophia*). Holy wisdom shines through in Homer's poetry as well, at least to the extent that divine will could be bestowed upon Homer, a heathen.

The fact that pagans, representatives of Greek and Roman antiquity, and others who never came into contact with the Bible, could have recognized God and divine truths was in line with the beliefs of Huldrych Zwingli. The reformer from Zürich believed that all truth comes from God, and everywhere that the truth is spoken, God is at work.[21] Obviously, this would not have replaced the importance of the Gospels' teachings and people's conversion to God. He found it quite plausible, however, to encounter a person from the Old Testament in Heaven, and right next to him a pagan from antiquity. He was possibly influenced by the *Quaestio de salvatione Aristotelis* from the pen of the Dutch educator Lambertus de Monte (1430/35–1499), which he had received as a gift from his colleague Heinrich Glarean around 1511.[22] Correspondingly,

18 Ammonios turned away from Christianity at an early age. The rumor that he remained a Christian goes back to the church father Eusebius (ca. 260–340).

19 Gessner (1542b), f. 16v.

20 This edition could not be found.

21 Pfister (1952), p. 124.

22 Von Moos (2014), pp. 42f.

Zwingli wrote in 1531 in his *Erklärung des christlichen Glaubens* to the French king Francis I: "Then you must expect to see the community and society interacting with the holy, wise, believing, upstanding, brave and virtuous people, all of whom have lived since the creation of the world. Here, you will see Adam, the saved and the savior; Abel, Enoch, Noah, Abraham, Isaac, Jacob, Judas, Moses, Joshua, Gideon, Samuel, Phineas, Elija, Elisa, Isaiah and the virgin Mother of God; those foretold by the prophets, David, Hezekia, Josiah, the Baptist, Peter, Paul, Hercules, Theseus, Socrates, Aristides, Antigonus, Numa, Camillus, the Catos, the Scipions, Louis the Pious, and his forebears named Louis, Philipp and Pepin and all other ancestors who have departed in faith."[23]

Given this assumption that antiquity was enlightened by God, it is understandable that Gessner and others in Zürich would turn to specific ancient texts for a selective interpretation of dreams. Nonetheless, they fundamentally rejected astrology and occultism.[24] Heinrich Bullinger also shared the belief that wisdom was the same throughout time, and across all cultures and religions. It was eternal wisdom (*philosophia perennis*), according to the title of a book by the Vatican librarian Agostino Steucho (1496–1549), which at least Theodor Bibliander and Heinrich Bullinger studied in Zürich.[25] Bullinger even went so far as to claim that the various ancient names of gods ultimately denote the God of the Bible, but that polytheism had taken over in the course of time: "And here one must speak up. If the heathens acknowledged the one and true God, then why was he given so many different names in books by the poets and philosophers? We wish to show that these names were given to the true God in order to emphasize his power and wisdom, and were not initially intended to signify many gods. In his book about the mysteries of the heavens, Iamblichus says that the Creator of understanding and master of truth and wisdom who brings hidden heavenly powers to light through his actions, is named *Annum* in the Egyptian language. However, because he is especially honest and full of truth, his name is also *Pythia*. The Greeks named him *Vulcanus* and believed him to be a good craftsman. Because he was the source of everything good, he was also called *Osiris*, and due to his various strengths and effects, he had even more names. It is most clearly written in Aristotle's book about the world with the following words: 'Even though there is only one God, he has many names, and this is because of the many powers that he displays to different groups of people.'"[26]

---

23 Zwingli (1995), p. 340.
24 Leu (1990), pp. 94f.
25 Christ-von Wedel (2005), pp. 42f.
26 Bullinger (1574), p. 39.

In 1544, Gessner released another title in Basel that dealt with the allegorical Homer exegesis. It was about the *Allegoriae in Homeri fabulas de diis* (*Allegorical Interpretations of Homer's Fables of the Gods*) by Heraclides Ponticus (ca. 390–320 BC) which Gessner translated into Latin.[27] Today, the otherwise little-known grammarian and philosopher Heraclitus (1st century AD) is considered to be the author who best represents the allegorical analysis of Homer. Dated January 1544,[28] the book's dedication is made to Hieronymus Fricker from Bern, who in 1542 became the bailiff of Mendrisio. Gessner wrote that this work was dedicated to him because his Zürich compatriot, Johannes Rhellicanus (ca. 1498–1542), had already dedicated the edition of Homer's *Vita* by Plutarch to him, and it had clearly pleased him. Gessner likely knew him from his student days in Bourges. According to a second short preface dedicated to Fricker, Gessner followed the Latin and Greek text by Heraclitus with a few Greek text fragments by various authors. These had been collected by Bishop Arsenios from the southern Greek town of Monemvasia and printed in Rome with errors. This was the reason for re-working and publishing them here once again.[29]

Gessner's preoccupation with Homer and interpretations of Homer did not come out of the blue. The great Greek poet enjoyed enormous popularity among Humanists and reformers in Zürich. Gessner commented about Homer in his *Bibliotheca universalis* of 1545 that if one were to summarize all of the international praise for him, it would produce a book thicker than both the *Iliad* and the *Odyssey* put together.[30] Furthermore, there was a strongly-held opinion among scholars that behind the primary content of the texts were additional hidden messages that needed to be uncovered. In 1528, Bullinger wrote about this in the study guide for his students in Kappel: "This is one thing that all completely agree upon. Those who understand antiquity know that all the creations by the ancient poets most surely have an added meaning in the background, whether about history, theology, natural science or morality. This was confirmed by Eustathius, an interpreter of Homer and Landinus, an interpreter of Virgil – and several times by Donatus and Servius as well. From those among us, Rhodiginus attests to this in the fourth book of his *The Fruits of Reading from Antiquity* in the first and second chapters. This also

27 Héraclite (1962), p. XLIII.
28 By the way: On January 24, 1544 a total solar eclipse took place. Cf. Pellikan (1892), p. 144.
29 Hieronymus (1992), pp. 160 and 264.
30 Gessner (1545), f. 335v.

goes for Erasmus in the second book of *Provisions* and in his *Adagia*, and also Melanchthon in his first book of rhetoric."[31]

Therefore, it is no wonder that Homer is included among the school authors in Zürich.[32] Another Greek work by Gessner the physician was an edition and very first Latin translation of a collection of sayings by Johannes Stobaeus (5th century) from Stobi, Macedonia. It was printed by Christoph Froschauer in 1543.[33] First released in Venice in 1535, it contained quotes from ancient works, some of which had been lost, and because it was teeming with errors, Gessner decided to create a new version. He cited it in the *Bibliotheca universalis* as his most useful publication to date.[34] In the foreword, dedicated on June 13, 1543 to the Bern politicians Hans Jakob von Wattenwyl and Hans Franz Nägeli, he explained that the work was especially suited to politicians. This was because it contained many moral, economic and political pieces of wisdom – the moral ones taught one about virtue, and that vices should be driven away. Politicians had a special responsibility to set an example in the community and to behave impeccably. The economic phrases by Stobaeus were also pertinent to statesmen because they would help them to properly and responsibly care for their families. For whoever cannot keep his own house in order, cannot govern a state. The book dealt with many political themes and further pieces of useful advice. This collection of quotes from 242 authors, who were listed at the beginning of the work, was organized into 123 thematic groupings, so-called *loci* or *sermones*. Once again, Gessner explained that the worldly wisdom presented in this work does not contradict divine wisdom. Instead, it is possible to find Christian values in Platonic dialogs, such as, that it is better to suffer injustice than to inflict it upon others. Based on their way of life, if Socrates, Pythagoras and others like them were alive today, who – even among Christians – would not consider them saints?[35]

In his subsequent foreword to the reader, Gessner praised the care taken by the book printer Christoph Froschauer and mentioned at this point that he had put together a sales catalog of Froschauer's printed books,[36] from which only one copy has survived in the University Library of Basel.[37] He also mentioned that he was preparing an index that comprehensively listed all literature by author and topic, by which he was referring to his *Bibliotheca universalis* of

31 Bullinger (1987), vol. 1, p. 37.
32 Ernst (1879), p. 92.
33 Cf. on the history of the Stobaeus editions: Curnis (2008).
34 Gessner (1545), f. 182v.
35 Stobaeus (1543), f. α5r.
36 Cf. Leu (2014a), p. 67f.
37 UB Basel, shelf mark: BE IX 6, no. 2.

1545 and *Pandectarum libri XXI* of 1548. Gessner added that he had verified the excerpts from Stobaeus using more modern editions wherever possible. He also included some brief information about this obscure scholar's life. In conclusion, he mentioned that further little-known collections of sayings by Greek authors were to be found above all in Italian libraries. The Gessner Stobaeus edition was also based on a manuscript from the private library of Diego Hurtado de Mendoza (1503–1575) in Venice, in addition to the very faulty edition printed in Venice in 1535.[38] The Zürich edition was very popular.[39] Shortly after Gessner's book was published, the Italian reform-refugee Celio Secundo Curione (1503–1569), who taught in Lausanne's Latin School, apparently wrote to Bullinger on March 4, 1543, requesting that this publication, and others by Gessner, along with a Bible in Latin, be sent to him.[40]

8.3 Martial Edition

In 1544, Gessner was busy with the publication of a dictionary of Latin names of places and people, the so-called *Onomasticon propriorum nominum*,[41] as well as the third printing of his *Lexicon Graecolatinum*[42] with a famous foreword about the use and venerability of the Greek language. In the same year, he also produced an edition of the Roman poet Martial (40–102 BC) that could be suitable for schools, because Martial was especially well-known for his obscene and lascivious poetry, which was often too erotic for school lessons. In the *Bibliotheca universalis* of 1545, Gessner cited this as his second most important work.[43] Not only Bullinger, but also Gessner and other scholars in the city on the Limmat strove to provide an optimal supply of literature for the younger generation. This was clearly reflected in Zürich's print production, since educational titles and teaching materials made up 13 percent of its publication

---

38 Gessner (1545), f. 456r. Gessner also used a fragmentary manuscript from the fifteenth century that Oporinus had sent to him. It is kept in the UB Basel, call number: F VI 8.

39 Cf. Bernhard (2009), p. 174f.

40 HBBW 13, pp. 97–99.

41 Gessner regarded the *Onomasticon* as a useful tool for commentators of ancient texts, since they would find in it not only the identification of ancient names, but also explanations of them, to which they could refer and would not have to research everything again themselves. The work should thus serve as a commentary (*communis commentarius*) on antiquity, common to all and accessible to all. Cf: Blair (2010), p. 123f.

42 Gessner noted by hand in his own copy of the *Bibliotheca universalis* (Zürich 1545) that this quarto edition of the *Lexicon Graecolatinum* had appeared in 1544 (ZBZ, call number: Dr M 3, f. 182v).

43 Gessner (1545), f. 182v.

output in the sixteenth century.[44] The discussion revolved around the questions: Who were the important authors? What is the proper reading material? Which ancient classics are most effective for learning the Latin and Greek languages? How should one deal with questionable authors from antiquity who are stylistically excellent but morally ambiguous? And specifically, what is to be done with Martial, who was stylistically brilliant but often morally questionable? Gessner did not discard the reading of Martial, but argued for a responsible approach, publishing a version appropriate for school usage that was cleansed of all obscenities. The polyhistor from Zürich omitted the most scandalous of the fifteen hundred epigrams, shortened others, and removed the most offensive passages. He also arranged the short poems into eighty-five thematic groupings (*loci*), which he said would make it easier to find certain passages.[45]

Gessner's edition of around 350 pages is followed by a short philological commentary by Jacob Micyllus, professor of Greek at Heidelberg University (1503–1558). The end of the book contains a barely noticed work written by Gessner and announced on the title page as *Dialogi tres*, comprising 72 octave pages.[46] The names of the persons who appear in it are reminiscent of the comedy *Adelphoe* by the Roman writer Terence. Gessner in fact casts the following figures:

Aeschinus: a youth
Demea: father of Aeschinus
Harpula: a pretty slave girl
Mitio: Demea's brother and foster father of Aeschinus
Radamanthus: a judge
Syrus: a teacher
Teleboas: a court usher

Five of these names also appear in the comedy *Adelphoe* written by the roman writer Terence, including the three main characters Aeschinus, Demea and Mitio, as well as Harpula and Syrus. While the first three have the same roles here that Terence gave them, the latter's version of the character Harpula is the hetaira *Bacchis* (a bacchanalian concubine), who is listed as Psaltria (a harp-playing slave). Instead of being a teacher, Syrus in Terence's play is a slave of Mitio. The names Teleboas and Radamanthus appear in Greek literature,

44 Leu (2008a), p. 230.
45 Martial (1544), f. *3r.
46 Leu (2014).

although no direct reference can be made to the former. Gessner's character of the judge, Radamanthus, is known from ancient mythology as "a just man," who after his death was appointed by the gods to be one of the three judges in the underworld.

Along with the characters, there are also correspondences in the story line. The main theme of *Adelphoe* is the question of whether a disciplined, strict upbringing is ideal for young persons, or if it is better for their development and more will become of them, if they are given a great deal of freedom and autonomy. For Gessner, the main subject of the discussion between the two exponents Demea and Mitio is whether young people should be exposed to obscene literature. Demea is strictly against it, while Mitio finds it unproblematic. In both pieces, Demea represents authoritarianism, and Mitio takes a liberal position. The outcome of Terence's comedy is that neither method of upbringing reaches its goal, and the readers must independently conclude that a compromise between the two extremes is the right path. Likewise, Gessner's *Dialogi* reaches the compromise conclusion that Martial should not be completely banned from school classes, nor should he be read uncensored. Instead, a compromise is called for here as well, and a responsible and cleansed version of his work would most properly serve the needs of the school.

Every sixteenth-century student and teacher who read Gessner's *Dialogi* was immediately aware of the similarities to Terence. While this observation may have caused some to smirk, it certainly made them curious as to which conclusion the author himself had drawn. The Terence comedies were too well known in the early modern school system[47] for Gessner's references to go unnoticed. For example, the Zürich school regulations of October 1532 prescribed the reading of Terence at the third-year level of the Latin School.[48] To satisfy the schools' needs, editions of his comedy were released in Zürich in 1547, 1552, 1555, 1561 and 1570. As mentioned above, Gessner felt that his editions of Stobaeus and Martial were important and useful, but not all of his contemporaries felt the same way. As a result, there were no subsequent printings of the Martial edition. Bern's treasurer, Eberhard von Rümlang, wrote to Bullinger on October 28, 1544 with greetings to Gessner, but also with the advice that he should involve himself with more serious subjects than poetry.[49]

It is well known that for reasons of morality, not all contemporaries were enamored of fine literature. However, in Zürich between 1530 and 1550, a high

47 Bullinger (1987), vol. 1, p. 42f.; Dietl (2005), pp. 20–23.

48 Egli (1879), no. 1896, p. 823.

49 HBBW 14, p. 495.

point was reached for school and citizen dramas,[50] which spawned both local and foreign acting troupes. For instance, from November 29, 1543 until early May 1544, an ensemble from Speyer (Germany) took up residence in the city on the Limmat.[51] By comparison, under the rigid regime in allied Geneva, even reading the knightly novel *Orlando furioso* by Ludovico Ariosto (1474–1533) was forbidden.[52] It seems that no one in Zürich was disturbed by such books. Even Rudolf Gwalther, who later became head of the church there, had a copy of Ariosto in his private library.[53]

8.4 Author Royalties

Various publications by Gessner were produced not only to satisfy his own interests, but also because the author sought to improve his income.[54] Nowhere in the existing documentation can one find how much the polyhistor from Zürich actually earned from this spare-time work. Nonetheless, a 1535/36 list from Heinrich Bullinger contains information about revenue, interest and loans regarding author royalties received from Froschauer: "In April 1535, I calculated with Mr. Christoph Froschauer, and I still owe him 90 guilders and 7 shillings. Against this, I have earned 36 guilders for my writing. Therefore, I still owe him 54 guilders and 7 shillings. For this, I have proofed *Acta Apostolorum* (*Acts of the Apostles*).[55] Additionally, I have made comments in four Epistles,[56] and comments in the Epistles to the Thessalonians, and the Epistles of Paul to Timothy.[57] Also, the report about the ill in German.[58] For this, he gave me one guilder for three printed sheets. Without the *Acta Apostolorum*, the three mentioned booklets add up to a total of 69 and one half sheets. This makes

50 Brunnschweiler (1989), p. 120.

51 We know this thanks to Wolfgang Haller (1525–1601), who by hand recorded somewhat randomly in individual printed annual calendars how the weather was on individual days or months and whatever happened that was worth mentioning. Cf. *Kalender oder Laaßbuechlin sampt der Practick uff das M.D.XLIIII. Jar*, Zürich, Eustachius Froschauer, 1544, [p. 4] (ZBZ, MS. D 269_1).

52 Jostock (2007), pp. 193–196.

53 ZBZ, call number: 15.152. Cf. Leu and Weidmann (2020), p. 93.

54 Cf. the letter to Bullinger September 1558: Hanhart (1824), pp. 139–147.

55 Heinrich Bullinger: *In Acta Apostolorum ... commentariorum libri V*, Zürich 1535.

56 Heinrich Bullinger: *In D. Apostoli Pauli ad Galatas, Ephesios, Philippenses et Colossenses epistolas ...*, Zürich 1535.

57 Heinrich Bullinger: *In Apostoli Pauli ad Thessalonicenses, Timotheum, Titum et Philemonem epistolas*, Zürich 1536.

58 Heinrich Bullinger: *Bericht der krancken ...*, Zürich 1535.

23 guilders, and more than 31 guilders are still owed. This is without factoring in the *Acta Apostolorum*."[59]

From these remarks, one can surmise that Bullinger received one guilder for writing or revising three printed sheets, which corresponds to 12 folio pages.[60] Gessner's *Bibliotheca universalis* of 1545 had 1,300 folio pages and the four volumes of *Historia animalium*, printed between 1551 and 1558, totaled 3,400 folio pages. If compensated in the same manner as Bullinger, this would mean a payment for his authorship of 108 and 280 guilders, respectively. While these are impressive sums, which Gessner was surely pleased to receive, the work for Froschauer often placed him under a great deal of pressure. This is exemplified in a letter to Benedictus Aretius (ca. 1522–1574) from August 1553 in which Gessner describes Froschauer as an annoying and relentless taskmaster.[61] Whether author or publisher, it was obvious that one earned more money from thick books than thin ones, which made voluminous works more welcome. Gessner wrote about this issue to the young Bauhin on August 1, 1563, noting that Froschauer did not like to print thin volumes, but that his cousin occasionally would.[62] We also learn from Gessner's correspondence that during times of war and crisis, it was difficult to find a printer for scientific and medical books, because opportunities for sales were limited.[63]

Gessner's remuneration from the Basel printers Episcopius and Froben was actually greater than the Froschauer payments. For the 72 folio pages of his comprehensive introduction to the large Galen edition in 1562,[64] this amounted to twice 15 pounds. This was equivalent to 15 guilders,[65] or one guilder for nearly five folio pages, which was more than double what Froschauer paid. A certain rate of inflation might have played a role, but never to this extent, since book prices remained quite stable during the sixteenth century.[66]

59 ZBZ, MS. K 40, f. 338r.

60 It is possible that he received more money for writing a new text than for revising an existing one with a view to a new edition, but Bullinger's summary account does not distinguish this.

61 Gessner (1577), f. 117r.

62 Bauhin (1591), p. 122.

63 Letter from Gessner to Crato von Krafftheim, October 18, 1561. Cf. Gessner (1577), f. 9v.

64 The introductory section comprises 77 folio pages, but five of them were not written by Gessner.

65 Wackernagel (1881), pp. 39 and 41.

66 Leu (1993), p. 34; Leu (2004), p. 86.

CHAPTER 9

Professor of Natural Sciences

Along with his less than lucrative medical practice and his earnings from publications, which have just been described, Gessner took a teaching position (*lector*) at the *Hohen Schule*. Partnered with Otto Werdmüller,[1] together they taught philosophy, mathematics, natural sciences and ethics until 1546. During the year 1542, they were also assisted by the previously-mentioned Melchior Wirz. Because there was no money budgeted for a new professor, Gessner and Werdmüller shared an augmented student stipend. This amounted to 14 pounds every quarter, plus six bushels of grain (worth about 30 guilders). Starting in 1541, they received 30 pounds per quarter, and in 1542 it was 40 pounds per quarter, which amounted to 60 or 80 guilders per year. In 1546, Werdmüller became archdeacon, and Gessner filled the position as a salaried *Professor Physicae* (professor of natural sciences) all by himself. His colleague, Josias Simmler, substituted for him on occasion.[2] Gessner's yearly salary from the student office (*Studentenamt*) remained constant at 160 pounds, or 80 guilders, (along with specified amounts of grain and wine) starting in 1542 until he received the canon prebend in 1558.[3] Added to this were his earnings as a physician and from publishing activities. His income was still much less than that of Bullinger, who earned 350 guilders per year, but he was indeed more well off than certain country pastors such as Conrad Clauser, who in 1542 received 36 guilders, 21 bushels of grain, 10.5 *Malter* of oats, 4 *Viertel* of legumes, 10 buckets of wine, plus firewood from the Zürich community of Elsau.[4] For dedicating the first volume of *Historia animalium* in 1551 to the city council, Gessner's compensation was raised to include 10 *Malter* of grain and 10 buckets of wine, and his promotion to city physician in 1554 brought an increase in salary to 142 guilders.[5] While this certainly improved his situation, due to his investments in research and family, it did not completely help him out of his financial predicament.[6] He would have had to spend more than a year's income

1 Gessner refers to Werdmüller as "primus apud nos philosophiae interpres," from which it must be concluded that he was already teaching before Gessner. Cf. Fabricius Montanus (1555), p. 121.

2 Milt (1947), pp. 33–35.

3 Cf. the so called *Studentenamtsrechnungen* in the StAZH, G II 39.2 and G II 39.3.

4 Leu/Weidmann (2004), pp. 26f.

5 Cf. p. 348.

6 Cf. p. 165.

© KONINKLIJKE BRILL NV, LEIDEN, 2023 | DOI:10.1163/9789004541696_010

in order to buy a house for 200 guilders, like his colleague Conrad Pellikan had done in 1544.[7] There was also not much left with which the bibliophile Gessner could purchase books, which at the time were relatively expensive. For instance, *De humani corporis fabrica* by Andreas Vesalius (Basel 1543) was priced at 5 guilders, or the complete 10-volume works of the Church Father Augustine, printed in Basel in 1529/29, went for 23 guilders.[8]

9.1 Teaching Material

The sources reveal little when it comes to the content taught by Gessner and his colleagues Werdmüller and Wirz. According to a description by the theologian Ludwig Lavater (1527–1586) in his book *Über die Gebräuche und Einrichtungen der Zürcher Kirche*, with a foreword dated January 1559, Gessner interpreted the scientific writings of Aristotle and other authors from 2:00 to 4:00 every afternoon (ill. 14).[9] There were no teaching activities on Thursday and Saturday afternoons, thus no physics class either.[10] The purpose of teaching this subject was to relate general knowledge of the natural sciences that could be applied to Bible study, but also for a personal understanding of nature and to reveal the greatness of the Creator. In addition, the subject provided an introduction to the basics of state-of-the-art medicine, which should facilitate the continuation of the actual specialized studies at a university.[11] The School Regulation of 1559 required that the *Professor physicus* should chose an appropriate scientific compendium and work through it with the students.[12] In his Gessner biography, Josias Simmler gives us an insight as to which textbooks Gessner used during his 24-year teaching career from 1541–1565. He mentions that Gessner frequently lectured on relevant works by Luther's ally Philipp Melanchthon (1497–1560), the Tübingen Aristotelian Jacob Schegk (1511–1587), the Wittenberg Professor Johannes Bernhardi, also known as Velcurio (1490–1534), the Silesian physician Hieronymus Gürtler von Wildenberg (1464/65–1558), the Italian Humanist Ermolao Barbaro (1454–1493), the Spanish philosopher Sebastian Fox Morcillo (1526/28–1559) and others. He focused in

7 Pellikan (1892), p. 143.
8 Leu/Weidmann (2004), pp. 24–28.
9 Lavater (1559), f. 19r. On Aristotle studies at other Reformed universities, cf. Rohls (2012/2013).
10 For the period around 1550, see Ernst (1879), p. 95. For the period after the school reform of 1559, see StAZH E II 476, f. 14r.
11 Göing (2003), pp. 86f.
12 StAZH E II 476, f. 13r.

ILLUSTRATION 14 Notes written by Gessner in his personal copy of Aristotle, *Opera omnia*, Basel 1539.

ZBZ, ABTEILUNG ALTE DRUCKE UND RARA: B 83: A

particular on Aristotle's *Nicomachean Ethics* and *Physics*.[13] Additionally, one finds in the files of the *Hohen Schule* that on April 24, 1561, he was required to read roughly 80 pages of *Descriptionis universae naturae ex Aristotelis pars prior* by Jacques Charpentier (Jacobus Carpentarius) with his students, with 10 auditors assigned to him.[14]

In the *Bibliotheca universalis*, Gessner underscored the importance of Schegk's *Philosophiae naturalis disputationes*, printed in Tübingen in 1538, which he used with his students, because he felt at the time that nothing could surpass it for scholarship, methods, brevity and style.[15] In 1548, he released *Naturalis totius compendium* by Ermolao Barbaro through the publisher Johannes Oporinus in Basel. The *Epitome in universam Aristotelis physicam* by Hieronymus Gürtler von Wildenberg was appended, which, according to Simmler, Gessner had also used in his teaching. Even after Gessner's death, this work was still used in teaching the natural sciences, as is evidenced by the personal copies of Johannes Collin (1553–1609)[16] and Johann Werner Guldinbeck, who gave his volume to Hans Ludwig Breisacher (1579–1645) as a gift in 1596.[17] After finishing at the *Hohen Schule*, Collin and Breisacher were later both involved in Zürich church activities. Gessner's partially hand-annotated, personal copy of Johannes Bernhardi's (Velcurio's) commentary on Aristotle's *De anima* (*On the Soul*), printed in Basel in 1538, is still in existence.[18] The book changed hands several times and continued to be used in class work for nearly 100 years, as is indicated by a list in the back of the book with 35 names of people who attended a related lecture at the *Hohen Schule* in May of 1628. Bernhardi's commentary on Aristotelian physics also seems to have found its way into the teaching plan, since an unnamed student noted in his 1566 copy that a lecture about it had started on September 11, 1570.[19] In 1563, Gessner also published

---

13 Simmler (1566), f. 8v.

14 StAZH E II 458, f. 19v; cf. Göing (2013), pp. 271 and 425.

15 Gessner (1545), f. 362r.

16 According to the ownership note on the front parchment cover, this copy belonged to an unknown Jakob Nüscheler in 1549, who possibly also visited the *Hohe Schule* and attended Gessner's classes. The decipherment of Nüscheler's ownership entry is only possible under UV light.

17 ZBZ, call number: Z 248.

18 ZBZ, call number: Dr M 877_2. Cf. Göing (2003), where, however, Bartolomaeus Bernhardi is erroneously named as the author.

19 Ioannes Velcurio: *Commentarii in universam physicam Aristotelis* …, Tübingen 1566, f. A_2r (ZBZ, call number: NP 2451_2). This work might have been used at an external teaching institution, but the volume belongs to the holdings of the library of the Naturforschende Gesellschaft in Zürich, founded in the eighteenth century, and therefore seems to have always been in Zürich and also used here. That the date, September 11, 1570 refers to the

his own textbook about the theory of the soul (*De anima*), which he discussed with his students[20] and which was still used by the student Caspar Waser (1565–1625) twenty years later.[21] Gessner had fears that due to its complicated language, the book would not sell well.[22] The described teaching materials and their long years of use show that the content of natural science lessons introduced by Gessner at the *Hohen Schule* continued to follow roughly the same path for decades to come.

9.2 Study Guide

During the course of the year 1560, Gessner put down on paper his thoughts about how prospective scholars, especially physicians, should go about their studies.[23] His remarks began with the suggestion that delving thoroughly into Aristotle's philosophy and knowledge of natural science was a must. Likewise, it would be necessary to study the Greek commentaries on Aristotle as well as pertinent texts by the Arab philosopher and physician Averroes (1126–1198). In case interest was lacking or one had alternative career goals, it would be sufficient to become familiar with compendia about Aristotle by contemporary authors. In addition, he recommended that those studying medicine should at least read Aristotle's works on natural science; the theologians, about metaphysics; and the jurists about his politics, ethics and economics. He also strongly recommended to his readers the study of Aristotle's logical and dialectical writings, jointly known as *Organon*, since they formed the basis for scientific work and argumentation. If the *Organon* were not introduced by an instructor, he suggested reading the Greek commentators Simplicius of Cilicia (ca. 490–560), Johannes Philoponus Grammaticus (6th century) and Alexander of Aphrodisias, as well as paraphrases by the French humanist Jacques Lefèvere d'Étaples (1450/55–1536). Should that also not be possible, or if these titles were not accessible, as an introduction, he suggested the works of the pedagogue from Strasbourg, Johannes Sturm (1507–1589), the theologian Johannes Hospinian (1515–1575), the reformer Matin Borrhaus (1499–1564),

---

beginning of lectures, I conclude from similarities of student entries to Glarean lectures, cf. Grafton/Leu (2014), pp. 11 and 19.

20 Gessner (1563), p. 721. Gessner dedicated the work to the emperor's personal physician Julius Alexandrinus (1506–1590), to whom he sent three copies on April 7, 1563. Burmeister (1975), p. 226.

21 Cf. the copy in the ZBZ, call number: 7.381.

22 Burmeister (1975), pp. 214f.

23 ZBZ, MS. C 50a, f. 104r–107v.

and regarding Aristotle, the previously mentioned Jakob Schegk. One should then advance from dialectics to ethics, where Aristotle's *Nicomachean Ethics* was the appropriate text. This subject also included economics, which was addressed better by Xenophon than Aristotle. Politics fell under the same category, about which the latter had written extensively. The natural sciences came next on the list. For foundational reading, Gessner cited the eight books *On Physics* by Aristotle, flanked by various commentaries from Themistius (ca. 317–388) up to Schegk. This was followed by the enumeration of further scientific treatises by Aristotle that covered meteorology, minerology, plants and the theory of the soul, and ending with books about the animal world. For this, he not only suggested works from ancient and Byzantine commentators, but also from contemporary authors of books about animals such as Edward Wotton (1492–1555) from England, the above-mentioned Guillaume Rondelet as well as his own works.

In the pages that followed, which are no longer worked out in full but consist of jottings, he recommended the Aristotle commentaries by the previously-mentioned Charpentier and by Thomas Aquinas, Averroes and Simplicius. For metaphysics, Theophrast (371–287 BC) and the Bolognese philosopher Giovanni Battista Camozzi (Camotius) (1515–1581) could be referenced. Regarding the immortality of the soul, which Aristotle famously rejected, more information was to be found in Platonic philosophers such as Hermes Trismegistus,[24] Iamblichus (ca. 240–320), Plotinus (205–270), Marsilio Ficino (1433–1499) and of course Plato (428–348 BC) himself. This addressed a theme that brought theology into the study plan. Gessner then rounded out the passage with an exhortation to read the Bible every day – without fail – along with the technical literature.

In the next step, there were a few random ideas about the methodology of study. Gessner noted that together with reading and writing, practical exercises should not be neglected either, which consisted of thinking about a topic for oneself, writing about it and discussing it with others. Furthermore, what one heard and read should be gathered into summaries, collections of statements and tables for memorization. For certain subjects, a sort of lexicon should also be created in which the terminology was listed in alphabetical order along

---

24 It may be surprising that Gessner cites the ancient esoteric Hermes Trismegistus, but he was regarded by contemporaries as the transmitter of ancient secrets since the rediscovery of his writings in Florence in 1462. The scholars in Zürich fell back on him and similar writings when it helped to substantiate their statements. Heinrich Bullinger, for example, also referred to Hermes Trismegistus when he had to explain that even the pagans in ancient times had held the view that there was only one God and not many gods. Cf. Bullinger (1574), p. 8.

with commentaries and references. As an example, he cited the *Dictionarium medicum* by the physician scholar from Brescia, Aloysio Mundella († 1565), who had recently done this for the works of Galen. This was the *Theatrum Galeni*, whose hithero unknown publication history is elucidated by Gessner's correspondence.[25]

9.3 *Theatrum Galeni*

On September 29, 1551, Mundella sent a letter asking what Gessner thought about him (Mundella) creating a medical index of the works of the Roman physician Aulus Cornelius Celsus (ca. 25 BC–50 AD).[26] In an additional handwritten note dated October 5, Mundella thanked Gessner for the index he had written for Mundella's *Dialogi medicinales*. It had been printed by Froschauer in Zürich in the same month, and Gessner had also contributed a Greek poem at the back of the book.[27] Unfortunately, Gessner's answer to Mundella has not survived. It was not until August 20, 1560 that Mundella appears again in Gessner's correspondence, when he wrote to Taddeo Duno (1532–1613), a physician in Locarno: "My dear Duno, I have received what you sent from Aloysio Mundella and thank you. Since I too am quite busy, I will write directly to him once I have more information about his *Theatrum*, such as when it will finally be finished, and if it is to be a stand-alone publication per his wishes. I will write to the book printer in Basel and include Mundella's letter."[28] At this point, the *Theatrum Galeni* was already in Basel for printing. Apparently, it was Gessner who had made arrangements with the Basel printer Nicolaus Episcopius the Elder (1501–1564) for the printing of the volume with nearly 1,400 folio sheets. In the surviving fragments of a letter written on October 18, 1560 by Mundella to Gessner, he asked whether Gessner had any news from the printer about his *Theatrum*.[29] On April 10, 1562, Mundella requested that Gessner ask the printer to work with special care.[30] There must have been a long period of inaction, since Mundella contacted Gessner two years later, prompting Gessner to write to Theodor Zwinger (1533–1588) in Basel on July 14, 1564, asking for information: "I am sending you the letter from Mundella and

25 Cf. Hieronymus (2005), vol. 2, no. 300.
26 ZBZ, MS. C 50a, no. 34.
27 The fact that Gessner made the register to this work as a commission for Froschauer is not mentioned anywhere else, not even in the book itself.
28 Gessner (1577), f. 88r; translation slightly revised according to Peine (1941), p. 24.
29 StAGR D V/37 C 36.06.03.
30 StAGR D V/37 C 36.06.15.

ask that you send me a brief answer as to whether Episcopius is possibly printing the work and when he might have it finished."[31] The years of delay probably arose because Mundella had produced his extensive alphabetical manual on Galen before the publication of the new definitive Galen edition of 1561/62, which had been published by Nicolaus Episcopius and Hieronymus Froben (1501–1563). Now thousands of references had to be adapted. On April 8, 1565 Gessner requested that Zwinger, his colleague in Basel, exert pressure to have the *Theatrum Galeni* finished.[32] On June 14, Gessner wrote again, communicating that Mundella had died.[33] His nephew, Giovanni Baptista Mundella, was staying with him in Zürich and would now come to Basel. Gessner urgently called upon Zwinger to do everything he could to see that the *Theatrum Galeni* would finally be published.[34] Zwinger's estate in the University Library of Basel contains two notes in the hand of Conrad Gessner, which he must have written sometime in the second half of 1565. In one, it states that he is sending the foreword to *Theatrum Galeni* to Giovanni Battista Mundella, which was to be printed in 1568.[35] In the other, Gessner writes shortly before his death that the Episcopius printing house should give him one or two copies of the book, since he had made such an effort on its behalf and had written at least 30 letters to the printer in Basel and to Mundella in Italy. In the case that he might not live long enough to see the final product, he wished for a copy to be sent in his name to Caspar Wolf (1532–1601) in Zürich.[36]

9.4 *Loci*-Collections

The reason why Gessner mentioned Mundella's work as an exemplary *Dictionarium* in his reflections on university studies is that it is a so-called medical collection of *loci*,[37] based on a working methodology that was also promoted at the *Hohen Schule* (ill. 15) and in particular by Gessner. Especially since the beginning of book printing, which had multiplied the number of available

31 Gessner (1577), f. 108v; translation slightly revised according to Peine (1941), p. 78.
32 Gessner (1577), f. 110r.
33 The year of death 1553, which is widespread in the literature, is definitely wrong.
34 Gessner (1577), f. 112r.
35 UB Basel, shelf mark: Frey-Gryn Mscr I 13, no. 83.
36 UB Basel, shelf mark: Frey-Gryn Mscr I 12, no. 170.
37 Accordingly, Mundella's subtitle reads: "Theatrum Galeni … quo vel indicis loco in omnes libros Galeni libros, vel locorum communium instar in re medica, lector eruditus … utetur."

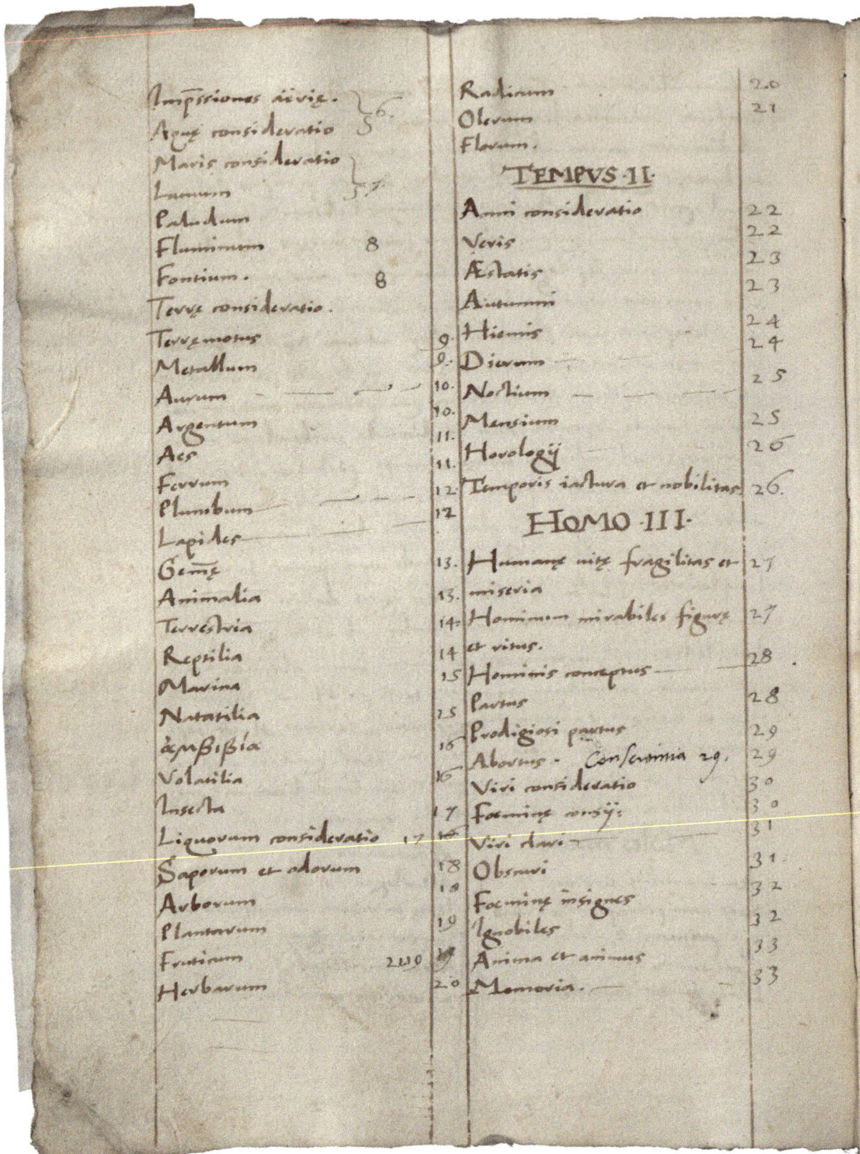

ILLUSTRATION 15 As a 16-year-old student at the *Hohen Schule* in Zürich, Rudolf Gwalther hand-wrote a *Loci*-collection in which he organized reading sources according to topics. Page from the table of contents with the keywords *Tempus* (time) and *Homo* (person) with corresponding subterms. ZBZ, HANDSCHRIFTENABTEILUNG: MS. D 129

texts,[38] scholars had been contemplating processes for easily storing and recalling information. To this end, a method was followed in Zürich that had been formulated by the famous humanist Erasmus[39] with its roots in antiquity.[40] Erasmus suggested that in order to preserve knowledge gained from reading a variety of books and to keep it accessible, every scholar should compile his own encyclopedia consisting of thematically sorted literature references: "Now, to derive maximum use from them [books], one should first set up a collection of positions, categories and forms in which everything noteworthy that is encountered may be entered in an organized manner. How such collections are to be organized will be explained later in *Foundations of the Abundant Style* [*De duplici copia verborum ac rerum*]. For those who do not have sufficient time or an adequate number of books, Pliny alone would have much to offer – likewise, Macrobius and Athenaeus as well as Gellius. Above all, however, one must turn without delay to the origins, that is, to the Greeks and ancient sources in general."[41]

The text about methodology mentioned by Erasmus, in which he explains how such encyclopedic collections of *loci* are created, was published in 1512 under the aforementioned title *De duplici copia verborum ac rerum*. In Zürich, it became part of the Latin School's curriculum.[42] In the second book of this work, Erasmus dedicated an entire chapter to describing the *loci* method.[43] First, one must make a list of important terms (*loci*). These should be augmented by other words that are related, in opposition, or associated in some other manner. Relative to the *locus*, Erasmus recommended that notes be taken while reading various authors to record literary sources and other important items such as turns of phrase, sayings, anecdotes, or illustrative examples. Through the act of writing them down, one will find them easier to recall. The collected knowledge will then be accessible for publication or presentation at any time.

In 1519, Erasmus applied the method of *loci* to theology in his *Ratio seu methodus*: "I cannot be certain whether that which I am about to relate will be of particular use to those who properly apply it. That is, you should either collect various theological themes [*locos aliquot theologicos*] yourself or take up those gleaned by others. Then you should gather all that you read in selected 'nests' [*nidulos*] to clarify what has been borrowed and what you wish to store. For

---

38 Blair (2010).

39 Leu (2007), pp. 327–342.

40 Moss (1996).

41 Erasmus (1963), p. 34.

42 Leu (2007), p. 332.

43 Erasmus (1978), pp. 279–659. The mentioned chapter is on pp. 635–648.

instance (to cite a few pertinent examples): about faith, about fasting, about enduring evils, about supporting those who are ill, about putting up with godless officials, about avoiding the annoyances of the weak, about studying the Holy Scriptures, about the responsibilities of piety for parents and children, about Christian love, how to honor the nobility, and about envy, resentment, chastity and the like. For there are countless combinations one may consider. After the subjects have been categorized as related or unrelated, they must be listed as to relevance to all the books of the Old Testament, the Gospels, and letters to the Apostles – and what conforms or does not (as we have already cited in our *Copia*). If it seems good to one, it is possible to add in that which one finds useful from ancient authors and books by heathens. I believe I can discern from his writings that St Jerome also used this method. If there were issues to discuss, content was readily available; if an explanation was needed, comparisons were easily made."[44]

As mentioned above, Gessner created various *loci* collections for individual authors and subjects. For instance, in 1544, he generated 80 *loci* (themes)[45] for the reorganization of Martial's poems. These background collections of material also showed through in his other works, and it is only a mild exaggeration to say that some publications can be compared to an overturned basket of notes or *loci* collections.[46] The same can be said for many works by the reformer Heinrich Bullinger.[47] The *Pandectarum libri XXI*, a key word index for thousands of authors and books in the 1545 *Bibliotheca universalis*, was an enormous *loci* collection printed in 1548. Gessner says in the foreword that all of the works from the *Bibliotheca universalis* were listed according to content and *locus*.[48] In the first book of the *Pandectarum libri*, which is devoted to the subject of grammar, he speaks more specifically about the *loci* method in parts 4 and 5 and recommends it to his students.

There are three expansive handwritten volumes from Gessner's estate entitled *Thesaurus medicinae practicae* that provide a good view into his working methods. They consist of thousands of paper clippings that Gessner cut from letters, books, manuscripts and his or other people's notes and organized under specific medical topics (*loci medici*) (ill. 16). Caspar Wolf, the administrator of

44 Erasmus (1967), p. 453.

45 Gessner's Italian colleague Ulisse Aldrovandi also compiled an 83-volume encyclopedia of readings, which he wrote down on slips of paper and pasted in alphabetical order. Cf. Miani (2004).

46 Particularly when certain passages consist of thematically related quotations. Cf. Colombat/Peters (2009), p. 47.

47 Leu (2007), pp. 330–332; Koch (1968), plate in appendix.

48 Gessner (1548), f. $^{*}{}_{3}$r.

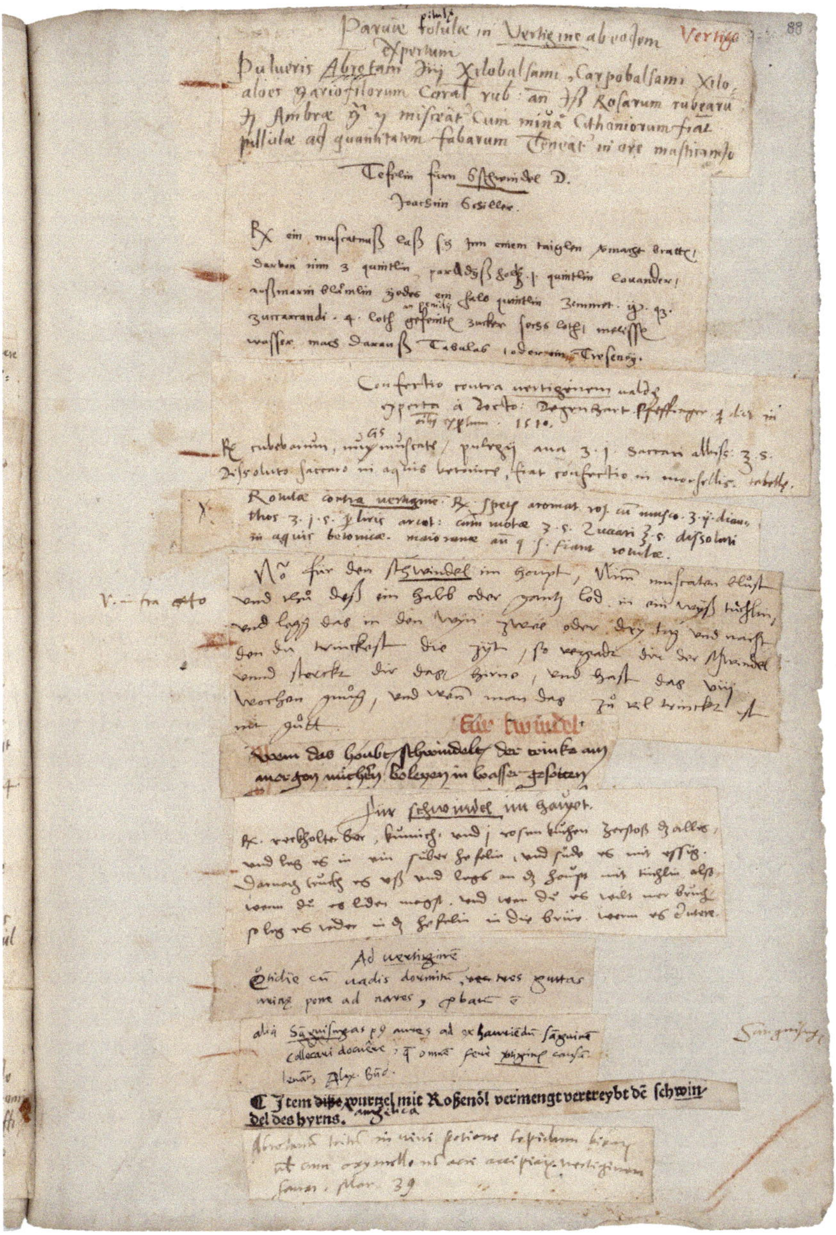

ILLUSTRATION 16 Caspar Wolf's medical *Loci*-collection with recipes, remedies and advice, in part from Gessner's possessions, that are organized by illness (here, vertigo-dizziness). The third and fourth entries from the bottom are in Gessner's handwriting. The paper clippings are annotated with German, Latin, handwritten and printed texts (second from the bottom).

ZBZ, HANDSCHRIFTENABTEILUNG: MS. S 204A, F. 88R

his estate, made the effort in 1596 to glue the notes into the three mentioned volumes. It is possible that he augmented them with some of his own material. The result was an enormous compendium of medical remedies in which recipes and medicines were organized according to illness and could be looked up by disease.[49] Volume 1 deals with head-related illnesses, volume 2 is dedicated to the breast and abdomen, and volume 3 delves into kidney, as well as venous and sexually transmitted diseases.

9.5 Lectures

With regard to Gessner's lecturing activities, only one manuscript fragment of 24 handwritten pages has survived. The rest of the corresponding estate was published by Caspar Wolf in 1586. This fragment[50] contains lecture notes about chapters 5–18 of the second book of the natural history by Pliny the Elder (23/24–79 AD), a Roman officer and natural scientist who died during the eruption of Vesuvius. The pertinent chapters[51] are primarily concerned with the seven traditional planets surrounding the Earth (ill. 17), with Gessner relying in part on Melanchthon in his explanations. One will find no mention of the name Copernicus here since, like Aristotle[52] and Melanchthon, Gessner believed in a geocentric worldview. This particular lecture, held in 1561, was a *lectio publica* that was open to attendance by all the students of the *Hohen Schule*.

Another of Gessner's lectures, however, was mentioned in some notes from a student in 1563. In the Zentralbibliothek Zürich there is a copy of a geography booklet by the Romanian reformer Johannes Honter (1498–1549), which has the following handwritten note on the cover page: "D. Conradus Gesnerus praeceptor meus plurimum observandus auspicatus est haec Cosmographica rudimenta 25 die Februarij Anno 1563" (My most esteemed teacher, Conrad Gessner, began with the *Cosmographica rudimenta* on February 25, 1563).[53] On the flyleaf, the unnamed student summarized the 58-page booklet that

49 Cf. Blair (2010), pp. 213–225.

50 ZBZ, MS. C 50a, f. 28r–34v and 39r–45v.

51 Gessner's chapter count corresponds to the Pliny edition published in Basel in 1539, which he acquired in 1543. His own copy is now in the UB Erlangen, MS. 2720. Cf. Leu et al. (2008), p. 202.

52 Flashar (2013), P. 281.

53 ZBZ, shelf mark: RR 1831. Gessner's involvement with Honter's work is also evidenced by his partly richly annotated copy of Honter's *Cosmographiae rudimenta libri duo*, which

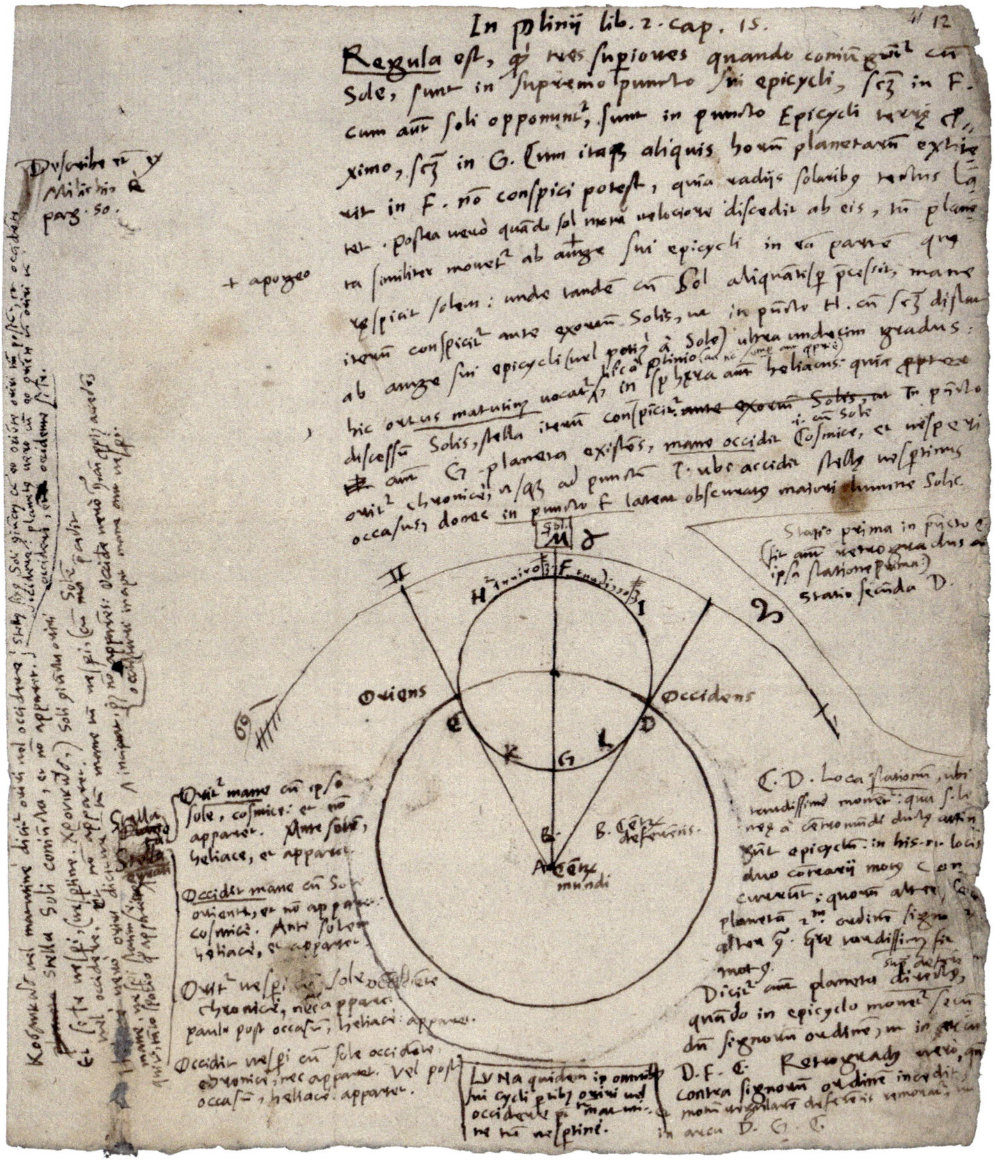

ILLUSTRATION 17 Gessner's lecture notes about the cosmological books by Pliny the Elder.
ZBZ, HANDSCHRIFTENABTEILUNG: MS. C 50A, F. 41R

was divided into four books: book 1 covers elements of the heavens, the stars, geographical longitude and latitude, climate zones, etc.; book 2 deals with the geography of Europe; book 3 deals with Asia, Africa and the islands;[54] and book 4 goes into human anatomy and illnesses, animals, plants, offices, occupations and more. The 15 supplementary maps give an interesting glimpse into the worldview that was taught in Zürich, as well as the level of geographical knowledge at the time. The first table provides a schematic interpretation of the cosmological system according to Ptolemy's worldview, in which the Earth is placed at the center with the sun being the fourth planet. The second illustration shows the Earth with the names of the 12 winds, as they were taught in ancient Greece. The third map shows the entire world with the four continents that were known at the time (ill. 18). The remaining woodcuts show maps of various countries. America appears very vaguely on the world map, but contrary to earlier versions, it is no longer shown as an island but as a significant land mass. In the text, the New World is only indirectly mentioned at the end of book 3. Nonetheless, one may assume that Gessner taught more about the new continent to his students, since he possessed wide-ranging botanical and zoological knowledge of the New World. The unknown student's copy of Honter's *Rudimenta cosmographica* is shot through with empty pages, which suggests that he intended to write his lecture notes directly into the book. Oddly enough, his handwritten entries and margin notes break off starting on page 5. Furthermore, in the booklet there are also seven handmade schematic sketches by the student that must have first been drawn on a wall panel by Gessner. This conclusion is based on two of these drawings being found among the posthumous scientific manuscripts by Gessner that were selected and published by Caspar Wolf.[55] This includes the previously mentioned 12 winds, where the drawing and the printed version are identical (ill. 19a, 19b). Also among the notes about the above-mentioned Pliny lecture, one finds 14 astronomical graphics, that Gessner must have displayed to his audience in some manner.[56]

Remarkably, included in Gessner's sketches is an obsolete medieval zone map of the Earth, which references the commentary by the ancient author Macrobius on Cicero's *Somnium Scipionis* (II,9).[57] The Earth was divided into

---

appeared in Basel in 1534 as an appendix to Dionysius Periegeta's work *De totius orbis situ* (ZBZ, call number: 20.1842).

54 America was also considered an island at the time, as evidenced by the title of Columbus' famous letter, first printed in 1493: *Epistola … de insulis … nuper inventis*.

55 Gessner (1586), pp. 128f.

56 Already Aristotle worked with graphics and drawings, cf. Flashar (2013), pp. 22, 79 and 288.

57 Edson et al. (2005), pp. 45 and 59f.

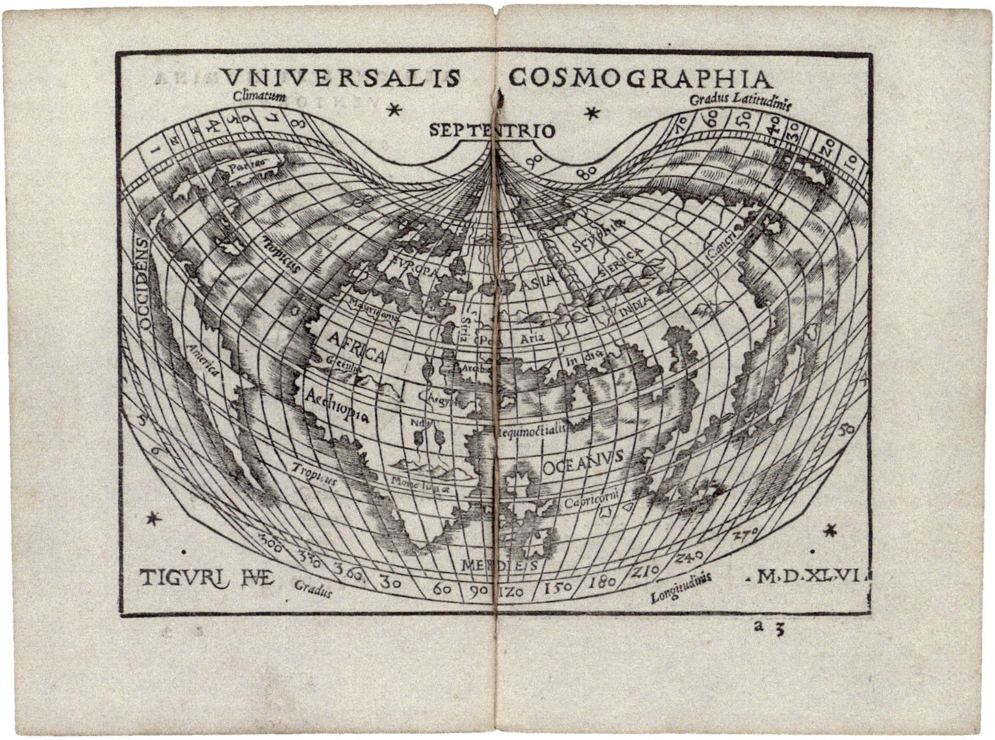

ILLUSTRATION 18 The world map, in: Johannes Honter, *Rudimentorum cosmographicorum libri*, Zürich 1552. According to a handwritten note on the cover page of a copy of the book, Gessner discussed the volume with his students in February of 1563.
ZBZ, ABTEILUNG ALTE DRUCKE UND RARA: RR 1831

five belts, in which only two were said to be habitable since the two at the poles are too cold and the one at the equator is too hot. Additionally, the equatorial zone was thought to be an impassable space separating the inhabitants of the northern and southern hemispheres. With the courageous explorations of the Portuguese seafarer Gil Eanes, this theory had been proved wrong, and all zones were subsequently known to be inhabited. Eanes passed the feared West African Cape Bojador in 1434, thus opening the door to worldwide journeys of discovery. The St Gallen reformer Joachim Vadian (1484–1551) had already taken a position on this in his comments on the Roman geographer Pomponius Mela (1st century AD), which were published in Vienna in 1518.[58] Gessner was also surely aware of these issues, and perhaps used the five-zone graphic as a crude climate model.

58 Näf (1944), vol. 1, p. 270. Cf. Suter-Meyer (2020), pp. 140–164.

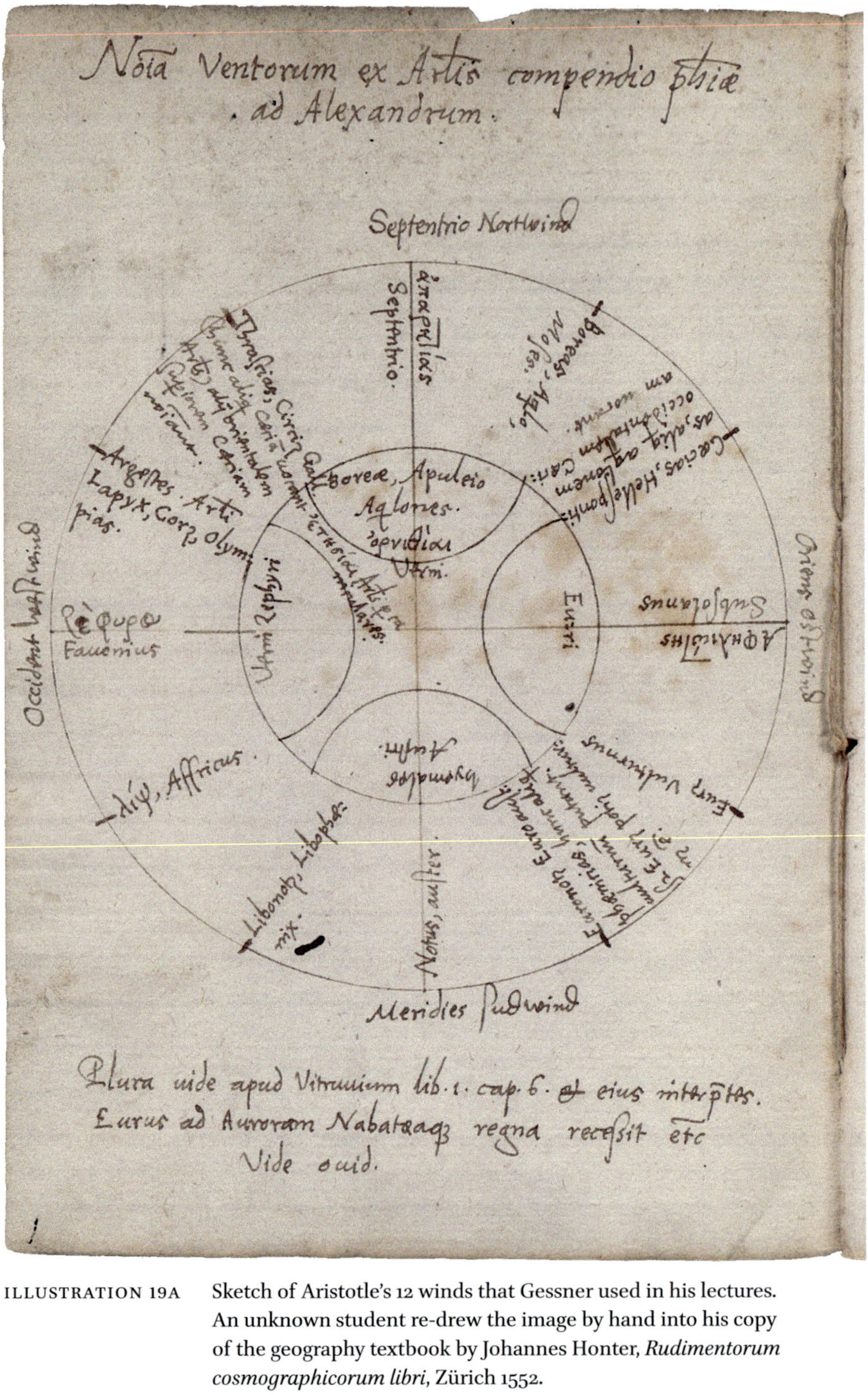

ILLUSTRATION 19A Sketch of Aristotle's 12 winds that Gessner used in his lectures. An unknown student re-drew the image by hand into his copy of the geography textbook by Johannes Honter, *Rudimentorum cosmographicorum libri*, Zürich 1552.

ZBZ, ABTEILUNG ALTE DRUCKE UND RARA: RR 1831

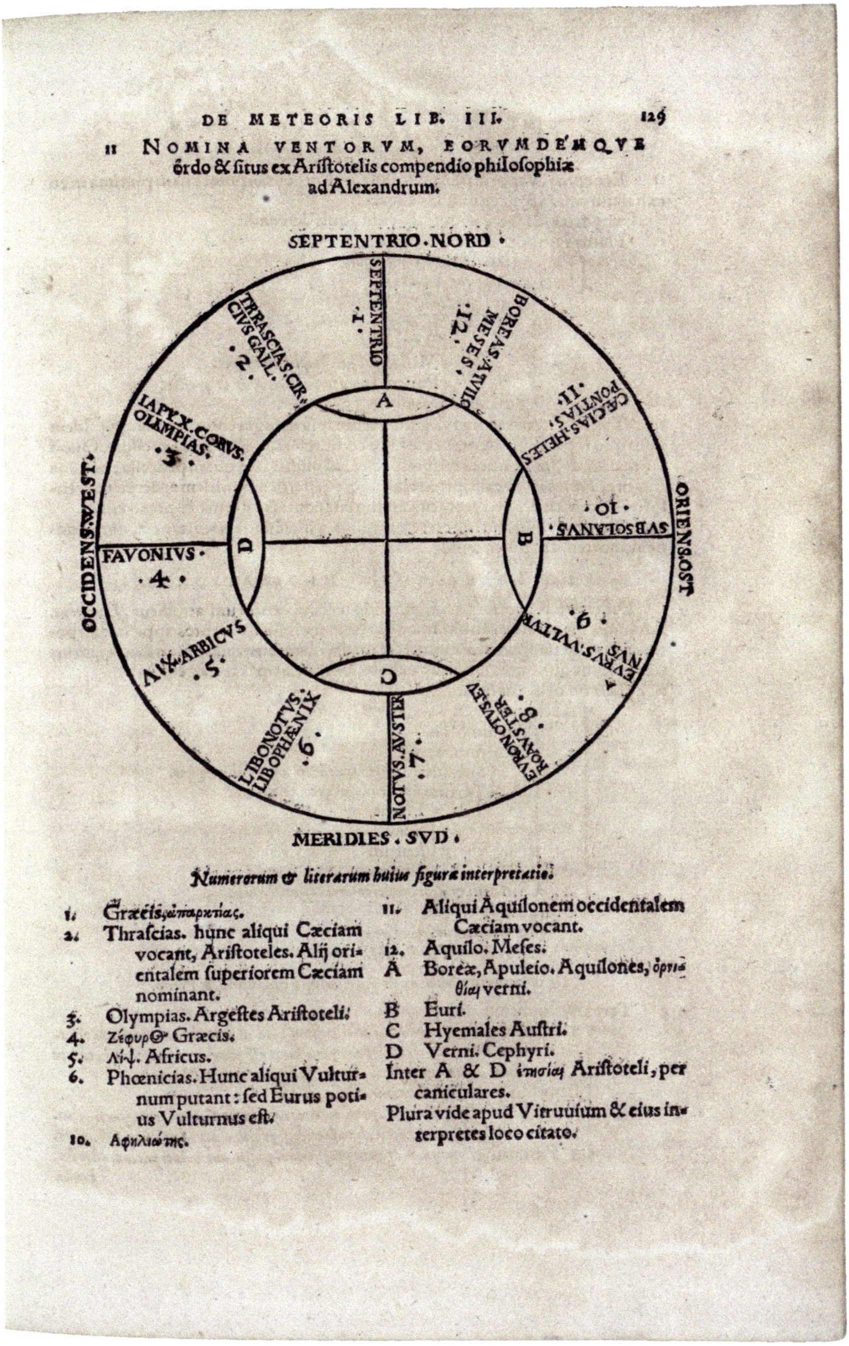

DE METEORIS LIB. III. 129

11 NOMINA VENTORVM, EORVMDEMQVE ordo & situs ex Aristotelis compendio philosophiæ ad Alexandrum.

Numerorum & literarum huius figuræ interpretatio.

1. Græcis, ἀπαρκτίας.
2. Thrascias. hunc aliqui Cæciam vocant, Aristoteles. Alij orientalem superiorem Cæciam nominant.
3. Olympias. Argestes Aristoteli.
4. Ζέφυρος Græcis.
5. Λίψ. Africus.
6. Phœnicias. Hunc aliqui Vulturnum putant: sed Eurus potius Vulturnus est.
10. Ἀφηλιώτης.
11. Aliqui Aquilonem occidentalem Cæciam vocant.
12. Aquilo. Meses.

A Boreæ, Apuleio. Aquilones, ὀρνιθίαι verni.
B Euri.
C Hyemales Austri.
D Verni. Cephyri.

Inter A & D ἐτησίαι Aristoteli, per caniculares.

Plura vide apud Vitruuium & eius interpretes loco citato.

ILLUSTRATION 19B After Gessner's death, Caspar Wolf edited his handwritten lecture notes, among which was a diagram of Aristotle's 12 winds. Conrad Gessner, *Physicarum meditationum libri*, Zürich 1586, p. 129. ZBZ, ABTEILUNG ALTE DRUCKE UND RARA: 5.86

Along with graphic renderings, which even included the first schematic sketches of various forms of earthquakes,[59] Gessner also employed astronomical instruments in his teaching. In his estate is an introduction explaining how to use an astrolabe, which he possibly spoke about or dictated to his students.[60] Unquestionably, maps were used in teaching at the *Hohen Schule*, since Gessner not only received his own personal map of Greece from Oporinus in September 1544,[61] but this map was also officially purchased by the *Hohen Schule* in 1547.[62] It is based on a drawing by the Greek humanist, librarian and collector of handwritten manuscripts, Nikolaos Sophianos from Corfu, who was especially active in Rome and Venice. It was published in Rome no later than 1542 and reprinted by Johannes Oporinus in Basel two years later. The famous woodcut artist Heinrich Vogtherr the Elder (1490–1556) provided the cut illustration. The map was reproduced on eight pages and measured 84 × 113 cm. Oporinus produced a second printing just one year later, and it was also published by Johannes Schröter in Basel in 1601. Today, all that remains of this is a single print from 1545 in Athens and one from 1601 in Basel.[63]

In his famous *Pandectarum libri* of 1548, Gessner published a compilation of 17 European maps that he was familiar with, including the aforementioned map of Greece. He also mentioned additional maps of regions, including Scandinavia, Poland, Prussia, Lithuania, Italy, France, Switzerland, Bavaria, Hungary, Germany and others.[64] Therefore, in Zürich there was clearly a familiarity with various maps and quite possibly some were used in teaching, besides the above-mentioned map of Greece. The need for map materials was also expressed in an undated letter written to Bullinger by Diethelm Keller in the name of the Latin School at the Fraumünster church sometime between 1537 and 1541.[65] He requested the purchase of geographic maps for classwork that were to be paid for with church funds. The request failed due to a lack of understanding by the canons, who had been educated in another spirit. Whether the maps were eventually purchased for the Fraumünster school is unfortunately not known. Bullinger might well have approved of Keller's

59 Gessner (1586), p. 132. The topic of earthquakes was more present in Zürich than one would think. On April 24, 1557, and July 9, 1559, the earth shook, as Wolfgang Haller (1525–1601) noted in his calendars printed by Froschauer or *Lassbuechlin* (ZBZ, shelf mark: MS. D 269_{12} and 270_{1}). Both quakes are missing in the earthquake catalogue of Switzerland (Schwarz-Zanetti/Fäh 2011).

60 ZBZ, MS. C 50a, f. 111r–112v.

61 Letter from Johannes Gast to Bullinger, September 5, 1544, in: HBBW 14, pp. 376f.

62 Cf. the so called *Studentenamtsrechnungen* in the StAZH, G II 39.2.

63 Muller (1990), pp. 202f.; Muller (1997), p. 324; Hieronymus (1992), pp. 429–431.

64 Gessner (1548), f. 111v.

65 HBBW, Ergänzungsband A, pp. 26f.

ILLUSTRATION 20 A reverse printed map of the Holy Land from 1525 with the Zürich fleet off the coast. It shows the first map that was bound inside a Bible.
ZBZ, ABTEILUNG KARTEN UND PANORAMEN: 31 NV 02:1

request. This is shown by the Bible that was printed in Zürich between 1524 and 1529 in small folio format. It was the very first to be published with a map of the Holy Land (ill. 20), which was meant as an aid for a better understanding of the text. Incidentally, it portrayed the Zürich fleet off the coast of the Holy Land![66]

Furthermore, Bullinger made recommendations in his 1528 study guide for students as to how to furnish one's personal work space with maps, globes, and astronomical instruments: "Globes or astronomical spheres should be hung high on the wall or from the ceiling. The walls should be decorated with maps by Ptolemy or other imagery related to the sciences."[67] The future head of the Zürich church still felt that the plates in various editions of Ptolemy were the best. The *Cosmographia* by the Greek scholar Ptolemy, who worked

66 The map in the Froschauer Bible is a copy of a map by Lucas Cranach the Elder. Cf. Dürst (1993), pp. 50–53.
67 Bullinger (1987), P. 137.

in Alexandria in the 2nd century, had been known to the Arab world since the 9th century. The Europeans first became aware of it in the early fifteenth century. It had been translated into Latin in 1406 by the Italian humanist Giacomo d'Angelo (or Angelus de Scarperia)[68] and was first printed in Bologna in 1462. A beautifully colored 1486 copy that had been published in Ulm was available to professors and students in the *Stiftsbibliothek* at the Grossmünster in Zürich.[69]

The greater part of Gessner's surviving lecture materials were published by Caspar Wolf in 1586 under the title *Physicarum meditationum, annotationum et scholiorum libri V*, more than 20 years after Gessner's death.[70] Book 5 has a separate title page and contains Gessner's comments on specific works by Aristotle. It concludes with a treatment dating from April 4, 1565 that spans more than two folio pages and covers the three-part nature of the human soul as a reflection of the three-part nature of God. Wolf compiled the nearly 300 folio pages of Books I–IV, together with the 250-page Book V, which contained excerpts, notes, collected materials, and handwritten annotations from books in Gessner's private library. He had not left behind a lecture folder, but rather a mass of loose papers and notes that Wolf had to sift through. Wolf commented on this in the foreword: "Because he was busy with so many other endeavors, Gessner tended to jot down on any handy slip of paper the thoughts that popped into his head. When he had time, he would return to the notes and work more thoroughly on the ideas he found worthwhile, then toss the memos in a box. While reading a book, if he came upon a hypothesis that he had read elsewhere and he felt that commentary was needed – whether in agreement or not – he would write it down above, below or on the next margin."[71]

The polyhistor often admitted in his correspondence that his study room looked chaotic. For instance, in a letter to Johannes Bauhin of February 15, 1564, he explained that he was not always able to expeditiously write down in a letter his opinion on certain matters because his observations were strewn across countless slips of paper, and it was impossible to look through each and every one of them.[72] A few months before on November 14, 1563, he let Bauhin know

68 Edson et al. (2005), pp. 41 and 57.

69 ZBZ, call number: Ink K 112.

70 Wolf was not sure whether the publication of Gessner's lecture notes was justified and asked Jakob Zwinger for advice on April 7, 1583. Cf. UB Basel, Frey-Gryn Mscr I 11, f. 130. I kindly thank Dr. Tilmann Walter (Würzburg) for this reference.

71 Fischer et al. (1967), p. 126. For the Latin original see: Gessner (1586), f. aa_5v.

72 Bauhin (1591), p. 139. On February 18, 1565, he wrote to Adolph Occo (Gessner 1577, f. 58r, cited in Tilemann, 1939, p. 44) about the description of a certain medicine: "I know that I have it at home too, but it lies under such a quantity of loose leaves about the plague that I have lost some hours to find it."

that he could not find his letter because after answering it, he had laid the letter on a pile, cut them up, and distributed the snippets by content among his collections of notes (ill. 21).[73] On May 5, 1565, he asked Adolph Occo to write letters on only one side of the paper because he tended to cut them into pieces.[74] This explains why there are relatively few letters to Gessner in his surviving correspondence. Elsewhere, he mentioned possessing a myriad of paper snippets.[75]

The largest of the four *Physicarum meditationum … libri* is the fourth, titled *De anima* (*About the Soul*). It is possible that Gessner devoted a great deal of attention to this subject in his lectures, especially since, as mentioned above, he had written a textbook on the topic. It was the fourth part in a published anthology that was printed by his nephew Jakob Gessner. In addition to the works of Juan Luis Vives (1493–1540), Philipp Melanchthon and Veit Amerbach (1503–1557), his nephew wished to publish something by his scholarly uncle and made a corresponding request.[76] In the epilogue, Gessner explained that he had limited himself to the theory of the human soul, which had not been addressed or adequately covered by others. Thus, he wrote here extensively about the five senses (touch, taste, smell, hearing and sight)[77] and in doing so, leaned more towards Galen and Vesalius than towards Aristotle.[78] Gessner's *De anima* was generated by a *loci* collection, and its content differs from Book IV of *Physicarum meditationum libri* of 1586.[79] The most unique elements in the latter are the graphic illustrations that attempt to portray certain psychological issues. While Gessner fundamentally remained an Aristotelian,[80] this did not hinder him from "including many elements of Renaissance Neoplatonism in his philosophical synthesis of his psychology (and in the drawings). The Neoplatonic body of thought appears in the text (almost always without quotes and not always detectable at first glance) and especially in the illustrations. Their attributes stem mostly from Renaissance symbols and hieroglyphics and clearly reflect humanistic imagery."[81] Gessner's considerable preoccupation with the theory of the soul was noticed abroad, which is why the Leipzig physician Wolfgang Meurer (1513–1585) wrote a dedicatory epistle to him in his 1564

73 Bauhin (1591), p. 126.
74 Gessner (1577), f. 62v.
75 Gessner (1577), f. 86v.
76 Gessner (1563), p. 721.
77 Gessner (1563), p. 951.
78 Göing (2013a), p. 226.
79 Göing (2013a), pp. 209–222.
80 Cf. on Aristotle's psychology: Flashar (2013), pp. 297–318.
81 Müllener (1965), p. 173.

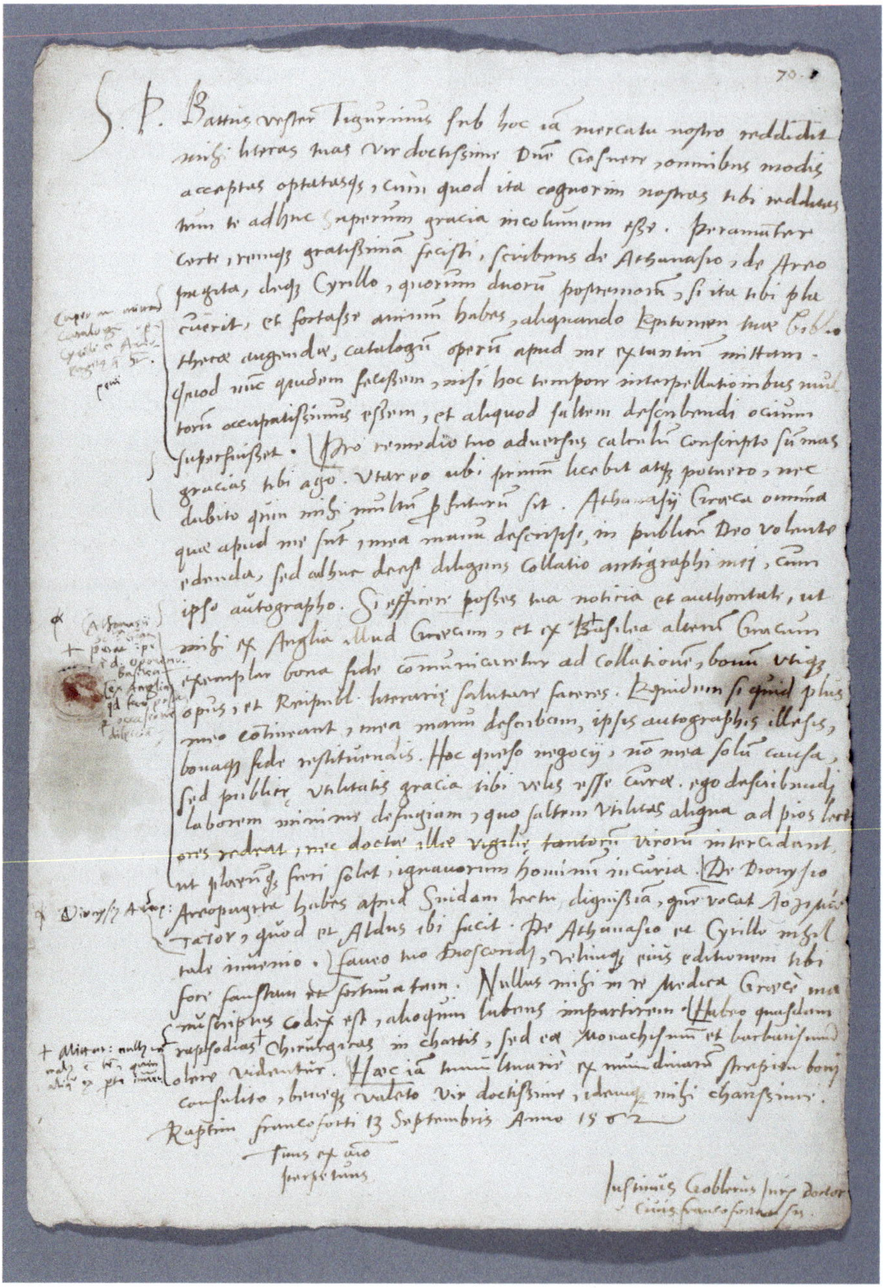

ILLUSTRATION 21 A letter from Justinus Gobler to Gessner, dated September 13, 1562. Gessner marked individual thematic sections with brackets, which he would then cut out and organize according to subject.
STAGR: D V/37 C 36.06.18

edition of Aristotle's *De anima*. He sent a few copies to Zürich on March 9 of the same year.[82]

9.6 Students

Gessner took his teaching responsibilities very seriously. When, on November 29, 1563, the school council discussed how little students were benefitting from ethics instruction and that they should therefore be given more natural philosophical instruction, something that the two young physicians Keller and Wolf could take upon themselves so as to provide some relief to Gessner, the latter did not think much of the idea. While he did indeed give up the teaching of ethics, he was not about to give up his academic seat, as is recorded in the minutes of the council meeting: "When this was brought to Dr Gessner's attention, our gracious gentleman believed that it would be disadvantageous to him; it would give the impression that he wished to be unburdened from his recent lectures and give up his canon prebend; and that he wished to push the work onto others. Moreover the *Historia stirpium* (*History of Plants*), which he had started to compile 30 years ago, was not such a burden to him, because he had also not promised anything to any printer who could press him for it. This is why he wishes to continue with his lectures until the next exam. If the gracious gentlemen, the school councillors, still clearly advised against this, which no one can blame them for, he would accept their verdict. The subject was therefore tabled and postponed until the exam."[83]

Various students, aged 16 to 20, had a more casual attitude about matters than Gesser. They frequently ignored attendance requirements and skipped classes. The school council was then compelled to address Gessner concerning school truancy on July 6, 1560, November 25, 1560 and April 19, 1561. He was not the only one afflicted by the students' disregard. Rudolf Gwalther wrote the following into the school council minutes on May 24, 1561: "So then in classes, I [Gwalther] experienced how the boys were lacking in ambition and wrote down very little. They also brought other books and spent their time reading them."[84] Additionally, there was the usual misconduct that every teacher is familiar with, such as theft, mischief with girls, fistfights and more.[85]

82 Letter from Meurer to Gessner. ZBZ, MS. S 204c, p. 12.
83 StAZH E II 458, f. 53v.
84 StAZH E II 458, f. 23v.
85 StAZH E II 458, f. 27r and 36v.

Naturally, there were also serious, hardworking students, who Gessner sometimes took on as assistants. In a letter to the Danzig physician Johannes Placotomus (1514–1577) on March 27, 1557, Gessner mentions an unnamed *amanuensis* (assistant) who had copied several notes for him by hand.[86] On January 6, 1560, Gessner wrote to Jacques Daléchamps that he sought a young, scholarly fellow with some understanding of medicine, who could reside with him and help him complete the *Historia plantarum*.[87] From an August 28, 1564 letter to Adolph Occo it becomes clear that Gessner often had the help of an unnamed young man for copy work.[88] Nearly four years later, on April 8, 1565, he asked his Basel colleague Theodor Zwinger, whether there might be a poor, young, modest medical student in Basel who could assist him with writing and copying. He was no longer able to handle all of the work himself. In exchange, this student could learn from Gessner and hear his lectures. And there might even be an opportunity to accompany him on an excursion to the Alps in the following summer.[89] It is not known whether Zwinger was able to find someone he could recommend and send to Zürich.

One pupil that remained in Gessner's memory was Anton Schneeberger, who later became a physician. He was born in Zürich, but enrolled at the University of Kraków in 1553, where he had contact with Swiss professors and students and began practicing medicine in 1558. With Gessner's *Catalogus plantarum* (1542) as an example, he wrote Poland's first botanical work, *Catalogus stirpium quarundam latine et polonice conscriptus* (Kraków 1557). He also passed along botanical and zoological information to his teacher in Zürich, who in 1565 cited Schneeberger as his treasured compatriot and former student.[90] There were Poles who were guests of Gessner in February and March of 1564. These included Jan Ostroróg (ca. 1545–1582), his father Jakub Ostroróg, who made his castle near Posen available to the Bohemian brothers (Hussites) for religious services, and Stanislaw Ostroróg (1519–1568). Gessner also hosted Stanislaw Starzechowski[91] (ca. 1540–1582) and his personal teacher, as well as Friedrich Scharff, who was Jan Ostroróg's instructor from Breslau.[92] Jan's brother Wenzel

86 Gessner (1577), f. 137v.

87 BNF, MS. lat. 13063, p. 264.

88 Gessner (1577), f. 66r.

89 Gessner (1577), f. 112r.

90 Hryniewiecki (1938), pp. 8, 31f. and 55.

91 Cf. Appendix II in this book, no. 11.

92 Cf. Gessner's *Liber amicorum*, entries numbers 181 and 187–190 (National Library of Medicine, Washington, shelf mark: MS. E 77); Gessner (1577), f. 17v; Wotschke (1908), p. 217; Hryniewiecki (1938), p. 55.

Ostroróg (1545,1574) was also in Zürich, but left early.[93] He had arrived a few months before, and on August 14, 1563, he wrote to Gessner with a request to send medicine that would make his beard grow. He was already 21 years old, but without suitable facial hair.[94] Ostroróg was in good company. The 20 year-old Felix Platter related his attempts to finally grow a beard: "I spent the whole night copying the medical ingredients from a booklet that claimed to be a special recipe to make hair grow. The book presented it as a secret for making a beard grow. Because my chin was hairless, and I wanted so much to improve my appearance, I frequently smeared it around my mouth at night, which made a mess of the pillow. Sometimes I also had my jaw scraped with a razor, but this did not help."[95]

Along with insufficient beard growth, Gessner's medical correspondence also included the topic of women with excessive eyelash growth (distichia or ectopic cilia), a problem that the surgeon Johannes Muralt tried to solve by burning out the hair roots.[96]

Not only Poles, but also Englishmen, Frenchmen, Italians, Hungarians and others – often including those who were persecuted for their faith – spent time in the city on the Limmat and got to know and appreciate Gessner.[97] The Englishmen Robert Horne (ca. 1519–1580) and Richard Chambers († ca. 1566) wrote a letter to Bullinger on February 3, 1556 in which they passed along greetings to Gessner, who had been their teacher and from whom they had endlessly benefitted.[98] There was also a group of students from Frisia who had studied under Gessner in 1562/63 and for whom he had written letters of recommendation to the University of Tübingen.[99]

With regard to the above-mentioned Poles, Gessner wrote to Crato von Krafftheim on March 5, 1564 that it had been several years since he had taken anyone in or housed any boarders.[100] In general, when reading his correspondence, one has the impression that Gessner and his wife lived a reclusive life in which he was constantly overloaded with work. He wrote to the director of the botanical gardens in Padua, Melchior Wieland (1520–1589), on May 27, 1562 to apologize for his tardiness: "My duties as municipal physician … force

93 Wotschke (1926), pp. 47 and 54.
94 StAGR, D V/37 C 36.06.30.
95 Platter (1976), P. 248.
96 Bauhin (1591), P. 111.
97 Cf. the Hungarians Gábor Perényi, Johannes Balsaráti, Petrus Lupinus, Lukács Szikszói, Titus Amicinus and Márton Berzeviczi: Bernhard (2009), pp. 169 and 172.
98 Letter from Horn and Chambers to Bullinger (StAZH, E II 375, 484).
99 Gessner (1577), f. 44v.
100 Gessner (1577), f. 17v.

me to neglect my studies and responsibilities to my friends, plus, I must teach the natural sciences every day and manage all my familial obligations. Indeed, my worries about those who are closest to me, most of whom are dependent on me, and other things too numerous to mention here, keep me at a remove from myself and especially my friends."[101] To the imperial physician, Crato von Krafftheim, Gessner's justification for haste was that he had to do everything by himself without any servants, something that Crato surely would not have said about himself.[102] On 21 February 1564, Gessner wrote a brief and succint letter to him: "I have the least amount of time of anyone."[103] Even in the year of his death, due to work, it seems that Gessner barely had time for his personal contacts. On February 25, 1565 he mentioned to Achilles Pirmin Gasser (1505–1577), whose sister Anastasia had been living in Zürich since the beginning of the year, that because of the poor weather, the plague and too little time, he had never seen her: "Although I have not met her yet, I understand that your Anastasia is well. This is because I am always at home like a sacrificial bowl, tied up with my papers."[104] In the end, it would be another month before Anastasia and her two daughters paid Gessner a visit on March 30.[105]

As Gessner's *Liber amicorum* (friend and guest book) shows, he did enjoy receiving visitors at his home, which by 1558 was a three-story house (ill. 22).[106] His friends were always welcome, for example, Adolph Occo, whom he informed on September 26, 1565 that a guest room was always available to him.[107] The more than two hundred entries in his *Liber amicorum* are witness to the fact that he was well known and that many scholars had paid their respects. The physician and religious refugee Hieronymus Massarius (1480/85–1564) stayed with him for at least two months in 1554,[108] and the Rhineland physician Johannes Ewich (1525–1588) spent a few days with him in 1557.[109] However, there is no evidence that foreigners or students stayed with him for longer

101 Gessner (1577), f. 139v.

102 Letter from Gessner to Crato von Krafftheim, April 24, 1563, in: Gessner (1577), f. 13r.

103 Gessner (1577), f. 16r.

104 Burmeister (1975), p. 345. Anastasia was married to the Zürich surgeon Felix Kerer, whom Gessner held in high esteem.

105 Burmeister (1975), P. 351.

106 Cf. p. 180.

107 Gessner (1577), f. 65r.

108 Cf. the letters from Wilhelm Gratarolus, January 5 and February 26, 1554, to Bullinger (StAZH E II 336, 60f.). Massarius possessed later a noteworthy garden in Strasbourg, cf. Cordus (1561), f. 238v.

109 Letter from Girolamo Zanchi to Bullinger, December 30, 1557 (StAZH, E II 356, 765). Ewich promised Gessner a Greek poem for the fourth volume of the *Historia animalium*

ILLUSTRATION 22 Gessner's residence at Frankengasse 6 in Zürich.
PHOTOGRAPH: DONAT STUPAN, ZÜRICH

periods after about 1555. It is possible that Gessner or even the church leader Bullinger wished to avoid unpleasant experiences with boarders from wealthy households who then complained about their inadequate reception at Gessner's home. This was the case when Sebastian Schertlin von Burtenbach (1496–1577) wrote to Bullinger in the spring of 1554 that it was not going well in Gessner's house for the three sons of Pankraz von Freyberg zu Hohenaschau (1508–1565). Schertlin had converted to Protestantism in 1546, commanded the infantry of the Upper German cities in the Schmalkaldic War and was the chief officer of the city of Augsburg from 1553. Pankraz von Freyberg, on the other hand, was made *Hofmarschall* (Lord Steward) in 1553 and as such was the head administrator at the court of Duke Albrecht V (1528–1579) in Munich and was a supporter of the Reformation. Schertlin complained to Bullinger on April 4, 1554: "Greetings. First off I wish you well and good. My dear sir and close friend, I have received a notification from the Bavarian Lord Steward Pangratz after he sent three young boys, his sons, to Mr. Gessner in Zürich. They were not only to be educated with assurances of proper Christian instruction, but it was also agreed that they would be fed familiar foods. It is worrisome that Mr. Gessner's wife is somewhat stingy and strict. To eat, the boys, who were well raised and still of tender age, often received only soup, coarse beef and the like to eat. For this, a payment of 100 guilders per year, etc. was agreed to and was received from Freyberg, who subsequently asked me for advice. I told him that I know of no better solution or plan than to write to you sir, my close confidant, and ask that you speak to Mr. Gessner and in particular to his wife, so that these boys can eat, drink and be well cared for in a manner wholly appropriate to their age and status. And if there is no hope for change regarding Gessner's wife, indeed, I would continue to request your assistance and advice so the boys can live in a different and better location, where they can eat and drink well, but nonetheless continue learning from the honored and eminent Mr. Gessner. I would like to kindly ask you, along with von Freyberg, a God-loving man, to help us."[110]

On the one hand it speaks well of Gessner's fame that members of the Bavarian aristocracy would decide to send their sons to study under him in Zürich. On the other hand, in Zürich one could ill afford to ignore such complaints. It was then decided to house the three boys with Rudolf Funck († 1584), who had been deacon of the Fraumünster church since 1552. The switch took place very quickly, and Pankraz von Freyberg himself wrote to Bullinger on August 22: "Venerable, learned and especially dear friend and gentleman,

on aquatic animals, but it probably never arrived in Zürich. Cf. Gessner's *Liber amicorum*, entry no. 108 (National Library of Medicine, Washington, call number: MS. E 77).

110 StAZH, E II 356a, 1016.

to you I readily offer my friendly services at any time. Your letter of June 13 arrived and was well received by my servant Cristoffer Werndlin. I should have answered it sooner, but I was unable to write a reply due to not being at home for an extended time. It was necessary to ride to other locations in the service of and under orders from my gracious prince and lord Duke Albrecht in Bavaria, etc. Now, regarding the issue of my three sons, whom I arranged to be moved from Dr. Gessner's to Rudolf Funck's, your eminence has assuaged my urgent concerns. Regarding the honored and highly acclaimed gentleman, Dr. Gessner, there have been absolutely no further complaints about the care of my sons. Due to being busy with other students and very important issues, he was not able to give my sons the attention they needed. I am, however, no less thankful and amicable toward Mr. Gessner for his substantial efforts. I would be unhappy, were he to withhold his previous good will toward me and my sons. Furthermore, I wish, wherever possible, to kindly repay him and ask that you be sure to let him know this on my behalf. I have also written to Mr. Rudolf Funck to inquire whether, according to my greatest wishes, he has managed to provide a good spiritual and material upbringing for my sons and has tended to their needs as requested. Further, I ask that you kindly tell me in confidence whether he specifically mentioned or noted that Wilhelm has misbehaved or disobeyed. Under no circumstances should this be allowed. Funck may employ appropriate measures or methods to force him to behave, to discipline him or make him obey. For as God is my witness, it is an especially grievous matter, and I do not like it when he or one of his brothers do not comport according to their birth and misbehave when being trained or taught. As their father, I try to live up to the same standards. Please follow up on this further with my continued, friendly confidence. I will do everything in my power for you and your family to deserve this. Incidentally, I would also ask that in your next message you write about my sons' progress in their lessons and other matters and send the letter to Dr. Gerion Seiller in Augsburg. Lastly, I am grateful for your continued diligence and effort. Not only have I heard of this from my servants, but I myself have found it to be the truth and hope to deserve this in the future as well. May God the heavenly father always bless you and your house with his holy mercy. I am hereby faithfully at his command."[111]

We see from these two letters that the problem was not seen to be the highly regarded and very busy professor Gessner, but rather that his household was not well managed. In particular, the boys' nourishment was wanting, which undoubtedly fell under the purvue of Gessner's wife. A decade later, Gessner himself complained to his friend Crato von Kraffheim on February 21, 1564 that

111 StAZH, E II 356, 132.

he could not take in any of the above-mentioned Poles because his wife was not well suited for domestic duties.[112] Whether this was due to her character or physical complaints is not clear from the letters. Nonetheless, he did end up taking in the Poles as a favor to Crato von Krafftheim. Although the food at Gessner's house apparently often left something to be desired, it seems as though he had nothing against a good glass of wine. Not content with the products from Zürich's vintners, he always got his wine from the canton of Graubünden.[113]

Gessner's wife suffered greatly from varicose veins at a relatively young age.[114] This began in 1546 and was a repeated topic in the correspondence.[115] In a letter to Cosmas Holzach in Schaffhausen, Gessner expressed regret that his wife was so headstrong and that she had rejected his recommended remedy for varicose veins from South American guaiac wood. Gessner himself had used frog and earthworm oil on pregnant women, which had also helped his wife.[116] On occasion, the condition of his wife's health was a hindrance, causing him to cancel a planned excursion to the Grisons[117] in the summer of 1560, and preventing him from giving as much time to his garden as he would have liked.[118] In the summer of 1562, she spent time at a spa in Baden, where Gessner kept her company for four days.[119] Despite all the complaints and efforts for his wife, it does not seem that he regretted his marriage. For on February 25, 1565, he advised the young Johannes Bauhin to take a wife since this would make his studies more calm and comfortable. However, he should make sure that his prospective wife was pious and well mannered.[120]

112 Gessner (1577), f. 16r.

113 Cf. the letters from Johannes Fabricius Montanus in Chur to Bullinger, November 10, 1560, and August 9, 1565 (StAZH E II 375, 621 and A 248.2).

114 Letter from October 10, 1561, to Cosmas Holzach. Gessner (1577), f. 85v.

115 Gessner (1577), f. 85r; Gessner (1584), f. C_3r; September 6, 1561, to Jacques Daléchamps: BN Paris, MS. lat. 13063, pp. 257f.

116 Gessner (1577), f. 85v.

117 Bauhin (1591), p. 96.

118 Gessner (1577), f. 115r.

119 Letter from Gessner to Gasser, June 28, 1562, Burmeister (1975), pp. 200f.

120 Bauhin (1591), p. 151 („Sed da operam, ut piam, et moribus tuis convenientem invenias").

CHAPTER 10

The World of Books

Having characterized Gessner's professorial activities, let us return to the year 1543 and examine some of the other, roughly chronological stations of his life. As previously described, he became familiar with the cities of Basel, Paris and Lyon and their print production during his time as a student. In March 1543, he had the opportunity to visit the Frankfurt book fair with Froschauer, which gave him further fascinating insights into the relatively young world of book printing. During this trip, he made the acquaintance of Arnoldus Arlenius Peraxylus (1510–1582), a scholar of Greek from Holland, who worked as a librarian for Diego Hurtado de Mendoza, the Spanish ambassador to Venice from July 1539 to December 1546. In the Spaniard's rich collection of handwritten documents, Arlenius found a manuscript by the Jewish historian Flavius Josephus (1st century), on the basis of which he produced the first printed edition in 1544.[1] After his death, Hurtado de Mendoza's library made its way to the Escorial near Madrid.[2] Gessner informed Vadian of his encounter with Arlenius and took the opportunity to let him know that Greek authors, especially those hidden away in Italian libraries, might be acquired through him.[3] Arlenius had previously provided Oporinus and other printers in Basel with handwritten manuscripts and books from Italy.[4]

10.1 Diego Hurtado de Mendoza

In the summer of 1543, most likely in August and September,[5] Gessner visited Arlenius in Venice and returned to Switzerland by way of Feldkirch[6]

1 Hieronymus (1992), pp. 337–340; Hobson (1999), p. 78.

2 Graux (1880), pp. 163–273.

3 Letter from Gessner to Vadian, April 4, 1543: Arbenz (1908), vol. 6, p. 211f.

4 Jenny (1964), pp. 9–18.

5 Arlenius probably wrote the letter to Vadian at the end of Gessner's stay in Venice. This means that he left in the first half of September 1543. According to Gessner (1548), f. 303r, he stayed in Venice for a month, which means that he must have arrived there at the beginning of August.

6 Gessner wrote to Gasser on August 5, 1554: "Greetings to you from our Johannes Fries, who was once allowed to experience your hospitality in Feldkirch when he returned from Venice, just as I myself had before." Cf. Burmeister (1975), p. 138.

© KONINKLIJKE BRILL NV, LEIDEN, 2023 | DOI:10.1163/9789004541696_011

with a letter for Vadian dated September 9.[7] In the letter, Arlenius described his guest from Zürich as highly intelligent.[8] During 1544, Gessner wrote both entries about Arlenius and Hurtado de Mendoza in his *Bibliotheca universalis*, which was printed in 1545.[9] He praised the former as a scholarly young man who had tracked down various Greek texts that had, as it were, been buried in Italian libraries. He mentioned that while in Venice during the previous year he had met the latter, who had a fantastic library with very rare Greek books. Furthermore, Arlenius had introduced him to the catalogue of Mendoza's library,[10] which he used in his *Bibliotheca universalis* at least 113 times as a reference to confirm the existence of certain works.[11] In the foreword of the second printing of Johannes Stobaeus's collection of phrases, which was published in 1549 by Oporinus in Basel (but paid for by Froschauer), Gessner wrote an addendum to the reader[12] about meeting Arlenius some years before, explaining that he "is currently in Florence publishing first editions with Lorenzo Torrentino for Cosimo de Medici. Among other works, he provided him with a very old Stobaeus manuscript from the library of Hurtado de Mendoza, his former employer and imperial emissary to Venice. After his return, he made a comparison to the 1543 edition and found passages in each that were missing in the other. He wrote all of these down so that the new publication now contains not only more sentences, but also entire sermons. In the end, he also used the manuscript to improve erroneous sections, fill in gaps and provide source information."[13]

Gessner must have been extremely impressed by the person as well as the collection of Hurtado de Mendoza (ill. 23). He dedicated the fourth edition of his *Lexicon Graecolatinum* to him, which included a three folio-page dedication dated 1545. Beside this singular hymn of praise, the title page spelled out the following: *Lexicon Graecolatinum ... cum praefatione ad illustrissimum virum D. Diegum Hurtadum a Mendozza* (Greek Lexicon ... with a preface to the illustrious D. Diego Hurtado de Mendoza). From other sources, it is also apparent that Hurtado de Mendoza must have been an exceptionally friendly, well-educated and tolerant person, who was particularly interested in all that

7 Gessner's stay in Venice in the summer of 1543 has repeatedly been erroneously dated to the summer of 1544.

8 Arbenz (1908), vol. 6, pp. 253f.

9 Gessner was busy completing the *Bibliotheca universalis* in February 1545, cf. Gessner (1545), f. 407v.

10 Gessner (1545), f. 92v and 205v.

11 Sabba (2012), pp. 93–104.

12 Stobaeus (1549), f. α_6v.

13 Hieronymus (1992), p. 460.

ILLUSTRATION 23 Portrait of Diego Hurtado de Mendoza, oil painting by an anonymous artist from the second half of the sixteenth-century.
PRADO NATIONAL MUSEUM, MADRID: P00295

was beautiful – not least female beauties. In 1536 he engaged in a variety of sexual escapades and in 1541 carried on an affair with a Jewish woman, which at the time was not thought to be proper for a good Catholic,[14] and all the less so since he was even being considered a possible candidate for the papacy and was responsible for the opening meeting of the Council of Trent four years

14 Samson (2006).

later, in which he represented the interests of Emperor Charles V. A foremost authority on the history of the Council of Trent, Hubert Jedin (1900–1980), summed up the Spanish diplomat with the following words: “In the history of the Council, Mendoza will not only be remembered as an imperial ambassador, but also as one of the strongest intellectual influences on the Council. The Spanish grandee from a large old family was a humanist of the highest caliber and a passionate collector of manuscripts and books. Even before the opening of the Council, he had his treasured library brought to Trent. It contained numerous manuscripts by classic Greek authors and church fathers, printed books covering all the sciences, and not least of all, many Lutheran books that he may have procured in Venice. Mendoza’s library was a substitute for the Council’s library, which failed to materialize, and was used as a working tool and point of attraction for the humanists in attendance at Trent.”[15]

Gessner would surely have agreed with the judgment of Juan Páez de Castro (ca. 1512–1570), a Jesuit priest and Hurtado de Mendoza’s secretary. He resided with Mendoza and on August 10, 1545 wrote to Jeronimo de Zurita y Castro (1512–1580), the secretary of the Spanish Inquisition since 1537: “My house is full of books, as many from him [Hurtado de Mendoza] as I wish to have: printed, handwritten and even his notebooks: Now we understand the *Mechanica* by Aristotle in which big ideas are explained. This is because he [Medoza] translated them *en romance* [into a romance language] and added short commentaries. I believe I can be of some assistance to him, but he only says: ‘We are studying, Mr. Ioan Páez.’ He is so good and so humane, what can your grace say? There is no one else like him. His knowledge is so many-faceted and unique; he is a great Aristotle scholar and mathematician; and no one is his equal in Latin and Greek. Ultimately, he is an extremely perfect human being. He brought me many books in three categories: a great abundance of Greek manuscripts, printed books on all subjects and others by the Lutherans. All of these are publicly accessible to those who ask, except for the Lutheran books, which will only be given out to people who need them for the Council [of Trent]. … Everyone believes that many decisions will be made at this Council, and that His Majesty will appoint him a bishop, and His Holiness will make him a cardinal. Pray to God that this will come to pass, for with him, all will be in good hands.”[16]

15 Jedin (1957), pp. 236f.
16 Andrés de Uztarroz (1878), p. 530.

10.2 The Hunt for Greek Manuscripts

Arlenius not only helped Gessner with the afore-mentioned Stobaeus manuscript, but with other Greek texts as well. These included, for example, a Byzantine manuscript of the collections of sayings by Antonius Melissa and Maximus the Confessor. As previously noted, he published it in Greek, and with assistance from Jean Ribit, also in Latin. To this work, Gessner added two further titles that were based on manuscripts from Hurtado de Mendoza's library. Through arrangements by Arlenius, these were brought back from a trip to Venice by Gessner's erstwhile fellow student, Johannes Fries, on June 2, 1545.[17] The works were *Oratio contra Graecos* (*A Speech Against Greeks*) by the Christian apologist Tatian (2nd century) and *De Deo et fide christianorum contra gentes ad Autolycum* (*On God and Christian Faith against the Heathens*) by Bishop Theophilus of Antioch (2nd century). In 1546, the anthology of all three pieces was published by Froschauer in Zürich.[18] Fries was responsible for the edition of the Greek texts, and he described Gessner in the foreword to the Greek version of Theophilus's text as his best and closest friend and praised Italy as a most highly respected and valued country.[19] Incidentally, this view was not shared by all of his contemporaries. For example, in a letter of December 26, 1555 to Bullinger, Gereon Seiler from Augsburg described Italians as deceitful, ambitious and greedy.[20] The Latin translation of Theophilus was undertaken by Conrad Clauser (1515/16–1567), who was a pastor in Elsau at the time. The Latin version of Tatian, complete with commentary, was done by Gessner himself.

Gessner's private library contained another Greek manuscript, about remedies with texts by Dioscorides and Stephanos of Athens. He had received this as a gift from the physician Agostino Gadaldini (1515–1575), whom he possibly met during his stay in Venice.[21]

The hunt for Greek manuscripts and thus often unknown texts as well as their editions preoccupied many humanists, since it was assumed that the

---

17 Unfortunately, the original manuscripts can no longer be found. The only evidence of Fries's trip to Venice is a copy of *Musica theorica* (Venice 1529) by Luigi Fogliani, which he had acquired for his private library. Leu (2003), p. 321f. The date of Fries' return to Zürich on June 2, 1545 is derived from a handwritten entry by Wolfgang Haller in a printed calendar that was in his possession: Jakob Rueff: *Kalender oder Laaßbuechlin ... uff das 1545*. Jar, Zürich, Eustachius Froschauer, 1544, [p. 34] (ZBZ, shelf mark: MS. D 269_2).

18 Cf. Gessner (1545), f. 60r, 607r and 614r.

19 Antonius Melissa et al. (1546), p. 244.

20 StAZH, E II 346, 300.

21 The manuscript is now in the ZBZ, MS. C 50. Cf. also Garofalo (2004).

text corpus handed down in Latin, in contrast to Greek, was largely known and that no major new discoveries were possible.[22] Furthermore, top-tier Italian humanists were very enamored of Greek culture. From the end of the fourteenth century, the lagoon city was home to many Greeks, including the famous Cardinal Bessarion (1403–1472). Using his own private library, in 1468 he established the core of the *Biblioteca Marciana*, as much as 60 percent of which consisted of Greek texts.[23] Not least, thanks to his efforts ancient Greek texts provided a new approach to natural science in Italy.[24] Additionally, the Ferrara-Florence-Rome Council (1439–1445), which was meant to resolve differences between the Eastern and Western Churches, brought hundreds of Greeks to Venice, as did the 1452 conquest of Constantinople by Sultan Mehmed II. This intensified cultural contact and elevated *La Serenissima* (Venice) to be the most important of markets for Greek manuscripts.[25] Therefore, it is no surprise that various Greek books were printed by none other than Aldo Manuzio (1449–1515), and that Greek, the language of Plato and Aristotle, was typically spoken in the academy he founded.[26]

Gessner's correspondence indicates that he was in good company with regard to his interest in Greek manuscripts. Not only did his younger colleague, the physician and natural scientist from Bologna, Ulisse Aldrovandi (1522–1605), collect Greek manuscripts,[27] but also many other scholars, including the Hungarian physician and philologist Johannes Sambucus (1531–1584). On January 18, 1560, the latter wrote from Padua to suggest a few titles[28] that Gessner should include in the new edition of his *Bibliotheca universalis*. He also informed Gessner that the Viennese court had purchased about a thousand manuscript codices from Constantinople.[29] Two years later, on February 13, 1562, he wrote from Paris that he had spent six months traveling to French libraries, which Gessner would surely also have enjoyed. He listed for Gessner the rarities that he had found, including an old codex that he had come upon, which contained medical notes from antiquity. Three days before, a commentary about Hippocrates's (ca. 460–370 BC) sixth book of *Epidemics* by the 6th

22 Cf. Sabba (2012), p. 116.

23 Ciccolella/Speranzi (2010), p. 443.

24 Cohen (2010), p. 59.

25 Sabba (2012), p. 83.

26 Magno (2012), pp. 43 and 90–97.

27 Letter of Tamás Jordán (1540–1585) from Transylvania to Gessner, 15 March 1563 (StAGR D V/37 C 36.06.22).

28 Cf. on Gessner and Hungary: Bernhard (2009); Sambucus gave Gessner his *Emblemata* printed in 1564. Cf. Leu et al. (2008), p. 220f.

29 ZBZ, MS. F 61, f. 99r/v. On Sambucus as a manuscript collector, see Gerstinger (1926).

century physician Palladios had fallen into his hands. As far as he knew, this text was not to be found anywhere else.[30]

Book printers shared the humanists' joy of discovery and were interested in acquiring and printing unknown Greek manuscripts. However, Greek books were not easy to sell, as Oporinus complained on April 13, 1555 to Caspar von Niedbruck (1525–1593), an imperial councillor in Vienna.[31] None other than Erasmus of Rotterdam praised the editorial efforts of humanist printers as a contribution to the *respublica literaria*. He wrote the following about Aldo Manuzio in Venice: "He has a singular and constant wish, striving with unwavering zeal and sparing no effort to produce genuine, unspoiled and precise literary editions as a foundation for discerning minds. If I may say so, it is obvious how many important contributions has he already made in this regard – and under difficult circumstances. If God is merciful to our Aldus in his undoubtedly incomparable and excellent projects, and he supports these fine arts and grants him a merciful fate, then I can promise to all working scholars that in just a few years, thanks to his efforts, all of the important authors in the four languages of Latin, Greek, Hebrew and Aramaic, of all branches of knowledge will become available in revised and complete form. No one will be left wanting for any literary resources. When this is done, it will become very clear how many good codices have been inaccessible until now, either because they have been passed over out of heedlessness or because they have been withheld from the public out of the craving for prestige of certain people who are only eager to stand alone in the odor of wisdom. Then it will become clear just how many fatal errors also those authors bristle of whom one considers today that their texts have been sufficiently corrected ... Heaven knows it is a Herculean task, requiring a highly confident spirit to redress such an almost completely and extremely flawed situation for everyone. The hidden must be found, the ignored brought forth, that which has been erased brought back to life, and that which has been mangled must be supplemented and improved, what is distorted by the fault mainly of common printers, for whom the profit of even a single gold coin is more important than the entire world of literature."[32]

Gessner wrote to Niedbruck on March 22, 1555 that Michael Isengrin (1500–1557), a printer from Basel, wanted him to inquire as to where there still might be good old works from antiquity hidden away. Gessner himself had little to offer on this front. Almost all Latin texts had already been printed, and many of the Greek manuscripts had by then been located in Italian libraries. This is

30 StAGR, D v/37 C 36.06.11.

31 Steinmann (1969), p. 145.

32 Erasmus (1983), pp. 187–189 (*Adagia*, Festina lente).

why he had searched through many catalogues and entered the corresponding Greek works into the *Bibliotheca universalis*. While in Venice with Arlenius, with whom Isengrin was familiar, Gessner had seen a Greek manuscript with works by the patriarch Photios I of Constantinople (ca. 820–891) that had been obtained from Mendoza's library. Arlenius was still in Venice, and surely this and other manuscripts could have been found in his possession.

If Gessner had received this long-desired book in Zürich, he would have had it translated into Latin, published, and dedicated to Niedbruck. He would not have translated it himself, as there were enough young scholars in Zürich who would have been eager do it under Gessner's supervision.[33] Niedbruck subsequently arranged for a young fellow from Padua to prepare a copy of the manuscript,[34] but the text was never published. The only Photios edition published in the sixteenth century was released in Basel in 1561 and was based on a manuscript from the estate of Bonifacius Amerbach.[35] Niedbruck had probably sent two Greek manuscripts to Gessner in the 1550s: the *Historia Barlaam et Iosaphat* by John of Damascus (7th/8th century AD) as well as a piece about the twelve jewels on the breastplate of the Jewish high priest from the pen of Epiphanius of Salamis († 403 AD).[36] The latter was published by Gessner in 1565.

10.3 Venice: Book Publishing Metropolis

While visiting Venice in the summer of 1543, the 27-year-old Gessner not only met financially powerful bibliophiles and collectors, he also gained insights into the city as a gate to the Orient and as one of the most important European cultural centers of the Renaissance. He saw various buildings by the Renaissance architect Jacopo Sansovino (1486–1570), such as the *Biblioteca Marciana* under construction (begun in 1537), the *Logetta* at the foot of the St Mark's bell tower and the *Palazzo Dolfin* (ill. 24). In the same year, Titian (ca. 1490–1576) created his famous painting *Ecce Homo*, and the most important book of Italian Protestantism, *Beneficio de Cristo*, was published in Venice.[37]

33 ÖNB, Cod. 9737i, f. 199r/v; partially published in: AK 9/1, P. 165f.
34 Letter from Niedbruck to Gessner, June 12, 1555: ÖNB, Cod. 9737i, f. 273r/v.
35 Hieronymus (1992), pp. 591–601.
36 Leu et al. (2008), pp. 286 and 290.
37 Partridge (2015), pp. 302f.

ILLUSTRATION 24 View of St Mark's Square in Venice from the Tower of the *Chiesa di San Georgio Maggiore*. The *Biblioteca Marciana* by Jacopo Sansovino (the white columned building left of *St Mark's Square*) was under construction during Gessner's visit.
PHOTOGRAPH: LUKAS KELLER, ZÜRICH

Gessner lingered in this pre-eminent early modern capital of humanistic book publishing. In the early era of book printing until 1500, there were 153 verified printing houses, and 3,075 of the 10,465 titles printed on Italian soil were published there.[38] In the fifteenth century, there was no other place in the world that produced so many books. In the sixteenth century, it was only Paris that quantitatively surpassed Venice, its 27,262 titles accounted for nearly half of the 63,953 Italian publications.[39] The lagoon city remained the most important printing center south of the Alps until the era of Napoleon.[40] A few statistics will clarify just how significant book production in Venice was. The most famous printing centers in which three fifths of the 27,000 known incunables (fifteenth-century book printings) were produced were:

38 Magno (2012), p. 21; Barbieri/Irace (2010), p. 525.
39 Infelise (2014), p. 61.
40 Cf. Eickhoff (2007), pp. 219–222.

TABLE 1 Printing centers of the fifteenth century

City	Number of titles published in the fifteenth century
Venice	3705
Paris	3026
Rome	2021
Cologne	1531
Lyon	1334
Leipzig	1210
Strasbourg	1121
Milan	1106
Augsburg	1073

During the fifteenth century, 764 books were published in Basel, but only nine in Zürich.

In the sixteenth century, print production multiplied. The relevant references for various cities and countries in Europe are still being developed, so that it is not yet possible to obtain a complete overview of European print production from the fifteenth to the eighteenth century. The following numbers have been established for the most important printing centers during the sixteenth century (with more than 8,000 titles each):

TABLE 2 Printing centers of the sixteenth century

City	Number of sixteenth-century titles
Paris	30,000 to 40,000[a]
Venice	27,262[b]
Lyon	10,000 to 15,000
Wittenberg	9,471[c]
Rome	8,446
Leipzig	8,088
Basel	8,075[d]

a Martin (1983), p. 442.
b The figure for Italy are based on the *Edit 16* database available on the Internet.
c The figure for German-speaking countries are based on the VD 16 database, which is accessible on the Internet.
d Leu (2014a), p. 54f.

Additionally, there were more print shops in hundreds of other cities that achieved significant print production. For example, Cologne printed 7,856, Strasbourg 6,620, Geneva 4,146[41] and Zürich 1,642.

10.4 *Bibliotheca universalis*

The numbers on this list convincingly show that, within a few decades, the available amount of published and accessible knowledge increased enormously. Suffice it to say, for instance, that in Venice alone, more titles were printed during the sixteenth century than in all of Europe during the previous incunable period. Because only a half century had passed since book printing had been invented (ca. 1450), the statistics speak for themselves. Scholars took various approaches to collecting, sorting, selecting and summarizing knowledge, using collections of *loci*, indices, directories and diverse reference methods.[42]

An old and established way to achieve an overview of authors was to create book references or bibliographies, organized by author or topic. Gessner went this route with his *Bibliotheca universalis*, which was printed in Zürich in 1545 and contained 1,264 folio pages (ill. 25). He was not the first to try to blaze a trail through the flood of texts, but his undertaking was the most extensive of its kind to date, which earned him the epithet "father of bibliography."[43] He also followed up on his first inventory from 1545, which included 5,031 authors,[44] with a second one that was classified by subject groups. In 1548, a volume covering the mundane sciences was published, and in 1549 a more narrow volume concerning theology was released. Fundamentally, Gessner included all manuscripts and printed works in Greek, Latin and Hebrew, although he occasionally added works that were written in the vernacular, such as those by Martin Luther. Gessner trod new ground not only with the number of titles organized, but also with the methods used. Wherever possible, he strove to provide the author, title, printing location, printer, year, format and number of pages for each work. Furthermore, he provided the individual author entries with short biographies, factual comments or excerpts from their writings. This took the *Bibliotheca universalis* beyond being just a bibliography, to being a universal literary history from antiquity to the Renaissance.[45]

---

41 The figure for Geneva is based on the *GLN* database of the Bibliothèque de Genève, which is accessible on the Internet.

42 Blair (2010), pp. 3 and 11.

43 Bay (1916).

44 Sabba (2012), p. 39.

45 Note that the Baghdad bookseller Muhammad ibn Ishaq al-Nadīm produced a bibliography of 3,500 authors as early as the 10th century. Cf. Wellisch (1986).

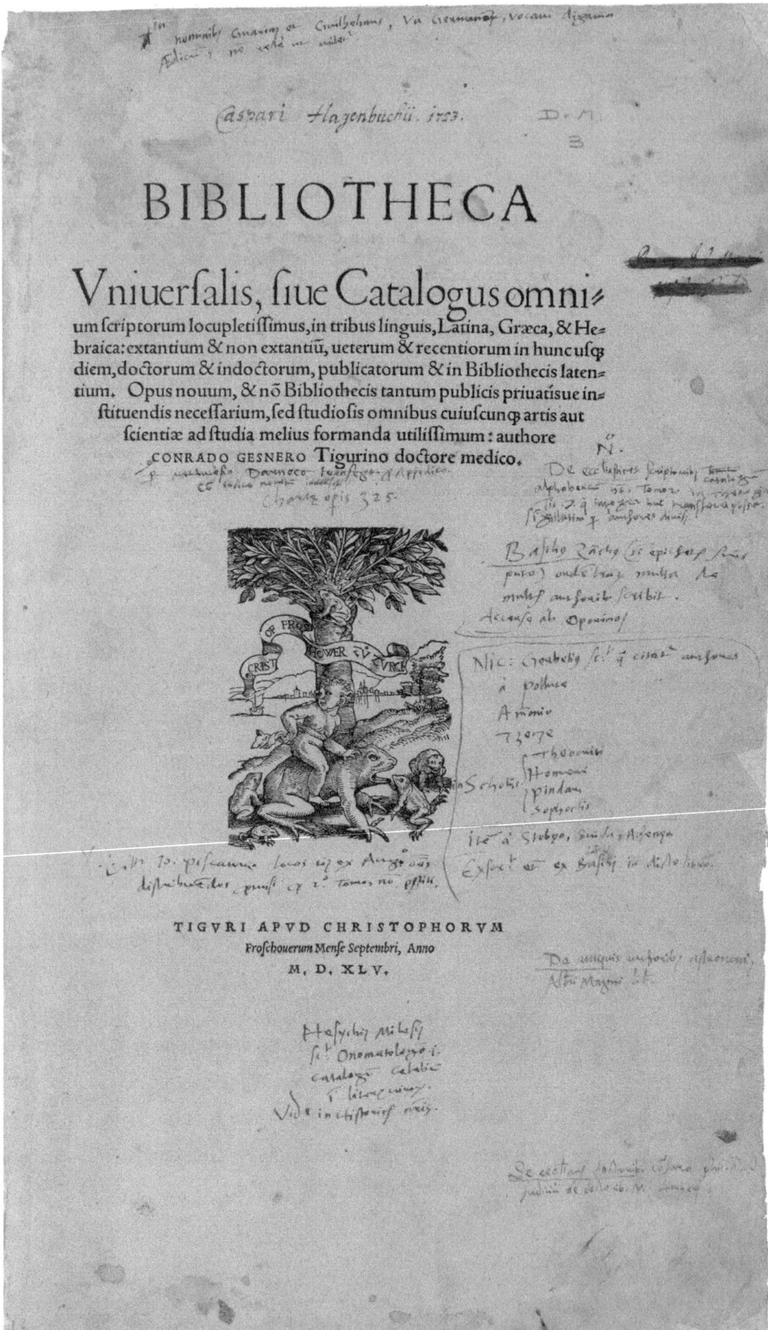

ILLUSTRATION 25 Gessner's personal copy of the *Bibliotheca universalis* with many annotations and additions intended for a new edition.
ZBZ, ABTEILUNG ALTE DRUCKE UND RARA: DR M 3

In July of 1545, Gessner dedicated his bibliographic work of the century to the Augsburg merchant and imperial counsellor Leonhard Beck von Beckenstein (1518–1575) with a nearly eleven folio-page foreword. Von Beckenstein himself owned a large library, and today his stately patrician house is home to the Maximilian Museum. Gessner praised him for more than a page and in passing also mentioned that he had begun work on the *Bibliotheca universalis* in 1542. He thanked God and wrote that he was happy to have found a way out of this literary labyrinth that had befuddled him for three years. Now, as one who had been shipwrecked and drifting out on the open sea, he was back in his home port once more.[46]

In the introduction, Gessner regretted the loss of many famous book collections since antiquity. In Alexandria, 700,000 manuscripts had been burned, as well as 120,000 more in Constantinople during Emperor Zeno's († 491) time. The contemporary era had also not been spared from book burnings and the destruction of libraries, as was evidenced by the fate of the Hungarian King Matthias Corvinus's (1443–1490) library when the Turks conquered Budapest in 1541. The Ottoman attack on Moravia and Silesia at the end of 1542 had also prompted much fear among the Swiss.[47] Should following generations be robbed of their precious books, and therefore their access to education and science, Gessner thought it possible for a decline in Europe to the same cultural level as the indigenous populations in the newly discovered America. For this reason, he had created a sort of index of Western cultural assets with his *Bibliotheca universalis*. Wealthy people could then use it at any time to build libraries and store knowledge to be passed on to future generations. Gessner's comments and book excerpts would have been helpful in the selection of suitable works. The book titles had been collected from printers' catalogues, public and private libraries he had visited in Italy and German-speaking countries, suggestions from scholars and friends and works from previous epochs. While he refrained from applying censorship, he did not consider all of them worth mentioning. He had held many of the books in his own hands and quoted word-for-word from them. This work could also serve as a library catalogue by placing the library signature next to the corresponding titles. Should there be books or new entries that were not yet registered, they could be written down in the margins. It was also possible for friends and neighbors to use the *Bibliotheca universalis* as a catalogue for their own private libraries. Each would enter not only their own books but also the books of their friends that they might not have. Thus, if need be, a person might be able to access a colleague's books, were he not able to afford many books himself. This was clearly an issue

46 Gessner (1545), f. $*_6$r.
47 Stettler (2014), pp. 42f.

due to the relatively high cost of books. Gessner himself maintained a library containing several hundred volumes,[48] and also used his personal copy of the *Bibliotheca universalis* in just this way by jotting down not only factual comments, but for more than 100 titles, the Zürich residents who owned them.[49] In 1551 the catalogue of the *Stiftsbibliothek* contained only 771 items, and there were only 10 guilders per year at disposal for purchasing new books, so scholars would buy a considerable number of books themselves and loan them out to each other.[50] Therefore, having a list with information as to who owned which books would have been especially useful. Nonetheless, no one is known to have diligently taken on the task of creating an index for the entire community of Zürich's scholars.

The librarian of the *Stiftsbibliothek*, Conrad Pellikan, also used the *Bibliotheca universalis* from time to time as a library catalogue and sometimes entered signatures into his copy.[51] It is also known that Philipp Eduard Fugger (1546–1618), son of Georg Fugger the Elder, also used Gessner's universal bibliography as a catalogue for his own library.[52] Otto Henry I, Elector Palatinate (1502–1559), even planned to create a new library with help of the *Bibliotheca universalis*. On October 31, 1544, Bucer wrote about this to Bullinger with a request to have the already printed portions sent to him: "I urgently request that you send me the already printed sheets from our Gessner's *Bibliotheca* that Duke Otto Henry seeks to obtain. He is creating a library and wishes to learn from Gessner's *Bibliotheca* which books are worth purchasing."[53] Notably, Josias Simmler dedicated his edition of the *Bibliotheca universalis* to the interested Palatine prince. In this version, the literary-historical remarks had been cut and it was published under the title *Epitome bibliothecae Conradi Gesneri*.

Continuing in his dedication to Beck von Beckenstein, Gessner explained that the authors are listed alphabetically by their first and not by their last names. Whoever was not familiar with them could search through a list of authors – organized by family names – at the back of the book. As an abbreviation, he used a "T" when citing an entry from the *Catalogus ecclesiasticorum scriptorum* by Johannes Trithemius (1462–1516), printed in 1494. Whenever he added anything to the latter, he placed an asterisk in the text. The abbreviation

48 Leu et al. (2008).

49 Gessner's own copy is preserved in the ZBZ, call number: Dr M 3.

50 Leu (2004), p. 82.

51 ZBZ, call number: IV O 2.

52 Widmann (1966), p. 5.

53 HBBW 14, p. 506; translated according to Hanhart (1824), p. 115. Georg von Anhalt owned a copy of the Bibliotheca universalis that Philipp Melanchthon had obtained for him. Cf. regest 4448, in: https://www.hadw-bw.de/forschung/forschungsstelle/melanchthon-briefwechsel-mbw/mbw-regest.

"cl." stands for *claruerit* (born) and "M" for *mortuus est* (deceased). At the end, Gessner listed some of his sources, which he often did in his works and for which he was much more conscientious than many of his contemporaries. This involved his consultation of library catalogues and reference works, or complete assimilation into the *Bibliotheca universalis* as in the case of Trithemius. A glance at the second part of the work, published in 1548 under the title *Pandectarum libri XXI*, shows that these references are incomplete, and that, according to his own words, he had used several others.[54]

Gessner's trip to Italy in the summer of 1543 led him to Bologna, Brescia, Como, Ferrara, Padua,[55] Pavia,[56] Venice and Verona.[57] As far as Pavia, he was accompanied by the German physician Johannes Dantz († 1546).[58] Once in Ferrara, he studied live plants and animals, including a great white egret, an avocet and a glossy ibis.[59] It is also possible that he met with a Jewish Portuguese doctor, Amatus Lusitanus (1511–1568), from whom he had at least three prints in his private library.[60] Lusitanus remained in the Jew-friendly city until 1547.[61] During the final decade of the sixteenth century, two thousand of the fifty thousand residents were Jewish.[62] In Bologna and Venice, Gessner visited various libraries, examined catalogues and works and wrote copious notes. While in Bologna, he visited the monastery of San Salvatore, but in Venice, he not only examined Hurtado de Mendoza's collection, but also the *Marciana* (*Bibliotheca Bessarionis*) and the libraries at the monasteries of San Giovanni e Paolo and Sant' Antonio del Castello.[63] Additionally, he also became familiar with at least the catalogue of Greek works in the *Laurenziana* in Florence as well as those in the *Vaticana* in Rome.[64] In the summer of 1545,[65]

---

54 Sabba (2012), S. 79f.

55 Gessner noted in the margin of his copy of Pietro Andrea Mattioli's *Commentarii ... in libros Pedacii Dioscoridis Anazarbei de medica materia* (Venice 1558), f. 381r, on mountain mint (*Calamintha*): "Hanc puto vidi Patavii. ..." ZBZ, call number: Dr M 438. I owe this reference to Prof. Dr. Jan-Andrea Bernhard (Ilanz).

56 Gessner herbarized in Pavia. Cf. Gessner (1555), p. 22.

57 Sabba (2012), p. 81.

58 Gessner (1545), f. 411r. Gessner noted the otherwise unknown date of death by hand in his own copy of the *Bibliotheca universalis* (ZBZ, call number: Dr M 3). In his *Tabulae simplicium medicamentorum*, published in Basel in 1543 (VD 16 D 109), he is named as "Medicus." Thanks to Dr. Tilmann Walter (Würzburg) for this reference.

59 Gmelig-Nijboer (1977), p. 89.

60 Leu et al. (2008), pp. 88f.

61 Arieti (2014), p. 52.

62 Modestino/Nascimbeni (2014), p. 71.

63 Sabba (2010), pp. 83–117.

64 Sabba (2010), pp. 117–123.

65 According to the preface in: Antonius Melissa et al. (1546), f. †2r, Gessner stayed in Augsburg in the summer of 1545.

while the *Bibliotheca universalis* was going to press,[66] he visited Augsburg and was a guest at the public library, founded in 1537. He referenced its collection twenty times in his universal bibliography.[67] He apparently added in information garnered from Augsburg during the printing preparations.

While, up to this point, Gessner's universal bibliography was the most extensive of its kind, it was by no means complete. Much of what he included was random in nature. The *Bibliotheca universalis* could not claim to deliver a representative cross-section of Western print production. When one attempts to quantitatively evaluate printing locations, the results show that Gessner's book did not represent individual printing centers proportionately to their actual production. Basel, for example, is the most frequently listed printing location with 38 percent of the indexed titles. It is followed by Venice with 18 percent, Cologne with 10 percent, Strasbourg with 10 percent, Wittenberg with 4 percent and Frankfurt with 2 percent.[68] In reality, Venice with 11,332 titles exceeded Basel's 3,514 printings by a factor of three. Up until 1545, Cologne with 4,736 and Strasbourg with 4,903 also sent more books to press than Basel.

Gessner was completely aware of this deficiency, which is why he continued to collect and expand his bibliography. While he himself continued to work on it, the 1555 publication of an *Appendix* with 2,000 additional authors was turned over to his colleague Josias Simmler.[69] In the foreword, Gessner praised the imperial councillor Caspar von Niedbruck as a knowledgeable jurist, who not only knew Greek and Latin, but had also studied Italian and Spanish and was a great bibliophile.[70] In 1551 the Basel theologian Conrad Lycosthenes (1518–1561) produced an abridged edition (*Elenchus*) of the *Bibliotheca universalis* that was thinner and less expensive. In 1555, Simmler and Gessner released the *Bibliotheca universalis* plus *Appendix* in a similar manner.[71] Gessner wrote to Niedbruck on February 2, 1556 that he wished to publish the *Bibliotheca*

66 Gessner's dedicatory epistle to Leonhard Beck von Beckenstein dates from July 1545 and was written in Augsburg. According to the printer's colophon, the entire work was available in September of the same year, although Gessner had still hoped in February that it would be available within four months. Cf. Thylesius (1545), f. a5r.

67 Sabba (2012), pp. 136–142.

68 Sabba (2012), p. 161.

69 Cf. the letter from Gessner to Bonifacius Amerbach, in which he informed him that he was working on the appendix and was looking for Richard de Bury's *Philobiblon* for this purpose. Cf. AK 6, no. 2761.

70 Simmler (1555a), f. $*_3$r.

71 Cf. Wellisch (1984), pp. 50–52: Conrad Lycosthenes: *Elenchus scriptorum omnium* … Basel 1551; Josias Simmler and Conrad Gessner: *Epitome bibliothecae Conradi Gesneri* …, Zürich 1555.

universalis in yet another form – organized according to Hebrew, Greek and Latin authors as well as by ancient and modern authors. However, this never materialized.[72]

From Gessner's correspondence we learn that he incessantly collected catalogues from book printers and private libraries, especially with an eye toward a new edition of his *Bibliotheca*. On February 3, 1548, he wrote to Jean Ribit in Lausanne that in case he might have a list of new books – not those printed in Basel, but rather more uncommon ones – he should send it to him.[73] On October 21, 1562, he turned to Theodor Zwinger in Basel with the request that the printer Pietro Perna (1522–1582) send him the books he had promised, and a catalogue as well.[74] A year and a half later, he received from Zwinger a list of books by the French physician and mathematician Jacques Peletier (1517–1582).[75] He expressed thanks to Joachim Camerarius the Younger on January 27, 1565 for having sent the list of works by the Northern Italian professor of medicine Giovanni Battista da Monte (1498–1551).[76] Then on April 3, he thanked Adolph Occo for the list of Greek books that the Augsburg council had donated to the public library[77] and also for the catalogue of valuable books in the private collection of Occo's father.[78]

In 1561, at the latest, Gessner was planning a new edition of the *Bibliotheca universalis*, as stated in a letter to Sylvius Caesar Scaliger (1530–1585), the eldest son of the famous Italian humanist Julius Caesar Scaliger (1484–1558). Gessner also asked for a biography of the father and a list of his works, making assurances that father and son would be included in the next edition.[79] Similarly, he wrote to the Cologne theologian Georg Cassander (1513–1566) on October 9, 1562 to inquire about his new publications.[80] At the same time, the second part with the title *Pandectae* should also come out again. On January 7, 1565, he let Occo know that the new edition of the *Pandectae* containing many printers' catalogues would soon be finished, but the book was never published.[81]

72 ÖNB, Cod. 9737k, f. 23r–25r.
73 BNF, Département des manuscrits, Latin 8641, f. 13v–14v.
74 Gessner (1577), f. 107r.
75 Letter from Zwinger to Gessner, February 5, 1564: StAGR, D V/37 C 36.06.36.
76 Rath (1950), pp. 159f.
77 Letter of Gessner to Occo of 3 April 1565: Gessner (1577), f. 72r.
78 Letter of Gessner to Occo of 5 May 1565: Gessner (1577), f. 62r/v.
79 Gessner (1577), f. 132v/133r.
80 ZBZ, MS. S 105, no. 132.
81 Gessner (1577), f. 60v.

The *Bibliotheca universalis* enjoyed broad approval and soon made a name for its author beyond the borders of the Swiss Confederation.[82] It has been verified that the Basel printer Hieronymus Froben consulted the book,[83] and Johannes Sambucus repeatedly sent Gessner information about obscure titles.[84] Bartholomäus Westheimer complained to Bullinger that Gessner had not included his works in the third part, the theological subject catalogue entitled *Partitiones theologicae*.[85] These and other facts indicate that Gessner's universal bibliography was one of the most important author encyclopedias of its time. It was comparable to today's *Who's Who*, in which one would gladly be included and immortalized.

Not everyone utilized the book with good intentions, as the offices of the Roman Inquisition document. Various compilers of *Lists of Heretical Books* (*Catalogus librorum haereticorum*) under Pope Paul IV (1476–1559) searched through the *Bibliotheca universalis* for suspect authors and publications.[86] It is likely that Pope Sixtus V also had a copy of Gessner's universal bibliography for the same purpose.[87] As of 1549, this and other works by Gessner were themselves counted among the forbidden books.[88] An attempt by Venetian booksellers to obtain a release for at least the natural science and medical titles by the polyhistor from Zürich was not approved. On October 15, 1558, the Roman ichthyologist Ippolito Salviani wrote to a colleague, Ulisse Aldrovandi in Bologna, that he absolutely had to have the new fish book by Gessner, even if it could only be read with permission from inquisition officials.[89] But in the 1563 *Index*, which was created as a result of the Council of Trent, Gessner continued to be counted among the forbidden Protestant authors.[90] Even in Antonio Sotomaior's *Index librorum prohibitorum et expurgandorum*, printed in 1667 in

82 Shortly after its publication, copies of the work reached as far as Bohemia, for example. Cf. Hejnic/Bok (1989), p. 16.

83 StAGR, D V/37 C 36.06.47.

84 ZBZ, MS. F 61, f. 99r/v. In 1561 or 1562, Sambucus wrote to the bibliophile and numismatic Henri de Mesmes in Paris, to enquire whether he could see certain coins and his coin catalogue, so as to determine whether it and the catalogue of his private library deserved inclusion in Gessner's *Bibliotheca universalis*. Cf. Gerstinger (1968), pp. 54f.

85 PD Dr. theol. Reinhard Bodenmann (Brugg) dates this letter to May 22, 1551; personal communication of October 11, 2014; StAZH, E II 377, 2690.

86 Moreni (1989), p. 131f.

87 Grendler (1977), p. 290.

88 Germann (1994), p. 92.

89 Pinon (2002), p. 490. Salviani writes „Gisnerus" instead of „Gesnerus."

90 Grendler (1977), pp. 99, 148 and 297. There were also Catholic dignitaries, such as Julius von Pflug (1499–1564), the last Catholic bishop of the diocese of Naumburg, who owned numerous Gessner works. Cf. Pollet (1977), p. 325.

Madrid, there are 27 folio-pages covering Gessner's works. Twenty-three pages are devoted to Gessner's bibliographic works alone, which have been covered with black ink to render them illegible for Catholic readers.[91]

The Roman Curia's measures imposed hardship upon both the book business and intellectual exchange. For example, by the 1540s, the importation of Protestant books from Basel to Northern Italy had become more difficult.[92] Indeed, the Index of Forbidden Books expanded even further; not only were specific authors and works banned, but also more or less the entire output of 15 print shops in Basel.[93] Added to this was the fact that the transport of books over the Alps, as opposed to the more easily navigated waterways, was already difficult enough. Whether by land or water, paper items were transported in barrels to protect them from moisture. The carrying capacity of a horse ran to about 615 pounds, but the majority of alpine passes were not easily traversable. For this reason it was necessary to switch to pack horses or mules, which could carry loads of up to 375 pounds. On flatland, a freight wagon could cover about 18.5 miles per day, but in the mountains, certainly much less. The most important and most used way to transit the Alps led over the Brenner Pass, but the main route from Basel to Italy remained the Gotthard Pass. The many toll charges made the goods more expensive and led to the cheapest routes being chosen instead of the shortest.[94]

10.5 *Pandectarum libri* and *Partitiones theologicae*

The second volume of the *Bibliotheca universalis* was published by the fall of 1548. Listed according to topic (*loci*), the entire literature from the first volume of 1545 was included as well as a further 2,000 authors and their works – the same as those in the previously mentioned *Appendix* of 1555. As a rule, Gessner only refers to the author, sometimes by naming the short title. More extensive information must be taken from the author entry in the *Bibliotheca universalis*. Among the titles referenced by Gessner are several works that today are considered very rare[95] or have even been lost, as is demonstrated by the example of musicology.[96]

91 Sotomaior (1667), pp. 195–221.
92 Leu (2014), pp. 75f.
93 Steinmann (1967), pp. 97f.
94 Wirtz (2006), pp. 215–221.
95 Furno (2008), pp. 52–54.
96 Bernstein (1973), p. 125.

Gessner's *loci* system is divided into 21 subject areas (ill. 26), which are further divided into about 250 sub-topics (*tituli*) with thousands of keywords. The 21 subjects consist of the extended medieval trivium (grammar, dialectics, rhetoric, poetics), followed by the quadrivium (arithmetic, geometry, music, astronomy/astrology) and the various newer humanistic subject areas of history, geography, fortune telling, magic and handicrafts. Gessner was not at all comfortable with all kind of esoteric and occult literature, but because of the amount of literature on these subjects, however, they could not be ignored in a universal bibliography. Strangely enough, this domain underwent considerable growth during the century of the Reformation.[97] After these preparatory sciences (*praeparantes*) came the significant sciences (*substantiales*), namely physics, metaphysics, heathen theology, ethics, economics and politics. The list ended with the three classic university subjects of jurisprudence, medicine and Christian theology. The last of these was the pinnacle of all knowledge since it led to God as the Creator of all things. The *Pandectae* contains the literature of subjects 1 to 19. For medicine, Gessner's own area of study, and for theology, the most important of the sciences, he wanted to follow up soon with a separate volume. However, only one, about theology, made it to publication, i.e., the *Partitiones theologicae* of 1549. Throughout his life, with standards of perfection set too high, Gessner never got far with medicine. He wrote to Theodor Zwinger on November 30, 1565 that Caspar Wolf was handling the medical *Pandectae*.[98] But Wolf, too, was unable to bring the work to publication.

Gessner knew the idea of multiple indexing of a book collection as well as the scheme of 21 subjects from the catalogue of the *Stiftsbibliothek* at Grossmünster church, which had been established by Conrad Pellikan. The sequence he used, however, was slightly different. Pellikan followed up the trivium and quadrivium with the three university subjects theology, medicine and jurisprudence. Thus, the subject theology encompassed the *Theologica biblica*, the *Theologica scolastica* and the *Theologica heretica*. Included under the area of medicine and the natural sciences were *Philosophia*, *Philosophia naturalis* and *Medica*. Jurisprudence was covered by *Philosophia moralis, Ius civile* and *Ius Pontificium*. The newer sciences came at the end and were listed as *Oratoria, Historia, Poetica, Geographica* and *Superstitio*.[99]

---

97 Cf. z. B. Midelfort (1982).

98 Gessner (1577), f. 114r.

99 Germann (1992), pp. 67–72 and 89. Cf. also concerning Pellikan and Gessner: Escher (1937).

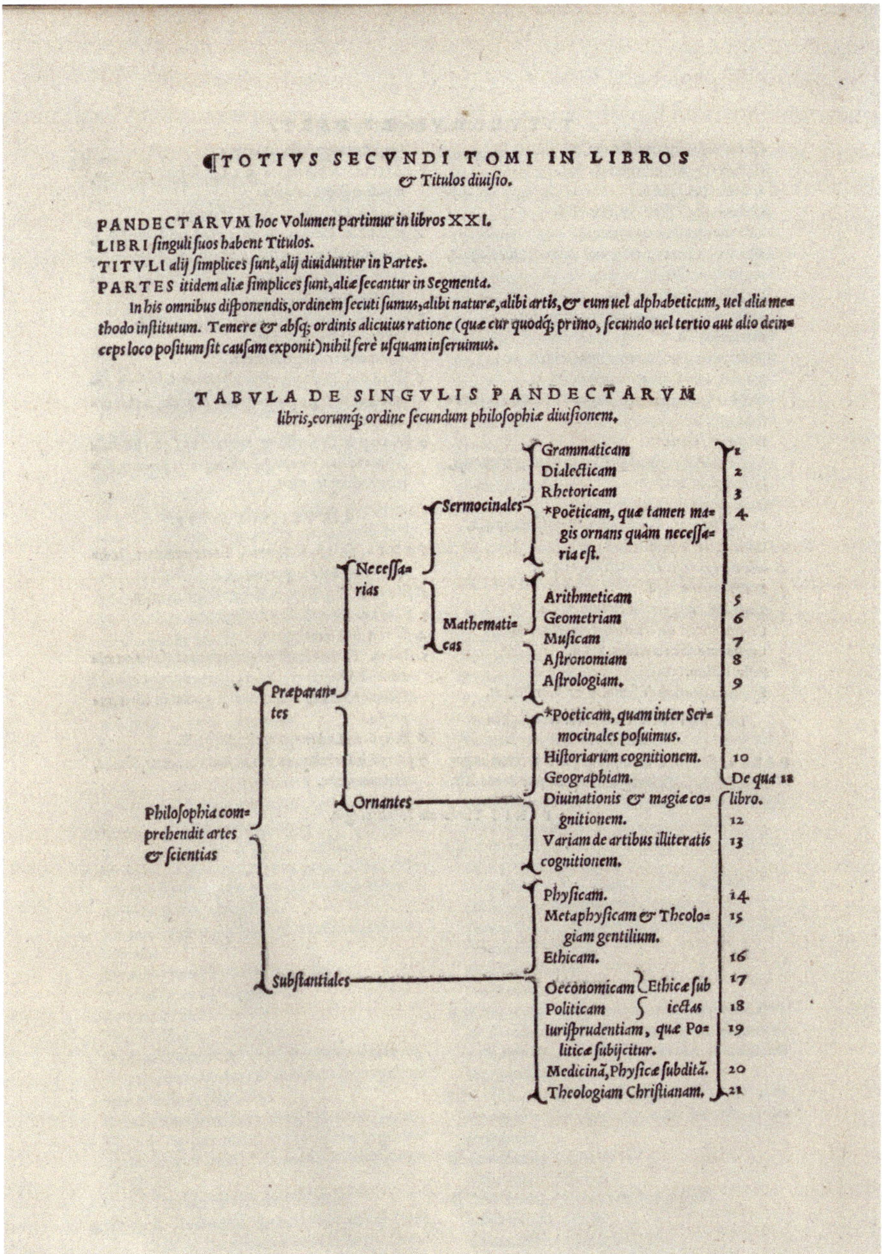

¶TOTIVS SECVNDI TOMI IN LIBROS
& Titulos diuiſio.

PANDECTARVM hoc Volumen partimur in libros XXI.
LIBRI ſinguli ſuos habent Titulos.
TITVLI alij ſimplices ſunt, alij diuiduntur in Partes.
PARTES itidem aliæ ſimplices ſunt, aliæ ſecantur in Segmenta.
In his omnibus diſponendis, ordinem ſecuti ſumus, alibi naturæ, alibi artis, & eum uel alphabeticum, uel alia methodo inſtitutum. Temere & abſq; ordinis alicuius ratione (quæ cur quodq; primo, ſecundo uel tertio aut alio deinceps loco poſitum ſit cauſam exponit) nihil ferè uſquam inſeruimus.

TABVLA DE SINGVLIS PANDECTARVM
libris, eorumq; ordine ſecundum philoſophiæ diuiſionem.

Philoſophia comprehendit artes & ſcientias
Præparantes
Neceſſarias
Sermocinales
Grammaticam 1
Dialecticam 2
Rhetoricam 3
*Poëticam, quæ tamen magis ornans quàm neceſſaria eſt. 4
Mathematicas
Arithmeticam 5
Geometriam 6
Muſicam 7
Aſtronomiam 8
Aſtrologiam. 9
Ornantes
*Poeticam, quam inter Sermocinales poſuimus.
Hiſtoriarum cognitionem. 10
Geographiam. De qua 11 libro.
Diuinationis & magiæ cognitionem. 12
Variam de artibus illiteratis cognitionem. 13
Subſtantiales
Phyſicam. 14
Metaphyſicam & Theologiam gentilium. 15
Ethicam. 16
Oeconomicam } Ethicæ ſubiectas 17
Politicam 18
Iuriſprudentiam, quæ Politicæ ſubijcitur. 19
Medicinã, Phyſicæ ſubditã. 20
Theologiam Chriſtianam. 21

ILLUSTRATION 26 Gessner divided the sciences into twenty-one subject areas. The most elevated of all sciences was theology. Diagram by Gessner (1549), f. a_6r.
ZBZ, ABTEILUNG ALTE DRUCKE UND RARA: 5.13

In the foreword to the *Pandectae*, Gessner cited Beck von Beckenstein from Augsburg not only as a patron but also as a motivator who encouraged him to complete the project. He also explained that he had given up on the original plan, which would have included a third volume with all works listed alphabetically according to their topic area or key words (*loci*). Instead, at the end of the *Partitiones theologicae*, there is a comprehensive subject index with roughly 4,000 terms.

Gessner's monumental bibliographic feat exceeded everything known to this point. Probably the most comprehensive bibliography at the time had been by Johannes Trithemius, which however listed only about 1,000 authors, a number that came nowhere close to the quality of Gessner's work. During the sixteenth century, two re-worked and supplemented editions entitled *Epitome bibliothecae* were published in Zürich. One was by Gessner's colleague Josias Simmler in 1574, and the other was released in 1583 by Johann Jakob Fries (1547–1611) with whom he had enjoyed good contacts as a youth and student.[100] The latter included ca. nine thousand authors. Still further, Gessner's pioneering efforts were taken up in other countries, continuing into the eighteenth century.[101]

10.6 Relationships with Printers

Gessner dedicated each of the nineteen books as well as the twenty-first book in the form of *Partitiones theologicae* to a printer. In doing so, he tried to take into account which subject was a specialty of each print shop. The first book, about grammar, was dedicated to Christoph Froschauer (ill. 27), although that had not been the original plan. Gessner writes about this in the foreword: "Among the famous book printers of our time, Froschauer, I have given you first place. However, this was not intentional, but purely a coincidence. I foresaw the last book of this work for you, which is about Christian theology, since you have printed many Latin and German works on that subject. However, I also wished to tread lightly and not arouse suspicion that I overtly favor my homeland and its citizens while giving little consideration to book printers abroad. Then, when it was clear that the last two books could not be published with the others, and that the first one was still without a dedication (at the start of publication, I had not even thought about dedications), I found this book to

100 Letter from Gessner to Fries, 14 May 1563. Fries was studying in Geneva at the time, and Gessner encouraged him to study Hebrew intensively. Cf. Gessner (1577), f. 129v. Cf. also on the continuation of Gessner's work by Fries: Christ (1922).

101 Wellisch (1982), pp. 53f.; Serrai (1990), pp. 204–262.

ILLUSTRATION 27
Portrait of Christoph Froschauer the Elder by an unknown sixteenth-century artist.
ORELL FÜSSLI AG (ZÜRICH)

be most fitting to none other than you. For it was you who first encouraged me early on to work on both parts of the *Bibliotheca*, and you have printed most of my other works at great personal expense and effort. For 26 years,[102] you have been tirelessly engaged in the publication of exemplary works, predominantly in Latin and German with a few in Greek. You execute these with the utmost fidelity and care, showing all the qualities that make up a great book printer. It is therefore appropriate that what resulted from happenstance, now appears to have been done with the best intentions and a sense of personal commitment. However, no one should conclude that the order of the books [of the *Bibliotheca universalis*], some of which are dedicated earlier and some later, signifies some kind of ranking. I did not actually take into account the

102 This would mean since 1522, but Gessner is wrong here. The currently oldest known Froschauer print is dated 1519. Cf. Vischer (2001), p. 26; Fueter and Gülland (2021).

order of the books, but rather their contents when considering which printer has distinguished himself by the number of works he has printed or by their importance. Sometimes, I would just commit the book to the first, best printer that occurred to me. Unlike gatherings of distinguished personages, there is no comparing or ranking; each printer is exceptional in his own way. One printer may have published in a single discipline or language, while another printer perhaps in two or more. Those who are outnumbered in such counting can make up for this with their meticulous work on just a few books. There are printers with only a single work to show for that will nonetheless outweigh several others. I realize only to well that there are other excellent printers. Would I have had enough books to dedicate, I certainly would not have passed over any who are praiseworthy."[103]

Different printers were personally acquainted with Gessner from his time spent in Basel, Lyon, Paris, Strasbourg and Venice. He worked together with others over extended periods of time, for instance, Froschauer or the print shop of his cousins Andreas and Jakob Gessner. He frequently helped the latter with welcome printing jobs, because their company never really got up on its feet. In 1554, Jakob had gotten into the printing business because there were no other opportunities,[104] which was surely not the best basis for running a successful business. He split from his brother Andreas in 1558 and took up scientific studies, receiving support from Joachim Camerarius the Younger from Nuremberg. Gessner wrote a thank you to him on August 5, 1558, but it is not clear from the letter what form the support had taken.[105] From 1560 on, Jakob ran his own printing business, but it too merely limped along. For this reason, and in contrast to Froschauer, he took on more slender and therefore less profitable publications.[106] It is likely that he gave greater attention to the iron trade since it was more lucrative.[107] For this reason, Gessner wrote to Adolph Occo in December of 1564 that, aside from Froschauer, almost no one was printing in Zürich any more.[108] In 1563 Gessner published the dietetic cookbook *Ars Magirica* by the East Prussian physician Jodocus Willich (1501–1552) through his cousin. He mentioned to Kentmann that this was done less for the content of the book and more for the sake of the printer.[109] Gessner's publication of

---

103 Fischer et al. (1967), p. 92.

104 Letter from Gessner to Joachim Camerarius, February 26, 1554, Staatsbibliothek Berlin, call number: Sammlung Darmstaedter/L Naturwissenschaften/Lb Botanik. Publiziert in: Freytag (1831), pp. 60–62.

105 Rath (1950), pp. 150f.

106 Letter from Gessner to Johannes Bauhin, August 1, 1563. Cf. Bauhin (1591), pp. 122–124.

107 Letter from Gessner to Melchior Wieland, May 27, 1562. Cf. Gessner (1577), f. 139v–140v.

108 Gessner (1577), f. 53v.

109 Letter from Gessner to Johannes Kentmann, August 27, 1563, cf. Gessner (1584), f. D1r–D2r.

his own book about minerals and fossils *De rerum fossilium, lapidum et gemmarum liber* in 1565 can probably also be seen as a favor to his relatives.[110] Gessner's untimely death was presumably a bitter blow to Jakob's business and most likely one of the reasons it had to be liquidated in 1566.

Gessner had an especially close relationship with Froschauer, who not only supported and paid him fairly well but also often pushed and pressured him.[111] There were other authors who could have sung a similar tune, such as Heinrich Bullinger, who wrote to Myconius: "Today, August 8, I am finally answering your letter of July 5. Up until today, as I complete my fourth decade,[112] I have been so chained to the printer that I have gotten nothing else done. I must immediately tend to church and domestic issues and lend my ear to people near and far. Every day, I must write as much as the printing presses swallow up. You know how stressful and demanding printers are."[113]

While Gessner and Froschauer not only worked together in the 1540s, and also stayed together in Augsburg[114] or Frankfurt, over the years it seems that they drifted apart. In any case, Gessner's acquisition of the canon's benefice was seen by him as a liberating blow against the printers' demands, which first and foremost meant Froschauer.[115] He also regretted having promised to publish the planned *Historia plantarum* with him.[116] Like many with an entrepreneurial spirit, it appears that Froschauer the Elder was a complex character. It was understandable that he refused to bring out Gessner's *Icones animalium* before the fall book fair of 1560,[117] or that he preferred to print theological books to medical titles because they earned more money.[118] However, it seems as though he had choleric characteristics as well.[119] All the same, Gessner was interested in a good professional relationship with Zürich's most important book printer, and for this reason he accompanied Froschauer's nephew, Christoph Froschauer the Younger (1532–1585), to Basel on September 29, 1561. In the previous month, the Basel city council had forbidden books from Zürich to be

110 Cf. Burmeister (1975), pp. 287–289.

111 Cf. p. 81.

112 The "fourth decade" refers to a series of sermons by Bullinger published by Froschauer under the title *Sermonum decas quarta* in 1550.

113 StAZH, E II 342, 242.

114 HBBW 15, pp. 416–420.

115 Cf. p. 81.

116 Cf. p. 60.

117 Letter from Gessner to Johannes Bauhin, July 14, 1560. Cf. Bauhin (1591), p. 97.

118 Letter from Gessner to Melchior Wieland, May 27, 1562. Cf. Gessner (1577), f. 139v–140v.

119 Letter from Johannes Pincier to Pellikan, ca. September 13, 1544, in which he complains that Froschauer became so indignant on the occasion of the Frankfurt Book Fair because of a financial matter that he had to be ashamed of himself (ZBZ, MS. S 55, no. 169). Cf. HBBW 14, p. 400.

sold in the city on the Rhine through the book seller Peter von Mecheln,[120] which prompted Gessner and Froschauer[121] to appear before them. However, the famous physician's presence in Basel was not enough to elicit a milder verdict.[122] By the summer of 1563, however, Froschauer sent another barrel of books to Peter von Mecheln in Basel.[123] It is possible that he did not sell the works locally but passed them further along instead.[124]

Gessner was not only familiar with Zürich's printing houses, but through his work on the *Lexicon Graecolatinum* of 1537 with Johannes Walder, he also became closely associated with various print shops in Basel. Gessner viewed the invention of the printing press as a gift from God[125] and was knowledgeable about the business. This was made especially clear in a famous passage in the *Pandectae*[126] in which he explains how to make an index for a book.[127] This did not involve developing new ideas, but rather a description of the working process of proofreaders in early modern print shops. It was not just about producing error-free texts, but also concerned other functions, such as creating registries.[128] Gessner suggested in simple terms that the person who wished to construct an index should write down a list of index terms on one side of a sheet of paper as they come up in the course of his reading. These would then be cut into snippets and generally organized by placing them in boxes. Later, they were to be more carefully sorted and fixed to full sheets of paper with glue, running threads or paper strips in their definitive alphabetical order. The last of these methods had the advantage of affording easy correction or shifting of the entries. The entire "album" then served as source material for a typesetter from which to compose lead-type matrices for a printed index. This method is also mentioned in a letter written from Pellikan to Myconius on August 30, 1535. At the time, Pellikan was putting together the index for his Bible commentary, which was published by Froschauer in six folio volumes from 1533 to

120 Grimm (1966), p. 1403.

121 Whether it was Jakob or Conrad or even another representative of the Gessner family is unclear. The summary of the document simply says: "Ist Froschower samt Gessner erschinen". However, there was probably no other Gessner who could have made a difference in Basel than Conrad, and whose presence would therefore have made sense.

122 Staatsarchiv Basel, "Handel und Gewerbe JJJ1," with old no. 34. I owe this reference to Dr. Beat R. Jenny (Riehen). On the increasing economic bottlenecks of Basel printers in the second half of the sixteenth century, see Leu (2014a).

123 Bauhin (1591), p. 119.

124 Gessner also obtained books from Basel from Peter von Mecheln, cf. Gessner (1577), f. 107r; Bauhin (1591), pp. 100f.

125 Gessner (1548), f. 19v.

126 Gessner (1548), f. 19v/20r.

127 Wellisch (1981).

128 Grafton (2011), pp. 26f.

1537. This amounted to 50,000 handwritten entries on little slips of paper. He informed his friend in Basel that he had already spent five weeks assembling the snippets and, so far, he was up to the letter "P."[129] Pellikan's *Index bibliorum* was ultimately released by Froschauer in August of 1537 as a separate volume.

The individual books of the *Pandectae* and the *Partitiones theologicae* were assigned to the following printers. Wherever possible, Gessner would add a catalogue of the printing output by each:

TABLE 3 Books dedicated to printers.

Book	Printer	Location	Catalogue	Date[a]
1. *Grammatica*	Christoph Froschauer Sr.	Zürich	Yes	8/21/1548
2. *Dialectica*	Joh. Bebel/Michael Isengrin	Basel	No	no date
3. *Rhetorica*	Johannes Oporinus	Basel	No	1/7/1548
4. *Poetica*	Nicolaus Brylinger	Basel	No	no date
5. *Arithmetica*	Robert Estienne	Paris	No	1/22/1548
6. *Geometria*	Johannes Petreius	Nuremberg	No	1/22/1548
7. *Musica*	Heinrich Petri	Basel	No	1/30/1548
8. *Astronomia*	Hieronymus Curio	Basel	No	2/1/1548
9. *Astrologia*	Johann vom Berg/ Ulrich Neuber	Nuremberg	No	2/2/1548
10. *Divinatio*	Wendelin Rihel	Strasbourg	No	2/5/1548
11. *Geographia*	Paolo Manuzio	Venice	Yes	2/13/1548
12. *Historia*	Sebastian Gryphius	Lyon	Yes	3/4/1548
13. *Artes illiterates*	Christian Wechel	Paris	Yes	4/1/1548
14. *Physica*	Johannes Herwagen	Basel	Yes	4/23/1548
15. *Metaphysica*	Johann Gymnich	Cologne	Yes	5/28/1548
16. *Ethica*	Jean Frellon	Lyon	Yes	6/16/1548
17. *Oeconomia*	Vincenzo Valgrisi	Venice	Yes	7/10/1548
18. *Politica*	Girolamo Scoto	Venice	No	7/15/1548
19. *Iurisprudentia*	Tommaso Giunta	Venice	No	7/30/1548
20. *Medicina*				
21. *Theologia*	Hieron. Froben/ Nic. Episcopius	Basel	Yes	Feb. 1549

a Meant is the date of the dedication preface.

129 StAZH, E II 358, 113.

The nine cases in which there are catalogues from printing houses provide an interesting source for the history of books that is not usually available. Along with Gessner's printed indices, the catalogues from Froschauer's print shop are available for the years 1543, 1562 and 1581. For Basel, the following catalogues have been preserved: Cratander for 1539; Froben, Herwagen and Petri for 1553/54; Oporinus for 1562 and 1567; as well as Petri and Henricpetri for 1579, 1583 and 1606.[130] Some of Gessner's lists may refer to unprinted catalogues that were assembled by himself from various sources. This is the case with the printer Sebastian Gryphius (1492–1556) from Lyon.[131] Others were possibly composed from memory.[132] The fact that he had assigned seven of the 20 books to printers in Basel and four to their professional colleagues in Venice reaffirms the earlier assertions that Gessner was especially familiar with these printing centers and their production.

The second book, on dialectics, was dedicated to the Basel printer Johannes Bebel († ca. 1550) and to his son-in-law Michael Isengrin, who sometimes worked together.[133] Along with their printing of the book on dialectics, Gessner also highlighted the complete Greek edition of Aristotle's works that they had produced in two volumes (Basel 1539), a heavily annotated copy of which he kept in his private library.[134] He emphasized the elegant font, the quality of paper and the removal of errors from the text. Praise was also given to the Greek commentaries on "the Stagyrite's" individual books. In light of this, one should not forget that only a few printers were practiced in typesetting Greek texts.[135] Finally, Gessner also mentioned other areas in which Bebel and Isengrin had produced noteworthy publications. This included ancient classics as well as works in law and medicine, among which the groundbreaking botanical work by the Tübingen physician Leonhart Fuchs, *De historia stirpium* (Basel 1542, German edition: *New Kreuterbuoch*, Basel 1543). Gessner predicted eternal fame for the book due to the outstanding quality of the text, the illustrations and the printing. Over the years, Gessner maintained close ties, especially to Isengrin. The latter sent him at least three works as gifts. In 1548, it was a book by the Italian doctor Giovanni Manardi (1462–1536). In 1554 (probably), it was a military treatise by the Byzantine emperor Leo VI (866–912). And finally, in 1556 there was the *Hieroglyphica* by the Italian humanist Giovanni

130 Leu (2014a), pp. 67f.
131 Furno (2008), pp. 44–47; Rozzo (2008).
132 Lutz (1937), p. 113.
133 Reske (2007), pp. 69f. and 73.
134 Leu et al. (2008), pp. 57f.
135 Steinmann (1969), p. 131.

Pierio Valeriano Bolzani (1477–1558).[136] In a letter from Gessner to Isengrin written on March 1, 1554, it is apparent that the printer from Basel had asked for help with the compilation of an index of rare handwritten manuscripts that were hidden away in various libraries. Unfortunately, the polyhistor from Zürich had to turn down the request because he was completely preoccupied with work on the *Historia animalium*. Gessner recommended the Italian religious refugee and bookseller Pietro Perna (ca. 1520–1582) who was in Basel and possibly available to help.[137] He also suggested that Isengrin seek information from Arlenius and other Italian scholars.[138]

Gessner dedicated the third book of the *Pandectae* to Johannes Oporinus, who was the most prolific and perhaps most diverse intellectual among sixteenth-century printers in Basel. Bullinger and Pellikan also maintained good contacts with him. Gessner first makes reference to the many teaching aids that Oporinus's print shop produced, especially the rhetorical titles. Textbooks as well as calendars, pamphlets, writs of indulgence, and occasionally, short publications were produced in great numbers during the fifteenth and sixteenth centuries. These amounted to a steady line of business for early modern printers, generating profits which permitted investment in more expensive and elaborate projects.[139] In his *Laudatio* to Oporinus, Gessner stressed the service rendered to philosophy (the editions by Porphyrius and Aristotle, among others) and to the natural sciences. He had printed Andreas Vesalius's (1514–1564) *De humani corporis fabrica*, one of the most famous works on anatomy and the human body. Oporinus also excelled in jurisprudence and theology. The first theological publication on Gessner's list is the 1542 Latin edition of the *Quran* with various critical commentaries. Additional works mentioned by Gessner range from titles by the reformed theologians Johannes Oecolampadius and Petrus Martyr Vermigli (1499–1562), to the *Loci theologici* by Philipp Melanchthon as well as to the heterodox writings by the Orientalist and Kabbalist, Guillaume Postel (1510–1581) or by the humanist Sebastian Castellio (1515–1563) from Savoy who broke with Calvin and ultimately chose Basel as his new home. This brief list alone shows that Oporinus exhibited enthusiasm for controversial topics of discussion, which more than once led to conflicts with the authorities.[140]

136 Leu et al. (2008), pp. 162, 170 and 245f.
137 Reske (2007), pp. 82f.
138 ÖNB, Cod. 9737i, f. 100r/v.
139 Leu (2008a), pp. 229–232.
140 Steinmann (1969), p. 177.

As gifts, Oporinus gave Gessner at least eight printed works, a portion of a hand-written manuscript[141] and a new version of the map of Greece that he had printed in 1556.[142] Gessner valued him as a skilled printer, which is why in 1548 he arranged to have him publish the afore-mentioned textbook by Ermolao Barbaro. Various letters reveal that beyond business issues, they maintained personal contacts and were very well acquainted. Gessner congratulated Oporinus on his third marriage – to Anna Holzach in May of 1565.[143] But he also knew enough about him to write to Crato von Krafftheim on August 16, 1561 that Oporinus was among the followers and friends of the reviled Paracelsus, and that the printer himself had reported that this fellow had interactions with demons.[144] Gessner was also aware of Oporinus's current projects, as shown in another letter to Crato von Krafftheim in which he writes that Oporinus was printing a comprehensive Greek-Latin dictionary by Robert Constantin for Heinrich Petri (1508–1579).[145] Apparently the printer from Basel occasionally required patience. For example, Gessner complained to Theodor Zwinger on December 13, 1564 that he had sent Cassiodorus's *Variae* to Oporinus from the *Stiftsbibliothek* Zürich, but he had never sent it back. Gessner had already been admonished several times by the head of the library, and he would be grateful if Zwinger would pass this on to Oporinus.[146] On another occasion, Gessner was trying to locate a cask which contained a sealskin that belonged to him, and that Oporinus had probably mistakenly packed at the Frankfurt book fair in 1564.[147]

The fourth book of the *Pandectae*, about poetics, was dedicated to the Basel printer Nicolaus Brylinger († 1565). In 1548, he had given Gessner a copy of his printing of the works of the Greek historian Xenophon (4th century BC) as a gift.[148] Gessner praised Brylinger's *Dictionarium poeticum* and his many annotated editions of ancient poets – especially the Roman poet-prince Virgil – that were specifically intended for the student market. In addition, he only had positive things to say about the Xenophon edition and the Greek-Latin versions of the New Testament. One criticism that he could not hold back, however, which applied not only to Brylinger but to all of Basel's printers, was their choice of ancient classics. Gessner argued for works of high regard, based either on their

141 Leu et al. (2008), p. 308.

142 Letter from Oporinus to Gessner, June 25, 1556. ZBZ, MS. C 50a, no. 29.

143 UB Basel, Frey-Gryn Mscr I 13: no. 83.

144 ZBZ, MS. C 50a, no. 40. Cf. also Kühlmann/Telle (2001/2004), vol. 1, pp. 598f.

145 Treviranus (1830), pp. 61f.

146 Gessner (1577), f. 109v.

147 UB Basel, Frey-Gryn Mscr I 13: no. 83.

148 Leu et al. (2008), p. 257.

language or content, and especially in the case of those meant for student use, but he felt that sexually excessive authors should be avoided, or that such passages should be purged. It was well known that Gessner himself had chosen this approach to his publication of the lascivious poet Martial.

The fifth book, about arithmetic, was dedicated to Robert Estienne (1499–1559) in Paris. Estienne later converted to Protestantism in 1550, moved to Geneva in 1556 and acquired citizenship there. Gessner explained his choice by citing Estienne's 1538 publication of the widely known arithmetical work *De arte supputandi* by the English bishop Cuthbert Tunstall (1474–1559). Furthermore, he had printed numerous other notable titles, and Gessner did not want him to be overlooked in the *Pandectae*. In 1548, Estienne presented Gessner with a textbook printed that year concerning the pathology and therapy of the Greek physician Alexander of Tralleis (ca. 525–605), who had been active in Rome.[149] This was a gift in return for the books that Gessner had previously sent to him.[150] His two sons, Henri (1531–1598) and Robert (1530–1571), who were also printers, dedicated the 1553 edition of various works by the Neoplatonist Bishop of Ptolemais, Synesius of Cyrene (4th/5th century AD) to Gessner personally. Still extant is a brief exchange of correspondence from 1549 to 1551 between Gessner and Henri Estienne that was written in Greek.[151] Gessner praised him as a publisher of reference books and reprimands the printing of senseless works such as love stories and low-grade poetry that he felt were counterproductive, not only to young people but people of all ages. He also mentioned that in Zürich there were no Greek manuscripts, but some could be found in Augsburg and Italy. He considered the people in Augsburg so petty, however, that there was little hope of acquiring them in Zürich on loan. In a following letter, Gessner lamented the moral deterioration in the city on the Limmat, which Estienne should not be surprised about, if he were to visit Zürich sometime. Years later, Gessner encouraged Franciscus Portus, a Cretan working in Geneva, to write a book that would teach Greeks about the main points of faith. He suggested Henri Estienne as a printer who might be won over for the project.[152]

---

149 Leu et al. (2008), pp. 47f.

150 Undated letter from Gessner to Henri Estienne. ZBZ, MS. C 50a, no. 54/1. Letter edited in: Orelli (1837), pp. 33f.

151 Orelli (1837). On December 4, 1565, Gessner thanked Adolph Occo in a letter written in Greek for his Greek letters and modestly remarked that he had not been able to cultivate Greek as he would have liked due to the lack of suitable teachers. In his whole life he had written Greek letters to friends only three or four times. Cf. Gessner (1577), f. 55v.

152 Gessner (1577), f. 133r/v.

Book six, about geometry, was dedicated to Johannes Petreius (1497–1550) from Nuremberg, whom Gessner knew personally. He praised his very attractive and carefully printed works, especially those in the areas of astronomy, mathematics and geometry. These included titles by Johannes Regiomontanus (1436–1476), Nicolaus Cusanus (1401–1464) and the groundbreaking 1543 work by Nicolaus Copernicus, *De revolutionibus orbium coelestium*. The last of these had already been listed in the *Bibliotheca universalis*, but with reference to the *De libris revolutionum Nicolai Copernici narratio prima* by his former Zürich schoolmate and later Copernican, Georg Joachim Rheticus (1514–1574). He cited the second printing of 1541 in Basel, but did not go into further detail.[153] Rheticus was an advocate for the Copernican system, and is known for having convinced Copernicus to publish his epochal works.[154] In his *Pandectae* of 1548, Gessner published two small treatises by him on astronomical and geometrical measuring. Rheticus had written them during his stay in Zürich the same year.[155]

The reason for devoting the seventh book of the *Pandectae* to Heinrich Petri in Basel was his 1547 printing of the *Dodekachordon* by the Swiss humanist and musicologist Heinrich Glarean. In it, the musical keys for medieval churches were supplemented by two additional keys, from which the major and minor scales were later derived. While Gessner held the work to be an especially beautiful volume from the Petri printing house, Glarean was not at all satisfied with the publication. He had a student assistant correct the many errors by hand in several copies before giving them as gifts to important people.[156] Gessner noted that along with publishing medical, theological and other scientific books, Petri had also printed many valued works in Latin, Greek, Hebrew and German. Because he did not have his own copy of a catalogue, he created a substantial list of them from memory. Gessner did not fail to thank Petri for the continual friendliness he had shown him over many years. This included giving Gessner a book in 1547 that was a compendium of cosmological and mathematical writings in Hebrew by Abraham bar Chya Albargeloni Hanasi Hasephardi (1065–1136) and Abraham Ibn Esra (1098–1164).[157] It is very likely that he also received other books from Petri, which may have included a copy of a work by the Persian physician Abū Bakr Muhammad ibn Zakarīyā

153 The ZBZ has preserved a copy of the afore-mentioned work by Rheticus, which was acquired in 1546 by the Basel teacher Christoph Pfäfferlin (ca. 1495–1565), who is known not least as the music teacher of Basilius Amerbach (call number: Gal Tz 1185_2).

154 Wolfschmidt (2010), p. 50.

155 Gessner (1548), f. 80r/v; cf. Burmeister (2015), p. 247.

156 Kölbl (2012), p. 91; cf. also Grafton/Leu (2014), p. 16.

157 Leu et al. (2008), pp. 62f.; cf. Hieronymus (1997), pp. 745–751.

ar-Rāzī (ca. 864–925), printed in Basel in 1544. He asked Theodor Zwinger on November 26, 1565 to send it to him: "I thank you and the famous doctor Huber (whom I honor and consider a teacher) very much for the volume by ar-Rāzī. I see that there is much in it that is necessary for my business. I don't have time to copy it all, and I don't dare put it in the hands of someone else who might sully such a beautiful book. Therefore, I will probably send it back just as I received it. I would appreciate it if you would ask the renowned councilman Heinrich Petri, my old highly esteemed patron, whether he might have an unbound or less elegant copy of this volume. He probably does have one since he published this and other works by ar-Rāzī. If he does, and will loan it to me, and is not adverse to occasional improvements (that would not disturb the flow), then I will make some corrections while reading. I can write down explanations of the many Arabic words found in the simple remedies as I did before in the second book by Avicenna.[158] If only some printer would publish this wonderful, and in my opinion, necessary volume for physicians, which has never been published in Germany."[159]

From 1543 on, Gessner made various contributions to Petri's efforts, especially by revising assorted dictionaries of ancient languages.[160] He also mentioned him occasionally in his letters. Although Gessner also corresponded with Petri, none of these letters has survived.[161] On July 24, 1564, he informed Crato von Krafftheim that Heinrich Petri had lost his wife, two daughters and a son-in-law to the plague,[162] an indication of how close he was to Petri's family.

The eighth book of the *Pandectae*, concerning astronomy, was dedicated to the Basel printer Hieronymus Curio († 1564). Gessner thanked him for following in the footsteps of his father, Valentin Curio (ca. 1500–1532) and his stepfather, Johann Walder († 1541) – both book printers – and also for producing estimable works. Of these, Gessner's preference was for ancient classical editions and astronomical titles. With a production of between 35 and 85 titles, all three of these printers belonged to the category of smaller printing houses in Basel.

The Nuremberg printers Johann vom Berg († 1563) and Ulrich Neuber († 1571) were honored with the dedication of the ninth book, about astrology. Gessner began his testimonial by noting that he had heard of their publication

158 Gessner owned an edition of Avicenna' *De animalibus* that had been published in Venice around 1500. Cf. Leu et al. (2008), p. 62.
159 Gessner (1577), f. 113v.
160 Hieronymus (1997), pp. 884 and 1272ff.
161 Gessner mentions in a letter to Theodor Zwinger that he also wrote to Petri. Cf. Gessner (1577), f. 105v.
162 Gessner (1577), f. 21r.

of various astrological and mathematical titles along with the writings of Luther. However, because of the great distance to Nuremberg, he was not that familiar with their print production. Gessner cautioned the two about astrology and its association with superstition and stressed that the more educated and devoutly Christian someone is, the more he will distance himself from this subject. It is even in dispute among heathens.

In the tenth book of the *Pandectae*, Gessner listed the literature about divination and magic. His dedication was to the Strasbourg printer Wendelin Rihel, who had taken private lessons in ancient Greek from Gessner[163] and had presented him with many books printed by his shop.[164] Gessner wrote that, as a rule, he had dedicated the other books to printers who had produced noteworthy works in corresponding fields. But in the tenth book he had done just the opposite by lauding someone who had not printed a great deal about magic and no occult works. Instead, he had printed texts that opposed the occult, along with good theological books. Gessner listed several titles by the reformers Wolfgang Capito, John Calvin and Martin Bucer along with works of Greek and Latin literature for which Rihel had received a level of fame. In conclusion, he mentioned the botanical work by Hieronymus Bock that Rihel had published in 1539, which became one of the standard botanical works of the sixteenth century. As an exception, Gessner felt a responsibility to write a foreword to the reader after the dedication notes in which he briefly justified including reprehensible superstitious and occult books in his bibliography. On the one hand, this approach was in keeping with his fundamental goal of indexing all works, without ideological considerations. In contrast to similar Catholic projects, it was to present a truly universal bibliography.[165] On the other hand, he pointed out that these books were listed with a warning – in a similar manner to theologians listing false doctrines, and physicians listing poisons. In this way, one could discern or look up whether or not a book was, at best, objectionable. If, against expectations, someone were to use his index as a starting point for misdeeds and devote themselves with interest to such condemnable writings, the cause would not lie with Gessner's work but rather with the moral wickedness of the person involved.

Gessner chose Paolo Manuzio (1512–1574), the son of the famous Aldus, for the dedication of the eleventh book of the *Pandectae*, which covered

163 Cf. p. 277.

164 Gessner (1577), f. 138r.

165 Cf. Zedelmaier (1992), pp. 213–224 and other passages. On the more liberal attitude of the Zürich Reformation in contrast to inquisitorial Catholicism, see Bibliander's polemic against the censorship decisions of the Council of Trent in: Bibliander (1551), pp. 66ff.

geography. Gessner praised the latter as the first printer to publish in Latin and Greek, and with particularly beautiful fonts.[166] He was considered to be a credit to the family name during the renaissance of the sciences (*renascentibus studiis*). While the invention of book printing had occurred in Germany, Aldus's perfection of the invention was no less remarkable. Finally, Gessner came to the print production of Paolo Manuzio, which included not only the impressive Cicero commentaries but also classics in geography such as the works by the Greek geographers Strabo (ca. 63 BC–23 AD) and Dionysius Periegetes (2nd century AD).

In his dedication of book twelve, on the topic of history, to the Lyon printer Sebastian Gryphius, Gessner vented his spleen about the many unfit and mediocre printers. He had actually intended to have all the books of the *Pandectae* dedicated only to printers in German-speaking lands. However, it became clear that many poorly corrected and erroneous texts were being released in the interests of profit instead of quality. Thus, he decided to consider other deserving printers, including Gryphius. It is hard to say how much personal contact the two had, but Gryphius's son, the bookseller Antoine Gryphius, was an important book supplier for Gessner.[167] The Parisian printer Christian Wechel was also considered by Gessner to be among the most important printers of the sixteenth century. For this reason, Gessner dedicated to him book thirteen, about various mechanical and other practical arts such as architecture, manual skills and agriculture.

Gessner dedicated the fourteenth book, about the natural sciences, to Johannes Herwagen (1497–1558) from Basel. Of interest to the history of science, Zürich's polyhistor declared in his dedication that, after Aristotle and his commentators, barely any works of note concerning the natural sciences had been published up to the sixteenth century. This made Georg Agricola's (1494–1555) *Bermannus* (1530) about earth sciences and Jean Ruel's *De natura stirpium* (1537) about botany all the more welcome. Gessner held Herwagen in high regard for publishing the commentaries by the Aristotle scholar from Tübingen, Jakob Schegk, and therefore wished to dedicate the fourteenth book of the *Pandectae* to him.[168] According to Gessner, he wrote the subsequent list of works from the Herwagen print shop from memory. How close the two men were on a personal level is not known. Gessner did note in his own copy of the *Bibliotheca universalis* that Herwagen had given him certain parts

166 All known 26 Greek incunabula were published in Italy, the first not in Venice but in Brescia in 1473, cf. Jones (2004), p. 22.

167 Cf. p. 61.

168 Cf. p. 139.

of Wolfgang Musculus's (1497–1563) commentary on the Gospel of John, which he passed along as a wedding gift to his relative Caspar Hirt.[169]

The fifteenth book of the *Pandectae* was about non-Christian philosophy and theology, which Gessner referred to as metaphysics. The dedication was addressed to the Cologne printer Johann Gymnich,[170] with whom he had no personal contacts. Gessner was aware that this subject matter, which according to the following table of contents (f. 237r) covered the basic questions of philosophy, human existence and pagan religions, was not among the topics in which Gymnich excelled. Nonetheless, he had hoped that Gymnich would be able to contend with it, especially since it concerned the basic questions upon which other sciences and scientific work depended. In this part of the *Pandectae*, Gessner also added literature about heathen religions, even though he knew quite well that they dealt with idolatry and were far removed from the fundamental issues examined by Aristotle, Socrates or Plato.

Book sixteen, about moral philosophy (ethics), was dedicated to the Lyon printer Jean Frellon (1517–1568), whom he had met personally in January of 1541.[171] After praising the meaning and significance of ethics for human coexistence, Gessner also thanked Frellon for his and his brother François's genial reception nine years before in Lyon.[172] This portion of the *Pandectae* was assigned to Frellon even though Gessner was well aware that this was not the primary field of his print production.

The last three books of the *Pandectae*, covering economics, political science and law, were distributed among three Venetian printers. The first went to Vincenzo Valgrisi, who worked in the lagoon city from 1540 to 1572. Gessner praised him as an outstanding printer and evoked fond memories of his hospitality, which he had enjoyed for a month in 1543.[173] Gessner dedicated the seventeenth book of the *Pandectae* to Valgrisi because of his reputation and friendship. He wrote that books concerning economics were not a particularly good fit with the printer's publishing program, and would be more relevant to his large family. Knowledge of domestic finances and economics were not only important for private households, but also for a statesman.

169 Leu (1993), p. 40.

170 The date of death given by Reske (2007), p. 433, as 1544 must be doubted due to this dedication epistle of 1548.

171 Cf. p. 59.

172 Either Gessner is mistaken in the year, or he made a side trip to Lyon from Lausanne in 1539, which is otherwise not further documented.

173 Cf. p. 113.

Gessner's dedication of the next book, which dealt with the literature on political science,[174] was directed to Girolamo Scoto (ca. 1505–1572), who, like his father Ottaviano (active until 1555), had printed many very good books in a variety of disciplines. He was eager to have the Scoto's catalogue in order to be able to include it in his own list. He could only remember a few titles, such as the commentaries to most of Aristotle's books. In particular, he had recently read the commentaries on Aristotle's zoological writings with profit.

The nineteenth and last book of the *Pandectae*, concerning the literature of jurisprudence, was dedicated to Tommaso Giunta (1494–1566) and the other descendants of the famous Luca Antonio Giunta (1457–1538), who had founded one of the great printing dynasties in Venice.[175] Gessner stressed the many legal and medical works that the elder Giunta had published, including an edition of Galen in Latin. Although he maintained no personal contacts with the Giunta family, Tommaso sent him a copy of a pharmaceutical publication by the Northern Italian physician Jacopo Dondi dall'Orologio (1293–1359) and one by Franciscus Alexander (1529–1587) as gifts.[176]

The twenty-first book of the *Pandectae*, which featured theological literature, was published separately as *Partitiones theologicae* in 1549 with a dedication to Hieronymus Froben (1501–1563) and Nicolaus Episcopius (1501–1564). From 1531 on, they had operated a print shop together in Basel. The focus of their output was medical and natural science texts, which would have been a reason for Gessner to devote the twentieth book, concerning medicine, to them. However, he did not want to wait any longer with his praise for the two, especially since their production of the four-volume Galen edition in Latin (1542) was of such significance. Gessner closed his dedication by noting that Froben and Episcopius were also accomplished in the area of theology, which was evident from the 1549 catalogue of their publications.

This index of 111 titles, printed on one and a half folio pages, listed not only medical works but also editions of the church fathers, numerous editions of ancient classics, a specially marked section with works by Erasmus of Rotterdam, various Hebrew prints and titles by the Italian humanists Flavio Biondo, Pietro Bembo and Raffaele Maffei (Volaterranus). Also still in existence is a second, single-paged index of publications from 1557 that lists 118 works.[177] In the 1549 catalogue, there are 93 titles which also appear in the 1557

174 On the subject of politics, it should be noted that in the ZBZ (MS. S 112, no. 128) are preserved two pages of Gessner's notes, which he entitled "Politica Tiguri." They contain incoherent jottings on guilds, mayors and prohibitions.

175 Nuovo (2013), pp. 51–71; Barbero (2022).

176 Leu et al. (2008), pp. 47 and 266.

177 Richter (1965), plate 3.

version. Only twenty-five new titles were added in the eight years between 1549 and 1557. The main topics are again: ancient classics, church fathers, Erasmus of Rotterdam and humanists in Latin and Greek. As we learn from a Juli 1, 1534 letter written by a Frisian scholar living in Basel, Viglius (Wigle) Aytta van Zwichem, it was, above all, the nearly countless works by the humanist prince Erasmus of Rotterdam that made Froben well known across country borders.[178] From preserved publishing records in print, it may be concluded that, over the years, the titles in question were produced in large numbers. They corresponded to a more or less stable demand, and the printer could proclaim these areas or individual authors as his intellectual property with which good money could be earned. The sometimes significant editions of several thousand copies, capable of guaranteeing sales over an extended period, required appropriate storage. This was not always easy to obtain or finance, as becomes evident from Oporinus's correspondence.[179]

Gessner remained in contact with Episcopius and Froben for decades and received many books from them as gifts.[180] In June of 1548, Froben informed him verbally about certain publication plans held by the Saxon mineralogist Georgius Agricola. This indicates that there must have been personal encounters between them, as must undoubtedly also have been the case with other Basel printers.[181]

Comparing the exchange of letters between Gessner and Episcopius on the one hand, and Froben on the other, it is evident that he maintained a collegial interaction with Episcopius. The correspondence is marked by technical questions that usually pertain to one's own book projects or those of others.[182] On September 30, 1560 Episcopius asked Gessner for advice. He and Froben planned to do a Latin translation of the popular chronicles of Johannes Stumpf that had been printed in Zürich in 1546/47. The project, however,

178 The letter was published in English translation in: Gerritsen (1991).

179 Steinmann (1969), pp. 148 and 181f.

180 Leu et al. (2008), pp. 73f., 117, 140f., 143, 161, 176f. and 232. According to a letter from Episcopius to Gessner of April 17, 1563, he also gave him the Isaiah commentary by Cyril of Alexandria (StAGR, D V/37 C 36.06.25).

181 Leu (1993), p. 39.

182 Cf. for example on the Hippocrates edition: letter from Episcopius to Gessner dated January 3, 1553 (ZBZ, MS. C 50a, no. 59). In Gessner's correspondence, Episcopius's name also occasionally appears in connection with book projects by others that might be realized at Episcopius's or that Gessner referred to him. See, for example, certain works by the Englishman John Caius (Gessner [1577], f. 111v), the Bavarian historian Johannes Aventinus (Burmeister [1975], pp. 291–293), or the Greek physician Dioscorides (Bauhin [1591], p. 110).

was never realized.[183] Episcopius's death was very painful for Gessner, as he related to Theodor Zwinger on April 8, 1564.[184] The relationship with Froben, however, seemed to place Gessner under some measure of pressure since he could hardly afford to turn down commissions from him. He complained of his enormous workload to Bullinger in a famous letter from the first half of September 1558.[185] Froben had added to it with his wish to match the Latin translation of the proposed large Galen edition to the Greek text. Thanks to financial support from the canons and a subsequent raise in salary, Gessner could afford to turn down the immense project. However, this did not stop Froben in 1561 from asking whether Gessner would contribute a foreword to the edition done in the same style as his article about Galen in the *Bibliotheca universalis*.[186] Gessner agreed, not only writing a foreword with high praise for Episcopius and Froben,[187] but also adding an extensive bio-bibliography about the celebrated Greek physician.[188] What at first glance seemed like a quite manageable bit of work, actually took up a great deal of time, as he lamented to Crato von Krafftheim on February 15, 1562.[189] Despite work-related tensions, however, it appears that Gessner kept good relations with the Froben family. For this reason, after Hieronymus Froben's death, the latter's son asked Gessner whether he would write a biography of his late father, or perhaps some Greek or Latin poems, as Gessner mentioned in a letter of June 26, 1563 to Achilles Pirmin Gasser.[190]

---

183 StAGR, Shelf mark: D V/37 C 36.06.01.

184 Gessner (1577), f. 107r.

185 Cf. p. 348.

186 Letter from Froben to Gessner before August 29, 1561 (StAGR, D V/37 C 36.06.47).

187 Edited in: AK 11, no. 4693, pp. 974–994.

188 Gessner also provided various other publications with detailed specialist bibliographical sections, such as the *De stirpium facultatibus libri* by Hieronymus Bock, published in 1552, the *Chirurgia*, which he edited in 1555, or the Strasbourg edition of Dioscorides from 1562.

189 Gessner (1577), f. 10v.

190 Burmeister (1975), pp. 236f.

CHAPTER 11

Augsburg Intermezzo

In July 1545,[1] Gessner accompanied the book printer Froschauer to Augsburg (ill. 28). By about 1500, the city of more than 25,000 residents was one of the most important financial centers of the Christian world as well as one of the most significant trading hubs in the German Empire. The wealth of the city was extensively tied to the Fugger enterprise with its financial businesses, trade in goods, and profits in silver and copper, all of which added up to an immense fortune. Jakob Fugger (1459–1525) was one of the wealthiest men of his time and is known to have financed the election of Charles V, a Habsburg, to become emperor.[2] Even Michel de Montaigne wrote in his travel journal in 1580 about the Fuggers: "The Fuggers – there are many branches of the family, but they are all well off – take the highest position in the social life of the city. We have seen two great halls in their palace, one is large and high-ceilinged with a marble floor; the other is lower with walls full of oval portraits, both old and new. In the back there is a small chamber. These were the most magnificent rooms I have ever seen."[3]

It is fairly certain that Gessner wrote the dedicatory foreword of the *Bibliotheca universalis* to Leonhard Beck von Beckenstein while visiting Augsburg. On February 1 of the same year, Gessner had already dedicated a work to him, which was an edition of various texts by the Italian humanist Antonius Thylesius (1482–1534). These included the famous *De coloribus libellus*, which Goethe was to integrate into his *Farbenlehre*.[4] Gessner explained that he wished to dedicate a work to Beckenstein that would continue to draw attention in the future – an accomplishment that, with respect to his own writings, he could not and should not hope for. He had collected various works by Thylesius and, for the first time, published some that had not been published before. Thylesius had been the teacher of the Zürich professors Johann Jakob Ammann and Rudolf Collin (among others) when they had studied in Milan. A few years

1 The *terminus post quem* is a letter from Georg Fröhlich (Laetus) from Augsburg to Bullinger dated June 20, 1545. Gessner was not yet in Augsburg at that time. A letter from Fröhlich to Bullinger dated August 18 must be considered as *terminus ante quem*. Gessner was then already back in Zürich. Cf. HBBW 15, pp. 371–375 and pp. 452–454.

2 Roeck (2005), pp. 98–104.

3 Montaigne (2007), p. 102.

4 Hennig (1987), pp. 174f.

© KONINKLIJKE BRILL NV, LEIDEN, 2023 | DOI:10.1163/9789004541696_012

earlier, the latter had published some of Thylesius's idylls through Froschauer[5] and left others for Gessner.[6]

The reason for Gessner's trip to Augsburg was Anton Fugger's wish for him to teach his two sons, Marx (1529–1597) and Hans (1531–1598) as well as his nephews – fatherless since 1535 – Ulrich (1526–1584), who had a gift for languages,[7] and Raymund (1528–1569).[8] The Augsburg town clerk, Georg Fröhlich, wrote to Bullinger on June 20, 1545 that Anton Fugger's intentions had surprised him, especially because as bankers to the Habsburgs, the Catholic Fuggers belonged to the bulwark of the traditional church in confessionally divided Augsburg.[9] Fröhlich continued that the nephews were leaning toward the evangelical faith, and that there was even concern about it in Augsburg.[10] One month later, on July 28, he let the head of the Zürich church know that he had extended a dinner invitation to Gessner and Froschauer, for whom he unfortunately had too little time. Also invited were the parish pastors Michael Keller († 1548), Wolfgang Musculus, Johannes Heinrich Held, the lawyer and municipal legal advisor Claudius Pius Peutinger (1509–1552) and the humanist, educator and librarian of the municipal library founded in 1537, Sixt Birck[11] (1501–1554).[12] They surely enjoyed good conversation, and Gessner

5 The *Cyclops* of Thylesius was published by Froschauer in 1537 at the end of an anthology of writings by Titus Calpurnius Siculus and Marcus Aurelius Olympius Nemesianus.

6 Thylesius (1545), f. a_4r–a_5r.

7 Gessner dedicated the *Icones avium* (1555) to him and described him in the dedication preface as the most educated among the nobles and the noblest among the educated. He was proficient in both languages (Greek and Latin) and knowledgeable in various sciences. Of the vernacular languages, in which Gessner also mentioned the individual animal names, he knew German, Italian and French, the last two languages very well.

8 In his letter, Fröhlich does not mention the first name of Fugger, who intended to bring Gessner to Augsburg. According to Nebinger/Rieber (1978), it only makes sense to see Anton Fugger behind it. His brother, who died in 1535, left behind eleven children, of whom Ulrich and Raymund were the only males who, in terms of age, would still have been considered in need of a a tutor.

9 Roeck (2005), p. 108.

10 HBBW 15, p. 373.

11 Gessner gave Sixt Birck (Xystus Betuleius) the four volumes of *Historia animalium* (1551–1558) with the following handwritten dedication: "Doctissimo viro D. Xysto Betuleio, amico longe charissimo. Conradus Gesnerus D. D." The volumes were very attractively colored, but the coloring does not correspond to the colored first volume preserved in the ZBZ, nor to the pictorial originals for the animal books found in the University Libraries of Amsterdam (MS. III C 23 & 24) and Basel (MS. K I 1), from which it can be concluded that the coloring without an original was done later in Augsburg, possibly even after the sixteenth century. The books are now kept in the Erlangen University Library (call number: Trew C 46–49).

12 Gessner contributed a Greek dedicatory poem to the *Novi Testamenti Concordantia* (Basel 1546) by Sixt Birck.

TA IVXTA
AVGVS
FIGVRAM QVAM HI
CE TEMPORIBVS

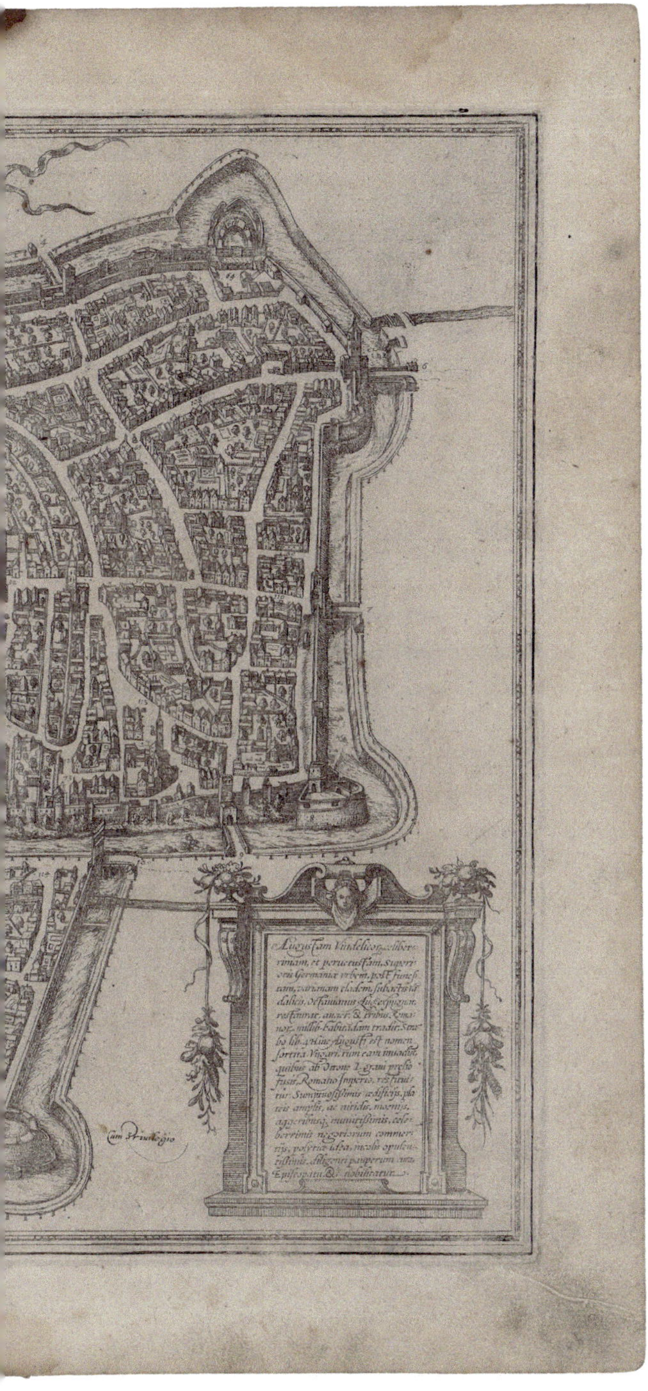

ILLUSTRATION 28
Map of the city of Augsburg with the Fugger palace and in the back the dwelling of Leonhard Beck von Beckenstein (red square): Georg Braun, *Beschreibung und Contrafactur der vornembsten Stätt der Welt*, Cologne 1582, plan no. 40.
ZBZ, ABTEILUNG ALTE DRUCKE UND RARA: T 45

would have been a standout for his earnestness, modesty and erudition. Anton Fugger seemed ready to invest in Gessner. He apparently considered sending him to Italy, which Fröhlich thought to be a dangerous undertaking. All the same, Gessner must have known what he was doing.[13] Then, Gessner and Froschauer departed unexpectedly. Fröhlich wrote to Bulllinger once more on August 18 that it was no wonder an agreement could not be reached between the extremely Catholic Fugger and the pious Gessner. He closed the letter by noting that he would be purchasing Gessner's *Bibliotheca universalis* as well as Zwingli's four-volume work of 1544/45 from the Frankfurt book fair.[14]

Gessner seemed to get along well with Fröhlich. On February 1, 1546, he dedicated his Tatian translation to him, which was published at the end of the above-mentioned *Sententiae* by Antonius Melissa and Maximus the Confessor.[15] During preparations for publication, Fröhlich had heard that Gessner wished to dedicate a work to him. This made him uncomfortable, so he asked Bullinger to prevent it from happening if at all possible.[16] In the dedication, Gessner mentioned the invitation to Fröhlich's home, and as an expression of thanks, he devoted the short work to him. In conclusion, he also sent greetings to the others who had attended the dinner, leaving out Peutinger and adding Bernardino Ochino to the list. The question must remain open as to whether Ochino had replaced Peutinger as a guest, whether he was perhaps an additional participant, or if he and Gessner had encountered each other elsewhere. One may be certain, however, that Gessner and Peutinger had indeed met. The latter leaned toward reformed Protestantism[17] and had sent his children's teacher, Johannes Kielmann, to Zürich in September 1546 to find a safe place for training during the Schmalkaldic War. Peutinger's children and Kielmann did in fact arrive in Zürich that October and were likely to have remained until the end of the war in April 1547.[18] According to a letter from Kielmann to Bullinger on September 28, 1550, they had lived with Gessner, since Kielmann referred to him as *antiquus hospes* (old host).[19]

---

13 HBBW 15, p. 417.

14 HBBW 15, pp. 453f.

15 Cf. p. 48.

16 HBBW 16, p. 193. The appearance of the dedicatory preface in the printed version evidently embarrassed Fröhlich, who protested that he felt entirely unworthy of it. HBBW 16, p. 300.

17 Roth (1928), p. 207.

18 Peutinger's letters to Bullinger of September 11 and October 24, 1546 (StAZH E II 346, 187 and E II 356, 991).

19 Kielmann to Bullinger, September 28, 1558 (StAZH E II 356, 100). Cf. also Kielmann to Bullinger of January 13, 1549 (StAZH E II 343, 401).

On the same day, February 1, 1546, Gessner not only wrote the dedication text for the Tatian translations, but also one for the Greek edition of the *Sententiae*, which was addressed to the Augsburg councilmen Johannes Welser (1497–1559) and Jakob Herbrot († 1564), who had served as mayors during Gessner's stay.[20] In it, Gessner praised the Augsburg library, which he had personally visited, and he underscored the importance of its stock of Greek manuscripts. He was certainly thinking of the 99 manuscripts that the city of Augsburg had purchased in Venice in 1543/44 not least through the efforts of Wolfgang Musculus. Unlike other such purchases, the buyer in this case was a city instead of a private person, something quite unconventional for the times and quite deserving of Gessner's outspoken praise.[21]

Gessner's dedication to the two previous mayors irritated the Augsburg city council. It saw itself as a chamber of equals from which two mayors were chosen to more or less be *primi inter pares* (first among equals) for one year. In this regard, Fröhlich wrote to Bullinger on April 1, 1546, that Gessner should write any similar future dedications to the entire council.[22] Ultimately, Gessner followed this advice on June 20, 1564 in his dedication to the Augsburg council in Johannes Moibanus's edition of Dioscorides.[23] His earlier animosity is probably the reason why no honorarium was given after he had sent two beautifully bound copies of the *Sententiae* to Augsburg, something which Gessner later complained about to Musculus. One was for the council and the other for the Augsburg city library.[24] The council thanked Gessner in writing and arranged, over the objections from Musculus, for two of the long sought-after Greek manuscripts to be loaned out to him.[25] One was a description of the Earth *De situ orbis* by Dionysius Periegetes, and the other was a commentary on the *Octateuch* by Procopius of Gaza (5th/6th century). The first contained only a quarter of the text, which made it of no value to Gessner. He subsequently asked Froschauer's business manager, Gregor Mangold,[26] to take it back to Augsburg by way of Lake Constance along with other books. Despite an

20 HBBW 16, p. 261.

21 Zäh (1997).

22 HBBW 16, p. 301.

23 Dioscorides (1565), f. a_{ii}r–a_{vi}v.

24 Letter from Gessner to Musculus before February 21, 1550; Stadtbibliothek Zofingen, Pa 15.1, no. 73 (1766/67). For the preface to the Galen edition addressed to the Basel council, he received 5 thalers as a gift, cf. p. 68. The council of Nuremberg also showed gratitude to the wound physician Jakob Baumann, who received 50 guilders for the dedication of his German translation of Vesalius's *De humani corporis fabrica* (Nuremberg 1551). Cf. Keller et al. (2008), vol. 4, p. 283.

25 HBBW 16, pp. 291f. Cf. Roth (1907), pp. 156 and 193.

26 Steinmann (2000), p. 11.

unfortunate shipwreck, the manuscript could be saved, and today it is located in Munich's Bavarian State Library.[27] He had the Procopius piece copied by Sebastian Guldinbeck and sold it to Conrad Clauser for 10 guilders. Since he paid Guldinbeck the same amount for the copy work, no profit was made, which Gessner pointed out to Musculus,[28] who spent several months in Zürich in 1548/49.[29] The Augsburg original was still in Gessner's possession in 1550, and when exactly it was returned is unknown. In any case, today it is kept in the Bavarian State Library.[30]

While Gessner had dedicated the first part of the *Sententiae* with its Greek text to two Augsburg politicians, the second part, translated into Latin, was dedicated on February 5, 1546 to the Protestant-leaning Johann Jakob Fugger (1515–1575, ill. 29). Gessner had spent a great deal of time in Fugger's library reading a commentary about Oppian's *Halieutica*.[31] He mentioned that he had wished the Latin portion also to have a patron and therefore thought of Fugger, who was known for his interest in literature and science. It was intended that he perceive the dedication as a sign of Gessner's approval of him. He likened the text's secular sagacity to jewels, and the theological wisdom of the church fathers to the stars. 20 years later, these played a role in Gessner's design of a family crest.[32]

In the following years, it seems that Gessner no longer traveled to foreign lands. He did, however, occasionally venture outside Zürich, for instance, to St Gallen in December of 1545.[33] Aside from his official duties, from 1546 to 1548 he was deeply preoccupied with work on the *Pandectae* and the *Partitiones*. From 1548 on, he devoted himself to the publication of his monumental animal encyclopedia, as he was to recall ten years later.[34] Other work included the publication of the above-mentioned natural science textbook by

27 HBBW 16, p. 292. BSB Munich, Cod. Monacensis graec. 529. Gessner seems to have continued to be interested in text editions of Dionysius, at any rate the Pole Florian Susliga announced in a letter to Bullinger of May 13, 1549, that he would soon send Gessner one. Cf. StAZH, E II 338, 1468.

28 Letter from Gessner to Musculus before February 21, 1550; Stadtbibliothek Zofingen, Pa 15.1, no. 73 (1766/67).

29 Bodenmann (2000), p. 574.

30 HBBW 16, p. 292.

31 Gessner (1558), f. b_4v.

32 Cf. p. 246.

33 Gessner undertook more travels within the Swiss Confederation than had been assumed so far or proven in the biographies. For example, he stayed in St Gallen at the end of December 1545, as can be seen from Bullinger's correspondence. Cf. HBBW 16, p. 81.

34 Gessner (1558), f. b_3r.

ILLUSTRATION 29 Augsburg memorial to Johann Jakob Fugger in front of the Maximilian Museum, formerly the residence of Leonhard Beck von Beckenstein.
PHOTOGRAPH: LUKAS KELLER, ZÜRICH

Ermolao Barbaro and a work of only a few pages about Hippocrates.[35] He was also busy with the eight-volume Galen edition in Latin, printed by Froben and Episcopius in 1549. The nearly complete edition of Galen in Greek, printed by Aldo Manuzio in Venice in 1525/26, instigated a large number of Galen printings up to 1560.[36] These included the Basel edition based on a translation by the German humanist Janus Cornarius (1500–1558). Gessner was responsible for the structure of the texts, chapter headings and summaries of individual sections, which required a thorough prior study of Galen's writings. In the 1540s, five complete Latin editions of the Greek physician were published in Europe.[37] In no decade before or after, up to 1600, were there more editions printed, which indicates the level of demand. The abundance of Galen and Aristotle titles published in the mid-sixteenth century leads to the conclusion that there was an intense preoccupation in Europe with these two authors.[38]

35 Wellisch (1982), A 22.
36 Nutton (2008), pp. 370f.
37 Durling (1961), p. 280.
38 Leu et al. (2008), pp. 21–25.

Bebel and Isengrin's 1550 Aristotle edition was based on their earlier 1539 edition, which Gessner kept in his personal library with intensive annotations and edits.[39] Gessner sent his copy to Basel, to permit the printers to carefully integrate his numerous corrections. Others who worked on this third Aristotle edition in Basel were Justus Velsius and Matthias Flacius Illyricus (1520–1575).[40] A short time later, a further Aristotle and a Theophrast edition were released, which on March 1, 1554 Gessner asked Isengrin to send to him.[41]

---

39 Leu et al. (2008), pp. 57f.

40 Hieronymus (1992), pp. 170f.

41 ÖNB, Shelf mark: Cod. 9737i, f. 100r/v.

CHAPTER 12

The Animal Books – A Renaissance Noah's Arc

In the eyes of his contemporaries, Gessner's greatest and most important accomplishment was his still famous five-volume encyclopedia of all the then-known animals.[1] Four volumes, covering four-legged livebearing animals, four-legged egg-laying animals, birds and aquatic animals were published between 1551 and 1558. The fifth book, about snakes and scorpions, was posthumously published in 1587 by his estate executor Caspar Wolf with assistance from Jacob Carron. Volumes one through four contain about 3,500 folio pages with nearly 1,100 woodcuts. With the exception of Volume 2, the illustrations tripled in number from book to book, so that the 1558 book about fish ended up with an impressive 737 images.[2] Gessner justified his focus on zoology by citing the many scholarly books on botany, and the fact that minerology had been thoroughly covered by *Bermannus*, the standard reference by Georgius Acricola. By comparison, little attention had been given to the animal world, and only in an incomplete manner.[3] Gessner's attention to this subject was appreciated and welcomed by his peers, as is shown in the above-mentioned biography by his colleague Josias Simmler, who praised no other work as highly as the *Historia animalium*.[4] According to the famous French anatomist Georges Cuvier (1769–1832), Gessner's achievement is the most important zoological work of the early modern period.[5] This feat is even more remarkable when one considers his many other responsibilities as a physician and professor, as well as Gessner's struggles with poor health. In 1552, the same year that his mentor Myconius and the city physician Clauser died, Gessner also fell severely ill and came close to dying. When his childhood friend, Theodor Beza in Geneva, heard about his serious illness, he wrote to Bullinger on May 15: "What should I write to Gessner? Were he to be taken from us, I should congratulate him rather than feel sorry. But when I think of his unique kindness, his erudition, his diligence and his friendship, nothing would be more bitter to me now than his death. I am too afraid to write to him. But you see, I'm imploring you to soon rid us of this fear. If he nonetheless dies – which I wish for least of all – be

1 The term "Arche de Noé de la Renaissance" comes from Pinon (1995), p. 34.

2 Numbers according to Wellisch (1984).

3 Gessner (1551), f. α_2r.

4 Simmler (1566), f. 8v–13v.

5 Cuvier (1841), p. 90.

© KONINKLIJKE BRILL NV, LEIDEN, 2023 | DOI:10.1163/9789004541696_013

prepared to befittingly console us. One hopes that this last undertaking will not be necessary."[6]

A month later, Gessner was in good health again,[7] but in the following year, the Worms lawyer Wolfgang Weidner reported to Bullinger: "When I was last in Zürich, Gessner must have been something over 40 [he was 37 years old!] and in the best years of his life, but his pallor spoke volumes about his excessive research projects."[8]

Work on the *Historia animalium* posed an immense challenge for Gessner. The material to be processed for such a comprehensive animal encyclopedia was immense and, considering the information and communication possibilities at the time, quite astounding. He was surely pleased and satisfied to be working on such a pioneering project, but it also frequently placed him under pressure, especially since Froschauer constantly badgered him to finish the books. For example, on March 1, 1554, he informed Michael Isengrin in Basel that he could not assist him with his bibliographic project because he was completely preoccupied with the animal history.[9] The situation remained this way for years, for on June 6, 1560 he told Theodor Zwinger: "Now I'm being tormented by two printing presses that are printing the fish and four-legged animals; I can barely lift a finger to scratch my head."[10]

12.1 Natural Theology

Unlike his contemporaries, Gessner did not see his work on the natural world to be the means to an end. It was not for providing food, clothing and medicine; instead, he pleaded for the study of nature itself. In the foreword to the first volume of the *Historia animalium* he stressed the unbelievable diversity of the animal world, he marveled at the songs of birds, and was astounded by the physical and organizational skills of ants and bees. There was not a single living creature, with its specific individual qualities or other wondrous characteristics, that did not ultimately suggest an ingenious Creator. He considered those who were only searching for benefits and profit from nature to be spiritually poor. How many things are there that are not directly useful, that nonetheless captivate and fascinate us, such as precious topaz, diamonds, chrysolite,

6 Calvin (1875a), p. 316.

7 Letter from Bullinger to Myconius, June 12, 1552 (StAZH, E II 342, 282).

8 Hanhart (1824), p. 140.

9 ÖNB, Shelf mark: Cod. 9737i, f. 100r/v.

10 Peine (1941), p. 67. See the Latin original in: Gessner (1577), f. 104v.

emeralds and the like? And if these lifeless objects can enthrall us, then what is living nature capable of?[11] On August 5, 1558, Gessner dedicated the fourth volume on aquatic animals to Ferdinand I, who had been elected emperor on February 28 of the same year and maintained a conciliatory religious policy. He emphasized that these mute creatures were deserving of attention and study and should not be passed over without a word. Sea creatures were unjustly feared and considered mindless, but they were also begotten by a magnificent Creator and worthy of research.[12]

Like Aristotle,[13] Gessner recognized a natural hierarchical ladder behind all creatures that ultimately led to God, the source of all life. The ladder began with stones and minerals, ascended to the plant and animal worlds, then to humans, up to the angels and ended with the Creator of all things. Avoiding any speculation, Gessner had little to say about angels and left that up to the theologians.[14] His first area of interest, and also the first volume of the *Historia animalium*, focused on livebearing four-legged animals. This was because they were highest on the hierarchical ladder in their similarity to humans with regard to body parts, feelings, intellect and behavior. He had intended to make a second subsequent volume about all the other living creatures, but this turned into four more books.[15]

For Gessner, the study of nature inevitably led to the glorification of the Creator. This is why in the first volume of the *Historia animalium*, the foreword to the reader included a long quote from the Old Testament Book of Job (Job 38.39–39.30). At its heart is the omnipotence of the Creator and the insignificance of man. The last page of volume one of the *Historia animalium* shows the same woodcut of the creation of the world as that found since 1531 on first page of Genesis in the large Zürich folio Bible. Above the illustration, Gessner added a Greek quote from the pre-Socratic philosopher Empedocles (ca. 495–435 BC), who had praised the elements of the earth that comprise all things. Gessner, however, diverted the tribute from the elements to God. The passage reads in translation: "As the naturalist Empedocles wrote, all things owe praise and glory to the one God, from whom exists all that is, was and later shall be – and from whom trees sprouted, men and women, beasts, birds, and water-nourished fish."[16]

11 Gessner (1551), f. α_3r/v.
12 Gessner (1558), f. a_3r.
13 Flashar (2013), pp. 348f.
14 Cf. on this whole complex of theology and especially natural theology: Leu (1990).
15 Gessner (1551), f. α_2v.
16 Cf. Gemelli Marciano (2009), fragment 27, pp. 178f.

Following the dedicatory preface to the Zürich city council in the first volume of the *Historia animalium*, for which Gessner received a yearly honorarium of 10 bushels of seeds and 10 buckets of wine,[17] there was a short printed note by Christoph Froschauer, which mentioned an abridged version of the extensive volumes. This short version was aimed at readers who could not afford such an enormous work or were unable to invest a great deal of reading time. Subsequently, in 1553, he published the *Icones animalium quadrupedum viviparorum et oviparorum* (*Images of Livebearing Four-legged and Egg-laying Animals*), and in 1555 the *Icones avium* (*Bird Images*), which contained animal illustrations but very little text. The topic of natural theology and the awareness of God in nature were echoed in the dedication of the *Icones animalium* to the English noblemen Thomas (1510–1555) and John Grey (ca. 1523–1564). They were the brothers of Henry Grey (1517–1554) to whom Bullinger dedicated the *Sermonum decas quinta* in 1551. Queen Mary I put Henry to death on February 23, 1554, after executing his daughter Jane (1536/37–1554) on February 12, 1554. Jane had been the Queen of England from July 10 to 19, 1553. Nine days after she was dethroned, Gessner wrote the dedication to her two uncles. It is not likely that he was already aware of the turmoil in England. His praise for the Grey family – especially Henry Grey and his daughters Jane and Catherine (1540–1568) – was because of their erudition and piety. He encouraged them to marvel at the animals and therefore the Creator of all things, for they lived through Him. And through them, God left traces of His existence and divinity (*vestigium divinitatis*) in the visible world.[18]

The second edition of the *Icones animalium quadrupedum* was published in 1560 with much more illustrative and textual material than the first. It was dedicated to none other than Queen Elizabeth I (1533–1603) of England, whom Gessner praised in a 28-line Greek poem. Unfortunately, this did not sit well with the Queen since her permission for such a dedication had not been obtained beforehand. She also found it inappropriate for such a work to be devoted to her when the first edition had been dedicated to the Grey family. The Greys were indeed related to Elizabeth I, but the Queen did not think that Jane Grey should have been crowned as the legitimate successor to Edward VI. Elizabeth's father, Henry VIII had intended the throne for her half-sister Mary I ("Bloody Mary"). Because they competed for succession to the throne and for other issues, relations were tense between the Greys and

17 Kapp (1886), vol. 1, p. 319. "Kernels" means rye or wheat. A "Malter" contained about 150 liters and a "Eimer" (bucket) about 100 liters.

18 Gessner (1553a), p. 2.

Henry VIII's daughters, Mary I and Elizabeth I.[19] Moreover, it was not proper to dedicate a second edition of a work to a person of higher social standing than the first edition.[20] Gessner had been in the dark about all this.

On March 7, 1561, Elizabeth's secretary of state, William Cecil, brought her the news from Zürich that, contrary to custom, Gessner had not received a penny from either the Greys or the court for his dedication.[21] He was short of money with 40 indigent relatives, and would appreciate at least being reimbursed for the cost of the books he had sent as gifts, which amounted to four crowns. He also enclosed one of his latest works. Cecil had been informed of Gessner's situation by John Caius,[22] and the court did not delay in sending the needy researcher in Zürich six pounds for his *De animalibus*.[23]

Gessner's dedication to the Queen was a singular celebration of creation and the Creator. First of all, he praised the English monarch's Christian virtues, intellect, and knowledge of ancient languages, referring to her as a most gracious heroine (*clementissima heroina*). Then, he explained how the observance of nature led to the awareness of God and ourselves. Humans are like a small world of their own, with an earthly body composed of natural elements, and the image of God in their souls. Between humans and the other creatures, He placed the four-legged animals, who unlike humans, possess no intellect. Gessner pointed out that God created not only visible things, but also invisible things such as the spirit and the angels. The latter are God's watchmen over the pious, but also His avengers against those who do wrong. But one should worship only God, just as we see there is only one sun. The moon is its reflection, shining not from its own light, but because the sun shines on it. So it is with God, the Lord of everything, who lends His power to all of nature. He who does not know nature does not know God. Furthermore, nature clearly shows how God cares for all things that He has created. All of the millions of animals have food, "clothing" and a place to live. How much more will He care for those He finds to be true to their faith?

With regard to the body and its movements, men are not different from animals. However, a difference is made apparent through our reason (*ratio*)

---

19 Doran (2015), pp. 43–60.

20 Blair (2021), pp. 74f.

21 Cf. Schottenloher (1953).

22 Stevenson (1866), p. 10, no. 17. On Caius and Gessner, cf. Funk (2017), Nutton (2018), see index, p. 149; Grafton (2019).

23 Calendar of the Cecil Papers in Hatfield House, vol. 1, 1306–1571. Originally published by Her Majesty's Stationery Office, London, 1883, pp. 257ff. Available online at: www.british-history.ac.uk/cal-cecil-papers/vol1/pp257-263. I owe this reference to Prof. Dr. Ann Blair (Harvard). For the prices of Gessner's animal books, see p. 216.

and intellect (*mens*), with which we come close to the angels. Furthermore, humans walk upright, so that the head can be lifted toward the sky to remember our commitments to God. And, not only our physical eyes, but also the eyes of our soul should look to the Creator of all things. Man can also learn from the animals and the qualities that God has given them. For animals will instinctively do many good deeds that would be difficult for humans. Jesus himself commanded men to be as wise as serpents and as innocent as doves (Matthew 10:16). Moreover, the inherent qualities of cranes, bees, groundhogs, martens, squirrels and other animals are amazing; how they live together socially, hide away, change to other locales and regions, create ingenious living spaces and remain cautious and circumspect relative to other creatures. All these are thoughts that Gessner may have acquired from his spiritual father Huldrych Zwingli.[24]

He also marveled at the discipline, restraint and chastity of elephants, for Aristotle had claimed that having mated once, the bulls would never touch the female again.[25] Gessner also pointed out that, unlike humans, no animal would kill another of the same species, with devotion to the herd taking priority. Note that this does not correspond to today's ethological knowledge. At any rate, the selflessness of bees also impressed Gessner, not to mention their industriousness, as they collected sweetness without setting aside any of it for themselves. His moral theology led to the assertion that in observance of the animal world, humans should recognize their own damnable, corrupt and sinful nature. One should admit this to God and ask Jesus Christ for His mercy and forgiveness.[26]

On June 13, 1560, the same day that he dated the previously described dedication to Queen Elizabeth I, Gessner also wrote one to King Maximilian II (1527–1576) of Bohemia for the illustration plates of aquatic animals (*Nomenclator aquatilium animantium*). The King was to succeed his brother Ferdinand I to be crowned emperor in 1564.[27] Maximilian was open to Protestantism and, among other things, collected books and manuscripts, especially with the support of Caspar von Niedbruck. Once again, Gessner wrote in the foreword extensively about the different levels of divine creation.[28]

---

24 Zwingli (1983), p. 104.

25 Aristoteles, *Historia animalium*, 9,46 (= 630b).

26 Gessner (1560), f. A_{ii}r–f. A_{iii}r.

27 Gessner gave Maximilian II a copy of his fish book, but the latter did not acknowledge receipt for months, which is why Gessner inquired with Paul Skalic on March 4, 1561, about what had happened. Cf. Glesinger (1950), p. 40f.

28 Gessner (1560a), f. aa_2r–f. aa_3v.

12.2 Autopsy

Gessner's voluminous *Historia animalium* was the first attempt to describe all the animals inhabiting the then-known four continents. Never before had anyone attempted such a bold project. Animal illustrations were only occasionally available, such as in the *Hortus sanitatis* by Johann Cuba, Bernhard von Breydenbach's (1440–1497) *Reise ins Heilige Land* or Michael Herr's *Grundlichem underricht ... seltzamer art, natur, krafft und eygenschafft aller vierfüssign thier* of 1546. It is remarkable that of all people Gessner should have taken on this project. He was not from a major trading nation with access to an extensive information and transportation network, and he did not live by the sea, where he would at least have had access to aquatic creatures. Furthermore, he did not reside in a major metropolis such as Paris or Venice, hubs from which he might have benefitted. On the contrary, Gessner dwelled with 7,000 inhabitants in Zürich, at the foot of the nearly untraversable Alps, far removed from the major European trading routes. As if these obstacles were not enough, he decided to go beyond confirming and supplementing information from Aristotle and Pliny and to include animals from regions unexplored by the ancients, such as the Alps or Northern and Eastern Europe. In addition, he strove his entire life to find the best possible illustrations, several hundred of which were included in his animal books. This set a new standard for the visual documentation of worldwide fauna. Gessner's zoological encyclopedia was considered one of the most important reference works until replaced in the eighteenth century by Georges-Louis Leclerc Comte de Buffon's (1707–1788) *Histoire naturelle*. Comparable works have been published more recently in German-speaking countries, including Alfred Edmund Brehm's (1829–1884) 10-volume *Illustriertes Tierleben* (*Illustrated Animal Life*), and in the twentieth century, Bernhard Grzimek's (1909–1987) 13-volume *Tierleben* (*Animal Life*).

While from a current standpoint, Gessner's work may look outdated, it still contains much information about animal history. For example, at the time, the Asian lion (*Panthera leo persica*) was well known, whereas today he is threatened with extinction. Only a few are still to be found in India's Gir National Park.[29] Another well-known example is the hermit ibis (*Geronticus eremita*), which used to inhabit Central Europe, but is now only found in Turkey, Syria and Morocco (ill. 30).[30] With regard to Gessner's book about birds, the biologists Springer and Kinzelbach write: "Among the sources, bird books from early scientific literature are of special significance. They provide a considerable

29 Enenkel (2007), p. 58.
30 Strohl (1917).

De Coruo ſyluat. Lib. III. 337

DE CORVO SYLVATICO.

AVIS, cuius hîc effigies habetur, à noſtris nominatur uulgo **ein Waldrapp**, id eſt coruus ſyluaticus, quòd locis ſyluoſis, montanis & deſertis degere ſoleat: ubi in rupibus, aut turribus deſertis nidificat. quare etiam **Steinrapp** uocatur, & alibi (in Bauaria & Stiria) **ein Claußrapp**: à petris ſeu rupibus, & pylis (nam pylas, id eſt anguſtias inter duos montes Germani **clauſen** appellant, hoc eſt loca clauſa) in quibus nidos ſtruit. Lotharingi, ut audio, corneille de mer, id eſt cornix marina: quam et in iuglandibus aliquando nidificare ferunt. ſed forte ea alia auis eſt. Circa lacum Verbanum coruus marinus dicitur. alibi in Italia coruus ſyluaticus, ut in Iſtria circa promontorium Polæ, ubi homine per funem demiſſo per rupes nidis eximuntur, & inter menſarum delicias habentur, ut apud nos quoq; in montium quorundam rupibus, ſic enim Fabarias thermas repertas aiunt, cum auceps quidam per altiſsimas rupes propter has aues ſe demiſiſſet. Alibi in Italia coruo ſpilato, id eſt coruus depilis, quoniam ſeneſcens caluescat. Germanicè quidam nuper conficto à ſe à ſono uocis eius nomine **Scheller** uocabat. ¶ Sunt qui phalacrocoracem hanc auem interpretentur, quoniam & magnitudine & colore ferè coruum refert: & calueſcit, ut uidi, cum adultior eſt. Turnerus Ariſtotelis coruum aquaticum & Plinij phalacrocoracem, & coruum ſyluaticum noſtrum auem unam eſſe arbitratur, tertium genus graculi. Coruus ſyluaticus Heluetiorum, inquit, auis eſt corpore longo et ciconia paulò minore, cruribus breuibus, ſed craſsis; roſtro rutilo, parum adunco (curuo) & ſex pollices longo; alba in capite macula, & ea nuda, ſi bene memini. Quòd ſi palmipes eſſet & interdum nataret, indubitanter tertium graculorum genus eſſe adfirmarem. uerũ licet auem in manibus habuerim, an palmipes fuerit, nécne, & calua, non bene memini, Sic ille. Sed cum nos certò ſciamus, palmipedem non eſſe coruum ſyluaticum noſtrum, non poterit eſſe coruus aquaticus Ariſtotelis, ſed neq; Plinij, qui (ut diximus) phalacrocoracem, id eſt coruum caluum, eundem & aquaticum facit. Noſter uerò ſyluaticus non eſt aquaticus, neq; in aquis degit: ſed in pratis, & locis paluſtribus uictum ſibi quæritat. Iam cum Ariſtoteles tertium graculi genus palmipes faciat, id quoq; coruus ſyluaticus noſter eſſe non poteſt.

¶ AVIS quam prius hæmatopodem eſſe putabam (inquit Bellonius) nunc potius ibin nigram eſſe conijcio: cuius Herodotus & Ariſtoteles meminerunt. Ea corporis mole auem à Gallis uulgò corlis (arquatam maiorem noſtram eſſe arbitror) dictã refert, uel paulò minor eſt, tota nigra, capite phalacrocoracis, roſtro iuxta caput plus quàm pollicari craſsitudine, inflexo modicè in arcum & in acutum deſinẽte, rubicundo, qui crurum etiam color eſt. Proceritas crurum ea ferè quæ in ardea

F

ILLUSTRATION 30 Hermet ibis, in: Gessner (1555a), p. 337.
ZBZ, ABTEILUNG ALTE DRUCKE UND RARA: 5.2

amount of material from the author's own viewpoint. They are archives for proxy data on bird life, which can be processed and evaluated with regard to their faunistic content. At present, however, as with other categories of sources, they have not yet been consistently indexed nor sufficiently published."[31]

With regard to providing scientific data, Gessner was much more precise than, for example, his similarly famous zoological successor Ulisse Aldrovandi.[32] The material that Gessner gathered, whereby the bulk was from Switzerland, Germany, Italy, England and France, is of particular importance, especially for the following reason: "It documents the condition of the Central European bird world during the climatic transition phase between the Medieval Warm Period and the advent of the Little Ice Age. It establishes an excellent basis for comparison during the subsequent considerable change in climate. The copius proxy data makes it possible to research the bird world. Both climatic influences are represented in the bird book with regard to the included species. The Medieval Warm Period of the late Middle Ages is documented by 'thermophilic' species (e.g. rufous-tailed rock thrush – *Monticola saxatilis* and the blue rock thrush – *Monticola solitarius*). Evidence of seabirds living far inland (gannets – *Sula bassana* and puffins – *Fratercula arctica*) suggests the occurrence of storms that were harbingers of the Little Ice Age."[33]

When contemplating the conditions under which Gessner executed this monumental work, the nature of his working methods takes on extra significance. He speaks openly in the forewords to the *Historia animalium* and the *Icones animalium* about his sources and the people who helped him. Furthermore, he usually specified where information came from, although it is not always clear what was acquired from others and what he discovered on his own. In addition, one must not overlook the fact that various creatures were researched, examined and illustrated by him personally. In the foreword to the reader in his book about aquatic animals, Gessner praises three ichthyologists, Pierre Belon (1517–1564), Guillaume Rondelet and Ippolito Salviani (1514–1572), whose results were obtained through *observare* (observing), *scribere* (describing), *peregrinare* (travelling) and *dissecare* (dissecting).[34] It is clear from various passages of the animal books that Gessner also used these methods, which he summarized with the term "autopsy." In his 1542 foreword to the *Historia stirpium*, the botanist Leonhart Fuchs had extolled the exemplary working methods of the ancient Greeks, who went to great lengths to personally view

31 Springer/Kinzelbach (2009), p. 3. Cf. Henrich (2006).

32 Springer/Kinzelbach (2009), p. 34.

33 Springer/Kinzelbach (2009), p. 48.

34 Gessner (1558), f. b_1r.

and examine plant species (αὐτοψία).[35] In the German translation of 1543, the shortened relevant passage says: "It is indeed known that the ancient physicians, especially Theophrast, Dioscorides and Galen, regarded nothing more worthy and necessary than the thorough examination of herbs. For this reason, they traveled through many lands at great cost and risk to life and limb to see with their own eyes and observe for themselves in order to know everything about herbs and where and how they grow."[36]

Gessner's animal books are rich with observations and descriptions, some of which he personally contributed, but also from conversations with hunters, fishermen, bird catchers and shepherds.[37] He also gained much from an abundance of literature and the letters from his correspondence partners.[38] A brief insight into the publications processed by Gessner can be found in the literature lists that he compiled in the first and fourth volumes. The first volume cites a total of 251 authors, respectively 3 Hebrew, 68 Greek, 49 ancient Latin, 15 Arabic translated into Latin, 90 Neolatin, 8 German, 3 Italian and 5 French. Regarding the 134 authors that he marked with an asterisk, Gessner had thoroughly worked his way through everything they had written about animals in the texts listed. Among these, 80 were modern contemporary authors, which meant that Gessner's zoological foundations were based mostly on "the modern." The bibliography in the fish book contained another three-digit number of authors. However, these were not sorted according to either time period or language, and referenced works that in the meantime were lost. Considering the listed and recognized texts, it may be assumed that his private library must have contained about twice as many as the 500 titles that have so far been identified.[39]

Gessner placed great importance on the relevant works of Belon and Rondelet, whose content he integrated in their entirety. He used 80 percent of the woodcuts in Rondelet's *Libri de piscibus marinis* (Lyon 1554/55), but also sent some illustrative material to the Frenchman, as for example the fish of the Danube.[40] Unfortunately, Gessner did not get a look at Salviani's book until late in the process. From its 100 copperplate illustrations, Gessner was only able to use one or two since he already had images of the other fish. He also used only a few woodcuts from Belon.[41] Just as Gessner always mentioned Rondelet and

35 Fuchs (1542), f. α_4r.
36 Fuchs (1543), f. 2_1r.
37 Gessner (1551), f. β_1r.
38 Delisle (2007).
39 Cf. Leu et al. (2008).
40 Egmond/Kusukawa (2016).
41 Gessner (1558), f. b_2v.

Belon with respect, they behaved similarly towards him. Rondelet, by contrast, had often criticized his compatriot Belon and the Italian Salvini for stealing his intellectual property. Gessner described this as a sign of the greatest ingratitude and something which he had never done.[42]

Work on the fish book was beset with particular difficulty since he lived so far from the sea. He actually wanted to delay publication of the volume until he had resolved a few remaining issues, but the printer had already invested a lot of money and did not tolerate further delays.[43] Printing preparations for the volume were extended from the winter of 1557[44] into the summer of 1558,[45] mostly due to Gessner compiling new material that he wished to integrate. Just six months after publication, he was already considering an appendix to the fish book,[46] which indicated some dissatisfaction with the results produced under time constraints.

Gessner would have gladly traveled more often, like his colleague Pierre Belon, who had visited Italy, Greece, the Near East and England. He did not, however, have the means to do so. During his time in Provence in the fall of 1540, he managed to take several excursions to the coast, and in Venice, he began with his fish illustrations during the summer of 1543.[47] For the rest, he had to abstain from further trips abroad. Whether or not he tried to organize an excursion to Scotland during his time in Paris cannot be confirmed.[48] In June 1546, he attempted to reach the sea at Rotterdam by way of the Rhine River in order to research the local fish fauna. Due to the outbreak of the Schmalkaldic War, however, he had to leave the ship in Strasbourg and cancel the trip.[49]

In the foreword to the first volume of the *Historia animalium*, Gessner comments on his travels: "It was clear to me that if I did not travel to foreign countries, and used only observations from my home territory and from reading reference books, I could accomplish little. Therefore, I traveled to Italy and other parts of Germany. Had I been fortunate enough to find a patron, or had

42 Letter from Gessner to Leonhart Fuchs, October 18, 1556, in: Gessner (1577), f. 137v–138v.

43 Gessner (1558), f. b_3r.

44 Letter from Gessner to Johannes Placotomus, March 26, 1557, in: Gessner (1577), f. 136v–137v.

45 Gessner sent one fish book each to Fabricius and Kentmann in Saxony on August 25, 1558. Cf. Gessner (1584), f. B_2v–B_3v.

46 Letter from Gessner to Kentmann, February 27, 1559, in: Gessner (1584), f. B_3v–C_1r.

47 Gessner (1558), f. b_1v. He did not draw certain fish illustrations himself, but had them drawn, see Gessner (1558), pp. 553, 1224 and 1237.

48 Salzmann (1965), p. 119.

49 Simmler (1566), f. 9r. Simmler does not date this undertaking, but speaks of a war that broke out and engulfed all of Germany, which can only mean the Schmalkaldic War.

my finances not been so limited, I would have gladly roamed the most distant lands and seacoasts with a burning desire to gain more knowledge. I did not have such luck, so I did the best I could. This meant making friends in various parts of Europe who kindly, generously and conscientiously passed along many accurate drawings of animals, along with their names and descriptions."[50]

On February 25, 1556, the Rostock theology professor David Chytraeus (1530–1600) wrote to Bullinger of his desire to invite Gessner to the Baltic Sea: "I and many fine scholarly gentlemen wish very much to see Mr Gessner walking along a beach on the Baltic Sea. He could conveniently journey here with the bookseller from Wittenberg after the Frankfurt book fair, from which a coach travels here once a month. Were he to present his books to Duke August, the Margrave of Brandenburg, and our princes and councils in the famous neighboring cities of Mecklenburg, he might very well cover his travel expenses. He need not fear the cold temperatures along the coast, for the winter here is undoubtedly milder than in the Alps. As our modest circumstances permit, we will pay for everything that the very learned, excellent and deserving man of church and science might need."[51]

Although he most certainly must have been pleased by the invitation, for unknown reasons, Gessner had to turn it down. He was possibly too busy, as he related to Johannes Culmann, personal physician to the Prince of Württemberg on June 25, 1563. The prince had invited him to Tübingen and had even made a horse and servant available.[52] This gesture on Culmann's part may have been motivated by the fact that some time earlier, he had been welcome to reside with Gessner for more than a year.[53] Gessner had already visited him once in Stuttgart in April 1559, during which stay he also used the opportunity to see Venerandus Gabler in Tübingen.[54]

Among the four working procedures (observe, describe, travel and dissect) of his colleagues that Gessner underscored, dissection was the area in which he strove to excel. He was not shy about examining the inner organs of animals he could get hold of in order to reach specific conclusions about them. This was how he disproved the claim by Aelianus and Pliny that the size of

50 Hanhart (1824), pp. 130f.

51 StAZH, E II 338, 1530. Chytraeus also sent Gessner the *Historia aquatilium* (Rostock 1517–1520) by Nikolaus Marschalk, but Gessner gave a scathing verdict on the images it contained. Cf. letter from Chytraeus to Bullinger (StAZH, E II 338, 1530) and Gessner (1558), f. b_4v.

52 Gessner (1577), f. 45r.

53 Letter from Gessner to Johannes Funck, January 21, 1565, cf. Gessner (1577), f. 96r.

54 According to entries no. 117 and 118 by Culmann and Gabler in Gessner's *Liber amicorum* (National Library of Medicine, Washington, MS. E 77).

a mouse's liver changed with the phases of the moon.[55] A systematic analysis of Gessner's bird book shows that from the 215 illustrated species, he had personally studied at least 65. Of these, he had dissected about 20, including the Eurasian hobby, snipe, kingfisher, curlew, great bustard, crane, Dalmatian pelican, teal, spoonbill, buzzard, red thrush, black stork, juniper thrush, wood ibis, smew and little grebe.[56] The subsequent insights he gained are exemplified by his description of the dissection of male and female buzzards. One of the buzzards had been examined during his stay in Lausanne, which shows that Gessner had been captivated by zoological issues already as a young man: "I once closely examined a female buzzard, that had been brought to me in Savoy. The beak was short, black and curved with the point being a mixture of green and yellow (which I had not found in others). The feet were yellow and covered with feathers nearly to the middle of the leg. The back was brown (*fuscum*), the outer feathers being mostly somewhat reddish. The inner parts of the feathers, which were covered and hidden on both sides were whitish. Because of many dense, ruffled and very soft little feathers, these were difficult to access. There were white feathers on the belly and breast, most of which had blackish spots through the middle. The feathers on the head had white edges. The primary feathers (*alarum maximae pennae*) were blackish, but all other wing feathers were whitish inside. The body length from the beak to the end of the feet, or even to the end of the tail, amounted to seven hand widths. On the whole, it was equivalent to the weight and size of an average-sized chicken (*mediocrem gallinam referre*). Upon cutting it open, I found no testicles and believe it to have been a female. Conversely, at another time, I cut open a male, who was of the same size as the earlier one with a weight of 30 ounces. It had different coloring on the forehead and in other places. The eyes had yellow irises. The legs, from the belly to the claws were about nine fingers long. The belly, sides and legs had white feathers that were separated by wide, red, transverse rings. The forward leaning part was a monochromatic dark chestnut brown. The tailfeathers were predominantly white, but were set off by areas of dark red color. A pair of equal sized testes (the label *triorchis*, or triple testes, does not correspond to reality), shaped like kidneys, were positioned on either side of the spine (*testes ad spinam dorsi inter utrosque*). The part in the middle was smaller than a testicle and whitish."[57]

55 Riedl-Dorn (1989), p. 56.

56 Springer/Kinzelbach (2009), pp. 403 and 426–546; Gmelig-Nijboer (1977), pp. 88–94.

57 Springer/Kinzelbach (2009), pp. 53f.

12.3 Zoo

As was common among princes and powerful men in the East and West, Gessner also owned his own private zoo. It was not, however, for recreational purposes but was used for observation and animal research. At times, individual bird species could be found there, such as great horned owls, jays, finches, wallcreepers, hoopoes, and crossbills.[58] According to Gessner, the young raven he fed with raw meat, little fish and moist bread[59] was not kept in a cage. Francesco Bonetti wrote to Bullinger on October 24, 1550 from Bergamo that he wished to send Gessner a dormouse. Apparently, it never ended up in his possession, or at least he never mentioned it in his animal books.[60] Among other pet animals, he owned snakes from Central France, which he received on November 17, 1564[61] and used for making theriac, a venom antidote. In a letter to Achilles Pirmin Gasser of December 3, 1565, he wrote that the snakes were in the best of health.[62] Additionally, it is possible that Gessner kept species of mice and rats, and perhaps even martens, beavers, gophers and groundhogs. The descriptions of them in the *Historia animalium* are so exact that he must have either owned or dissected them.[63] Gessner was not unique in this regard. The Basel Hebraist and cosmographer, Sebastian Münster, also made observations of groundhogs in his home. He wrote to Conrad Pellikan on May 11, 1544: "I am still expecting the Valais description that the bishop promised me. As proof, he is going to send me two living groundhogs. In order to research the nature of this animal, I wrote to the Bishop's nephew."[64] Without doubt, a special feature of Gessner's zoo were the two guinea pigs from the New World (ill. 31), colloquially called "Indian bunnies." Johann Heinrich Munzinger, a doctor from Augsburg, gave him a male and female. Gessner wrote in the German translation of the first volume of his animal book, printed in 1563: "Known as an 'Indian bunny' or *Seüwle*, they have come to our part of the globe in recent years from the New World and are now quite common. This is because it is a quite fecund animal, often giving birth to eight or nine offspring at a time. In size, it is somewhat smaller than the common rabbit and not as long, with short feet – six fingers in front and five toes in the back like mice. It is without a tail and comes in all sorts of colors. Its voice is like that of young pigs, and

58 Springer/Kinzelbach (2009), pp. 52, 455, 459 and 499.

59 Gessner (1555a), p. 322.

60 Letter from Bonetti to Bullinger, October 24, 1550 (StAZH, E II 361, 284).

61 Letter from Gessner to Adolph Occo, November 18, 1564, cf. Gessner (1577), f. 66r.

62 Burmeister (1975), p. 368.

63 Leu (1992), pp. 295f.

64 Burmeister (1964), p. 75.

Appendix. 63

LATINE Biſon. GERMANICE Wiſent.
Aliqui hoc genus bouis feri pro Vro accipiunt.

AD ORDINEM IIII. EORVNDEM.

LATINE Cuniculus Indicus.
ITALICE Conigli de l' India.
GALLICE Connin d' India.
GERMANICE Indianiſch Künele.

4

ILLUSTRATION 31 Guinea pig, in: Gessner (1553a), p. 63.
ZBZ, ABTEILUNG ALTE DRUCKE UND RARA: AW 1

for this reason they are also called Indian pigs. They eat all kinds of vegetation, fruit, bread and oats, preferring cabbage above all. Their young are born after wintertime and are not blind like other rabbits. They are very wanton and lustful, which is why they are so prolific. Males fight over the females. They require no liquids, and it suffices to spread a little water on their dry food. Otherwise, they can become swollen with excess fluid. Supposedly, they are good to eat, but I have found this not to be the case since they have a great deal of yellow fat or grease like a pig and also have moist, unwholesome flesh."[65]

12.4 Museum

Along with observation, description, travel and dissection, empirical access to nature could also be gained through collecting. Although he did not explicitly mention it, it was something Gessner cultivated throughout his life. Nonetheless, unlike many of his contemporaries, he was not partial to the bizarre and the exotic, and did not compile collections of objects from nature to create *Wunderkammern* (cabinets of curiosities). Gessner's systematic collecting activities were meant to expand morphological knowledge about every object from the three branches of nature (minerals, plants, animals). Naturally, he tried to acquire everything possible, but he did not collect simply out of curiosity or cupidity. Instead, he was bound by the goal of pushing deeper into the treasures of nature in order to determine characteristics, links and relationships. Unfortunately, we have neither an inventory nor reports from his contemporaries about Gessner's collection of natural objects. Therefore, the following list cannot be considered to be complete. From his letters and books we know that aside from albums of botanical, geological and zoological illustrations, which we will discuss later, Gessner also possessed many physical objects. This included a broad variety of minerals and fossils, which he has depicted in his *De rerum fossilium liber* of 1565. Of these, six objects are now found in Basel's Museum of Natural History, including a fossilized crab and a coin made of sealing clay containing Arabic writing.[66] Of special interest in his earth science collection was a 30 cm high *Mons metallicus* (mountain of metal).[67] It was made from different metals and stones and was admired by Gessner's friends as well as craftsmen and goldsmiths from Zürich. Gessner

65 Gessner (1563a), f. 73v.
66 Leu (1994); Etter/Schneider (2016).
67 Gessner (1584), f. B_1v.

had received the precious piece together with insects crafted from silver[68] from Valentinus Gravius, a city councilman from the mountain town of Freiberg in Saxony.[69] It was a reciprocal gift for the dedication of the second volume of the *Historia animalium*. Gessner had dedicated the book to him in return for the numerous illustrations he had sent of animals from the North and Baltic Seas.[70] The gift from Gravius was probably a so-called *Handstein*, an artistically arranged and decorated piece of ore or mineral specimen, an item often collected and displayed by princes or people of wealth.

Additionally, Gessner possessed a four-digit number of dried plants, seeds and fruits. His museum also contained items found on beaches such as corals, mollusks and snails, including the genera *Strombus, Pecten, Tellina* and *Nerites*.[71] Scholars from England to Italy even sent each other dried fish specimens, some of which also made their way to Gessner.[72] Among these were two small spinenose horsefish (*Congiopodus sp.*), which are found in the South Atlantic and South Pacific. Gessner's examination correctly concluded that they had nothing to do with the fish *Porcilius* cited by the Greek geographer Pausanias (ca. 115–180).[73] How the Augsburg doctor Gereon Seiler († 1562) sent Gessner a sturgeon by way of Bullinger was not detailed in his letter to the head of the Zürich church.[74] Gessner also possessed the skeleton of a reddish flying fish that he probably disposed of since unmacerated bones do not keep well.[75] Along with other aquatic animals, he was also in possession of several bird skins. More than a third of the illustrations in his bird book were based on such bird skins,[76] some of which were found in his collection, such as the pink pelican (*Pelecanus onocrotalus*), which he observed on Lake Zug.[77] Gessner had a presumably stuffed specimen sent to his friend Johannes Fabricius Montanus (1527–1566) in Chur, which was returned to him on March 20, 1561.[78] It is likely that Gessner himself had stuffed several birds after dissecting them. He had probably observed the process as performed by his father, the furrier Urs Gessner. From Belon, he received the skeleton of a night swallow;[79] from

68 Kusukawa (2014), p. 335.
69 Gessner (1577), f. 131r. Cf. Kusukawa (2012), pp. 170–173.
70 Gessner (1554), f. $*_2$r/v.
71 Letter from Gessner to John Caius, August 29, 1561, in: Gessner (1577), f. 133v–136v.
72 Pinon (2002), pp. 485f.
73 Gessner (1577), f. 2v.
74 Letter from Seiler to Bullinger, October 19, 1557 (StAZH, E II 346, 354).
75 Gessner (1558), p. 516.
76 Springer/Kinzelbach (2009), p. 403.
77 Gessner (1555a), p. 605.
78 Gessner (1577), f. 90r.
79 Pinon (1995), p. 15.

the Grison Alps, the skin of a bearded vulture, complete with skin and legs;[80] from John Caius, a hawk's foot;[81] and from Johannes Ferrerius, an astronomer and philologist from the Piedmont, the beak of a Brazilian tucan.[82]

There were also items from animals of the four-legged kind to be found in Gessner's collection. These included a sealskin,[83] the horns of a Polish bison (*Bison bonasus*),[84] a boar's tooth[85] and moose antlers.[86] He asked John Caius in England for the antlers of a fallow deer and promised him the horns of a Swiss ibex in return.[87] He would have loved to have the antlers of a reindeer,[88] but the delivery did not work out.[89] He nurtured justified doubts about the illustration of an animal with three horns (ill. 32) that he received from the Swedish bishop and geographer Olaus Magnus (1490–1557). At some point after the first volume of the *Historia animalium* was published, Anton Schneeberger informed him that the animal only had two antlers.[90] In this case and others, Olaus Magnus was following old reports and legends, but his work on northern peoples contained more or less the only tangible information available concerning these remote areas. He wrote about the reindeer's three antlers: "Around midnight in both the Bothnian region and in greater Lappland, there is an animal with three antlers that is a species of deer. However, it is taller, stronger and faster and is called a *Reyniger* (reindeer) ... the two antlers (similar to that which deer have) are larger than the other, that is, longer with more points, as many as 15. The third antler, however, which is located in the middle of the head, has shorter points and provides ample defense of the head against wild animals, especially wolves."[91]

In his efforts to possess reindeer antlers, Gessner wanted to definitively settle the issue with an autopsy of such an animal. It was certainly the same empirical impulse that drove his interest in a unicorn horn that was hanging from

80 Gessner (1555a), p. 750; Springer/Kinzelbach (2009), p. 537.

81 Fischer (1966), p. 45.

82 Leu (1992), pp. 298f.

83 Letter from Gessner to Theodor Zwinger from 1565 (UB Basel, Frey-Gryn Mscr I 13: no. 83).

84 Gessner (1560), p. 31. Cf. also the use of the *Carmen de Bisonte* by Nicolaus Hussovianus (Choptiany [2013]).

85 Gessner thanks Felix Platter on May 16, 1561, for the boar's tooth. Cf. Gessner (1577), f. 97v.

86 Gessner (1560), p. 55. The donor was the Polish nobleman Hans Boner the Younger (1516–1562), who was a friend of Gessner's student Anton Schneeberger. Cf. Fischer (1966), pp. 114f.

87 Letter from Gessner to Caius, August 29, 1561, in: Gessner (1577), f. 133v–136v.

88 Letter from Gessner to Crato von Krafftheim, November 12, 1563. Gessner (1577), f. 15r.

89 Letter from Gessner to Crato von Krafftheim, August 1, 1563. Gessner (1577), f. 13r.

90 Gessner (1551), f. 950–952; Gessner (1560), p. 59.

91 Magnus (2006), pp. 277f.

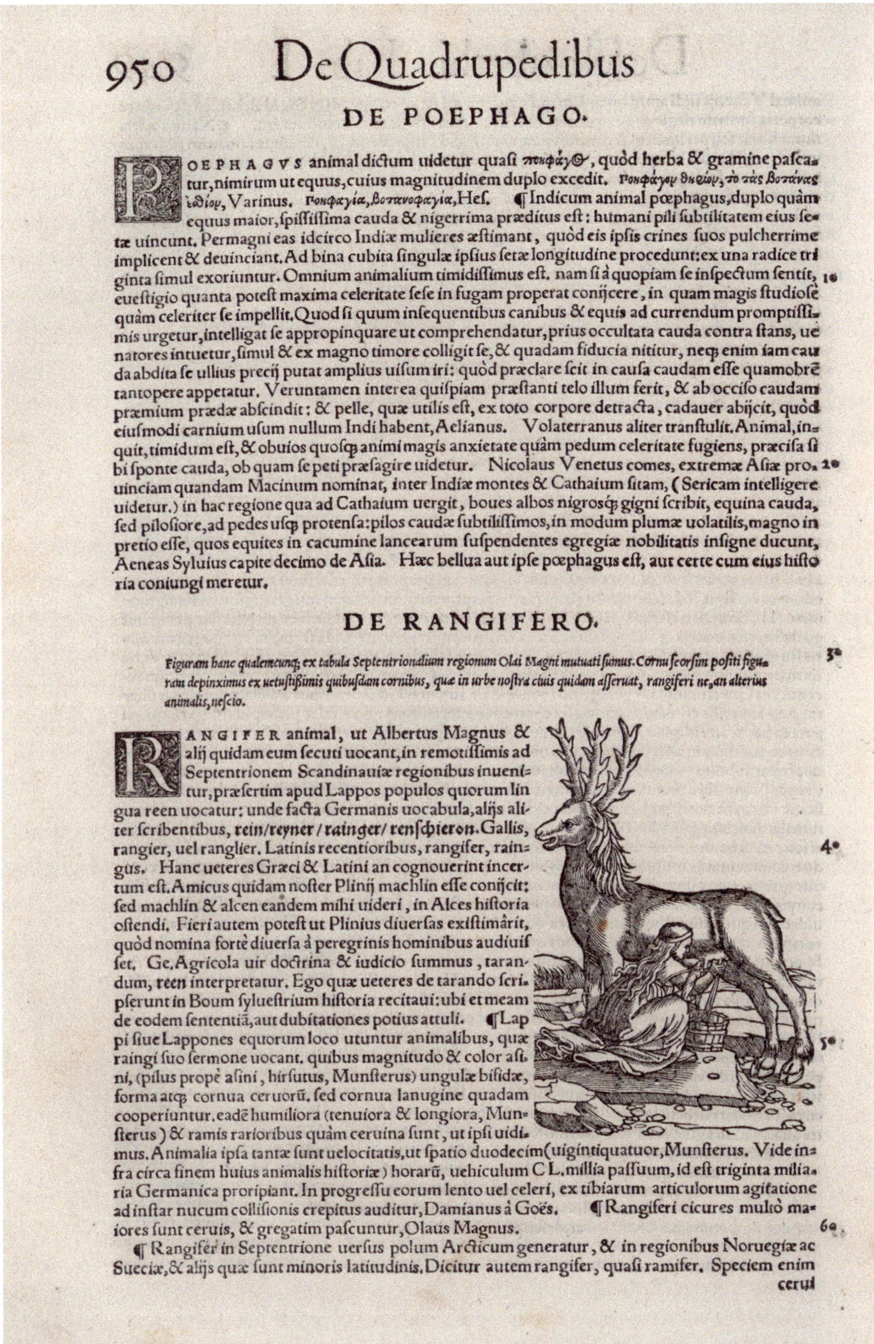

950 De Quadrupedibus

DE POEPHAGO.

POEPHAGVS animal dictum uidetur quaſi ποηφάγος, quòd herba & gramine paſcatur, nimirum ut equus, cuius magnitudinem duplo excedit. ποηφάγον θηρίον, τὸ τὰς βοτάνας ἐσθίον, Varinus. ποηφαγία, βοτανοφαγία, Heſ. ¶ Indicum animal pœphagus, duplo quàm equus maior, ſpiſſiſſima cauda & nigerrima præditus eſt: humani pili ſubtilitatem eius ſetæ uincunt. Permagni eas idcirco Indiæ mulieres æſtimant, quòd eis ipſis crines ſuos pulcherrime implicent & deuinciant. Ad bina cubita ſingulæ ipſius ſetæ longitudine procedunt: ex una radice triginta ſimul exoriuntur. Omnium animalium timidiſſimus eſt. nam ſi à quopiam ſe inſpectum ſentit, e ueſtigio quanta poteſt maxima celeritate ſeſe in fugam properat conijcere, in quam magis ſtudioſè quàm celeriter ſe impellit. Quod ſi quum inſequentibus canibus & equis ad currendum promptiſſimis urgetur, intelligat ſe appropinquare ut comprehendatur, prius occultata cauda contra ſtans, uenatores intuetur, ſimul & ex magno timore colligit ſe, & quadam fiducia nititur, neq; enim iam cauda abdita ſe ullius precij putat amplius uiſum iri: quòd præclare ſcit in cauſa caudam eſſe quamobrē tantopere appetatur. Veruntamen interea quiſpiam præſtanti telo illum ferit, & ab occiſo caudam præmium prædæ abſcindit: & pelle, quæ utilis eſt, ex toto corpore detracta, cadauer abijcit, quòd eiuſmodi carnium uſum nullum Indi habent, Aelianus. Volaterranus aliter tranſtulit. Animal, inquit, timidum eſt, & obuios quoſq; animi magis anxietate quàm pedum celeritate fugiens, præciſa ſibi ſponte cauda, ob quam ſe peti præſagire uidetur. Nicolaus Venetus comes, extremæ Aſiæ prouinciam quandam Macinum nominat, inter Indiæ montes & Cathaium ſitam, (Sericam intelligere uidetur.) in hac regione qua ad Cathaium uergit, boues albos nigroſq; gigni ſcribit, equina cauda, ſed piloſiore, ad pedes uſq; protenſa: pilos caudæ ſubtiliſſimos, in modum plumæ uolatilis, magno in pretio eſſe, quos equites in cacumine lancearum ſuſpendentes egregiæ nobilitatis inſigne ducunt, Aeneas Syluius capite decimo de Aſia. Hæc bellua aut ipſe pœphagus eſt, aut certe cum eius hiſtoria coniungi meretur.

DE RANGIFERO.

Figuram hanc qualemcunq; ex tabula Septentrionalium regionum Olai Magni mutuati ſumus. Cornu ſeorſim poſiti figuram depinximus ex uetuſtiſſimis quibuſdam cornibus, quæ in urbe noſtra ciuis quidam aſſeruat, rangiferi ne, an alterius animalis, neſcio.

RANGIFER animal, ut Albertus Magnus & alij quidam eum ſecuti uocant, in remotiſſimis ad Septentrionem Scandinauiæ regionibus inuenitur, præſertim apud Lappos populos quorum lingua reen uocatur: unde facta Germanis uocabula, alijs aliter ſcribentibus, **rein/reyner / rainger/ renſchieron**. Gallis, rangier, uel ranglier. Latinis recentioribus, rangifer, raingus. Hanc ueteres Græci & Latini an cognouerint incertum eſt. Amicus quidam noſter Plinij machlin eſſe conijcit: ſed machlin & alcen eandem mihi uideri, in Alces hiſtoria oſtendi. Fieri autem poteſt ut Plinius diuerſas exiſtimârit, quòd nomina fortè diuerſa à peregrinis hominibus audiuiſſet. Ge. Agricola uir doctrina & iudicio ſummus, tarandum, **reen** interpretatur. Ego quæ ueteres de tarando ſcripſerunt in Boum ſylueſtrium hiſtoria recitaui: ubi et meam de eodem ſententiā, aut dubitationes potius attuli. ¶ Lappi ſiue Lappones equorum loco utuntur animalibus, quæ raingi ſuo ſermone uocant. quibus magnitudo & color aſini, (pilus propè aſini, hirſutus, Munſterus) ungulæ biſidæ, forma atq; cornua ceruorū. ſed cornua lanugine quadam cooperiuntur. eadē humiliora (tenuiora & longiora, Munſterus) & ramis rarioribus quàm ceruina ſunt, ut ipſi uidimus. Animalia ipſa tantæ ſunt uelocitatis, ut ſpatio duodecim (uigintiquatuor, Munſterus. Vide infra circa finem huius animalis hiſtoriæ) horarū, uehiculum C L. millia paſſuum, id eſt triginta miliaria Germanica proripiant. In progreſſu eorum lento uel celeri, ex tibiarum articulorum agitatione ad inſtar nucum colliſionis crepitus auditur, Damianus à Goës. ¶ Rangiferi cicures multò maiores ſunt ceruis, & gregatim paſcuntur, Olaus Magnus.

¶ Rangifer in Septentrione uerſus polum Arcticum generatur, & in regionibus Noruegiæ ac Sueciæ, & alijs quæ ſunt minoris latitudinis. Dicitur autem rangifer, quaſi ramifer. Speciem enim

ceruí

ILLUSTRATION 32 Three-horned reindeer, in: Gessner (1551), p. 950.
ZBZ, ABTEILUNG ALTE DRUCKE UND RARA: NNN 41

a column of the Strasbourg Cathedral. Having heard of it, he asked Nikolaus Gerbel (ca. 1485–1560) on February 15, 1551 to inform him as to how it came to be there, what it was made of, whether it was hard or soft, how heavy it was and how it smelled. Unfortunately, Gerbel did not fulfill the request, which is why during a trip to Strasbourg in the spring of 1559,[92] Gessner had a look at the tusk for himself. He published[93] the fine specimen in the *Icones animalium* of 1560 and noted (ill. 33) that with the help of a string, he was able to measure along its curve, the length being nearly four Roman *Ellen* (about two meters). He scratched it with a knife, and it appeared to be a genuine horn. He wrote, however, that it was not from a unicorn, but rather from a very large and old aurochs.[94] In reality, it was a mammoth tusk. The relics from such animals were often interpreted as the remains of unicorns and were superstitiously revered in churches.[95]

Gessner also possessed various four-legged animals from the New World, that surely caused a sensation in Zürich. Adrianus Marsilius, a pharmacist from Ulm, sent him the shell, tail and claws from a nine-banded armadillo,[96] and Johannes Ferrerius sent him the skin from a whiptail lizard (without the head).[97] Wherever his personal collection was lacking, Gessner may have borrowed certain natural items for teaching or study purposes. For example, in 1559, Felix Platter in Basel sent him at his request a swordfish head,[98] and in 1560 a nautilus shell (*Nautilus pompilius*).[99] Following the example of Rondelet, Gessner erroneously labeled it a "Greater Argonaut" (*Argonauta argo*) in his animal books.

In 1558, Gessner had the upper floor of his two-storey house at Frankengasse 6 enlarged for use as a study (*coenaculum*). It had 15 windows with stained glass images of aquatic fauna.[100] An impression of how they might have looked is

92 Gessner (1584), f. C_1r/v.
93 Gessner (1560), p. 34.
94 Leu (1999), p. 18.
95 Leu (1999), p. 18.
96 Gessner (1560), p. 104.
97 Leu (1992), p. 299.
98 Gessner (1577), f. 97v.
99 Dill (2010), pp. 193–195.
100 Gessner's letter to Kentmann, 27 February 1559. Gessner (1584), f. B4r. On the 1576 city view by Jos Murer, Gessner's *Haus zum Sonnenzeit* has four floors, but the top third floor under the roof does not appear to have been used as living space. The 15 stained glass windows must also have been located on the second floor due to space constraints. Cf. Guyer (1973), p. 79f.; Ruoss (2016).

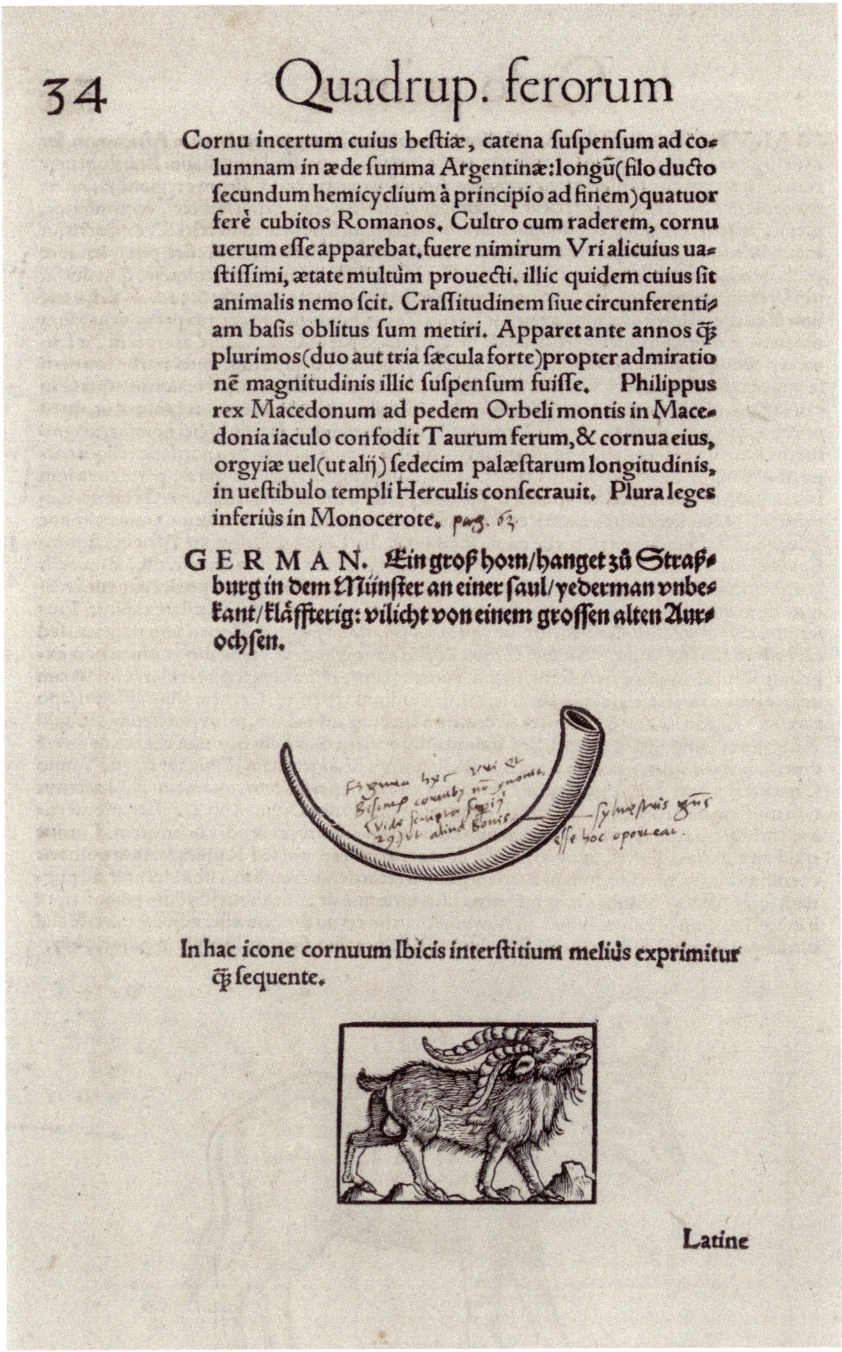

34 Quadrup. ferorum

Cornu incertum cuius beſtiæ, catena ſuſpenſum ad columnam in æde ſumma Argentinæ: longū (filo ducto ſecundum hemicyclium à principio ad finem) quatuor ferè cubitos Romanos. Cultro cum raderem, cornu uerum eſſe apparebat. fuere nimirum Vri alicuius uaſtiſſimi, ætate multùm prouecti. illic quidem cuius ſit animalis nemo ſcit. Craſſitudinem ſiue circunferentiam baſis oblitus ſum metiri. Apparet ante annos ꝙ plurimos (duo aut tria ſæcula forte) propter admirationē magnitudinis illic ſuſpenſum fuiſſe. Philippus rex Macedonum ad pedem Orbeli montis in Macedonia iaculo confodit Taurum ferum, & cornua eius, orgyiæ uel (ut alij) ſedecim palæſtarum longitudinis, in ueſtibulo templi Herculis conſecrauit. Plura leges inferiùs in Monocerote.

GERMAN. Ein groß horn/hanget zů Straßburg in dem Münſter an einer ſaul/yederman vnbekant/kläffterig: vilicht von einem groſſen alten Aurochſen.

In hac icone cornuum Ibicis interſtitium meliùs exprimitur ꝙ ſequente.

Latine

ILLUSTRATION 33 Mammoth tusk, in: Gessner (1560), p. 34, with Gessner's handwritten notes.

ZBZ, ABTEILUNG ALTE DRUCKE UND RARA: NNN 44

ILLUSTRATION 34 Stained glass design, probably for Gessner's study room.
ZBZ, HANDSCHRIFTENABTEILUNG: MS. P 66

provided by the design drawings attributed to the artist Grosshans Thomann (ill. 34).[101] Additionally, there were various gardens outside of Gessner's home in which he grew plants for research purposes. He also took a certain amount of pride in his natural history collections. On March 18, 1560, he invited Didymus Obrecht from Strasbourg to visit him in Zürich and view his garden along with the many metals, stones, gems, animals, images and the stained glass windows depicting aquatic animals.[102]

12.5 Organization

Gessner systematically organized his massive collection of materials into six main groups based on Aristotelian classifications. These were live-bearing four-legged animals, egg-laying four-legged animals, birds, aquatic animals, serpents and insects. Unlike modern zoological classification, he published bats together with birds,[103] mammals living in watery habitats (dolphins, fish otters, hippopotamuses, whales, walrusses, etc.) as aquatic animals, and scorpions were grouped with insects.[104] The book about snakes was published posthumously in Zürich in 1587 to which a fascicle about scorpions with its own title page was added. The planned work about insects remained unfinished. Gessner's material for this made its way through Joachim Camerarius the Younger to London, where Thomas Penny (1532–1589) continued to work on it together with notes from Eduard Wotton (1492–1555). When Penny died, his entomological estate went to his young friend Thomas Mouffet (1553–1604), who collected everything in a complete manuscript that was published in London in 1634 (ill. 35 and 36).[105]

Within the individual volumes, Gessner ordered the animals alphabetically according to their Latin names and presented some that were closely related in summary form. For example, he sub-categorized the aurochs, buffalo, cow, steer and bison under the Latin title *Bos* (buffalo), or had the hawks immediately followed by the falcons in the bird book.[106] The alphabetical arrangement

101 Naegeli (1981), p. 115.

102 Gessner (1577), f. 114v–115r.

103 Cf. Hinkel (1999).

104 The scorpions are depicted with eight legs instead of six, which makes them look like spiders. Cf. Gessner (1587), p. 3.

105 Fischer (1966), pp. 50–54. The magnificently illustrated manuscript of this is kept in the British Library in London, call number: Sloane MS. 741 or 4014 (some manuscripts from the Sloane collection have two call numbers).

106 Gessner (1555a), f. a_3v.

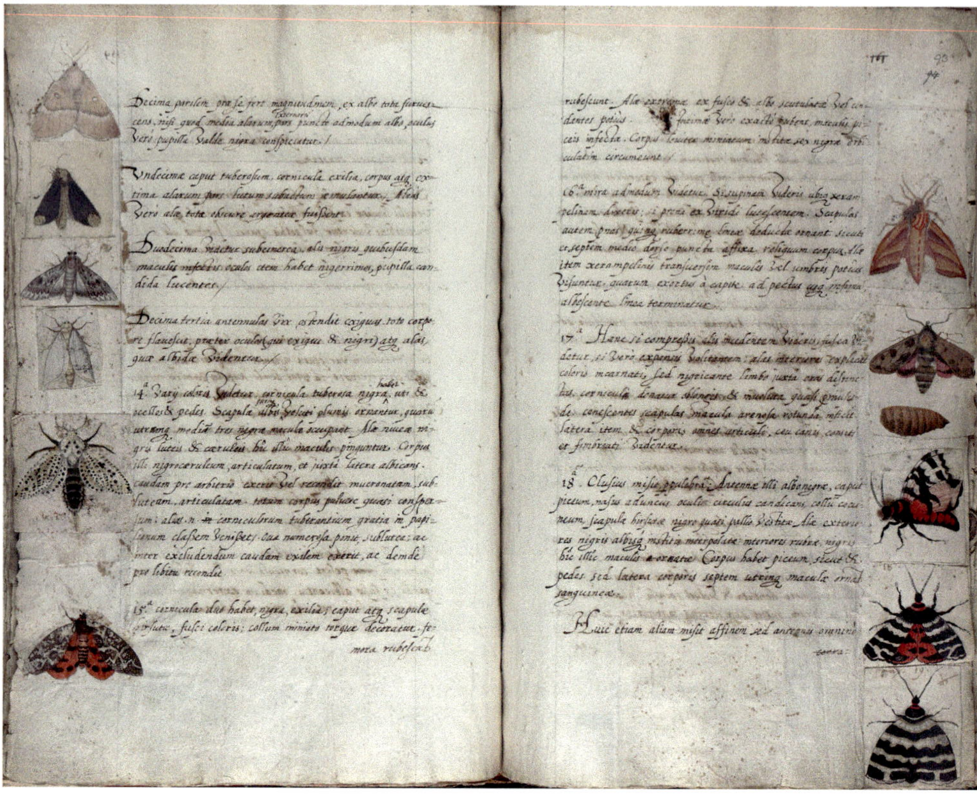

ILLUSTRATION 35 Print manuscript of Thomas Mouffet's work on insects.
BRITISH LIBRARY LONDON: SLOANE, MS. 741, P. 93

and the extensive lists in various languages made it easier to find a specific animal. They also show that Gessner's animal books were meant to be a comprehensive reference work. He wrote in the foreword to the reader in the first volume that those who wished to continue learning more about the animal world should read Aristotle or similar authors. His work was intended for use as a lexicon or reference book.[107]

In the illustration books, with their highly abbreviated texts, he set up a more sophisticated system, which was structured according to specific morphological similarities. For example, he distinguished between animals with or without horns. It resulted in the four-legged animals being organized into

107 Gessner (1551), f. β_2r.

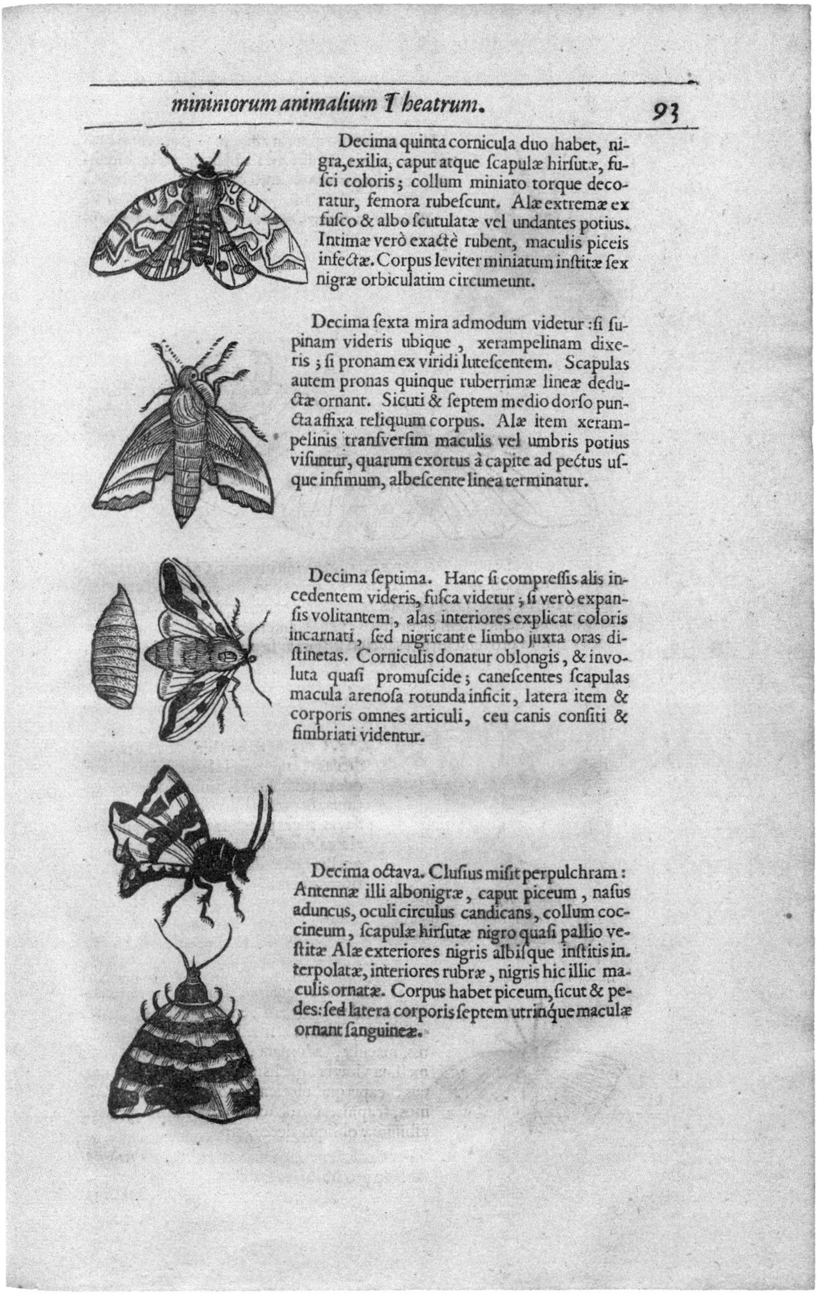

minimorum animalium Theatrum. 93

Decima quinta cornicula duo habet, nigra, exilia, caput atque ſcapulæ hirſutæ, fuſci coloris; collum miniato torque decoratur, femora rubeſcunt. Alæ extremæ ex fuſco & albo ſcutulatæ vel undantes potius. Intimæ verò exactè rubent, maculis piceis infectæ. Corpus leviter miniatum inſtitæ ſex nigræ orbiculatim circumeunt.

Decima ſexta mira admodum videtur: ſi ſupinam videris ubique, xerampelinam dixeris; ſi pronam ex viridi luteſcentem. Scapulas autem pronas quinque ruberrimæ lineæ deductæ ornant. Sicuti & ſeptem medio dorſo punctaaffixa reliquum corpus. Alæ item xerampelinis tranſverſim maculis vel umbris potius viſuntur, quarum exortus à capite ad pectus uſque infimum, albeſcente linea terminatur.

Decima ſeptima. Hanc ſi compreſſis alis incedentem videris, fuſca videtur; ſi verò expanſis volitantem, alas interiores explicat coloris incarnati, ſed nigricante limbo juxta oras diſtinctas. Corniculis donatur oblongis, & involuta quaſi promuſcide; caneſcentes ſcapulas macula arenoſa rotunda inficit, latera item & corporis omnes articuli, ceu canis conſiti & fimbriati videntur.

Decima octava. Cluſius miſit perpulchram: Antennæ illi albonigræ, caput piceum, naſus aduncus, oculi circulus candicans, collum coccineum, ſcapulæ hirſutæ nigro quaſi pallio veſtitæ Alæ exteriores nigris albiſque inſtitis interpolatæ, interiores rubræ, nigris hic illic maculis ornatæ. Corpus habet piceum, ſicut & pedes: ſed latera corporis ſeptem utrinque maculæ ornant ſanguineæ.

ILLUSTRATION 36 Thomas Mouffet, *Insectorum sive minimorum animalium theatrum*, London 1643, p. 93. Four of the five butterflies from the print manuscript have been reproduced on this page (the second from the bottom has been omitted).

ZBZ, ABTEILUNG ALTE DRUCKE UND RARA: TZ 92

seven categories,[108] the birds into eight[109] and the aquatic animals into 19.[110] Gessner's colleague, Johannes Fabricius Montanus, attempted to establish a system for the animal world that was as natural as possible in his 358-page work *Differentiae animalium quadrupedum secundum locos communes* (*Variations Among Four-legged Animals According to General Criteria*), published in 1555. Fabricius Montanus mentioned that Gessner had helped him, but it was not a zoological system in the modern sense. It was rather an accounting of the differences and similarities with regard to nutrition, types of habitats, character and other aspects. Part two of this work with the title *Similitudinum ab omni animalium genere desumptarum libri VI* (*Six Books about the Similarities Among All Species of Animals*) was written by Otto Werdmüller, who died in 1552. Gessner wrote the dedication to Werdmüller's son Abel, in which he compared the work to a collection of zoological *loci*.[111] Fabricius Montanus's work possibly corresponded to the sixth part of the *Historia animalium* that Gessner had originally intended to write on the topic of *De animalibus secundum genera et species*.[112]

The *Icones* volumes were not just a strategic commercial move by Froschauer, in which some or all of the already published illustrative material was used again to turn a second profit. To an extent, it was also a pictorial summary of the then-known animal world. Gessner's plan was to follow the extensive volumes with short texts and woodcuts for students. He explained the function of the volumes with comprehensive texts such as the *Historia animalium* and referred to them as *Pandectae*. He labeled the planned summaries *Compendia*. The following is from the foreword in the third volume of the *Historia animalium* about birds: "It would be a worthwhile effort to write a large, all-inclusive book about every area of knowledge, which I would give the title '*Pandectae*.' … Such volumes would be intended for use in libraries as reference books for scholars and professors. They are not suitable for continuous reading, but can be used more as lexicons in which everything of value might be found that has ever been written on a given subject. Their greatest application is to save the interested party from having to look things up in many books and do their own analytical research. In other words, they are intended to replace an entire library. Not to be discounted is the fact that students would also have access

108 Gessner (1560), p. 8.

109 Gessner (1555b), p. 127.

110 Gessner (1560a), f. aa_4r/v.

111 Fabricius Montanus (1555), pp. 121–126.

112 Gessner (1548), f. 221r, planed to write a *Historia animalium* with six parts: I. De quadrupedibus viviparis et oviparis; II. De avibus; III. De piscibus et aquatilibus; IV. De serpentibus; V. De insectis; VI. De animalibus secundum genera et species.

to all the vocabulary necessary for working on a specific topic. Based on such a *Pandectae*, a *Compendium* should be created to more briefly summarize that area of knowledge. Here, only the results are to be presented, and everything related to philology is omitted with no authors being credited. For this, only Latin should be used, and adequate attention should be given to the elegance of style. Such a *Compendium* is meant primarily for student use. For more exact information, it is also possible to refer back to the *Pandectae*, which to an extent can also serve as a kind of annotation to the *Compendium*."[113]

Since the illustration books fulfilled a great deal of the *Compendia*'s function, Gessner may have deliberately refrained from releasing them. In any case, they were never published. He did attribute a certain entertainment value to the *Icones* volumes. They were to offer a respite from more demanding studies and stimulate more appreciation for nature and the Creator.[114]

While Gessner's zoological classification was replaced by other concepts no later than the seventeenth century, some of his Latin names for animals have survived among specialists up to the present. Carl Linnaeus (1707–1778), who introduced the modern binominal nomenclature in the 10th edition of his *Systema naturae* of 1758, utilized many of Gessner's Latin names either in part or in full.[115] The same problem existed in the animal world as in the plant world that was addressed by Gessner in his 1542 *Catalogus plantarum*. He tried to summarize the countless confusing botanical names, terms and expressions from various scholarly languages and vernaculars into a Latin-Greek-German-French concordance. Plato famously said that the Greeks had settled all around the Mediterranean like frogs around a pond, but the Earth is much larger.[116] This led to the Greeks and Romans giving descriptions and names to the flora and fauna of the Mediterranean, but not the plants and animals in the rest of the world, not to mention the many new discoveries from overseas.[117] Thus, in Gessner's time, there were no universal designations, a problem that was first ingeniously solved by Linnaeus with the internationally recognized Latin binary nomenclature.

Gessner strove intensively to gather the different names of plants and animals in order to compare them and find out what exactly they might refer to, and which labels from both ancient and modern times would be appropriate to each organism. This terminological chaos is the reason that each animal book

---

113 Fischer et al. (1967), pp. 128f.

114 Gessner (1555b), [p. 2].

115 Springer/Kinzelbach (2009), pp. 76–79; Ohl (2015), p. 55.

116 Plato, Phaidon, 109b.

117 Cf. the corresponding statement by Gessner in his letter to Johannes Fabricius Montanus in Chur of August 2, 1558 (ZBZ, MS. C 50a, no. 25/3). So also in: Gessner (1560a), p. 279.

begins with an index of names in Hebrew, Greek and Latin. It is followed by an alphabetical directory in various vernaculars such as Arabic, German, English, French, Illyrian,[118] Italian, Persian, Polish or Spanish so that those with some degree of education would be able to find the desired animal. Additionally, each article began with an extensive explanation of the designations in various languages. Gessner even took into account the animal symbols in Egyptian hieroglyphics. His Egyptological discussion was based on *Hieroglyphica* by the late ancient philosopher Horapollo as well as a work with the same title from the pen of the humanist Giovanni Pierio Valeriano Bolzani. Gessner examined their sometimes bold claims with a critical eye, which led him to correct these authors in certain passages. For instance, Horapollo and others were of the opinion that moles had no eyes and were blind.[119] Gessner contradicted them, citing his own examinations.[120]

A good example of Gessner's studies in nomenclature is found in a passage from the foreword of the book of fish illustrations. There he writes: "My primary effort was directed at giving every animal a specific, intrinsic name, and this had to occur comprehensively in the book of aquatic fauna. In part, this is because there is such a large number of fish species and also because their names have often been wrongly identified. Frequently, a fish would be given multiple names in the same language. For river and lake fish, I sometimes had to come up with my own names, especially in German, which I usually indicated. One who is knowledgeable in natural history must be allowed to invent names when none are to be found."[121]

A little known early effort in his work on nomenclature for the fish book was a piece published in 1556 titled *Teütsche nammen der Fischen und Wasserthieren* (*German Names of Fish and Aquatic Animals*). This book is renowned as Gessner's only German publication, but here as well, most pages were written in Latin. This was the third title in his anthology *De piscibus et aquatilibus omnibus libelli III*, which was published by his cousin Andreas Gessner the Younger.[122] The book not only contained the oldest record of Saxon fish fauna by Johannes Kentmann,[123] it also clarified more than 100 fish names that were common to the German and English-speaking regions. Gessner correctly surmised that, to a degree, English and German were related: "We found

118 "Illyria" refers to the geographical area of Dalmatia and Albania.

119 According to current knowledge, the mole has eyes, but they probably only allow it to distinguish between light and dark.

120 Mauelshagen (2003), especially p. 229. Cf. for example Gessner (1560a), p. 355.

121 Gessner (1974), p. 21.

122 Bäumer (1990). Cf. Blair (2019).

123 Zaunick (1915).

English to be similar to the German language, so where German names were missing, we also used many English names."[124] Furthermore, Gessner says for example that the so called *Egli*, a food fish popular in Switzerland, is also named *Triechter Egle* when caught in the deeper parts of a lake. If the animal is brown on its sides, it is called a *Landegle, Roregle* or *Kraebegle*. Around Lake Constance, the smaller variety is called *Hürling* and the larger ones are named *Kretzer, Schaubfisch, Egle* or *Renckernegle*, depending on their weight.[125] One could also read that the German *Flunder* was called a *Flownder* in England and a *Flesum* in France.[126]

Gessner accumulated much of this information through a wide-ranging correspondence with hundreds of European scholars. Probably because of an urgent need to finish the fish book, on October 8, 1557, he wrote in somewhat of a rush to the Basel philosophy professor Johannes Hospinian (1515–1575): "I hear that in your area there is a fish that is larger than a lamprey and is called a *Semperin* (or *Lemperin*). I implore you to make inquiries on my behalf about this name with a fisherman. At the same time, please give me the names of all the fish that are caught in the Rhine by meeting with an old expert fisherman or having him come to call. Do not skip over even the smallest, least important ones, and if he can recall some that go by other names in other locales, please include these as well."[127]

Hospinian fulfilled the request expeditiously, and Gessner apologized on March 5, 1558 for taking so long to thank him.[128] A few weeks after the appeal to Hospinian, John Caius sent Gessner the English names of various fish.[129] Johannes Kentmann sent documentation about the fish of the Elbe River to his colleague in Zürich,[130] and the Hannover physician Burkhard Mythobius (1501–1564) did likewise for the Baltic Sea.[131] Gessner was also not shy about quizzing his visitors about common animal names in their home areas.[132] His *Liber amicorum* (guest book) collected the signatures of 227 visitors from

---

124 Gessner derived German fish names from English, cf. the dedicatory epistle to Caius in: Gessner (1556), p. 13.

125 Gessner (1556), pp. 115f.

126 Gessner (1556), p. 119.

127 Gessner (1577), f. 102r.

128 Gessner (1577), f. 102r.

129 Letter from Caius to Gessner, November 1, 1557, printed in: Gessner (1558), pp. 1294–1296.

130 Cf. e.g. E.g., Gessner (1584), f. C_1r/v. Cf. also Hackethal (1994), pp. 284f., and Zaunick (1915). Gessner agreed to find a printer for Kentmann for his work on the Elbe, but he advised him that the work would be better printed near him because of the corrections that would have to be made. Cf. Gessner (1584), f. B_2v–C_3r.

131 Gessner (1556), p. 94.

132 Gessner (1974), p. 15.

1555 to 1565. The entries included 27 Swiss, 57 Germans, 22 Frenchmen, 17 Dutchmen, 13 Poles, 8 Hungarians, 7 Italians, 4 Bohemians, 4 Englishmen, 4 Spaniards and a few Scots, Irishmen and Ukrainians.[133]

Gessner divided each chapter about an animal into eight subchapters that were labeled 'A' through 'H.' If necessary, due to a surplus of material, the eighth chapter would be further divided into eight more subdivisions labeled 'a' through 'h.' He characterized the main sections as follows:[134]

A: Naming the animal, if possible, in 13 languages with a list of synonymous labels. If a German or Latin name were missing, he would often create one himself.

B: The geographic range of the animal; the occurrence of possible varieties of the species and their differences. The morphology and anatomy.

C: Habitat, lifestyle, physiology, signs of good health and its maintenance. Reproduction, birth, rearing, life expectancy.

D: Emotional life, customary behavior, instincts, positive and negative characteristics, positive and negative dispositions toward members of the same species, humans and objects.

E: Use of the animal by humans (excluding for food or medicine). How to hunt, catch and tame the animal, how to care for it, feed it and how to use it as either a pet or work animal. How to use individual parts such as its pelt, hooves or waste.

F: Food products to be derived from the animal.

G: Medicines to be derived from the animal. Bites, blows and injuries caused by the animal and how to cure them.

H: Philological questions regarding the animal, etymology and metaphoric application of its name. Illustrations and sculptures of the animal. People and geographic areas named after it. This is followed by sections which deal with all the topics from A to G in a purely philological way. The final sections are related stories, fables and proverbs.[135]

Unfortunately, Gessner never said how he arrived at this structure. Certain aspects are a result of the historical preoccupation with the animal world since antiquity. The hunting and use of animals had been a topic for centuries with regard to pets, work animals, resources for clothing, nutrition and

133 Durling (1965). Gessner's *Liber amicorum* is preserved in the National Library of Medicine in Washington (USA).

134 Gessner (1551), f. γ_1v–γ_3r.

135 Fischer (1967), pp. 129f.

medications.[136] However, Gessner deserves credit for comprehensively presenting the material in this form. Remarkably, nearly 200 years earlier, an Arab theologian in Cairo named Muḥammad Ibn Mūsā ad-Damīrī (ca. 1341–1405) had written a two-volume animal lexicon, titled *Ḥayāt al-hayawān*.[137] His discussion of the following seven quite similar points about individual animals is much shorter, less scientific and covers considerably more superstitious and mythological aspects than Gessner's:

A: The animal's name and its meaning.
B: Description of the animal and its behavior.
C: Reports from Islamic tradition.
D: Legal regulations regarding the animal as food or in everyday use.
E: Sayings.
F: Medicinal uses.
G: Interpretation of the animal when it appears in a dream.

Indeed, there are structural correspondences in ad-Damīrī's and Gessner's works, but there is no evidence that the polyhistor from Zürich had any prior knowledge of *Ḥayāt al-hayawān*. Nor does ad-Damīrī's name appear in the Arab sources that Gessner listed in the first volume of the *Historia animalium*.[138]

12.6 Mythical Creatures

When leafing through Gessner's animal books, there are some especially eye-catching illustrations of peculiar creatures with origins in mythology or folklore. Because they are included next to illustrations of authenticated animals, they have become elevated iconographically to the same level. Taking a closer look and reading the Latin text, however, reveals Gessner's doubts about some of these fabled creatures or his outright denial of their existence.[139] Only a few were accepted,[140] for example, the unicorn, which still had believers as late as 1758 after its identification by Carl Linnaeus.[141] Gessner relegated the seven-headed hydra to the realm of fiction,[142] and he had grave doubts about the

136 Delaunay (1962).
137 Somogyi (1950).
138 Gessner (1551), f. β6r.
139 Reyer (2021), pp. 82–85.
140 Gmelig-Nijboer (1977), p. 119.
141 Reichholf (2015), p. 175.
142 Gessner (1558), pp. 541–543.

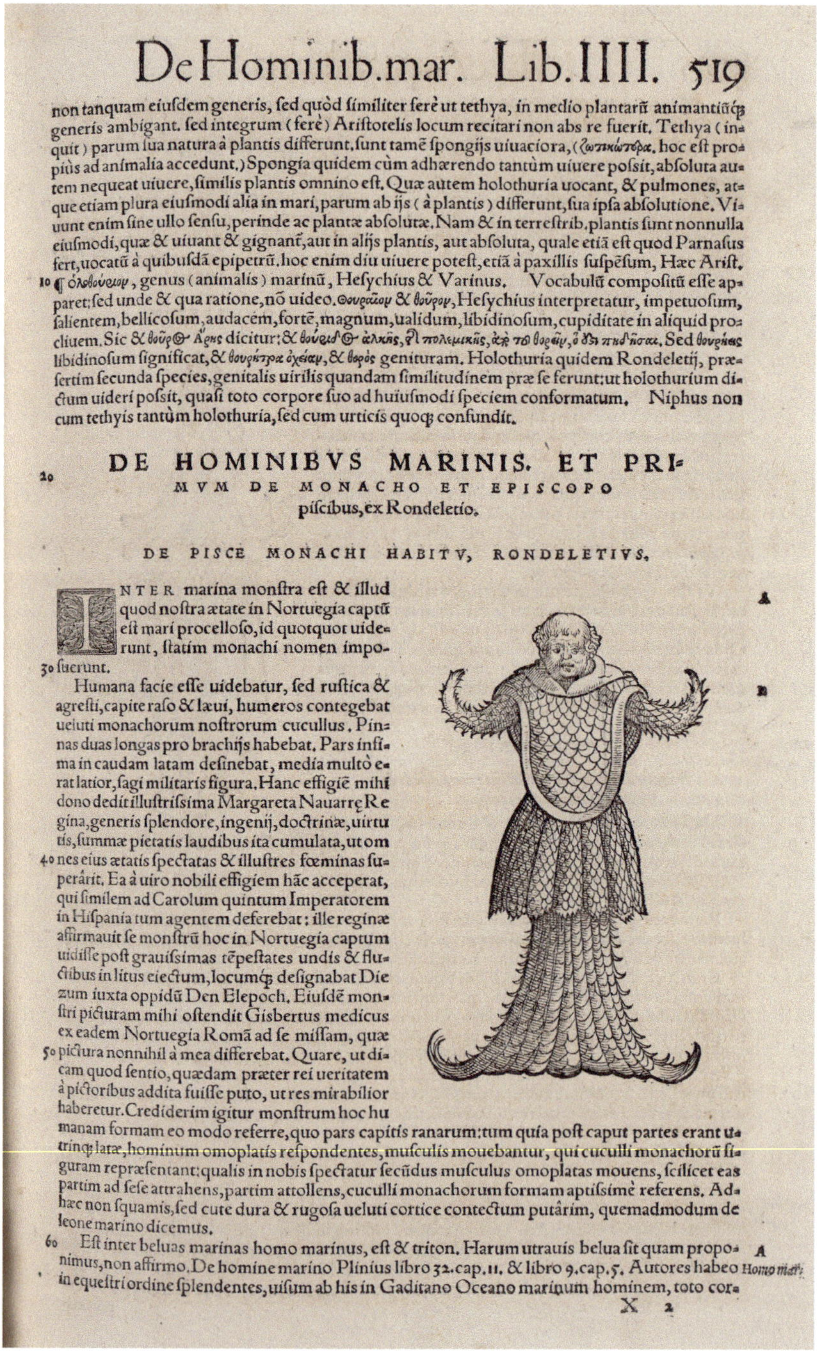

De Hominib. mar. Lib. IIII. 519

non tanquam eiusdem generis, sed quòd similiter ferè ut tethya, in medio plantarũ animantiũq; generis ambigant. sed integrum (ferè) Aristotelis locum recitari non abs re fuerit. Tethya (inquit) parum sua natura à plantis differunt, sunt tamẽ spongijs uiuaciora, (ζωτικώτερα, hoc est propiùs ad animalia accedunt.) Spongia quidem cùm adhærendo tantùm uiuere poßit, absoluta autem nequeat uiuere, similis plantis omnino est. Quæ autem holothuria uocant, & pulmones, atque etiam plura eiusmodi alia in mari, parum ab ijs (à plantis) differunt, sua ipsa absolutione. Viuunt enim sine ullo sensu, perinde ac plantæ absolutæ. Nam & in terrestrib. plantis sunt nonnulla eiusmodi, quæ & uiuant & gignant, aut in alijs plantis, aut absoluta, quale etiã est quod Parnasus fert, uocatũ à quibusdã epipetrũ. hoc enim diu uiuere potest, etiã à paxillis suspẽsum, Hæc Arist.

¶ ὀλοθούριον, genus (animalis) marinũ, Hesychius & Varinus. Vocabulũ compositũ esse apparet; sed unde & qua ratione, nõ uideo. θουριαῖον & θοῦρον, Hesychius interpretatur, impetuosum, salientem, bellicosum, audacem, fortẽ, magnum, ualidum, libidinosum, cupiditate in aliquid procliuem. Sic & θοῦρος Ἄρης dicitur; & θούριδος ἀλκῆς, τῆς πολεμικῆς, ἀπὸ τοῦ θορεῖν, ὅ ἐστι πηδῆσαι. Sed θουρήεις libidinosum significat, & θουρήτορα ὀχείαν, & θορὸς genituram. Holothuria quidem Rondeletij, præsertim secunda species, genitalis uirilis quandam similitudinem præ se ferunt: ut holothurium dictum uideri poßit, quasi toto corpore suo ad huiusmodi speciem conformatum. Niphus non cum tethyis tantùm holothuria, sed cum urticis quoq; confundit.

DE HOMINIBVS MARINIS. ET PRIMVM DE MONACHO ET EPISCOPO piscibus, ex Rondeletio.

DE PISCE MONACHI HABITV, RONDELETIVS.

INTER marina monstra est & illud quod nostra ætate in Nortuegia captũ est mari procelloso, id quotquot uiderunt, statim monachi nomen imposuerunt.

Humana facie esse uidebatur, sed rustica & agresti, capite raso & læui, humeros contegebat ueluti monachorum nostrorum cucullus. Pinnas duas longas pro brachijs habebat. Pars infima in caudam latam desinebat, media multò erat latior, sagi militaris figura. Hanc effigiẽ mihi dono dedit illustrißima Margareta Nauarrę Regina, generis splendore, ingenij, doctrinæ, uirtutis, summæ pietatis laudibus ita cumulata, ut omnes eius ætatis spectatas & illustres fœminas superârit. Ea à uiro nobili effigiem hãc acceperat, qui similem ad Carolum quintum Imperatorem in Hispania tum agentem deferebat; ille reginæ affirmauit se monstrũ hoc in Nortuegia captum uidisse post grauißimas tẽpestates undis & fluctibus in litus eiectum, locumq; designabat Diezum iuxta oppidũ Den Elepoch. Eiusdẽ monstri picturam mihi ostendit Gisbertus medicus ex eadem Nortuegia Romã ad se missam, quæ pictura nonnihil à mea differebat. Quare, ut dicam quod sentio, quædam præter rei ueritatem à pictoribus addita fuisse puto, ut res mirabilior haberetur. Crediderim igitur monstrum hoc humanam formam eo modo referre, quo pars capitis ranarum: tum quia post caput partes erant utrinq; latæ, hominum omoplatis respondentes, musculis mouebantur, qui cuculli monachorũ figuram repræsentant: qualis in nobis spectatur secũdus musculus omoplatas mouens, scilicet eas partim ad sese attrahens, partim attollens, cuculli monachorum formam aptißimè referens. Adhæc non squamis, sed cute dura & rugosa ueluti cortice contectum putârim, quemadmodum de leone marino dicemus.

A
B

A Est inter beluas marinas homo marinus, est & triton. Harum utrauis belua sit quam proponimus, non affirmo. De homine marino Plinius libro 32. cap. 11. & libro 9. cap. 5. Autores habeo in equestri ordine splendentes, uisum ab his in Gaditano Oceano marinum hominem, toto cor-

Homo mar.

X 2

ILLUSTRATION 37 Sea monk, in: Gessner (1558), p. 519.

ZBZ, ABTEILUNG ALTE DRUCKE UND RARA: NNN 48

existence of the basilisk.[143] Fish-human creatures such as the sea monk (ill. 37) were discussed, but here, Gessner quoted the opinion of Rondelet, whom he regarded highly. The French colleague could not match the drawing to any living creature, but with admirable reserve, refrained from commenting on whether such monsters actually existed.[144] Rondelet received a drawing of the fabled being from none other than Queen Margareta of Navarre (1492–1549).[145] Reading between the lines, however, it was clear that neither Rondelet nor Gessner believed in the existence of sea monks.

In an effort to be thorough, he included some fantastic animals in his animal encyclopedia that were passed down from tradition, of which only about a half dozen withstood his critical examination.[146] Among those confirmed by Gessner was the *Cynocephalus* (wild man)[147] from Bernhard von Breydenbach's *Reise ins Heilige Land*, which Gessner identified as an ape.[148] The unicorn, perhaps the most famous of mythical creatures in the Western world, was also confirmed by Gessner.[149] He possessed not only an illustration of it – whose accuracy he questioned[150] – but he had also seen a horn of the animal with his own eyes. He mentioned this in a letter of March 19, 1564 to his Augsburg colleague Achilles Primin Gasser.[151] He wrote to Adolph Occo over a year later: "I do not doubt that this animal exists among the Indians and Egyptians, and that it is Aristotle's Indian donkey and Aelian's *Cartazonum*. I have seen an entire horn myself and possess a small piece of one. In Africa, there are also one-horned ibexes and oryx antelopes,[152] and the rhinoceros is also a unicorn."[153]

The horn that Gessner had seen was actually a narwhal tooth. The tusks of these sea mammals and those from mammoths were interpreted as relics from unicorns, and were pulverized for pharmaceutical purposes.

Various illustrations by Gessner show fantastic creatures that were actually genuine animals that were poorly drawn or distorted, such as the *Cyprinus*

143 Ley (1929), pp. 81–85.
144 Cf. Gmelig-Nijboer (1977), p. 111; Riedl-Dorn (1989), pp. 113f.; cf. too Michel (2015).
145 Gessner (1558), p. 519.
146 Gmelig-Nijboer (1977), pp. 97–120; Riedl-Dorn (1989), pp. 74–94.
147 Gessner (1551), pp. 970–973.
148 Fischel (2009), pp. 64–67.
149 Gessner (1551), p. 781; Riedl-Dorn (1989), pp. 81–86.
150 Fischel (2009), p. 62.
151 Burmeister (1975), pp. 299–301.
152 If you look at an oryx antelope in profile and the two horns are congruent behind each other, it looks as if the animal had only one horn.
153 Letter from Gessner to Adolph Occo, August 26, 1565, cf. Gessner (1577), f. 75v.

fish (carp) with a human face.[154] This probably resulted from "an exact observation, since carp are sometimes born deformed with roundish heads."[155] There were also bizarre looking dried-out rays that were sold as dragons, which, of course, did not fool Gessner.[156] The provenance of an exotic creature from the New World called a *Su* (ill. 38) was very likely the giant anteater (*Myrmecophaga tridactyla*).[157] When thumbing through the bird book, there is a woodcut depicting a female pelican using its beak to cut open its breast in order to feed its hungry young. This is a medieval symbol for charity and self-sacrifice. Gessner unequivocally states in his text that such a bird does not exist in nature.[158]

On the other hand, Gessner was also confronted by illustrations of animals that were entirely the product of human fantasy. Despite visible illustrative "proof," his profound physiological knowledge led him to the conclusion that some of these creatures could not possibly have existed.[159] For example, he noted in the case of a woodcut depicting the monstrous sea boar (*Hyaena cetacea*), taken from the *Carta marina* (1539) by Olaus Magnus, that he could not accept the positioning of the pig-like ears and three eyes that were located on the sides of the stomach.[160] Therefore, Gessner deemed the illustration to be inaccurate and rejected the existence of the animal (ill. 39).[161]

Occasionally, he gave warnings about erroneous representations, as he did in the case of the salamander with little stars on its back that had been widely circulated in Breydenbach's previously mentioned travel descriptions.[162] Yet, it appears that the starred salamander continued to trouble him. After Easter 1557, he wrote to Caspar Wolf, who was studying medicine in Montpellier, that he had heard of a salamander in Languedoc with a pattern of stars on its back. Gessner asked him to make a comparison between this salamander and the one he had published in the second volume of the animal history. In case there

154 Gessner (1558), p. 373. Gessner had received the illustration from Achilles Pirmin Gasser. A second one was sent to him by Raphael Sailer, son of the physician Gereon Sailer, from Augsburg, which Gessner mentions but does not use. Cf. the letter from Raphael Sailer to Bullinger of February 26, 1557 (StAZH, E II 346, 363).

155 Gantenbein (1997), p. 192.

156 Gessner (1560a), p. 139.

157 Leu (1992), pp. 296f.

158 Riedl-Dorn (1989), pp. 99f.

159 Springer/Kinzelbach (2009), p. 33.

160 Gessner (1558), p. 247.

161 Cf. for the history of the picture: Faust (1998–2010), vol. 5, no. 761.

162 Gessner (1553a), p. 58; Gessner (1554), p. 27; Gessner (1560), p. 119. Cf. Harms (1989), p. 361; Kusukawa (2010), pp. 325–327.

Additiones. 127

Lynx modò occultat, ut hîc: modò exerit, pro arbitrio, ut cum irascitur, pugnat, scandit aut se retinet, & quidem præ longos, ut illic.

GERMAN. Ein Vntz / ist ein Luchß oder Luchsen art.

Lyncum genera duo solus Oppianus facit (lib. 3. De uenatione) maius & minus. Lynces paruæ (inquit) Lepores uenantur: maiores uerò in Ceruos atque Oryges facilè insiliunt. Ambæ corporis figura similes: & similiter oculi utrisque suauiter fulgẽt, &c.

NOVI Orbis regionem quandam Gigantes dicti (lingua ipsorum Patagones) incolunt. & quoniam cœlo non admodum calido fruuntur, uestiunt se pellibus animantis illius, quam Su appellant, id est, aquam: ab eo nimirum, quòd magna ex parte circa fluuios degat. Est autẽ omnino rapax hæc fera, & formæ monstrosæ, qualem hîc exhibeo. Cum à uenatoribus urgetur, suæ pellis gratia, catulos suos in dorsum admissos cauda ampla longaque tegit, & fuga elabitur. Itaque dolo scrobe effossa, & frondibus obtecta, unà cum catulis capitur. Cum autem ita inclusam se uidet, rabie quadam suos catulos obtruncat & occidit: & clamore horribili ipsos etiam uenatores terret: à quibus tandem sagittis confossa excoriatur, Andreas Theuetus cap. 56. Descriptionis Americæ.

AD ORDINEM IIII. DE QVADRVpedibus feris minimis.

ASPREOLVM in Germanicis cõmentarijs lego: & quid legam, nescio, Scaliger. Ego Sciurum Ponticum, siue Varium esse conijcio. Quoniam Sigismundus Liber Baro in Cõmentarijs rerum Moscouiticarum, ubi de pellibus diuersis è Moscouia adferri solitis mentionem facit, Sciuros Ponticos prætererit, quorum pelles inter præcipuas sunt, nisi Aspreolorum nomine eos intelligamus. Aspreolorum pelliculæ (inquit) ex diuersis partibus adferũtur, &c. ex nõnullis semper decem numero colligatæ, quarum in singulis fasciculis duæ sunt optimæ, quas Litzschna appellant. tres aliquanto deteriores, quas Crasna uocant: quatuor, quas Pocrasna: & ultima Moloischna dicta, omnium uilissima. Harum singulæ una aut duabus dengis emuntur. Meliores & selectas in Germaniam & aliò mercatores magno suo cõmodo portant. Sic ille. Aspreoli quidem nomen, ad Pirolum uel Spiriolum (sic enim barbari quidem scriptores Sciurum uocant) accedit.

FINIS.

ILLUSTRATION 38 Su, in: Gessner (1560), p. 127, with Gessner's handwritten annotations.
ZBZ, ABTEILUNG ALTE DRUCKE UND RARA: NNN 44

De Cetis diuerſis. Lib. IIII. 247

neq; repræſentat icon. non binis enim ſed quatuordecim ceu cornibus radiatum caput eſt. cornua utrinq; ab oculo incipiunt, & per occipitium tranſeunt, neſcio quàm rectè. Albertus enim (cuius uerbis paulò antè in cetis Oceani Germanici deſcribendis uſus ſum) cetis noſtrorum marium appendices quaſdam eſſe ſcribit (circa oculos) ciliorum inſtar, corneas, octo ferè pedes longas, plus minus, pro magnitudine piſcis, ſpecie magnæ falcis fœniſecum, ducentas & quinquaginta ſuper oculum unum, & totidem ſuper alterum, ex lato in acutum deſinentes, non rigidas, ſed iacentes, & diſpoſitas uerſus tempora piſcis, ita ut ueluti os unum latum magni uanni inſtar efficiant, quo belua aduerſus tempeſtates oculos muniens operiat. Oculi unius foueam homines quindecim, aliquando uiginti capere.

DE CETO CAPILLATO.

CETVM Capillatum uel crinitum hunc piſcem nomino, cuius caput tantum Olaus delineauit in tabula, paulò infra A.l. in explicatione uerò nullam eius mentionem fecit.

DE HYAENA CETACEA.

HANC feram marinam Olaus exhibuit in D.k. porco ſimilem eſſe ait, & apparuiſſe in mari iuxta Thylen inſulam, quæ ſupra Orchades Septentrionem uerſus ſita eſt, anno Domini 1537. Ego hyænam uoco, à ſimilitudine, ſiue ſuis, ſiue hyænæ, quadrupedum. De ueterum hyænis piſcibus aut cetis dicam in H. elemento. In hac Olai icone non probo auriculas animali marino appictas roſtrum quoq; nimis porcinum uidetur. Hoc mirum quod in eius latere tres uelut oculi apparent expreſsi.

DE CETIS DVOBVS, IVBATO, ET MONOCEROTE.

MONSTRVM hoc potiùs quàm cetum, iubatum ſimul barbatumq; in mari glaciali pingit Olaus, paulò infra Gruntlandiam, remotiſsimam ad Septentriones regionem, intra B. literæ complexum: facie uiri ferè, in explicatione uerò tabulæ nuſquam eius meminit: ut neque Monocerotis, id eſt, unicornis, cuius caput paulo infra A.m. ex æquore prominet.

x 4

ILLUSTRATION 39 Sea boar, in: Gessner (1558), p. 247.

ZBZ, ABTEILUNG ALTE DRUCKE UND RARA: NNN 48

was a difference, he asked Wolf to please have an illustration made of it.[163] Wolf apparently found no such animal, which is why Gessner did not wait on a new illustration for the *Icones animalium* of 1560. He also published Breydenbach's false description in order to point out its inaccuracy.[164]

Gessner was always looking for well-executed and lifelike imagery, giving credit to his sources and admitting that not all illustrations turned out to be elegant and exact.[165] Nonetheless, the Northern Italian physician and botanist Pietro Andrea Mattioli (1501–1577) repeatedly attacked him for publishing fictitious illustrations. Gessner did not take this sitting down. He defended himself against the charges at length in the foreword to the reader in the *Icones animalium* of 1560. He had attempted to settle the issue in private correspondence, but was unsuccessful. Therefore, he responded to the public accusations with a public retort. Although he never mentioned Mattioli's name, some scholars were able to discern to whom he was referring. In his personal copy of the book, Gessner wrote in the margins that he was rebuffing Mattioli.[166]

12.7 Network and Images

As explained above, Gessner was preoccupied with zoological issues during his time in Lausanne and Venice, during which he collected items from nature and made drawings of sea life. The earliest written evidence of his plan to write an animal encyclopedia is a letter sent to Bonifacius Amerbach on November 18, 1545.[167] He wrote: "I am sending a list of animals that I am writing about in which I have noted those that have no acceptable image. With help from friends, I hope they can be obtained for the leopard, hyena, tiger, aurochs, bison, chamois, bison, lynx, fallow deer, Indian mouse or Pharaoh's mouse, which I believe is the Egyptian mongoose, and some others. I also have quite a few images of animals whose names are not listed in the enclosed register. In all cases, I have made sure that they were drawn true to life and in color. The

163 Gessner (1577), f. 122r/v.
164 Gessner (1560), p. 119.
165 Gessner (1558), f. b_2v: „Hoc non diffiteor: picturas nostras non ubique elegantes aut accuratissimas esse: sed eam culpam (si quae est) in typographum, qui sumptus sustinebat, inquam pictores et sculptores reycio. Qualescunque sint autem, verae sunt, hoc est, vel ad naturam factae, vel ad archetypum alterius authoris qui semper nominatur."
166 Gessner (1560), p. 7 (ZBZ, Shelf mark: NNN 44_1).
167 The letter from Gessner to David Chytraeus of September 2, 1543, in which he thanks him for the fish illustrations sent to him, must have been dated incorrectly. It is probably not 1543, but 1553. Cf. Omont (1891).

planned work is divided into the following elements: the names of the animals in various languages, physical description, activities and characteristics etc., as well as uses for medicine, food and other things."[168]

Amerbach was not exactly an animal enthusiast or zoologically inspired contemporary. Nonetheless, Gessner hoped to receive the desired illustrations through his extensive network of connections. Aside from the fact that Gessner's wish list has been lost, the lawyer from Basel apparently could or would not be of assistance. Ultimately, the illustrations Gessner published of the Egyptian mongoose and the hyena were two woodcuts that he found in an old Greek Oppian manuscript in Venice.[169] Gessner explicitly mentioned his dislike of the hyena illustration.[170] The half-page image of the aurochs was copied from a land map (*Tabula chorographica*),[171] and the tiger drawing had been sent by his Northern Italian colleague Giovanni Ferrerio.[172] The lynx illustration had been rendered from a pelt with the head missing, which, as with the South American whiptail lizard, the artist had to draw in.[173]

Many illustrations were taken from various printed works, such as the *Cosmographia* by Sebastian Münster,[174] or from woodcuts by artists such as Albrecht Dürer (1471–1528)[175] (ill. 40). However, he was still missing many images, which is why he created a second wish list. This was not only sent off in letters, it was also published in the 1548 *Pandectae*.[176] Of the nearly 1,000 animal names[177] on his list, about a third were marked with little stars, which meant that he already had illustrations for them. Accordingly, he possessed illustrations for about 70 of the 120 four-legged animals listed, 120 of the 350 birds, 50 of the 200 sea fish, 20 of the 50 river fish, 10 of the 50 so-called "whalefish" (*cetaceus*), 30 of the 70 other sea animals, 20 of the 50 insects and 3 of the 36 snakes. At the age of 32, Gessner had knowledge of twice as many

168 AK 6, no. 2761, p. 202.

169 Both animals are depicted, for example, in a Greek Oppian manuscript kept in the Biblioteca Marciana in Venice, which belonged to the famous collector of Greek manuscripts Cardinal Bessarion (Marc. gr. 479, f. 51v and 52r). A similar work must have been the model for Gessner. Cf. Furlan (1997), ill. 18b and 19a.

170 Gessner (1551), pp. 624 and 635.

171 Gessner (1551), p. 157; Gessner (1560), p. 30.

172 Gessner (1551), p. 1060; Gessner mentions the source in his handwritten annotations in his own copy of the *Icones animalium* (1560), p. 66f.

173 Gessner (1551), p. 769; Leu (1992), p. 299.

174 Pinon (2005), p. 252.

175 The most famous Dürer illustration in Gessner's works is his armored rhinoceros. Gessner (1551), p. 953.

176 Gessner (1548), f. 221r–233v.

177 Most of the figures mentioned here are rounded.

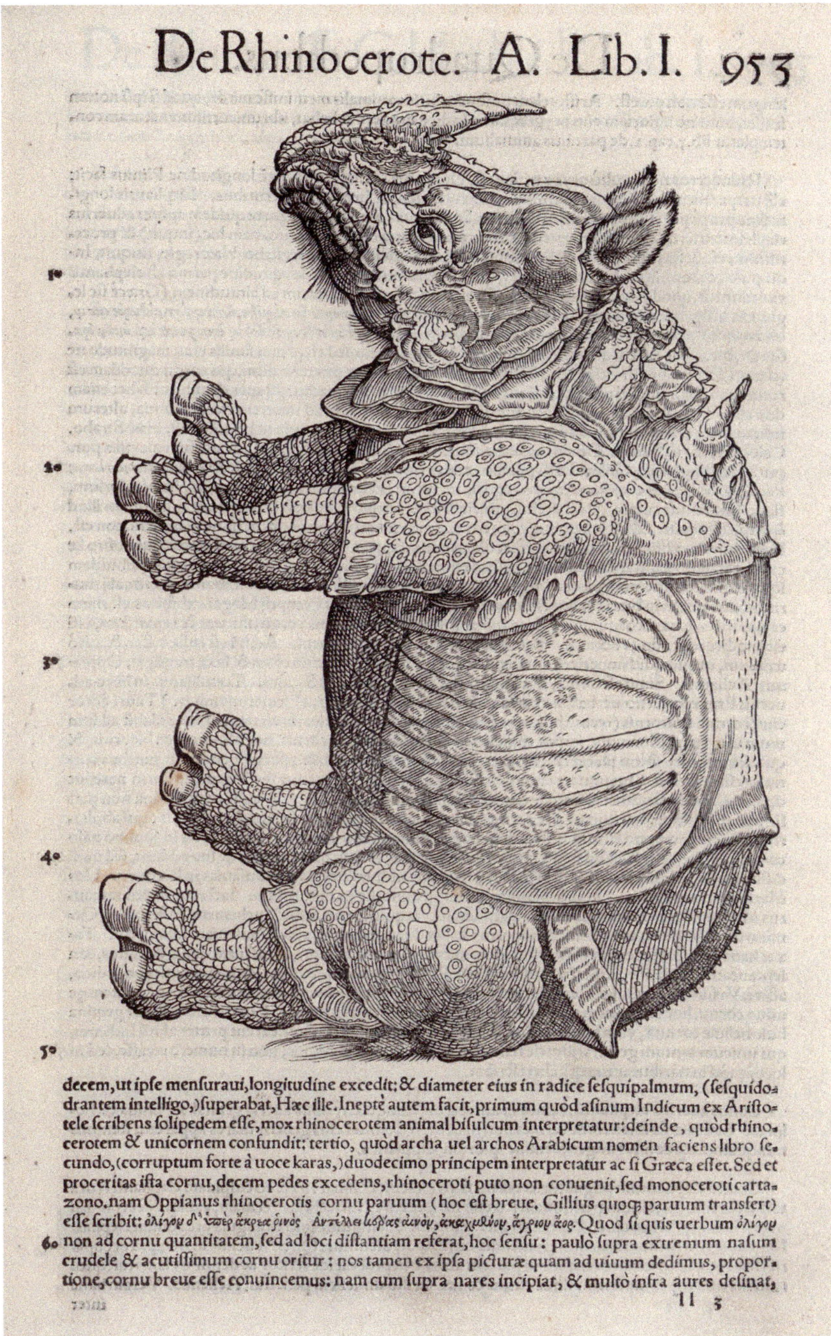
De Rhinocerote. A. Lib. I. 953

decem, ut ipſe menſuraui, longitudine excedit; & diameter eius in radice ſeſquipalmum, (ſeſquidodrantem intelligo,) ſuperabat, Hæc ille. Inepte autem facit, primum quòd aſinum Indicum ex Ariſtotele ſcribens ſolipedem eſſe, mox rhinocerotem animal biſulcum interpretatur; deinde, quòd rhinocerotem & unicornem confundit; tertio, quòd archa uel archos Arabicum nomen faciens libro ſecundo, (corruptum forte à uoce karas,) duodecimo principem interpretatur ac ſi Græca eſſet. Sed et proceritas iſta cornu, decem pedes excedens, rhinoceroti puto non conuenit, ſed monoceroti cartazono. nam Oppianus rhinocerotis cornu paruum (hoc eſt breue, Gillius quoq; paruum transfert) eſſe ſcribit; ὀλίγον δ' ὑπὲρ ἄκρατα ῥινὸς Ἀντέλλει κέρας αἰνόν, ἀκαχμένον, ἄγριον ἄορ. Quod ſi quis uerbum ὀλίγον non ad cornu quantitatem, ſed ad loci diſtantiam referat, hoc ſenſu: paulò ſupra extremum naſum crudele & acutiſſimum cornu oritur; nos tamen ex ipſa picturæ quam ad uiuum dedimus, proportione, cornu breue eſſe conuincemus: nam cum ſupra nares incipiat, & multò infra aures deſinat,

11 3

ILLUSTRATION 40 Rhinoceros, in: Gessner (1551), based on Albrecht Dürer's illustration, p. 953.
ZBZ, ABTEILUNG ALTE DRUCKE UND RARA: NNN 41

animals as Aristotle, who had listed around 500 species of animals.[178] Three years prior to printing preparations for the first volume of the *Historia animalium*, he was still missing two thirds of the desired illustrations. Gessner appealed to his readers to send him images of the animals with the promise that the contributor would receive a mention in the published volume.[179] Since Gessner's works had a broader distribution than that of the reformer Heinrich Bullinger,[180] this would have surely been attractive bait. It was well known among scholars how well received the *Bibliotheca universalis* had been in Europe, and that it would bolster one's reputation to be mentioned in a publication by Gessner.

It appears that several scholars responded to Gessner's call in 1548, for he noted that 25 of the 96 illustrations in the first volume had been sent to him.[181] Among them was Francesco Bonetti, who had wished to send him a dormouse illustration that he had commissioned from an artist in Bergamo.[182] Apparently, Gessner received the drawing from Gullielmo Grataroli, a Bergamo doctor, and not from Bonetti himself, and included it in the first volume of the *Historia animalium*.[183] Likewise, Johannes Echtius, a doctor and botanist working from Cologne, offered his assistance in 1550.[184] In the same year, Johannes Kentmann sent him his animal and plant albums,[185] which are preserved today in the Duchess Anna Amalia Library in Weimar.[186] It is not known when Gessner received the billy goat image drawn by Hans Asper (1499–1571) in 1547[187] that was published in Volume 1 of the *Historia animalium* in 1551[188] (ill. 41a and 41b).

Various illustrations were obtained through correspondence even before he made his appeal to the scholarly world in 1548. For example, there is

178 By way of comparison, around 1.4 million animal species are known today.

179 Gessner (1548), f. 221v.

180 Cf. Bernhard (2009), pp. 173f.

181 Gmelig-Nijboer (1977), p. 83.

182 Letter from Bonetti to Bullinger, October 24, 1550 (StAZH, E II 361, 284).

183 Gessner (1551), p. 619. Perhaps he never received the illustration from Bonetti, and Grataroli made one himself.

184 Letter from Johann Birckmann to Bullinger, January 18, 1550. Cf. Calvin (1875), pp. 513f.

185 Letter from Gessner to Kentmann, December 1, 1550, cf. Gessner (1584), f. A_1v–A_2r.

186 Herzogin Anna Amalia Bibliothek Weimar, Codex Kentmannus, Fol. 323. See online: http://ora-web.swkk.de/digimo_online/digimo.entry?source=digimo.Digitalisat_anzeigen&a_id=1025.

187 The original with monogram HA (Hans Asper) and the year 1547 is in the animal albums of Felix Platter, which contain drawings from Gessner's estate and are kept in the University Library of Amsterdam, MS. III C 23, f. 31r.

188 Gessner (1551), p. 302.

ILLUSTRATION 41A Goat with monogram H[ans] A[sper] from the animal album of Felix Platter. Many drawings from the album like this one was previously in Gessner's possession.
UNIVERSITY LIBRARY AMSTERDAM: MS. C III 22, P. 31

documentation for a fish image from Jean Ribit[189] and the offer of a bird-of-prey image from Johannes Gast, a pastor from Basel. He wrote to Bullinger about the exceptionally large animal: "*Avis mirae magnitudinis*, that eats 3 pounds of flesh at a time, has a curved beak, large feet. This one was caught by a farmer when it was still young and could not fly. Some believe it to be a bustard, and some say it is a large, young gyre falcon. An artist painted it. The picture should also be included in Dr Gessner's book of wild animals, if he does not yet have such a species of bird."[190]

With reference to Gessner's sources for graphic images, there is an interesting passage in an April 22, 1550 letter from Gessner to Kentmann. Gessner thanked him for the generous gesture of offering him access to his animal illustrations, for which Gessner had already received a list of names. Gessner explained: "I already have many pictures of animals that are named on the list,

189 Letter from Ribit to Gessner dated October 1, 1547 (BNF, Département des manuscrits, Latin 8641, f. 9v).

190 Letter from Gast to Bullinger, June 30, 1548 (StAZH, E II 366, 185).

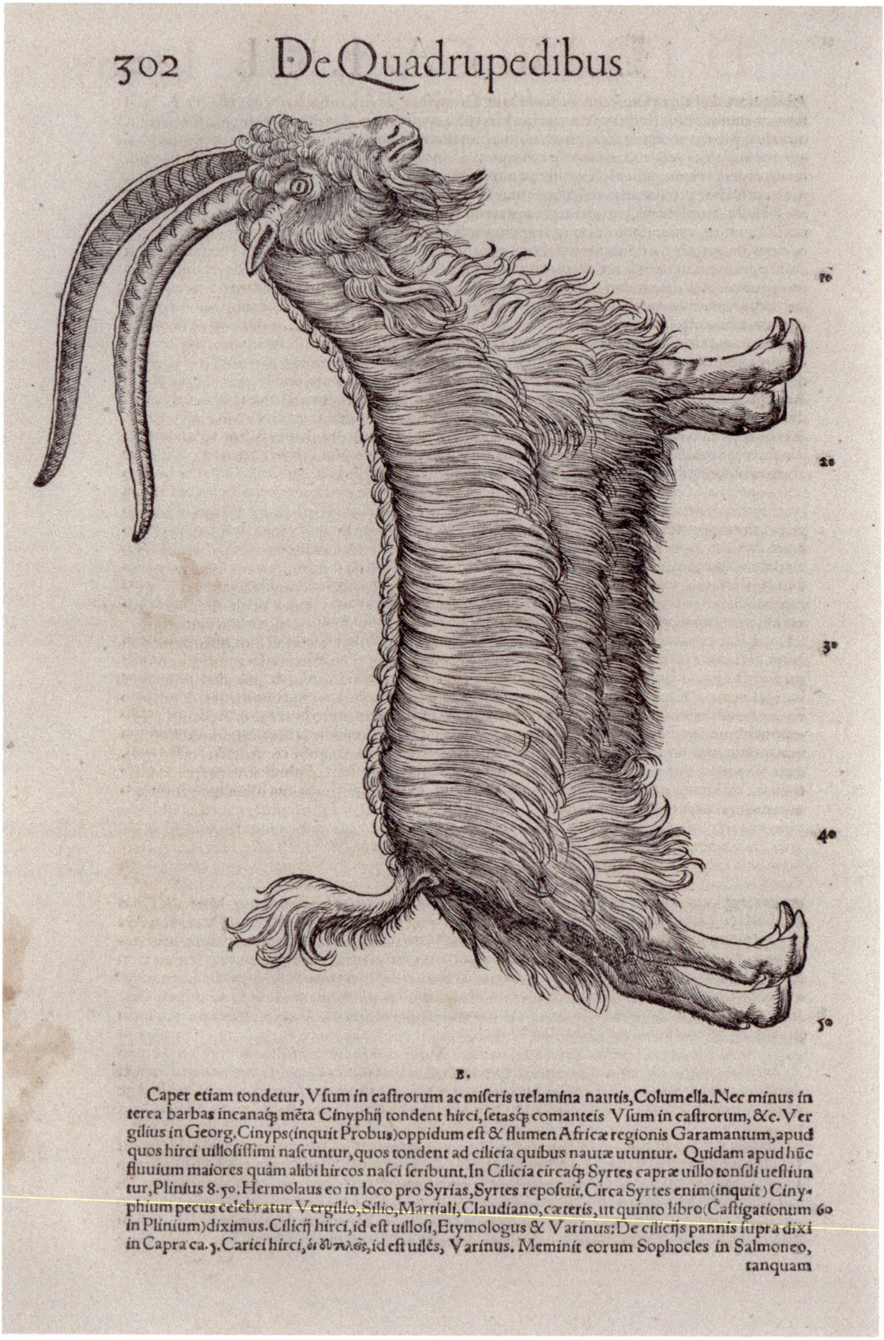

302 De Quadrupedibus

B.

Caper etiam tondetur, Vſum in caſtrorum ac miſeris uelamina nautis, Columella. Nec minus interea barbas incanaſq; mēta Cinyphij tondent hirci, ſetaſq; comanteis Vſum in caſtrorum, &c. Vergilius in Georg. Cinyps (inquit Probus) oppidum eſt & flumen Africæ regionis Garamantum, apud quos hirci uilloſiſſimi naſcuntur, quos tondent ad cilicia quibus nautæ utuntur. Quidam apud hūc fluuium maiores quàm alibi hircos naſci ſcribunt. In Cilicia circaq; Syrtes capræ uillo tonſili ueſtiuntur, Plinius 8.50. Hermolaus eo in loco pro Syrias, Syrtes repoſuit. Circa Syrtes enim (inquit) Cinyphium pecus celebratur Vergilio, Silio, Martiali, Claudiano, cæteris, ut quinto libro (Caſtigationum in Plinium) diximus. Cilicij hirci, id eſt uilloſi, Etymologus & Varinus: De cilicijs pannis ſupra dixi in Capra ca. 3. Carici hirci, αἱ εὐτελεῖς, id eſt uilēs, Varinus. Meminit eorum Sophocles in Salmoneo, tanquam

ILLUSTRATION 41B Hans Asper's goat as published, in: Gessner (1551), p. 302.
ZBZ, ABTEILUNG ALTE DRUCKE UND RARA: NNN 4₁

in part from other sources, and in part from the Nuremberg doctor Cornelius Sittardus [† 1550], who kindly sent them to me."[191] Sittardus was mentioned by Gessner in both the first and fourth volumes of the *Historia animalium* as a contributor, and he also supplied Gessner with illustrations even after the fish book was printed.[192] In the preface to the book about aquatic animals, Gessner mentions that Sittardus had received most of the images from the Dutch physician Gysbert van der Horst (ca. 1491–1556) from Rome, and that he owed him a debt of thanks.[193] Sittardus had studied in 1541/42 in Northern Italy, traveling together with the German doctors Valerius Cordus (1515–1544) and Nikolaus Friedewald as well as the Dutch physician Pieter van Foreest (1521–1597) to Rome. There, Cordus, who was a very promising young botanist, fell ill and received care in the hospital of Hieronymus Schreiber († 1547), then unexpectedly died on September 25. Several days before, Sittardus and Joachim Moller the Younger (1521–1588) had traveled on to Naples.[194]

In Rome, which had about 55,000 residents at the time, Sittardus met with van der Horst, who worked in the Santa Maria della Consolazione Hospital.[195] Sittardus copied roughly 40 fish illustrations from van der Horst's album of sea creatures and sent them to Gessner.[196] These drawings, which served as references for Gessner's woodcuts, are preserved in the University Library Amsterdam along with others.[197] When compared, some of the Kentmann animal drawings stored in Weimar, which are mostly of fish, bear remarkable similarities to the Amsterdam material. One might conclude that during his time in Rome in 1549, he also made drawings or had drawings made from van der Horst's images. In the letter, Gessner informed him that he was already familiar with a portion of the animals on the list sent by Kentmann, and that he already had illustrations of some that were sent by Sittardus. Therefore, Van der Horst, Sittardus, Gessner and Kentmann all possessed the same, repeatedly copied

191 Gessner (1584), f. A_1r.
192 Gessner (1551), f. γ_1r; Gessner (1558), f. b_5r; Gessner (1604), Paralipomena [appendix], pp. 6, 15 and 27f.
193 Gessner (1558), f. b_5r. Cf. Egmond/Kusukawa (2019).
194 Cf. the letter from Hieronymus Schreiber to Wolfgang Meurer of December 1, 1544 in: Cordus (1563), f. b_{iii}v–b_vv.
195 Egmond/Kusukawa (2016).
196 Where Sittardus got the illustrations, which he procured from Italy after Gessner's fish book was printed and sent to Gessner, cannot be said at present. They were published by the Frankfurt bookseller Andreas Cambier in 1604 as an appendix to Gessner's fish book. They come from Gessner's estate, as can be seen from the preface to the bird book that Robert Cambier had published with Johannes Wechel in Frankfurt in 1585. Cf. for example Gessner (1604), *Paralipomena* [Appendix], pp. 6 and 15.
197 Egmond (2013) and Egmond (2018).

ILLUSTRATION 42A Blue butterfish (*Stromateus fiatola*) from Johannes Kentmann's animal album that was copied from an illustration by Gysbert van der Horst in Rome. Johannes Kentmann, *Animalium aquatilium icones*.
DUCHESS ANNA AMALIA LIBRARY WEIMAR: FOL. 323, P. 30

ILLUSTRATION 42B
Cornelius Sittardus sent Gessner a copy of the same picture from Gysbert van der Horst, which together with other animal images made its way to Felix Platter in Basel after Gessner's death.
UNIVERSITY LIBRARY AMSTERDAM: MS. C III 22, P. 110

1110 De Aquatilibus.

B Latus eſt ac planus, caudam lunatam habens: Linguam carnoſam, delphini modo, ad humanam accedentem, ſuisq; partibus omnino abſolutam: quod præter cæteros branchijs præditos piſces ſolus habere mihi uiſus eſt, quemadmodum & hoc ſibi etiã peculiare, quòd præter omnium piſcium morem, pinnis ſub uentre careat, eſtq; prorſus inermis: Branchias utrinq; quatuor gerit, ſi quidem eam, quam capiti coniunctam habet, annumeres. Superiorem maxillam fixam ac immobilem, ut Scarus & Exocœtus habet: atq;, ut ait Oppianus,

Dentibus obtuſis eſt illi rictus inermis.

Cantharo quidem ſimilis eſt, ſed eo maior euadit, non tamen æquè craſſus. Lineam habet à capite ſub uentre ad umbilicum extenſam, atq; utrinq; in ceruice nigrã maculam. Squamis undecũq; caret, prę terquàm in eo lineę ductu, qui piſcis huius latera diſtinguit: uentrẽ quoq; paruum habet, ut & Lechia & Lampuga. Quamobrem & ei arctè inteſtina inter ſe cohærent, quæ tam numeroſis ſpiris in orbem circunductis ita inter ſe complicantur, ut globulũ eſſe dicas: hepar unius tantùm lobi nullo felle præditum: ſtomachũ in V literam efformatũ, cuius infima pars in mucronẽ turbinatur: atq; ad eius pylorum tam numeroſas uideas appendices, ut (quemadmodum nec in Thynno, Pelamide, ac Scõbro) nullis artibus numerari poſsint. Inteſtina fulcit glãduloſum pancreas. Lien illi ſtomacho ſubeſt.

COROLLARIVM.

FIATOLAE iconem à Bellonio exhibitam omiſimus, quòd forma & circunſcriptione corporis, ac pinnis cum ea, quam Rondeletius dedit, conueniat. Lineæ tamen ſeu ſtriæ & maculæ longè aliæ ac aliter in utriuſq; pictura ſunt: ut piſces eſſe diuerſos aliquis exiſtimaret, ſicuti etiã ex utriuſq; deſcriptione idem nõ uidetur. Rurſus uerò propter formam & pinnas, ut dixi, & quoniam uterq; circa Romam frequentẽ eſſe & fiatolam uocari ſcribit, unus uideri poteſt. A nobis poſita pictura, quam à Cor. Sittardo cum alijs piſcibus Romæ depictis olim accepi, cũ Belloniana ferè per omnia congruit, niſi quòd rudimenta quędam dentiũ oſtendit, cũ Bellonius omnino edentulam faciat: & lineam mediam rectã à branchijs ad caudã non exprimat, (cuius tamen in deſcriptione Bellonius non meminit,) nec aliã ſuperiorem Bellonio memoratam, incuruã, & ſquamoſam, cum reliquũ corpus ſquamis careat. De ea doctiſsimus Sittardus ſic ad me ſcribebat: Hic piſcis Fetolo uulgò dicitur, Romanis lampeca. ferẽ naſci ex pulmone marino, delicatiſsimus piſcis, nec exce-
(F) dit hanc magnitudinem. Erat autem miſſa ab eo pictura, uno alteró ue digito lon-

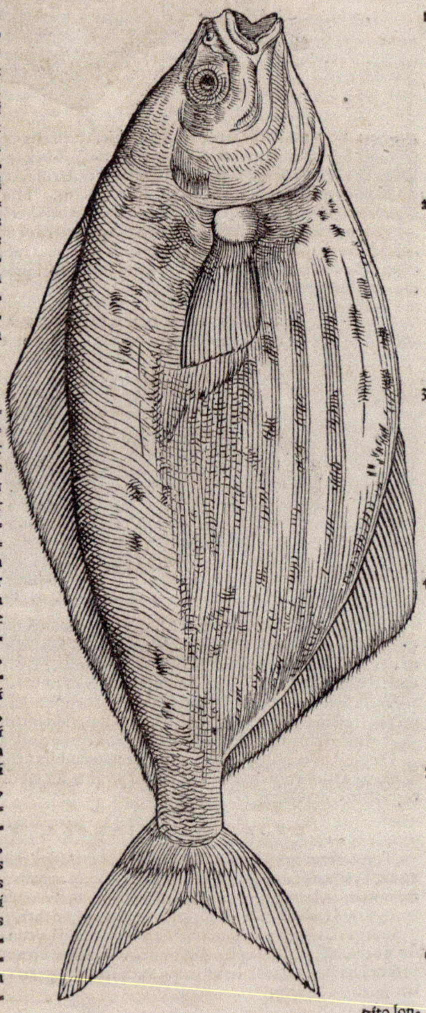

ILLUSTRATION 42C Publication of the blue butterfish by Gysbert van der Horst, in: Gessner (1558), p. 1110.

ZBZ, ABTEILUNG ALTE DRUCKE UND RARA: NNN 48

ILLUSTRATION 43A Torpedo ray from the animal album of Johannes Kentmann after an illustration by Gysbert van der Horst in Rome. Johannes Kentmann, *Animalium aquatilium icones*.
DUCHESS ANNA AMALIA LIBRARY WEIMAR: FOL. 323, P. 23

ILLUSTRATION 43B In Rome, Sittardus copied the same picture of the torpedo ray that Kentmann had and sent it to Gessner. After Gessner's death, it was added to Felix Platter's animal album.

UNIVERSITY LIBRARY AMSTERDAM: MS. C III 22, P. 122

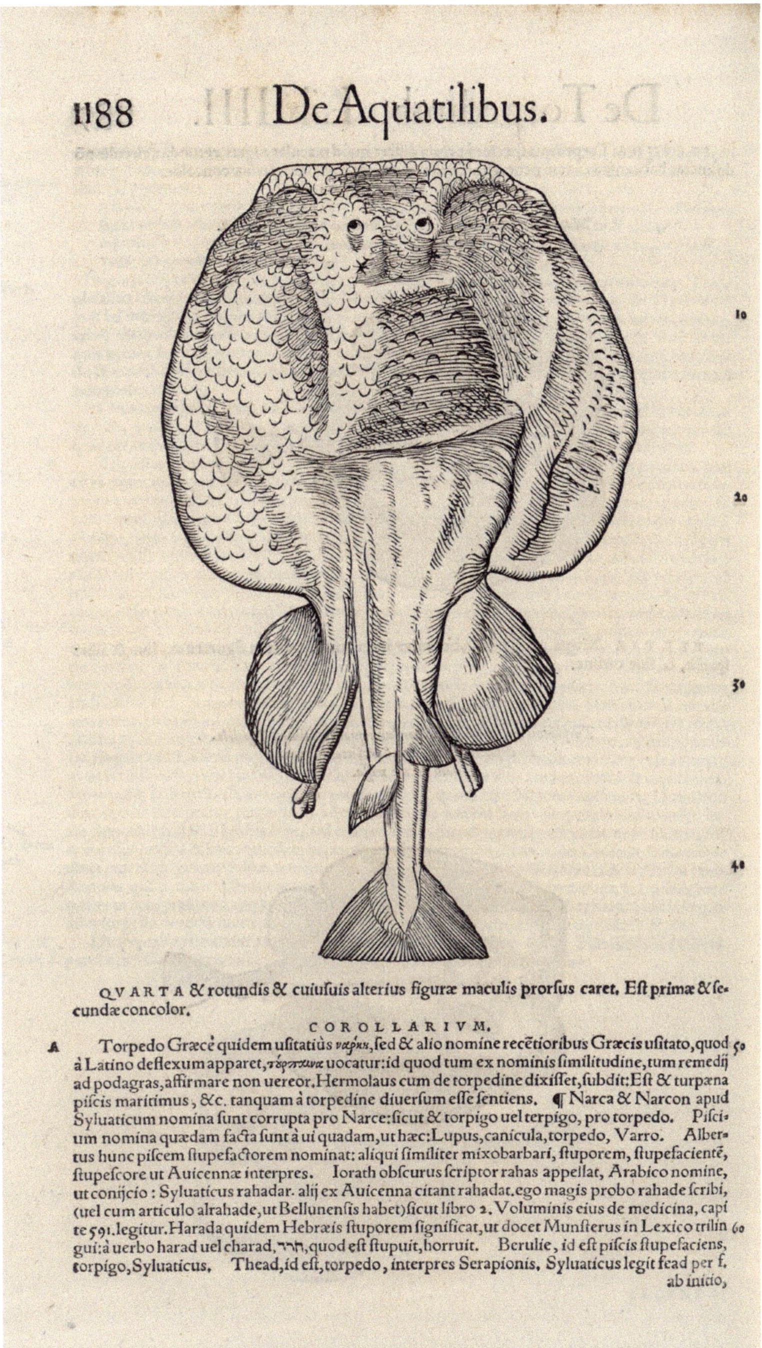

1188 De Aquatilibus.

QVARTA & rotundis & cuiuſuis alterius figuræ maculis prorſus caret. Eſt primæ & ſecundæ concolor.

COROLLARIVM.

A Torpedo Græcè quidem uſitatiùs νάρκη, ſed & alio nomine recẽtioribus Græcis uſitato, quod à Latino deflexum apparet, τόρπινα uocatur: id quod tum ex nominis ſimilitudine, tum remedij ad podagras, affirmare non uereor. Hermolaus cum de torpedine dixiſſet, ſubdit: Eſt & turpæna piſcis maritimus, &c. tanquam à torpedine diuerſum eſſe ſentiens. ¶ Narca & Narcon apud Syluaticum nomina ſunt corrupta pro Narce: ſicut & torpigo uel terpigo, pro torpedo. Piſcium nomina quædam facta ſunt à ui quadam, ut hæc: Lupus, canicula, torpedo, Varro. Albertus hunc piſcem ſtupefactorem nominat: aliqui ſimiliter mixobarbari, ſtuporem, ſtupefacientẽ, ſtupeſcore ut Auicennæ interpres. Iorath obſcurus ſcriptor rahas appellat, Arabico nomine, ut conijcio: Syluaticus rahadar. alij ex Auicenna citant rahadat. ego magis probo rahade ſcribi, (uel cum articulo alrahade, ut Bellunenſis habet) ſicut libro 2. Voluminis eius de medicina, capite 591. legitur. Harada quidem Hebræis ſtuporem ſignificat, ut docet Munſterus in Lexico trilingui: à uerbo harad uel charad, חרד, quod eſt ſtupuit, horruit. Berulie, id eſt piſcis ſtupefaciens, torpigo, Syluaticus. Thead, id eſt, torpedo, interpres Serapionis. Syluaticus legit fead per f. ab initio,

ILLUSTRATION 43C Publication of the torpedo ray by Gysbert van der Horst, in: Gessner (1558), p. 1188.

ZBZ, ABTEILUNG ALTE DRUCKE UND RARA: NNN 48

representations (ill. 42a, 42b, 42c; ill. 43a, 43b, 43c),[198] which were possessed by other scholars, as well as eight watercolors from a private collection that were described for the first time in 1996.[199]

Two of the eight illustrations – a ray egg case and a needlefish – can also be found in the Amsterdam album,[200] although Gessner published the needlefish (*Acus*) in his illustrated book of aquatic creatures.[201] Just as the fish images from Gysbert van der Horst could be seen in various Central European collections, the same bird pictures in Gessner's animal books also appeared in Jacques Daléchamps's animal album in Lyon. They had both received copies from John Caius in England.[202] It would appear that there was a lively exchange of animal images among scholars, within which certain well-done or favorite subjects were repeatedly copied and sent to various recipients.

In the above mentioned letter of April 22, 1550 to Kentmann, Gessner named a half dozen fish images from his list that he was interested in. He proposed that Kentmann should either have them drawn for him, and be reimbursed by Froschauer at the next Frankfurt book fair, or he should send the entire volume, so that he could copy everything of interest. Kentmann decided on the latter, for which Gessner later thanked him on December 1 and returned the material via Frankfurt on March 3, 1551.[203] It is not known whether Gessner did the drawings himself. Various surviving animal and plant studies prove that he was an outstanding draftsman with an exceptional feel for details. This is less evident in the fossilized horse's tooth from the countryside of Zürich[204] than in the case of the three aspects of a sea urchin. Here, for example, he made a precise copy of a woodcut published by Rondelet in his above-mentioned book on fishes and sea animals (ill. 44).[205]

Some of the bird images also trace back to Gessner, although conclusive proof is not always possible without written evidence.[206] As Gessner mentioned, Lukas Schan, a painter and bird catcher from Strasbourg, sent him

198 Egmond/Kusukawa (2016).

199 Holthuis (1996).

200 University Library Amsterdam: MS. C III 22, f. 107r.

201 Gessner (1560a), p. 92.

202 Van den Abeele (2002), p. 22. On these early modern picture albums, see Kusukawa (2011), pp. 194–198.

203 Gessner (1584), f. A_1v–A_2v.

204 Gessner (1560a), p. 356. Gessner had received the piece from Christian Hospinian, who had found it in the Zürich countryside, copied it and published it in original size.

205 Gessner wrote in his own copy of the *Nomenclator aquatilium animantium* (Zürich 1560), p. 257, in the margin: "Pinxi" (I have drawn it). ZBZ, call number: NNN 44_3.

206 Gmelig-Nijboer (1977), pp. 94f.

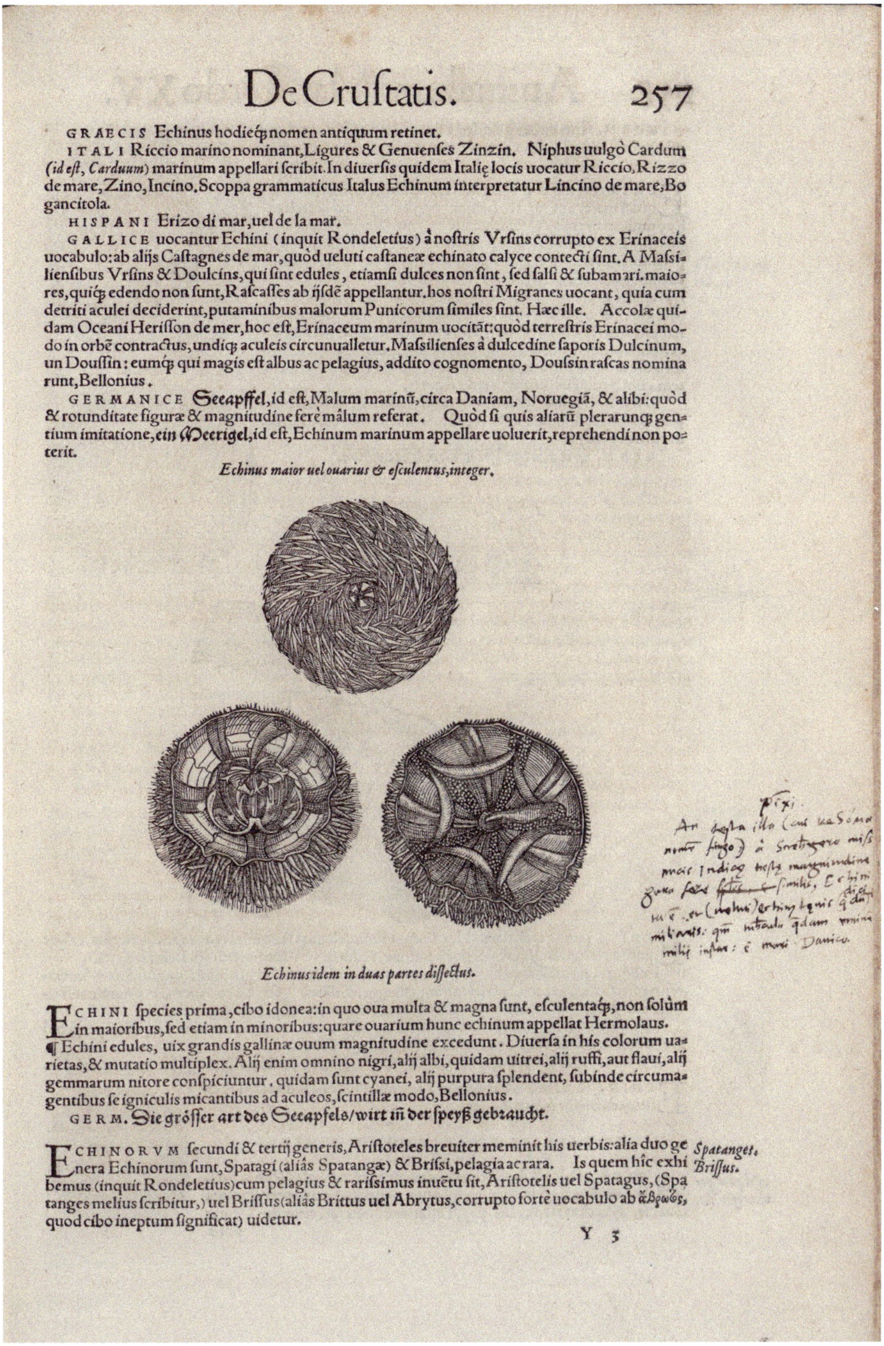

De Cruſtatis. 257

GRAECIS Echinus hodieq́ nomen antiquum retinet.

ITALI Riccio marino nominant, Ligures & Genuenſes Zinzin. Niphus uulgò Cardum (*id eſt, Carduum*) marinum appellari ſcribit. In diuerſis quidem Italię locis uocatur Riccio, Rizzo de mare, Zino, Incino. Scoppa grammaticus Italus Echinum interpretatur Lincino de mare, Bogancitola.

HISPANI Erizo di mar, uel de la mar.

GALLICE uocantur Echini (inquit Rondeletius) à noſtris Vrſins corrupto ex Erinaceis uocabulo: ab alijs Caſtagnes de mar, quòd ueluti caſtaneæ echinato calyce contecti ſint. A Maſsilienſibus Vrſins & Doulcins, qui ſint edules, etiamſi dulces non ſint, ſed ſalſi & ſubamari. maiores, quiq́ edendo non ſunt, Raſcaſſes ab ijſdē appellantur. hos noſtri Migranes uocant, quia cum detriti aculei deciderint, putaminibus malorum Punicorum ſimiles ſint. Hæc ille. Accolæ quidam Oceani Heriſſon de mer, hoc eſt, Erinaceum marinum uocitāt: quòd terreſtris Erinacei modo in orbē contractus, undiq́ aculeis circunuallet ur. Maſsilienſes à dulcedine ſaporis Dulcinum, un Douſſin: eumq́ qui magis eſt albus ac pelagius, addito cognomento, Douſsin raſcas nominarunt, Bellonius.

GERMANICE Seeapffel, id eſt, Malum marinū, circa Daniam, Noruegiā, & alibi: quòd & rotunditate figuræ & magnitudine ferè mâlum referat. Quòd ſi quis aliarū plerarunq́ gentium imitatione, ein Meerigel, id eſt, Echinum marinum appellare uoluerit, reprehendi non poterit.

Echinus maior uel ouarius & eſculentus, integer.

Echinus idem in duas partes diſſectus.

ECHINI ſpecies prima, cibo idonea: in quo oua multa & magna ſunt, eſculentaq́, non ſolùm in maioribus, ſed etiam in minoribus: quare ouarium hunc echinum appellat Hermolaus. ¶ Echini edules, uix grandis gallinæ ouum magnitudine excedunt. Diuerſa in his colorum uarietas, & mutatio multiplex. Alij enim omnino nigri, alij albi, quidam uitrei, alij ruffi, aut flaui, alij gemmarum nitore conſpiciuntur, quidam ſunt cyanei, alij purpura ſplendent, ſubinde circumagentibus ſe igniculis micantibus ad aculeos, ſcintillæ modo, Bellonius.

GERM. Die gröſſer art des Seeapfels/wirt iñ der ſpeyß gebraucht.

ECHINORVM ſecundi & tertij generis, Ariſtoteles breuiter meminit his uerbis: alia duo genera Echinorum ſunt, Spatagi (aliâs Spatangæ) & Briſsi, pelagia ac rara. Is quem hîc exhibemus (inquit Rondeletius) cum pelagius & rariſsimus inuētu ſit, Ariſtotelis uel Spatagus, (Spatanges melius ſcribitur,) uel Briſſus (aliâs Brittus uel Abrytus, corrupto fortè uocabulo ab ἄβρωτος, quod cibo ineptum ſignificat) uidetur. *Spatanges. Briſſus.*

Y 3

ILLUSTRATION 44 Gessner drew the sketch himself for this woodcut of a sea urchin, which is noted in the margin of his private copy ("pinxi"). Gessner (1560a), p. 257.

ZBZ, ABTEILUNG ALTE DRUCKE UND RARA: NNN 44

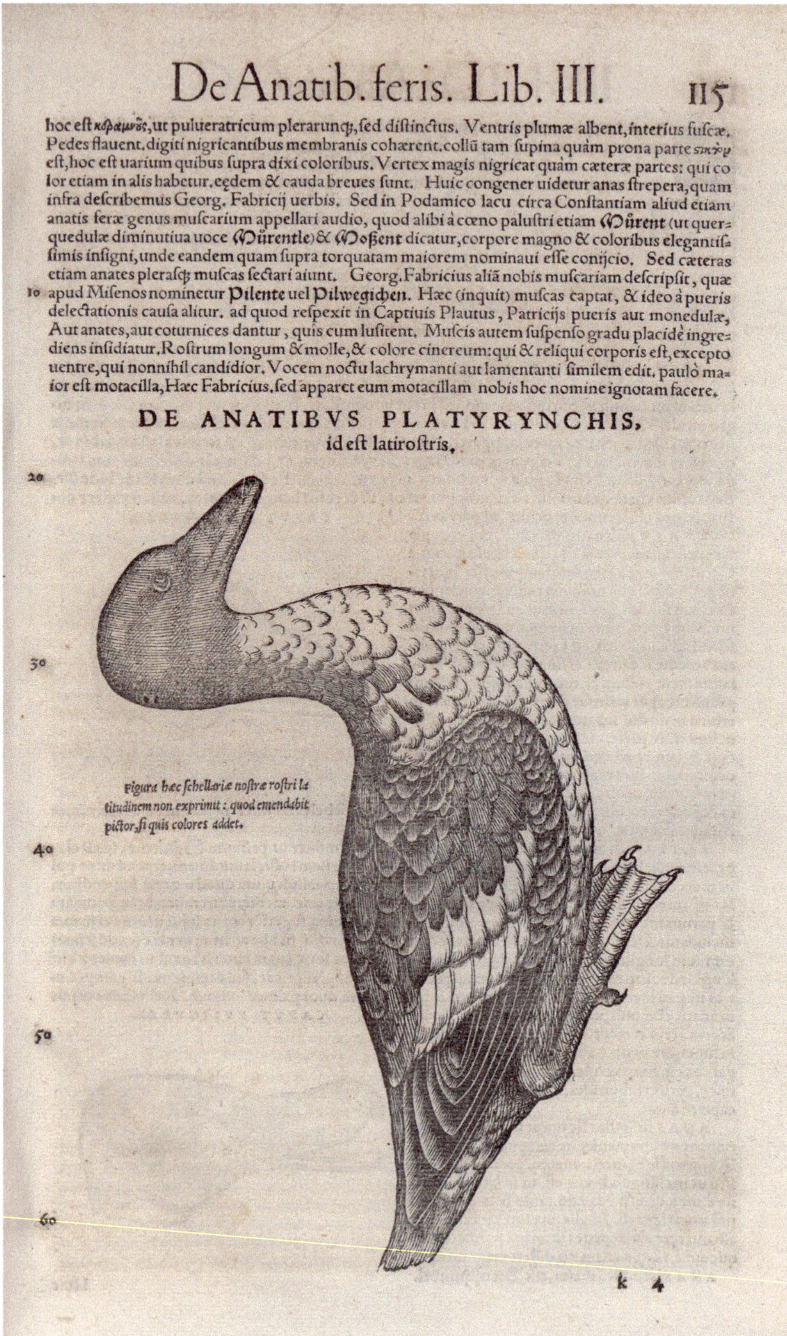

De Anatib. feris. Lib. III. 115

hoc est κορφαμνὸς, ut puluerat ricum plerarunq;, sed distinctus. Ventris plumæ albent, interius fuscæ. Pedes flauent. digiti nigricantibus membranis cohærent. collũ tam supina quàm prona parte στικτὸν est, hoc est uarium quibus supra dixi coloribus. Vertex magis nigricat quàm cæteræ partes: qui color etiam in alis habetur. eędem & cauda breues sunt. Huic congener uidetur anas strepera, quam infra describemus Georg. Fabricij uerbis. Sed in Podamico lacu circa Constantiam aliud etiam anatis feræ genus muscarium appellari audio, quod alibi à cœno palustri etiam Mürent (ut querquedulæ diminutiua uoce Mürentle) & Moßent dicatur, corpore magno & coloribus elegantissimis insigni, unde eandem quam supra torquatam maiorem nominaui esse conijcio. Sed cæteras etiam anates plerasq; muscas sectari aiunt. Georg. Fabricius aliã nobis muscariam descripsit, quæ apud Misenos nominetur Pilente uel Pilwegichen. Hæc (inquit) muscas captat, & ideo à pueris delectationis causa alitur, ad quod respexit in Captiuis Plautus, Patricijs pueris aut monedulæ, Aut anates, aut coturnices dantur, quis cum lusitent. Muscis autem suspenso gradu placidè ingrediens insidiatur. Rostrum longum & molle, & colore cinereum: qui & reliqui corporis est, excepto uentre, qui nonnihil candidior. Vocem noctu lachrymanti aut lamentanti similem edit. paulò maior est motacilla, Hæc Fabricius. sed apparet eum motacillam nobis hoc nomine ignotam facere.

DE ANATIBVS PLATYRYNCHIS, id est latirostris.

Figura hæc schellariæ nostræ rostri latitudinem non exprimit: quod emendabit pictor, si quis colores addet.

k 4

ILLUSTRATION 45 Common goldeneye, in: Gessner (1555a), p. 115.
ZBZ ABTEILUNG ALTE DRUCKE UND RARA: NNN 48

various animal illustrations.[207] Sometimes, Gessner would correct artists' drawings, as in the case of the common goldeneye (*Bucephala clangula*). In the printed legend, he noted that the duck's bill was too narrow (ill. 45). In case someone might wish to color the image, the artist was recommended to take this into account and widen the beak.[208]

After the publication of each animal book, Gessner also continued to search for better and, whenever possible, lifelike illustrations with an eye toward future editions. The first, richly illustrated addition to the first volume of the *Historia animalium* of 1551 came in the second volume of 1554 in the form of a comprehensive 27-page appendix. Sometimes he would print more than one picture of an animal in order to give the best possible impression of its form. In his personal copy of the *Icones animalium*, he often noted that he had alternate images for some animals. This might have been an extra picture of a tiger that was made at the court of Suleiman I in Istanbul, or a better illustration of the marmoset monkey, or an accurate drawing of a certain kind of buffalo.[209] Siegmund Freiherr of Heberstein, Neyperg and Guttenhag (1486–1566) was the provisional Austrian envoy at the Russian court and author of the repeatedly published *Rerum Moscoviticarum Commentarii*, which first appeared in Vienna in 1549. Gessner dedicated part of his book of aquatic animal plates to him and related in the dedication epistle of May 23, 1560 that he would be very grateful if the Freiherr had pictures painted of the Hungarian fish that lived in Lake Neusiedl and Lake Balaton, and provide their names in the German, Hungarian and Slavic vernacular.[210] On August 29, 1561, he thanked John Caius for five new illustrations, including a chamaeleon.[211] He also asked Crato von Krafftheim on August 1, 1563 for the fish illustrations by Johannes Moibaus (1527–1562). Six months later he asked whether there were animals at the imperial court in Vienna that he had not yet published.[212] Gessner's correspondence reveals to us a scholar with a discerning eye who had spent his entire life hunting for pictures of animals. He mentioned in a letter to Caius that to prevent the images from getting lost in his chaotic world of notes and snippets, he glued them into an album. Gessner wrote that the color illustrations Caius had sent were

207 Gessner (1551), f. γ_1v. Cf. Springer/Kinzelbach (2009), p. 421. Four bird drawings by Schan are kept in the UB Basel, cf. Leemann-van Elck (1935), p. 9.

208 Gessner (1555a), p. 114.

209 Gessner (1560), pp. 33, 67 and 96 (ZBZ, call number: NNN 44_1).

210 Gessner (1560a), p. 338.

211 Gessner (1577), f. 133v–136v.

212 Gessner (1577), f. 12v and f. 17r.

placed and glued into his book of animal images, and should Caius want them back, Gessner would have to have new drawings made.[213]

Gessner placed a high value on the quality of the illustrations, which he arranged to be drawn either from life or from a suitable sketch. For this reason, he was sceptical about those received from Olaus Magnus.[214] He had previously passed judgment on Magnus's illustrative material in the *Bibliotheca universalis*, with little confidence that it was based in reality.[215] In the introduction to the first volume of the *Historia animalium*, he wrote that he had made sure that all illustrations either corresponded to the living animal, or that he had drawn them himself. He had also received some from trustworthy friends. It was difficult to obtain good pictures of some creatures that had either a dangerous bite or were poisonous. He also wished for the woodcuts to represent the original colors of the animals. Since that was not possible because of technical reasons, the printer had the illustrations tinted according to Gessner's sketches for those who wished to have a colored version of the book. Furthermore, to the extent that page size permitted, the smallar animals among the birds, fish and insects were depicted at their actual size.[216]

The desire for and emphasis on true-to-life (*ad vivum*)[217] illustrations, arose from the *Zeitgeist* of the Renaissance, which influenced numerous painters of nature from the fourteenth to the sixteenth century. Many famous plant and animal images came from Albrecht Dürer, but he was not the first in Germany to make exact drawings from nature. This should be credited much more to Hans Pleydenwurff (ca. 1420–1472), Martin Schongauer (1445–1491) and others: "Dürer was not the first artist north of the Alps to draw from nature, but from what we know today, his is the earliest work that has been preserved. This is especially due to Dürer's own personal efforts to conserve *memoria*."[218] The new observance of nature might have had its roots in the nominalism of the thirteenth century, which took an early foothold in Northern Italy and Flanders.[219] As expressed in a June 22, 1554 letter to Georg Fabricius (1516–1571), Gessner also strove for true-to-life images that captured the illusion of reality. In the letter, he refers to a dispute between the two ancient artists, Zeuxis of

---

213 Gessner (1577), f. 135r. Also in Gessner's own copy of the *Icones animalium* (ZBZ, Shelf mark: NNN 44_1) there is a reference to his animal album on p. 15, where he noted to the illustration of a sheep: "In libro iconum meo 47." Apparently, the master drawing for the woodcut was on page 47 or bore the number 47.

214 Gessner (1558), f. b_2v.

215 Gessner (1545), f. 526v.

216 Gessner (1551), f. γ_1v.

217 For a discussion of the term painting "ad vivum," see, among others, Gessner: Kusukawa (2014), pp. 338–341.

218 Hess (2012), pp. 119f.

219 Wolf (2002), pp. 19–22.

Herakleia and Parrhasius to decide who was the better painter. Pliny the Elder recounted: "The latter [Parrhasius] is said to have engaged Zeuxis in a competition; Zeuxis had painted some grapes so realistically that birds were attracted to them. Parrhasius, however, had set up a realistically rendered painting of a curtain. Zeuxis was so proud of having deceived the birds, he insisted that the curtain be pulled open to display his painting. With embarrassment, he realized his error and declared himself the loser. He had fooled the birds, but Parrhasius had fooled him – an artist."[220]

After this reference to Pliny, Gessner followed up in his letter to Fabricius with praise for lifelike illustrations. This was the highest level that an artist could achieve, to emulate nature and make nonexistent things appear as if they were really there.[221] Of course, neither Gessner nor any of his contemporaries could live up to such a high standard.[222] This did not stop Caspar Wolf from posthumously praising Gessner for his plant drawings as another Zeuxis.[223]

12.8 Price and Distribution

Gessner's animal books can be found in many European libraries, both large and small, from Helsinki to Rome and Madrid to Moscow. They are even in the secluded libraries of aristocrats, such as the copy owned by the Pianetti family in Jesi, Italy.[224] Since professionally tinted woodcuts would have demanded a very high price, they are most often un-colorized copies. Froschauer tried to keep the price down by having his painter colorize twelve or more copies at a time. However, Gessner found the palette to be unnatural. If the artist had carefully painted each book one at a time, the price would have been too high.[225]

In a July 27, 1561 letter from the Heidelberg theologian Zacharias Ursinus (1534–1583) to Crato von Kraffheim, we learn that the total price for all four volumes of the *Historia animalium* at the Frankfurt book fair was 6 guilders and 17 shillings. The three books of illustrations (*Icones animalium quadrupedum, Icones avium* and *Nomenclator aquatilium animantium*), on the other hand, were priced at 1 guilder and 17 shillings. If colorized, the animal books could be bought for a total of 18.5 guilders. Ursinus listed the prices for individual volumes of the *Historia animalium* as follows:[226]

220 Plinius, *Naturalis historia*, 35,64.
221 Gessner (1577), f. 131r–132r.
222 Springer/Kinzelbach (2009), p. 412.
223 Wolf (1566), f. 48r.
224 Bigliardi Parlapiano (1997), p. 64.
225 Gessner (1577), f. 135r.
226 Becker (1892), p. 85.

TABLE 4 Prices for Gessner's animal books.

	Un-colorized	Colorized
Vol. 1 (1153 pp., 82 fig.)	2 guilders	4 guilders
Vol. 2 (143 pp., 43 fig.)	7 schillings	1 guilder 10 schillings
Vol. 3 (813 pp., 217 fig.)	1 guilder 10 schillings	6 guilders
Vol. 4 (1337 pp., 737 fig.)	3 guilders	7 guilders 10 schillings
Icones animalium	7 schillings	
Icones avium	10 schillings	
Nomenclator aquat. anim.	1 guilder	

Thus, a colorized copy was priced at two to seven times as much as one that was not colorized, depending on the number of tinted woodcuts. In a little less than a year the books had become somewhat more expensive. Gessner wrote on May 27, 1562 to Melchior Wieland that the bookseller's price for the animal books was 7.25 guilders and about 1 batzen. The books of illustrations were priced at 1 guilder and 10 batzen.[227] Two years later, as mentioned in a February 6, 1564 letter to Crato von Krafftheim,[228] the four volumes of the *Historia animalium* cost 7 guilders. As Gessner informed Joachim Camerarius the Younger on January 27, 1565, the colorized version sold for 7 to 8 guilders. In the same letter, he stated that due to higher demand for the volumes with illustrations, their price had multiplied, and even booksellers had to pay 6 guilders for them.[229] This amount was confirmed in a letter from Gessner to Achilles Pirmin Gasser on March 4, 1565.[230] Gessner procured colorized copies for his Augsburg colleague Gasser that Froschauer had not stored in Zürich, but in Frankfurt instead. This meant that Gasser had to wait. On November 6, 1564 Gessner wrote that in the meantime Gasser should have received the books of illustrations about four-legged animals and birds. On January 15, 1565, he sent Gasser the last of the three volumes, a colorized copy of the aquatic animals costing 6 guilders.[231] Today, Gasser's private copies of the *Icones animalium* and the *Icones avium* are kept in the Vatican Library in Rome. Gasser entered a price for the first of these at 6.5 guilders.[232] Therefore, all un-colorized books

227 Gessner (1577), f. 140v. Cf.: 1 guilder = 2 pounds = 15 batzen = 40 shilling.
228 Gessner (1577), f. 20r.
229 Rath (1950), p. 160.
230 Burmeister (1975), pp. 345f.
231 Burmeister (1975), pp. 288, 305, 332 and 338.
232 Kusukawa (2012), p. 61.

of illustrations were available for 6 guilders, and each colorized book of illustrations also cost around 6 guilders. This amounted to a total of 18 guilders for all three. Gessner wrote to Camerarius on August 27, 1565 that the illustrated volumes were out of print and no longer available.[233]

It is clear from these amounts that books in the sixteenth century were expensive, especially the illustrated or colorized titles. This is even more obvious when the prices are compared to incomes. Heinrich Bullinger earned about 350 guilders per year as head of the church in Zürich. A country pastor, however, earned only 40 guilders, plus some in-kind produce.[234] For his professorship, Gessner received 80 guilders per year, plus produce. He also earned 20 guilders after being selected to be the municipal physician, as well as varying sums for his publishing activities.[235] It is no wonder that in 1558 he turned to Bullinger with the urgent request for a raise in salary. This resulted in an appointment as canon with the receipt of a benefice that relieved him of his financial worries.[236] A look at the earnings of a craftsman shows that it was nearly impossible to afford books or education. In the mid-sixteenth century a master carpenter, mason or roofer in Zürich earned 6 guilders per month. His journeyman earned only 5.25 guilders per month. A half guilder could buy about 10 kilograms of beef or veal.[237]

Obviously, both Froschauer and Gessner wished to protect their investments in these magnificent works and prevent others from earning money from them by producing cheap copies elsewhere. During the early modern era, the only possibility was to be granted printers' privileges by a ruler who forbid unauthorized reproduction within his domain. The highest level of protection came from imperial and royal privileges, which is exactly what Gessner sought to obtain. Copy prohibitions that were granted by Emperor Charles V (1500–1558) and France's King Henry II (1519–1559) appear in print after the title pages within the two most lavishly illustrated animal books – the bird book of 1555 and the fish book of 1558. This was equivalent to copyright protection covering a large portion of early modern Europe. Gessner also attempted to obtain such a privilege from the king of France for the second printed volume of the history of animals in 1554. This is gleaned from a letter sent by the French ambassador Antoine Morelet du Museau to Bullinger on October 14, 1551 in which he wrote of having passed along the request to the court. He

233 Rath (1950), pp. 163f.
234 Leu/Weidmann (2004), pp. 26f.
235 Cf. p. 348.
236 Cf. p. 353.
237 Hauser (1962), plate in the appendix.

was to communicate with Zürich as soon as he had any news.[238] The entire effort took a long time,[239] which is why the royal decree first appeared in the third and fourth volumes. At the Habsburg court, the imperial personal physician Julius Alexandrinus (1506–1590) repeatedly lobbied for privileges for Gessner.[240] Being able to secure such a protective document more easily may well have been a consideration when choosing the location of a printer. This was expressed by Gessner in a letter to Crato von Krafftheim on March 17, 1561: "I am asking and urging you to finish the book *De destillationibus* that is still in progress. I hope I can find a good and respectable printer for you either in Basel or preferably Lyon. In addition to your privilege, it will be easy for a Lyon-based printer to also obtain one from his king, and thus he will be more inclined to do the printing."[241]

12.9 The History of Its Impact

Gessner's *Historia animalium* and the books of illustrations enjoyed great popularity and broad acceptance. This work by the Zürich polyhistor was not only considered the first encyclopedia of animals in the Western World, but it was possibly the high point of the "zoological decade" between 1550 and 1560, when numerous relevant works were published: "Zoology took off twenty years after botany; it was during the decade 1550–1560 that a real explosion of original publications on animals took place."[242] It is no wonder that by 1557 the bird book had already been translated into German followed by the other volumes in 1563. Whereas the Latin texts targeted scholars, the folios in German were meant for the aristocrats and patricians who had the financial means but not sufficient Latin language skills.

Josias Simmler wrote in the *Epitome Bibliothecae Conradi Gesneri*, which was printed in March 1555, that the Basel proofreader and translator Johannes Basilius Herold (1514–1567) had begun to translate the first volume of the *Historia animalium* into German.[243] Because Herold unabashedly took liberties with the text and even inserted autobiographical information,[244] the work was apparently taken away from him. The Winterthur doctor and pastor,

238 Letter from Morelet de Museau to Bullinger, October 14, 1551 (StAZH, E II 359, 2878f.).
239 Letter from Morelet de Museau to Bullinger, February 3, 1552 (StAZH, E II 359, 2881).
240 Cf. for example: Burmeister (1975), pp. 226 and 240–242.
241 Helmich (1938), p. 13; Gessner (1577), f. 8r.
242 Pinon (1995), p. 15.
243 Simmler (1555), f. 101r.
244 Burckhardt (1967), p. 97; Friedrich (1995), pp. 173–188.

Conrad Forrer (ca. 1530–1594), who finished the translation, wrote in the foreword to readers: "What concerns the translation, please take note and understand. Initially, the experienced Mr Johannes Herold took on this assignment and made it to the letter 'F' where the description of the fox begins. However, he elaborated too much in doing so, which would have been too cumbersome to the buyer and too tiresome for the reader. Furthermore, he was too taken with other things. I agreed to undertake it at the urging and request of both the most erudite Dr Conrad Gessner, who first wrote the book in Latin, and the printer. With as much diligence as possible, I have carried it through to the end. Afterwards, I also prepared the book on fishes for print in the same way."[245]

In Zürich, Forrer had enjoyed Gessner's classes in natural science, and later he studied medicine together with Felix Platter in Montpellier and obtained his doctorate in Avignon in 1555.[246] He was under the impression that not everyone was capable of such high-level work without the required knowledge of natural science: "Since such work of translation cannot be done by just anyone unless they have acquired a thorough knowledge of medicine, the characteristics of diseases and the like, such as dealt with in this book and in the customary old writings. In particular, he should love his neighbor and the German-speaking nation, which until now has been deprived of such a good. For what if so many natural histories had been collected from the ancients in a single book by the good and dear Dr Conrad Gessner with much insight, effort, work, incredible diligence and good presentation, but be of no use and benefit to the German-speaking nation? However, in order that many people, including the common man, might benefit from this important work, it was necessary not to delay this translation any longer."[247]

In the foreword of the fish book, which was also translated by Forrer, he states that he had translated it "some years ago," and that now in 1563 "it should be published."[248] Froschauer had probably hired Johannes Basilius Herold to translate the first two volumes of the *Historia animalium* about four-legged animals, Rudolf Hüsli (see below) to translate the bird book in 1555 (published in the same year) and Conrad Forrer for the 1558 translation of the fish book. Forrer then granted Froschauer's wish that he work on the torso of text that Herold had left behind, which he probably brought to a satisfactory conclusion in 1562. This made it possible for the animal and fish books to be published in 1563. It is highly likely that the translation work helped Forrer to keep

245 Gessner (1563a), f. a_{iii}r.
246 Schmid-Forrer (1983), pp. 18f.
247 Gessner (1563), f. a_{ii}r/v.
248 Gessner (1563b), f. aaa_{ii}v.

his head above water financially. Ambrosius Blarer wrote on February 17, 1562 to Bullinger that medical services in Winterthur were underappreciated and presumably poorly paid. In the course of that same year, Forrer changed professions and with Blarer's support began work as a pastor.[249] Nonetheless, it seems that he continued to practice medicine, which is evidenced in the six medically related letters sent to him by Gessner from 1563 to 1565.[250] Still benefitting from the plant specimens Forrer had sent to him, Gessner wrote advice to his colleague in Winterthur as to how to deal with epilepsy and postnatal psychosis.

Parallel to Herold and Forrer, Pastor Rudolf Hüsli (1524–1600) began translating the bird book in the summer of 1555. He had studied in Zürich and Oxford with Josua Maler (1529–1598) and Johann Rudolf Stumpf (1530–1592) and had been ordained in 1553. In the same year, he married Katharina Sprüngli with whom he raised 14 children. After stays in Albisrieden and Rafz, he began work on September 23, 1554 as a pastor in Töss, where he lived in modest circumstances. His residence was an extremely run-down house, "that no one wished to be living in during a strong wind for fear of collapse."[251] On September 22, 1555 he delivered a sermon about the giving of alms in which he criticized the administration of Zürich's church properties. He felt that the state was skimming off too much money for itself from the abolition of monasteries. Most of the wealth from former church property should have actually gone to the new reformed church. The use of these monies "became the most important topic of discussion between the church leadership and the council during Bullinger's time in office."[252] It did not take long for the council to hear about Hüsli's critical remarks. On December 10, he was unceremoniously arrested, thrown in the Wellenberg tower jail, and forbidden to give any further sermons. This caused Bullinger and the pastorship to get involved and express their displeasure about how their colleague had been treated. Discussions about the *Causa Hüsli* went on for months.[253] From December 21, 1555 Hüsli filled the pastor position in Weiach, and on March 30, 1557 he was hired in Zollikon with a raise in pay on September 24. From April 1559 until his death, he worked as the pastor in Dinhard.[254]

---

249 Schmid-Forrer (1983), pp. 26f.

250 Gessner (1577), f. 125r–128v; translated in: Schmid-Forrer (1983), appendix, pp. 18–25.

251 Bächtold (1982), p. 173.

252 Bächtold (1982), p. 143.

253 Cf. Bächtold (1982), pp. 168–180. Cf. also the handwritten *Collectanea ecclesiastica et politica Tigurina* by Hans Rudolf Steiner the Younger from 1667 (ZBZ, MS. J 305, pp. 68r–74r).

254 Hans Wilpert Zoller: *Schriftstücke zur schweizerischen Kirchengeschichte des 16. und 17. Jhs.* (ZBZ, MS. J 59_{14}, f. 123v).

In the foreword of August 1, 1557 about his translation of Gessner's bird book, Hüsli wrote that the work had taken him nearly two years to complete and was dedicated to the Zürich council as proof of his loyalty. Moreover, because of the uneventful time spent in Weiach, rather than remain idle, he hoped to prove himself useful to the general public by translating this book. It is very likely that Froschauer's financial backing helped to offset the poor salary of a country pastor in the little community of Weiach. Hüsli went on to say that the work was also important to him as a theologian. It pointed out creation and thus the Creator, and there was much to be learned from the animals about living life as a Christian.

In the foreword to the reader that followed, Hüsli explained that he had decided to make cuts everywhere to make the book thinner, easier and more readable compared to the Latin text. Initially, in chapter "A" he removed all the foreign names and their explanations, condensed the following sections and completely omitted the last chapter "H." It was also possible that he might have incorrectly translated some of the technical terms because he was neither a painter, world traveler nor doctor. With apologies, he also offered an explanation about the translation of plant names: "The herbs and medicinal plants have so many different names, that even the most educated people in this area are not in agreement."[255]

It was not until 22 years after Gessner's death that the fifth volume of the *Historia animalium*, covering snakes and scorpions, was published in 1587. The larger portion, about snakes, was compiled from Gessner's estate by Jacob Carron from Frankfurt, while Caspar Wolf put together the section on scorpions from Gessner's notes. Carron dedicated the foreword to the English ambassador and wealthy businessman Horatio Palavicino († 1600), who was knighted by Queen Elizabeth I in the same year, 1587. In it, he mentioned that Wolf had purchased Gessner's private library for a sizable sum before his death. Included were a large number of pages and notes, which, among other things, contained the research about snakes. Through arrangements by Christoph Froschauer the Younger, Carron's relative, the bookseller Robert Cambier, then bought them from the overburdened Wolf. Carron also added in his own research of the literature, which makes it not only Gessner's but also Carron's snake book.[256] Once again, it was Conrad Forrer who translated the work into German so that it could be published in 1589 by Froschauer in Zürich. Forrer writes in the foreword to the reader that it amounted to a pioneering achievement by Gessner since no one had ever written a book about snakes. He emphasized that this

255 Gessner (1557), f. aa_4r.
256 Gessner (1587), f. $*_2v$.

creature also glorified God's wisdom and was fascinating and worthy of study. Making the statement that researching nature must not be primarily beneficial was ahead of its time and of special interest: "Because describing snakes is part of the exploration and study of nature (disregarding fancy and usefulness) some scholarly men have also explored their gender and their characteristics with great enthusiasm and written about them when possible. But no one has ever been found who would have described and compiled everything and anything in this field with such effort and work as Conrad Gessner of blessed memory, if God had given him the time to complete it."[257]

The snake book, completed by Carron and printed by Froschauer the Younger in 1587, was part of the bookseller Robert Cambier's plan to reissue Gessner's *Historia animalium*, which he cited in 1585 as no longer available.[258] As a prelude, Cambier had the bird book printed at his own expense in the same year by Johannes Wechel, who ran one of the nine Frankfurt printing houses at the time.[259] Cambier wrote in the foreword to the reader that for the edited and extended edition of the *Historia avium*, a private copy belonging to Gessner had been made available with corrections, annotations and an appendix of additional notes.[260] Moreover, all bird illustrations and descriptions that could be found in his estate that did not appear in the first edition were incorporated in this one. The Frankfurt edition had 25 new images, plus textual additions and corrections, some of which were not added to the original text but were compiled in the appendix (*paralipomena*).[261] It was similar for the new edition of the fish book that Adreas Cambier published in Frankfurt in 1604. However, the new versions of the second volume in 1586 and the reprinted first volume of the *Historia animalium* in 1603 did not have separate appendices.

At this point, it would be a step too far to discuss also the further Latin editions, Edward Topsell's (ca. 1572–1625) English variant, Geoffroi Linocier's French version, the revival of Gessner's animal books in Johannes Heiden von Dhaun's book on Pliny, or in the works of other writers.[262] Likewise, at this juncture the publications that resulted from collaborations with Gessner will not be explored, such as *De canibus britannicis liber* (London, 1570) or

257 Gessner (1589), f. $*_{ii}$r.

258 Gessner (1585), f. $*_{5}$v.

259 Richter (1986), pp. 133f.

260 Gessner was wont to add annotations and corrections to copies of his own works with a view to a new edition, as he informed Achilles Pirmin Gasser on December 3, 1565. Cf. Burmeister (1975), p. 364.

261 Springer/Kinzelbach (2009), pp. 10 and 395.

262 Cf. Wellisch (1984), pp. 59–71; Friedrich (1995), pp. 188–246. On the survival of Gessner's animal images in manuscripts, cf. exemplarily Pyle (1984).

the *De rariorum animalium atque stirpium historia* (London, 1570) by John Caius.[263] According to Gessner, the previously mentioned work on dogs was to be printed in Zürich by Froschauer, who had the appropriate illustrative material.[264] For unknown reasons, this did not occur.

It is remarkable that Gessner's work and his iconographic program endured over decades and centuries and also sometimes appeared in completely unexpected contexts. Gessner complained to Crato von Krafftheim in a letter of August 1, 1563 that he was pestered by sculptors and painters who sought accurate plant illustrations to be used as resource images for their own work.[265] The "afterlife" of his animal illustrations in the visual arts has not been systematically researched.[266] However, it was clearly no coincidence that the potter and oven painter Hans Heinrich Pfau (1559–1636) hand copied the 1563 German translation of the first two volumes of the *Historia animalium* about four-legged animals. He also copied the woodcuts, while adding to and improving upon a few illustrations, as one can see from the Elephant imagery (ill. 46a and 46b).[267] Unfortunately, no oven tiles have been found with Gessner's animal images on them, although it is believed that some did exist.

There was also interest in Gessner's animal illustrations in foreign lands, including by none other than Queen Mary Stuart (1542–1587), who copied them into tapestries. Her embroidery brought a measure of diversion to the dull day-to-day existence of her 18-year imprisonment. Altogether, she took at least 60 animals from Gessner's books and stitched them with other subjects into tapestries that can still be admired today.[268] In addition to Gessner's works, she also used the woodcuts of plants that she found in Pietro Andrea Mattioli's (1501–1577) *Commentarii in Dioscoridis de medica materia* for reference. The source images for embroidery of animals from the New World came not only from Gessner, but also from André Thevet's *Singularitez de la France antarctique, autrement nommée Amerique*, which had been published in Antwerp and Paris in 1558.[269] Both editions were also to be found in Gessner's private library and were heavily annotated by him (ill. 47a and 47b).[270]

The French Franciscan Thevet (1516–1590) participated as pastor in the 1555 expedition of Vice Admiral Nicolas Durand de Villegagnon (1510–1571) to Brazil.

263 Perfetti (2000), p. 192.
264 Letter from Gessner to Theodor Zwinger, April 8, 1565, cf. Gessner (1577), f. 111v.
265 Gessner (1577), f. 13r.
266 A nice start for such studies was made by Ruoss (2019).
267 The manuscript is kept in the Stadtbibliothek Winterthur (MS. Fol. 228).
268 Bath (2008), p. 70.
269 Bath (2008), pp. 77–85 and 113–123.
270 Leu (1992), pp. 285f.

Von dem Helfant.

also die fliegen/auch andere überlägne thier vnn jr läben. Gillius spricht er habe haar von den Elephanten gesähen zweyer spangen lang. Was weyter von der gestalt vnn form der Elephanten erforderet wirt/mag auß der fürgestelten figur ersähen werden.

Von innerlicher gestalt der Elephanten.

Vier

ILLUSTRATION 46A Elephant, in: Gessner (1563a), f. 74v.
ZBZ, ABTEILUNG ALTE DRUCKE UND RARA: NS 4

ILLUSTRATION 46B Hans Heinrich Pfau hand copied the German version of Gessner's animal book (Vol. 1), replacing Gessner's elephant with its "screw-like" trunk with an improved illustration.
STADTBIBLIOTHEK WINTERTHUR: MS. FOL. 228, F. 80V

DE LA FRANCE ANTARCT.

gue Ahouaï, portant fruit veneneux et mortel, lequel
eſt de la groſſeur d'vne chaſtaigne moyēne, et eſt vray
poiſon, ſpecialement le noïau. Les hommes pour legere
cauſe eſtant courroucez cōtre leurs femmes leur en don
nent, & les femmes aux hommes. Meſmes ces malheu
reuſes femmes, quand elles ſont enceintes, ſi le mary les
a faſchées, elles prendront au lieu de ce fruit, certaine
herbe pour ſe faire auorter. Ce fruit blāc auec ſon noïau
eſt fait comme vn Δ delta, lettre des Grecs. Et de ce
fruit les Sauuages, quand le noïau eſt dehors, en font
des ſonnettes qu'ils mettent aux iambes, leſquelles font
auſsi grand bruit comme les ſonnettes de par deça. Les

Sauuages pour rien ne donneroiēt de ce fruit aux eſtra
gers eſtant fraiz cuilly, meſmes defendent à leurs en-
fans y attoucher aucunement, deuant que le noïau en

K

ILLUSTRATION 47A One of Gessner's personal copies of: André Thevet, *Les singularitez de la France antarctique, autrement nommée Amérique*, Anvers 1558, p. 67. Gessner noted on the center margin of the page that Petrus Coldenbergius in Antwerp had sent him a specimen of the illustrated delta-formed fruit from the ahouai tree (*Thevetia ahouai*), and that he had made a drawing of it.

UNIVERSITY LIBRARY BASEL: HX VI 30

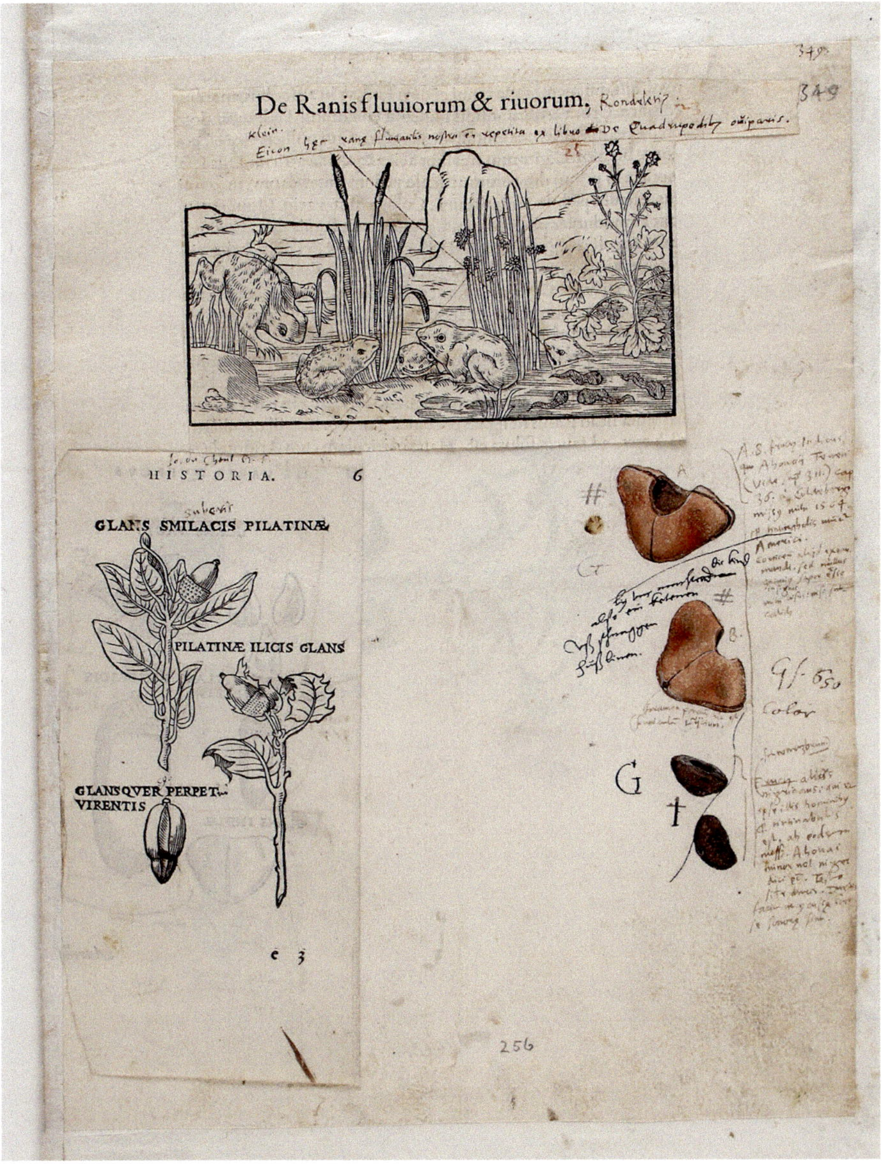

ILLUSTRATION 47B Gessner's drawing of the *Thevetia ahouai* fruit in the second volume of the *Historia plantarum* has been preserved in Erlangen. His drawing is far better that the woodcut by Thevet. Conrad Gessner. *Historia plantarum.*

UNIVERSITY LIBRARY ERLANGEN, MS. 2386/2, F. 349R

They reached Rio de Janeiro on November 10, but Thevet could not go ashore due to illness. Ten weeks later, he returned to France on the next ship, having never personally experienced the New World. An interesting fact, and possibly the reason for Queen Mary's interest in Thevet's work, is that Villegagnon was in command of the ship that brought her from Scotland to France in 1548 when she was five years old, and then back again in 1561.[271]

Gessner's animal images did not remain confined to the European continent and managed to make their way overseas as far as China. In the second half of the seventeenth century, the Flemish Jesuit Ferdinand Verbiest (1623–1688) reproduced 15 animals that were clearly taken from Gessner's work on his printed world map *K'un-yü ch'üan-t'u* and the accompanying illustrated description of the world *K'un-yü t'u-shuo* (ill. 48a and 48b). From there, some of Gessner's animal illustrations made their way into the Chinese encyclopedia *Gujin tushu jicheng* in 1725.[272]

271 Bath (2008), p. 77.

272 Walravens (1973); Lai (2018), pp. 148–158.

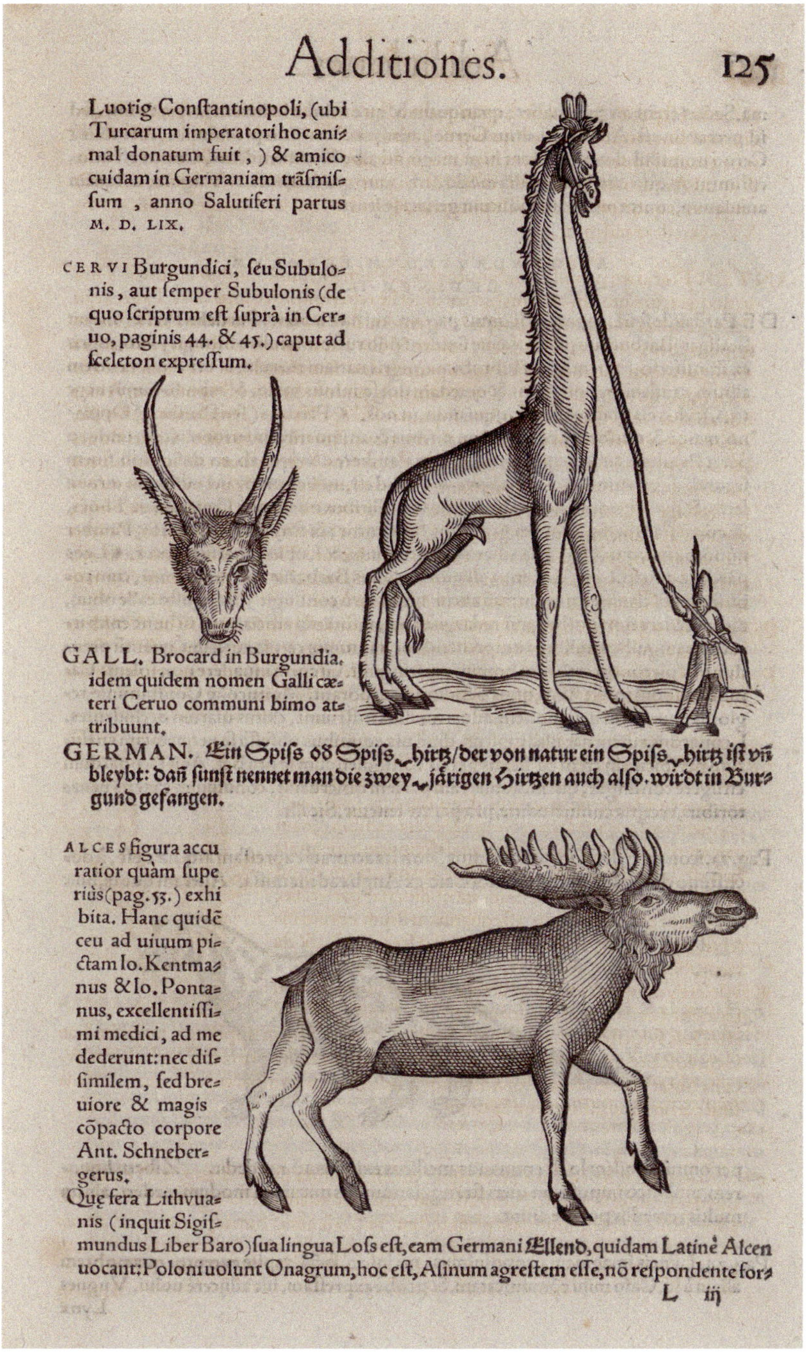

Additiones. 125

Luotig Constantinopoli, (ubi Turcarum imperatori hoc animal donatum fuit,) & amico cuidam in Germaniam trãsmissum, anno Salutiferi partus M. D. LIX.

CERVI Burgundici, seu Subulonis, aut semper Subulonis (de quo scriptum est suprà in Ceruo, paginis 44. & 45.) caput ad sceleton expressum.

GALL. Brocard in Burgundia. idem quidem nomen Gallicæteri Ceruo communi bimo attribuunt.

GERMAN. Ein Spiss od Spiss-hirtz/der von natur ein Spiss-hirtz ist vñ bleybt: dañ sunst nennet man die zwey-järigen Hirtzen auch also. wirdt in Burgund gefangen.

ALCES figura accuratior quàm supe riùs (pag. 53.) exhibita. Hanc quidẽ ceu ad uiuum pictam Io. Kentmanus & Io. Pontanus, excellentissimi medici, ad me dederunt: nec dissimilem, sed breuiore & magis cõpacto corpore Ant. Schnebergerus.

Quę fera Lithvuanis (inquit Sigismundus Liber Baro) sua lingua Loss est, eam Germani Ellend, quidam Latinè Alcen uocant: Poloni uolunt Onagrum, hoc est, Asinum agrestem esse, nõ respondente for-

L iij

ILLUSTRATION 48A Giraffe, in: Gessner (1560), p. 125.

ZBZ, ABTEILUNG ALTE DRUCKE UND RARA: NNN 44

ILLUSTRATION 48B Gessner's giraffe and other creatures from his animal books were printed by the Flemish Jesuit Ferdinand Verbiest on his world map.
UNIVERSITY LIBRARY UPPSALA

CHAPTER 13

Philological Works

In the 1550s, parallel to his work as a physician and professor of natural sciences, Gessner published not only the gigantic *Historia animalium*, but also a number of other botanical, medical and linguistic studies. As a supplement to the book about fish, he published an ichthyological work by the Moravian bishop Jan Dubravius (1486–1553) in 1559, with which he had not been familiar while writing the *Historia aquatilium animantium*. He merely noted there in the bibliography that it had come to his attention that the work existed in manuscript form and was located in Bratislava.[1] Gessner edited the text and added a short piece by the Greek physician Xenocrates of Aphrodisias (1st century) about edible fish and sea creatures. He dedicated the booklet to the Cologne doctor Johannes Echtius in gratitude for the fish illustrations he had provided. In the dedication, he wrote that he had received a manuscript fragment of the Greek text by Xenocrates from Guillaume Rondelet. He also included an Italian translation of the entire work by Giovanni Battista Rosario (1517–1578) and wrote a commentary comparing the Latin version by Rosario to the Xenocrates passages published by Edward Wotton in his *De animalium differentiis*.

13.1 *Mithridates*

The main philological work during these years was the so-called *Mithridates* of 1555, which was highlighted by Josias Simmler in his obituary of Gessner.[2] In this pioneering work on linguistics, Gessner examined the relationships among languages and assigned them to various families, even including Indian languages from the New World. Gessner mastered Hebrew fairly well[3] and had a good command of Greek and Latin. His French and Italian[4] were acceptable,

1 Gessner (1558), f. b_4r.
2 Simmler (1566), f. 7r. Cf. Peters (2019) and Sergeev (2019).
3 On August 7, 1563, he wrote to Johann Jakob Fries (1546–1611) in Geneva that he did not fully understand his Hebrew letter, partly because he did not know Hebrew very well and partly because of the unfamiliar shapes of the letters. Cf. Gessner (1577), f. 129v–130r.
4 Simmler (1566), f. 7r. Johannes Fabricius Montanus wrote to Bullinger on March 5, 1565, that Gessner could translate the enclosed Italian news for him (StAZH: E II 378, 1709).

© KONINKLIJKE BRILL NV, LEIDEN, 2023 | DOI:10.1163/9789004541696_014

and he had learned some Italian, Spanish[5] and Flemish.[6] Gessner was also familiar with botanical, pharmacological and zoological terms in Arabic.[7] His colleague Theodor Bibliander was of the opinion that someone who had a sufficient grasp of Hebrew would also be able to understand Arabic within two years.[8]

Gessner's title *Mithridates* referred to King Mithridates VI of Pontos (134–63 BC), who, according to Pliny the Elder, reigned over twenty-two different nationalities and spoke the language of each.[9] This work cannot be evaluated on its own, but must be viewed within the context both of the Renaissance emergence of linguistics and more specifically also the work by his colleagues Conrad Pellikan and Theodor Bibliander, with whom he donated a stained glass window to Heinrich Bullinger in 1545 (ill. 49).[10]

The humanists were known to have held the *tres linguae sacrae* (Hebrew, Greek, Latin) in high regard, which, according to Erasmus of Rotterdam, Christ had sanctified on the cross.[11] They were also interested in the vernaculars; let us mention here the Italian poems by Boccaccio, Dante and Petrarch or the studies on the German language and literature by Enea Silvio Piccolomini, Johannes Trithemius or Conrad Celtis. Along with antiquity and the history of nations, linguistic-historical documents, the relationships among languages and their mutual dependence were also subject of investigations. Sigismund Gelenius (1498–1554), a proofreader working in Basel, published in 1537 an etymological dictionary entitled *Lexicon symphonum, quo quatuor linguarum Europae familiarium, Graecae scilicet, Latinae, Germanicae ac Slavinicae concordia consonantiaque indicatur*. He made comparisons among Greek, Latin, German and Slavic words to show that their roots were related. One year later, the French polymath and cabbalist Guillaume Postel examined twelve different alphabets in his *Linguarum duodecim characteribus differentium alphabetum introductio*.[12]

These language-related studies were met with a great deal of interest, especially in Zürich, the source of a strong linguistic movement in the sixteenth

5 Leu/Weidmann (2004), p. 44.
6 Leu et al. (2008), p. 16.
7 Cf. the indices to the individual volumes of the *Historia animalium* and Gessner (1577), f. 113v.
8 Peters (1984), p. 9.
9 Plinius, *Naturalis historia*, 7,24.
10 Christ (2005), p. 116.
11 Cf. Gospel according to Luke 23,38.
12 Peters (1984), pp. 1f. and 7.

ILLUSTRATION 49 Stained glass window from 1545 that was donated by Conrad Pellikan, Theodor Bibliander and Conrad Gessner for Heinrich Bullinger's study. It depicts Jesus's last supper with his apostles. In the center lies the Passover lamb, which suggests the allegorical Lamb of God (John 1:29).
LANDESMUSEUM ZÜRICH, INV. NO. AG-1168

century.[13] In this city, home of the outstanding philologists Pellikan and Bibliander, every effort was made to translate the Hebrew and Greek Bible into German. In 1548, Bibliander released his *De ratione communi omnium linguarum et literarum commentarius*, in which he researched the commonalities among languages. His comparative language study was theologically motivated, since he was convinced that the linguistic chaos, a punishment for the attempt to construct the Tower of Babel, and the resulting multiplicity of languages were a result of sin. If it were possible to establish a single tongue as the world language, this would constitute the basis for peace and understanding among nations, as well as a common religion, given that arguments related to questions of faith had mostly been due to misunderstandings. In order to accomplish this, Bibliander thought Latin to be a good candidate for such a universal language.[14] At the end of his *Commentarius*, he printed important Christian texts in 14 different languages, including the Ten Commandments and the Lord's Prayer, among others.[15] This was his way of expressing that Christianity "had expanded beyond linguistic boundaries and can serve as a unifying factor."[16]

Unlike Bibliander, whose work he had thoroughly studied, Gessner's research was philologically, not theologically, motivated. He went well beyond the seventy-two established languages from the descendants of Shem, Ham and Japhet and named one hundred ten different languages and dialects[17] from which he briefly mentioned the following fifty-five:

Abyssinian
Arabic
Aramaic (Chaldean)
Armenian
Attic
Basque
Bithynian
Capadocian
Carian
Celtic (old Gaelic)

13 Peters (2002), p. 113.
14 Bibliander (1548), pp. 30f.; Peters (1984), pp. 8 and 13; Christ-von Wedel (2005), p. 53.
15 Bibliander (1548), pp. 224–235.
16 Peters (1984), p. 14.
17 Gessner (2009), p. 29.

Czech
Doric
English
Gaelic (Scottish)
Galatic
Georgian
German
Germanic languages
Gothic
Greek
Gypsy language
Hebrew
Hungarian
Indian
Ionic
Irish
Italian
Laconic
Latin
Lithuanian
Livonian
Lydian
Manx
Macedonian
New Greek
Old Egyptian
Oscan
Pelagic
Persian
Polish
Punic
Romansh
Rumanian
Russian
Sabine
Samaritan
Sardinian
Scythian
Slavic oder Illyrian languages

Slovene
Spanish
Tatar
Turkish
Welsh

Gessner discussed each of these languages in (Latin) alphabetical order. He began with Egyptian and ended with the language of the Vandals before finishing with a look at various exotic and mostly unknown languages. This work was also based on a corresponding collection of *loci* that Gessner had compiled from 154 authors. The most frequently sources were ancient authors such as Tacitus, Julius Caesar or Herodotus, but the most extensive sources were the books of humanists such as Sebastian Münster and Guillaume Postel. Merely two fifths of the text came from Gessner's own pen.[18]

He identified the language groups as Semito-Hamitic, Greek, Italian, Celtic, Slavic and Germanic[19] and considered Hebrew to be the oldest language, regarding it as pure and unadulterated. He correctly recognized Greek as being older than Latin: "It also seems clear to me that Greek is older than Latin, especially because the Romans based their alphabet on that of the Greeks. This is particularly so for the older forms. Additionally, they appropriated many Greek words, while conversely, the Greeks used next to no Latin words."[20] Gessner explained that the Romance languages came about as Roman Latin mixed with local tongues.[21]

Aside from the Balto-Slavic languages, he gave extensive space to the Germanic languages, which, all together, took up two fifths of *Mithridates*. It was not only humanists who had an interest in the national history and language. The wide-ranging use of the German language brought on by the Reformation also promoted cultivation of the mother tongue (ill. 50). A passage in Conrad Pellikan's house chronicles documents this clearly: "[The] sciences can be as clearly communicated and exactly expressed in the German language – the finest and richest of them all – as in any other language, including Greek, which is considered to be more prolix than Latin. Thus, if you wish to teach philosophy, you will be able to clearly lecture on Aristotle's concepts and you will feel compelled to use the more convenient new text books to cover the entire field of philosophy in German, out of love for the fatherland! When doing so, it is truly

18 Gessner (2009), pp. 47, 73–77.
19 Gessner (1974a), pp. 17f.
20 Gessner (1974a), p. 21.
21 Gessner (1974a), p. 23.

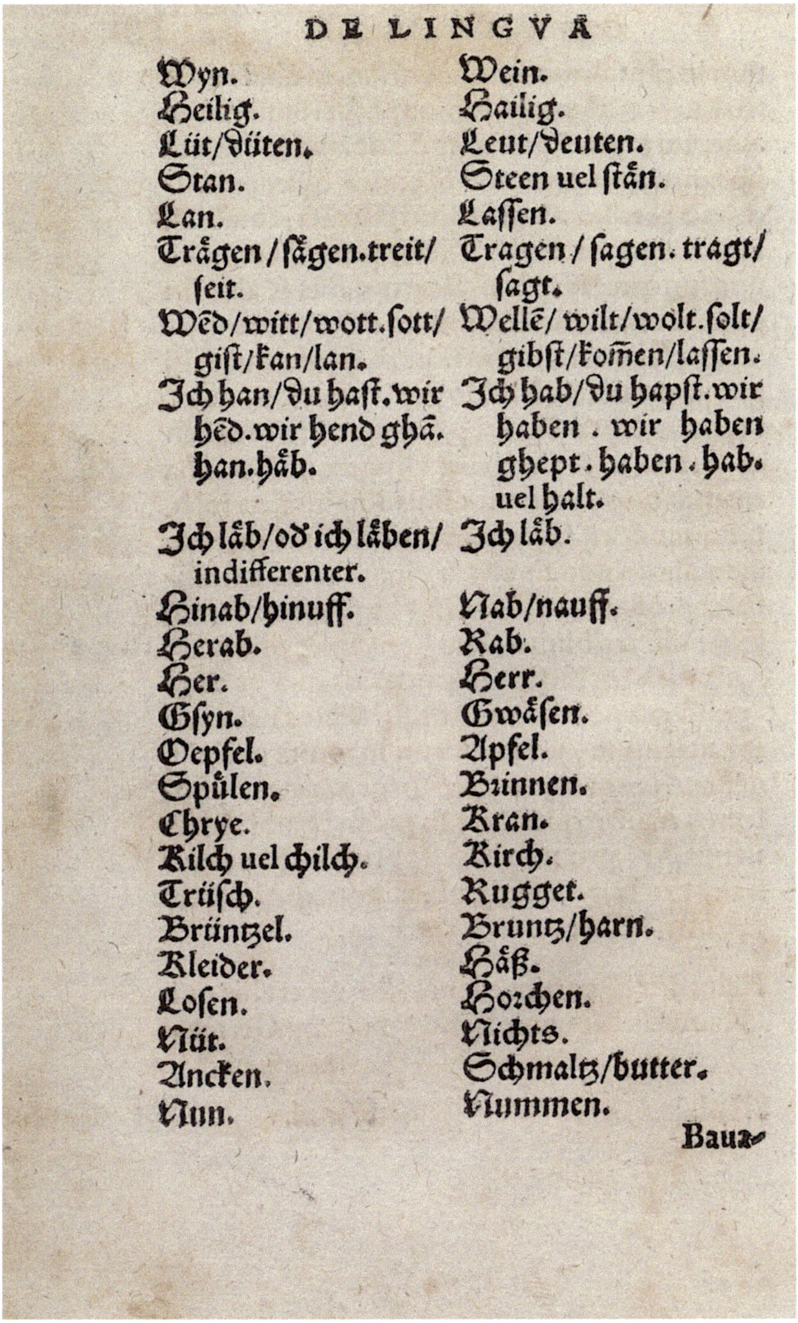

DE LINGVA

Wyn.	Wein.
Heilig.	Hailig.
Lüt/düten.	Leut/deuten.
Stan.	Steen uel ſtān.
Lan.	Laſſen.
Traͤgen / ſaͤgen.treit/ ſeit.	Tragen / ſagen. tragt/ ſagt.
Wēd/witt/wott.ſott/ giſt/kan/lan.	Wellē/ wilt/wolt.ſolt/ gibſt/kom̄en/laſſen.
Ich han/du haſt.wir hēd.wir hend ghā. han.haͤb.	Ich hab/du hapſt.wir haben . wir haben ghept . haben . hab. uel halt.
Ich laͤb/od ich laͤben/ indifferenter.	Ich laͤb.
Hinab/hinuff.	Nab/nauff.
Herab.	Rab.
Her.	Herr.
Gſyn.	Gwaͤſen.
Oepfel.	Apfel.
Spuͦlen.	Brunnen.
Chrye.	Kran.
Kilch uel chilch.	Kirch.
Trüſch.	Rugget.
Brüntzel.	Bruntz/harn.
Kleider.	Haͤß.
Loſen.	Horchen.
Nüt.	Nichts.
Ancken.	Schmaltz/butter.
Nun.	Nummen.

Bau-

ILLUSTRATION 50 Comparative lists of Swiss-German and Schwabish words, in: Conrad Gessner, *Mithridates*, Zürich 1555, f. 38v.
ZBZ, ABTEILUNG ALTE DRUCKE UND RARA: RR 1402

God-pleasing piety and a form of worship, when one does not neglect faith in God and the doctrines of the church! These days, certain pious men have already set out and forged ahead in this direction. However, there are arrogant and boastful linguistic buffoons who cannot write ten words without weaving in three Greek words and the eleventh in Hebrew! In that case, it would actually be more suitable and important to learn the language of the Turks, who are unfortunately and inevitably inching closer and closer to us. Then, perhaps with instructive words, we could make them more humane and receptive to Christian teachings, instead of provoking them with our immoderate vanity! And even without Latin and Greek, what would our German congregations fail to understand about the truth? It has now been almost 30 years that faithful and pious preachers have wisely favored German for their teachings. German is spoken in the churches, and everything to do with piety and Christian discipline is taught in this language, which is indeed why church doctrine is so clearly presented and interpreted! Now, Moses speaks directly to our people; men and women, boys and girls understand him far better than Parisian doctors and monks have, although they are called themselves Scotists and Thomists. Likewise, the Prophets, Christ, the Evangelists and Apostles speak only in German to our people. And there is no passage in the biblical canon so dark that Bullinger cannot relate it in the brightest light with blessings to be understood by all! This man of God who succeeded Zwingli as a church teacher with the most thorough command of all holy languages, nonetheless always makes use of the common vernacular, yes, even of the dialect, following the example set by Chrysostom, Gregory or Augustine."[22]

Gessner believed German to be the oldest vernacular,[23] and was particularly interested in its linguistic vestiges from the past. On October 9, 1562, he asked the Cologne theologian Georg Cassander to send him the alphabet, Lord's Prayer and other texts written in Gothic.[24] Additionally, he was preoccupied with the *Evanglienharmonie* (gospel harmony) by Otfrid of Weissenburg (ca. 790–875), the first famous name among old high German poets. His colleague from Augsburg, Achilles Pirmin Gasser, sent Gessner a manuscript copy of the work, which he had found in Ulrich Fugger's library[25] along with a dictionary of old German words. Gessner intended to publish the work but

22 Pellikan (1892), pp. 125f.
23 Gessner (1974a), p. 34.
24 ZBZ, MS. S 105, no. 132.
25 Lehmann (1956), p. 149.

was unable to find a willing printer.[26] This was not the end of it, however, and he wrote to Gasser on April 22, 1563: "If I am still alive, and my *Mithridates* were to be reissued, I will attach a page from your Otfrid as an example. Since its [*Mithridates*] publication, I have accumulated a great deal about various languages that I would like to add to the book."[27] Ultimately, Matthias Flacius Illyricus was able to publish Gasser's manuscript in 1571.

Gessner's interest in the Germanic languages and their dialects is reflected in the foreword to Josua Maler's German-Latin dictionary, which Gessner and Fries had instigated. A student of both, Maler took on the project to be published by Froschauer in Zürich in 1561. Added to the dictionary were numerous expressions, which, according to a modern judgement, "were common in earlier Swiss-German writings, but today are mostly forgotten. They can only be understood through the indirect route of Maler's Latin translation. None of the other German-speaking provinces have a similar linguistic-historical source from such an early period."[28] In the foreword to Maler's dictionary, Gessner once again paid tribute to German as the oldest of vernaculars, with traces of it found from Moscow to Milan.[29] He had hoped that someone would publish a universal bibliography of German literature, and if a person were to be found for such an endeavor, Gessner would have gladly made all of his relevant material available.[30]

At the end of *Mithridates* was a large-format fold-out table with the Lord's Prayer in 22 languages. The translations were compiled by Gessner from various sources. For instance, it has been assumed that the Hungarian version had been copied from János Sylvester's *Uy Testamentum mag'ar n'elevenn* (Sárvár 1541) or Dévai Bíró's *Orthographia Ungarica* (Krakow 1549). Only recently has it become clear that it actually came from a Hungarian student who had been living in Zürich.[31] However, Gessner was not satisfied with this version, as he expressed in a letter of February 2, 1556 to Caspar von Niedbruck. He sent von Niedbruck a copy of *Mithridates* with the request that he have the passage corrected by a Hungarian and to inform him accordingly.[32]

26 Burmeister (1975), pp. 230–244.
27 Burmeister (1975), pp. 233f.
28 Fischer et al. (1967), p. 80.
29 Maler (1561), f. *4r.
30 Maler (1561), f. *7v.
31 Bernhard (2015), p. 270.
32 ÖNB, Cod. 9737k, f. 23r/25r.

13.2 Classic Editions

In 1556, a year after the publication of *Mithridates*, Gessner published the first critical edition of the works of the Roman sophist and high priest Claudius Aelianus (2nd/3rd century), who wrote in Greek. Gessner was primarily interested in his 17 books about the animal world (*De animalium natura*). Frequently deriving models for moral behavior from zoology, these books were extensively quoted by Gessner in his *Historia animalium*. Aelian believed that animals, although unable to reason, were often more reasonable than humans and that they also avoided shameful acts that people sometimes willfully committed, despite being gifted with reason.[33] The folio volume of over 700 pages contained besides Aelian's animal stories with Pierre Gilles' (1490–1555) Latin translation, improved by Gessner, also Aelian's *Variae historiae libri* XIII in Latin, translated by the German philologist Justus Vulteius (1529–1575) as well as *De militaribus ordinibus instituendis*, translated by the Italian humanist Francesco Robortello (1516–1567) and the *Epistolae rusticanae*, whose Latin version was provided by Sebastian Guldenbeck (d. 1565). Together with Gessner, Guldenbeck had also been involved in the performance of Aristophanes's *Ploutos* in 1531. Gessner dedicated the bilingual Aelian edition to Johann Jakob Fugger, who had provided Gessner with a Greek manuscript by Aelian from Augsburg's public library, and made another available from his own private library. Gessner acknowledged him as an exceptionally well-educated man and praised his 12,000-volume Library,[34] declaring him to be one of the greatest founders of libraries in the history of mankind.

In the subsequent year, Gessner followed up with another Greek-Latin publication. This time it dealt with the only two surviving works by the Greek apologist Athenagoras of Athens (2nd century), that is, his petition on behalf of the Christians (*Apologia pro Christianis*) to Emperor Marcus Aurelius (121–180) and to his son Commodus (161–192) as well as his explanation of the resurrection of the dead (*De resurrectione mortuorum*). Athenagoras defended the Christians against the common accusations of that era, which included fornication, pedicide, cannibalism and other unfounded assertions. Additionally, he gave explanations for the insolubility of marriage, life after death and the nature of God. Robert Estienne, the Parisian printer who had become a citizen of Geneva in 1556 and to whom Gessner had dedicated the fifth book of the *Pandectae* (about arithmetic), had located a Greek copy of these two texts in the French king's library at Fontainebleau. He was shown yet another manuscript, which came from Rome, by the Italian bishop Guglielmo Sirleto

---

33 Leu (1990), pp. 84–90.

34 Dauser/Ferber (2010), p. 107.

(1514–1585), who took care to have a copy made of it.[35] Estienne took over the collation of the Greek texts and convinced Gessner to do the Latin translation and commentary on the *Apologia*, since no one else in Zürich had taken an interest in it.[36] Gessner also proofed the Greek text, for which Estienne thanked him, declaring Gessner to be a giant of literary accomplishments.[37] The Dutch philologist Petrus Nannius (1500–1557) was responsible for the translation of the *Resurrectio*. While Gessner also wrote a short commentary on Athenagoras' petition to the emperor, Estienne wrote a commentary on the second work.

Gessner dedicated his portion of the book to his patron Heinrich Bullinger, with whom he maintained a good and friendly relationship. A dedication of a work by Gessner to the head of the Zürich church was long overdue, but until then, Gessner had not yet found one that was appropriate to Bullinger's profession. He asked Bullinger to inform him if something in the book was theologically erroneous or incorrectly translated, so that he could correct it in this or a later edition.[38]

The work must have been published between the end of May and the end of August, since Gessner's dedicatory foreword was dated February 3, 1557 and Estienne's May 23. Furthermore, Beza wrote to Bullinger from Lausanne on September 4 that he had received several author's copies for Gessner but did not know how to send them to him.[39] This delay may have been the reason that Gessner was unable to send a copy of the book to Johannes Fabricius Montanus in Chur until December 3.[40]

The Athenagoras edition was released in Zürich again two years later, by Andreas Gessner the Younger, together with various short texts by Greek theologians in a Greek-Latin anthology with the title *Theologorum aliquot Graecorum veterum orthodoxorum libri*. The printer justified anthologies in his foreword to the reader, noting that neither booksellers nor printers were fond of slender volumes, because they were easily overlooked and easy to lose, a topic that Gessner also brought up in the foreword to his extensive *Chirurgia*.[41] Here, he chose a path similar to that of the Basel printer Heinrich Petri, who likewise

---

35 Athenagoras (1557), p. 190.

36 Letter from Gessner to Caspar von Niedbruck dated January 12, 1557 (ÖNB, Cod. 9737k, f. 183r).

37 Athenagoras (1557), p. 190, literally: "Ille literarum laborum Atlas." According to ancient mythology, Atlas was a son of the Titans.

38 Athenagoras (1557), pp. 79f.

39 CThB, vol. 2, p. 96.

40 Letter from Gessner to Fabricius Montanus dated December 3, 1557 (ZBZ, MS. C 50a, no. 25/2).

41 In the preface to the anthology *Chirurgia* (Zürich 1555), Gessner states that printers and booksellers preferred thick books because they were more likely to attract attention. Moreover, students would keep thick books more carefully than thin ones.

combined several old theological writings for publication in a single book (see his *Mikropresbyticon* or the *Orthodoxographa*).[42] Andreas Gessner the Younger also mentioned that the Greek manuscript in this edition, featuring 12 letters from the church father Ignatius of Antioch (2nd century), was from the private library of Caspar von Niedbruck.[43] Aside from the aforementioned texts, there were also old council decrees and some works by the Christian Neoplatonist Aeneas of Gaza (5th/6th century), the Byzantine theologian Demetrios Kydones (1324–1397), the apologist Hermias (3rd century) and the East Roman cleric Agapetos (6th century). As translators, the book mentioned, beyond Gessner and Nannius, Johannes Brunner († 1575), Johannes Wolf (1522–1571), both from Zürich, and Raphael Seiler (ca. 1544–1574) from Augsburg.

With the critical Greek first edition of the *Meditationes* by the Roman emperor and philosopher Marcus Aurelius, Gessner secured for himself a place in the history of classical philology. The former Latin teacher from Brugg, Michael Toxites (1514–1581) sent the now lost manuscript to Gessner from the library of the Palatinate Elector Otto Henry I.[44] The Augsburg humanist Wilhelm Xylander (1532–1576) did the translation and wrote a short commentary. Gessner sent him his Greek version. According to a letter from July 30, 1558, Xylander sent Fries and Gessner some fresh salmon and a marten pelt, possibly as thanks.[45] Along with the Latin translation, Xylander suggested edits to the Greek text, some of which Gessner took into account for the printing. Xylander's foreword, dedicated to the Augsburg merchant Georg von Stetten (1520–1572), is dated October 1, 1558, an indication that he was probably finished with his part. Xylander sent the manuscript plus translation and commentary to Gessner, who then completed the work. He added the Greek version of the *Vita Procli* by Marinus of Neapolis (5th century), and in mid-February 1559, he penned the dedication foreword to Anton von Werthern von Beichlingen (1528–1579), who had been a correspondent of Bullinger's since 1558.[46]

At about the same time, Andreas Gessner the Younger published a Latin edition of Leo Africanus's (ca. 1490–ca. 1550) description of Africa. He invited his relative Conrad to contribute something to the small book. In response, Gessner wrote a Latin translation of the Greek travel report *Periplus* by the Carthaginian admiral Hanno the Navigator (ca. 440–480 BC). Gessner dedicated

42 Hieronymus (1992), no. 436 and no. 437.

43 Gessner (1559/60), p. 4.

44 Salzmann (1960).

45 StAZH E II 356, 1033. Cf. Bührer (2002), pp. 178f.

46 Bullinger, *Diarium*, sub anno 1558.

the booklet on January 30, 1559 to the French scholar Jean du Choul, the son of a well-known jurist and researcher of antiquity, Guillaume du Choul (ca. 1496–1560). The latter's description of *Mont Pilat* in the Massif Central was published in the same year (1555) as Gessner's description of Mount Pilatus near Lucerne.[47]

47 Cf. Longeon (1979).

CHAPTER 14

Second Stay in Augsburg and the Imperial Grant of Arms

From March to August 1559, Emperor Ferdinand I assembled an Imperial Diet in Augsburg, the first and only one that he presided over. The agenda was full of important business: "First of all was Ferdinand's continual effort to overcome the division of faith within the empire, an ambition, so to say, that preoccupied him until his last breath. Additionally, there was the defense against the Ottomans, i.e., securing the borders of his domains. It was in part a Habsburg problem, the solution for which was afforded only conditional support by the imperial estates. Furthermore, there was the securing of peace and justice within the empire, as well as the everlasting topic of the *Reichsmünzordnung* (imperial coinage policy). But also of concern was the restitution issue regarding the dioceses and cities of Toul, Metz and Verdun in Lorraine. The French king had taken possession of them in 1552, and Ferdinand had raised this demand for the first time at the Imperial Diet of Augsburg in 1555. Not to mention, the new political situation in the Baltic states was marked by the Russian tsar threatening the Livonian state."[1]

The newly elected emperor wished to use this opportunity to meet the natural scientist from Zürich (ill. 51), who a year earlier had dedicated the impressive fourth volume of his *Historia animalium*, about aquatic animals, to him. Gessner recalled in his testament: "Among other books that I have written and published, the first one that I dedicated and sent to Emperor Ferdinand – a large book in Latin about fish and all aquatic animals – was received by His Majesty with extraordinary graciousness. Soon afterwards, I was invited by his personal physicians to the Imperial Diet in Augsburg, because they appreciated my work, and the emperor wanted very much to see me. When I arrived, they led me to His Imperial Majesty in his chamber, where the Emperor highly praised my fish book as pleasing his Majesty. We discussed many natural subjects for more than an hour, and he made a promise to me, that he would take me under his care. Shortly thereafter, his personal physicians wrote to me again, noting that they had not seen any coat of arms on my letters and sigils. On behest of the Emperor, they would therefore, if I pleased, issue and grant

1 Kohler (2003), pp. 272f.

© KONINKLIJKE BRILL NV, LEIDEN, 2023 | DOI:10.1163/9789004541696_015

ILLUSTRATION 51 Gessner at the age of 39. Portrait by Jos Murer (monogram IM above the frame), in: Gessner (1555a), back side of the title page.
ZBZ, ABTEILUNG ALTE DRUCKE UND RARA: 5.2

me a coat of arms as a recognition, which I could also bequeath to my blood relatives, honouring them."[2]

The initiative for the creation of a grant of arms, which was the first step towards ennoblement, thus seems to have come from the court physicians, but Gessner seemed to have no sense of urgency.[3] As is evidenced by a heraldic panel from 1561, which Gessner together with Heinrich Bullinger and Georg Keller gave to an unnamed person, Gessner was still using a self-designed coat of arms two years after the Augsburg stay.[4] It was not until the winter of 1564 that Gessner for the first time explicitly mentioned the subject of the imperial coat of arms, picking up on a theme that the personal physicians had first suggested. In a letter of 21 February 1564, he inquired in a letter to Crato von Krafftheim, with an eye on his kinship, about the grant of arms: "In the other two [letters], I have asked you, among other things, to make a request for an distinction from His Majesty the Emperor on behalf of myself and my family. You, Mister Julius and Mister Lauraeus, if he is at court, could add this as the culmination of your merits for me. I have enclosed a painted sketch showing how I would like it to appear, if you have no objections."[5]

Crato von Krafftheim took up the matter for his friend in Zürich, but he did not agree with the heraldic design in the sketch, which can be seen on the March 5, 1564 portrait of Gessner by Tobias Stimmer (see ill. 67). Gessner sent a new sketch to Crato von Krafftheim on March 26 with the following explanation: "I am sending an improved crest design according to your input. I have adorned the swan's wings (in the first sketch, the helmet decoration was a stork) with stars instead of flowers. The crown of ivy makes sufficient reference to my botanical work. The mouth of the snake no longer holds a stone, but, instead, a ring with a gem to signify that I have also written about gems, stones and ores. The stars scattered on the swan's wings shall represent the heavenly philosophy (theology), which rightly holds the highest place, as well as the works of the church fathers that I have published. This interpretation is for the rest of the world, but for myself, the symbol means that I should renounce the study of earthly things as soon as the works at hand are completed, as befits an old man like myself."[6]

The animals that appear in the four fields of the coat of arms, lion, eagle, dolphin and snake represent Gessner's books respectively about four-legged

2 Hanhart (1824), pp. 285f.

3 Frey (2017), p. 42.

4 I owe this reference to Dr. phil. Mylène Ruoss (Schweizerisches Landesmuseum, Zürich). Cf. Wartmann (1908), pp. 72–74; Weber (2016).

5 Helmich (1938), p. 43; Latin text in: Gessner (1577), f. 16v.

6 Fischer et al. (1967), p. 232.

FERDINANDVS
DIVINA FAVENTE CLEMENTIA
ELECTVS ROMANORVM IMPERATOR SEMPER AVGVSTVS AC

ILLUSTRATION 52 The imperial grant of arms for Conrad Gessner.
ZBZ, HANDSCHRIFTENABTEILUNG: FA GESSNER 1000

animals, birds, fish and snakes, although the last of these was published posthumously. The dolphin wears a crown because of Gessner's dedication of the book on aquatic animals to the emperor.[7]

The granting of arms did not take long. The document (ill. 52) was written on April 3 and signed by the Emperor on April 25. On April 26, the seal

7 Weber (2016).

was affixed,[8] and at Whitsun (May 21) 1564, Gessner expressed his thanks to Crato von Krafftheim.[9] He also inquired about von Krafftheim's expenses, which he wished to compensate – 30 guilders were usual for the a grant of arms.[10] The fact that the entire procedure was wrapped up so quickly was likely due to the Emperor being quite taken with the universal scholar from Zürich. Crato von Krafftheim told Gessner on February 28, 1564 about dining with His Majesty: "Then His Imperial Majesty turned to me and asked whether I know you personally. I answered that I find it better to know exactly how you feel about things from the many letters I've received from you, than if I had only met you face to face. His Majesty the Emperor, in the presence of the many admirable gentlemen, then began to speak highly of your wide-ranging scholarship and tireless diligence, and ended with the words: 'Gessner is integrity itself [*Gesnerum totam probitatem*].' These were the Emperor's own words. He added further praise of your family that I cannot elaborate on at this time."[11]

Gessner also used the time in Augsburg to intensify his personal contacts to various physicians. This included Crato von Krafftheim, with whom he had begun corresponding in the fall of 1559, and Achilles Pirmin Gasser, with whom correspondence had begun in 1554. In April, he visited the botanical garden of the Augsburg patrician Johann Heinrich Herwarth, where he saw a tulip that had been brought from Turkey. In 1561, he published a drawing of a tulip that he had acquired in 1557, which was the very first printed image of a tulip in Europe and the earliest published evidence of the flower that eventually became so popular (Ill. 53a and 53b).[12]

8 Letter from Crato von Krafftheim to Gessner dated April 27, 1564 (ZBZ, FA Gessner 1516: 1).

9 Gessner (1577), f. 18r.

10 Burmeister (1975), p. 345.

11 Hanhart (1824), p. 257; original in: ZBZ, Familienarchiv Gessner: 1516: 1.

12 Cordus (1561), f. 213r. The woodcut shows a tulip, the drawing of which Gessner had already received in 1557, as can be seen from the manuscript of the *Historia plantarum*. Erlangen University Library, MS. 2386/2, f. 220r.

ILLUSTRATION 53A Drawing of a tulip that Gessner acquired in 1557, in: Conrad Gessner, *Historia plantarum*.
UNIVERSITY LIBRARY ERLANGEN, MS. 2386/2, F. 220R

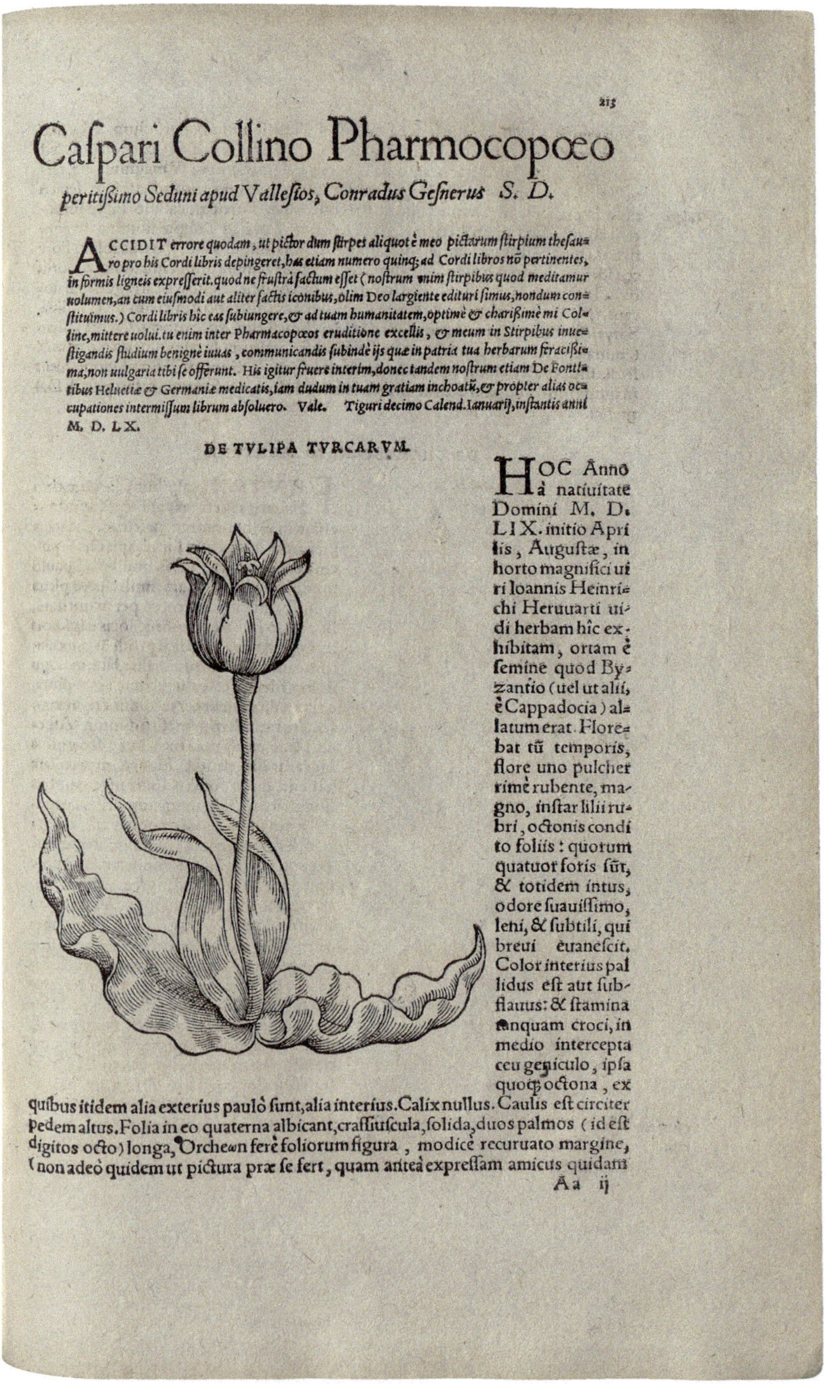

213

Caſpari Collino Pharmocopœo

peritiſsimo Seduni apud Valleſios, Conradus Geſnerus S. D.

ACCIDIT errore quodam, ut pictor dum ſtirpes aliquot è meo pictarum ſtirpium theſauro pro his Cordi libris depingeret, has etiam numero quinq; ad Cordi libros nõ pertinentes, in formis ligneis expreſſerit. quod ne fruſtrà factum eſſet (noſtrum enim ſtirpibus quod meditamur uolumen, an cum eiuſmodi aut aliter factis iconibus, olim Deo largiente edituri ſimus, nondum conſtituimus.) Cordi libris hîc eas ſubiungere, & ad tuam humanitatem, optimè & chariſsimè mi Colline, mittere uolui. tu enim inter Pharmacopœos eruditione excellis, & meum in Stirpibus inueſtigandis ſtudium benignè iuuas, communicandis ſubindè ijs quæ in patria tua herbarum feraciſsima, non uulgaria tibi ſe offerunt. His igitur fruere interim, donec tandem noſtrum etiam De Fontibus Heluetiæ & Germaniæ medicatis, iam dudum in tuam gratiam inchoatũ, & propter alias occupationes intermiſſum librum abſoluero. Vale. Tiguri decimo Calend. Ianuarij, inſtantis anni M. D. LX.

DE TVLIPA TVRCARVM.

HOC Anno à natiuitate Domini M. D. LIX. initio Aprilis, Auguſtæ, in horto magnifici uiri Ioannis Heinrichi Heruuarti uidi herbam hîc exhibitam, ortam è ſemine quod Byzantio (uel ut alij, è Cappadocia) allatum erat. Florebat tũ temporis, flore uno pulcherrimè rubente, magno, inſtar lilii rubri, octonis condito foliis: quorum quatuor foris ſũt, & totidem intus, odore ſuauiſſimo, leni, & ſubtili, qui breui euaneſcit. Color interius pallidus eſt aut ſubflauus: & ſtamina tanquam croci, in medio intercepta ceu geniculo, ipſa quoq; octona, ex quibus itidem alia exterius paulò ſunt, alia interius. Calix nullus. Caulis eſt circiter pedem altus. Folia in eo quaterna albicant, craſſiuſcula, ſolida, duos palmos (id eſt digitos octo) longa, Orchewn ferè foliorum figura, modicè recuruato margine, (non adeò quidem ut pictura præ ſe fert, quam anteà expreſſam amicus quidam

Aa ij

ILLUSTRATION 53B The first published image of a tulip, in: Cordus (1561), f. 213r.
ZBZ, ABTEILUNG ALTE DRUCKE UND RARA: I R 5

CHAPTER 15

Botany

Throughout his life, Gessner's first love was botany, which he was introduced to as a young boy by his great uncle, the chaplain Johannes Frick.[1] Because of his untimely death, Gessner's botanical oeuvre has survived only as an unfinished manuscript torso, which unfortunately deprives him of the recognition he deserves. His knowledge in this area was obtained from the botanical literature, from raising various plants in his garden, botanical forays and excursions as well as the creation of both a physical and an illustrated herbarium. The latter has been largely preserved in three volumes and contains sketches for his planned, comprehensive *Historia plantarum.*

15.1 Technical Literature

A glance at Gessner's private library[2] reveals that he possessed various important botanical titles, for instance, an edition of the works of the Greek botanist and student of Aristotle, Theophrast of Eresus (371–287 BC). It had been published in 1541 in Basel by Oporinus, and Gessner's copy was full of handwritten notes. Today, this volume is preserved at the University of Tartu Library (Estonia).[3] The Zentralbibliothek Zürich possesses instead Gessner's copy of Pietro Andrea Mattioli's *Commentarii in libros sex Dioscoridis Anazarbei de medica materia* (Venice 1558), in which there are not only extensive annotations by Gessner, but in three places there are also plant drawings with detailed studies of flowers and fruits. These demonstrate his outstanding skills as an observer and illustrator (ill. 54).[4] Two other noteworthy botanical volumes in Gessner's private library are the French and Dutch editions of a plant book by Rembert Dodoens (1517–1585), which he also heavily annotated. Altogether, there are about a dozen botanical works that have been found among his possessions.[5] However, he actually studied several times as many relevant books, which may have been included in his private library. In a December 12,

1 Cf. p. 24.
2 Leu et al. (2008).
3 The volume was already discussed in 1821 in the *Doerptische Beytraegen* (p. 394) as a holding of the University of Dorpat/Tartu Library, cf. Morgenstern (1821).
4 ZBZ, call number: Dr M 438. Cf. Hess (2019).
5 Leu et al. (2008), p. 21.

© KONINKLIJKE BRILL NV, LEIDEN, 2023 | DOI:10.1163/9789004541696_016

154 And. Matthioli Comm.

uocantur, nos uerò Lomie dicimus: tametſi Limonia, cùm ſint guſtu omnibus ijs ſpeciebus acidiora, ob id ſuccò & frigidiora, & ſicciora eſſe, argumento duci poßit. Quin & Aurantia cortice calidiora cenſenda ſunt, quandoquidem is reliquis acrior, atq; amarior guſtatu deprehenditur. Aurantiorum (quemadmodum etiam de punicis ſuperius diximus) in Italia tria obſeruantur genera, dulcia, acida, & uinolenta, ſiue miſta. Dulcia in omnibus partibus excalfaciunt. Aliorum uerò ſuccus refrigerat, prout maiore, ſeu minore referta ſunt aciditate. Quocirca febricitantibus ad reſtinguendam ſitim, acida quidem, & uinoſa, non autem dulcia ſunt exhibenda. Fit è Limoniorum ſucco, ſicut & citriorum, ſerapium ad reſtinguendum bilis feruorem non inutile. Item & ad peſtilentes febres, & eas, quæ contagione non uacant. Aqua, quæ è Limoniorum ſucco calentis aquæ balneo, & uitreis organis extrahitur, præterquàm quòd mulieribus expetitur pro fuco ad faciem expoliendam, uitiligines ubicunq; ſint, & totius corporis maculas emendat, uaros delet, & acaros interimit. Serapijs admiſta febribus acutis, et contagioſis mirè ſuccurrit. Epota uentris tineas enecat: id quod etiam præſtat ſuccus recenter expreſſus, unciæ pondere potus. Cæterùm non parum laboris impendiſſe uidetur Braſauolus in præfatione ſui de ſimplicium examine libri, ut doceret, unde emanauerit Arantiorum uocabulum. Sed licet ibi plura de eius origine referat; ea tamen, meo quidem iudicio, non multùm nomini quadrant. Quandoquidem ego potius exiſtimauerim, uulgò Arantia, quaſi Aurantia, altera expuncta litera, dicta fuiſſe. Nam cùm hæc poma auri colore reluceant, ab auro meritò nomen accepiſſe uidentur: unde, quantum equidem reor, Aurantia, hoc eſt, aurea mala dicta fuere.

Nomina. Reliquum eſt, ut ſingulis malis ſua reddamus nomina. Mala ita Latinis dicta, Græcis μῆλα uocantur: Arabibus, Tuffa, ſeu Tufaha: Italis, Mele: Germanis, Oepffel: Hiſpanis, Manſanas: Gallis, pomes. Ea uerò, quæ Græcè κυδώνια μῆλα, Latinè Cotonea, & Cydonia mala dicũtur: Arabicè Saffargel: Italicè, Mele cotogne: Germanicè, Quitten, aut Kutten: Hiſpanicè, Membrilhos, & Marmelos: Gallicè, Conting. Quæ autem Græcum nomen habent μελίμηλα, ἢ γλυκύμηλα, eorum Latinum eſt Mellita, uel Dulcia mala: Arabicum, Melomella, & Galopomella: Italicum, Mele dolci. Mala, quæ περσικὰ μῆλα, καὶ ῥοδάκινα Græco nomine, Latino Perſica mala appellantur: Arabico, Sauch, ſeu Chauch: Italico, Peſche: Germanico, Pferſich: Hiſpanico, Pexegos: Gallico, Peſches. Μῆλα ἀρμενιακὰ, πραικόκκια, καὶ βερικόκκια quæ Græci, Latini Armeniaca mala, et Præcocia nominant: Mauritani, Mermex ſeu Mirmix, Mex, Meſmes, ſiue Miſmis: Itali, Armeniache, Bacoche, Moniache, et Griſomele: Germani, S. Iohans pferſich: Hiſpani, Albiricoques, Albarchigas, et Aluaricoques: Galli, Abricot. Mala deniq; Μηδικὰ μῆλα, ἢ κεδρόμηλα à Græcis uocata, Latinis ſimiliter Medica mala, aut Citromala, et Citria dicuntur: Italis, Cedri, et Citroni: Germanis, Citrin oepffel, Iuden oepffel, et Citronaten: Hiſpanis, Cidras: Gallis, ung Citron.

Ἄπια. PYRA. CAP. CXXXII.

PYRORUM multa genera. Quæ tamen omnia adſtringunt: proinde adduntur conuenienter in repellentia cataplaſmata. Decoctum è ſiccis, & cruda ipſa aluum ſiſtunt: ieiunos eſitata lædunt. Achras pyraſtri genus eſt, quod tardiſsimè matureſcit. Vim pyro adſtringentiorem habet, & ad eadem conuenit. ſtringunt etiam eius folia. Ligni pyrorum cinis efficaciter his auxiliatur, quos fungi ſtrangulant. Aiunt fungos, ſi ſyluestria pyra cum eis coquantur, innoxios fieri.

Pyrorum cōſideratio.

QVONIAM Pyra uulgatißima ſunt non modò in Italia, ſed ubiq; in tota Europa; ideo ſuperuacaneum duxi eorum arbores per hiſtoriam explicare. Multa, ac uaria apud nos Pyrorum numerantur genera, quemadmodum & apud antiquos. Qui, ut Plinius refert libro, & capite XV. celebrauere ſuperba, falerna, decumiana, dolobelliana, pompeiana, liceriana, ſeueriana, tyranniana, fauoniana, lateriana, anitiana, tiberiana, ſignina, purpurea, ſementina, laurea, amphorina, coriolana, cucurbitana, & alia, quorum enumeratio longa foret. Quas ſanè uarias appellationes mutuata ſunt, uel ab hominibus, qui primùm ea in ciborum uſum reuocarunt; uel à locis, unde uenerant; aut à notis, quas aliquibus fructibus acceptas referebant; aut à colore proprio; aut à tempore, in quo ab arboribus decerpebantur. In hoc itaq; noſtri quoq; Hetruſci antiquorum conſuetudinem ſecuti, uarijs nominibus à diuerſis inditis cauſis ſua Pyra uulgò appellant moſchatelle, giugnole, ciampoline, roggie, ghiacciuole, ſpinoſe, quadrane, carouelle, papali, ſan Nicolo, durelle, zuccaie, campane, uernareccie, gentili, porcine, ſementine, & plura alia habent alijs nomenclaturis appellata. Cæterùm ſi quis uellet in noſtratibus ueterum Pyra collatione aſſequi, operæpretium eſſet, ut unumquodq; eorum genus peculiarem, perſpicuamq; habuiſſet deſcriptionem: quandoquidem fieri non poteſt, ut ex ſola nomenclatura innoteſcant.

Pyrorum facultas. Porrò in uniuerſum de eorum uiribus

ILLUSTRATION 54 Detailed drawing of pear blossoms by Gessner, in: Pietro Andrea Mattioli, *Commentarii secundo aucti, in libros sex Pedacii Dioscoridis Anazarbei de medica materia*, Venice 1558, p. 154.

ZBZ, ABTEILUNG ALTE DRUCKE UND RARA: DR M 438

1563 letter to Johannes Bauhin, he mentioned having studied more than 200 plant books for his *Historia plantarum*.[6]

An insight into Gessner's profound knowledge of the technical literature can be found in the Latin edition of Hieronymus Bock's book about herbs, published in 1552. The Strasbourg physician and Hebraist David Kyber (1525–1553) translated the work into Latin, and Gessner added a foreword, an extensive botanical bibliography[7] and five indices of plant names in Greek, Latin, German, Arabic and Hebrew. Gessner remarked in the foreword that different sciences, including botany, had experienced a renaissance (*stirpium historia renata*) and ascended to new heights.[8] The subsequent 32-page comprehensive bibliography of botanical authors was organized according to Greek (13 pages), Latin (2 pages), Arabic (2 pages), and modern writers (9 pages) as well as those who had made incidental contributions to the knowledge of plants (3 pages) and those who had concentrated on pharmaceutical applications (3 pages). Here, he also praised the Arabs, who had not only adopted a great deal of knowledge from the Greeks, but also produced their own important research.[9] He left out authors who had written about magic and superstition in relation to plants, such as Orpheus and Democritus, whom already Pliny had ridiculed.[10]

15.2 Nomenclature and Classification

Gessner adopted the literature, but through his own empirical studies, he went far beyond his predecessors. This is clear to all who study his notes describing plants for the planned *Historia plantarum*, which are currently preserved at the university libraries at Erlangen and Tartu.[11] The fact that he soon exceeded the ancient authors whom he revered is also clear from the tri-lingual plant lexicon (*Lexicon rei herbariae trilingue*) by his colleague David Kyber, which Gessner published in 1553. He added a nearly 100-page, comprehensive summary (*Tabulae collectionum*) with information citing the months in which plants bloomed – a matter of special interest to pharmacists. Here, we also find information about plant morphology that was explicitly derived from Theophrast's first book of his *Historia plantarum*. Gessner surpassed the Greek

6 Bauhin (1591), pp. 133–136.
7 Cf. also the plant lists with references in: Gessner (1548), f. 196r–209r.
8 Bock (1552), f. a_{viii}v.
9 Bock (1552), f. b_{iii}r.
10 Bock (1552), f. b_{i}r.
11 Cf. p. 290.

author,[12] however, with his descriptions of flowers and fruits. Looking over Gessner's morphological tables from 1553 – especially those that were posthumously reworked and published in 1587 by Caspar Wolf[13] – it is striking that Gessner distinguished the plants according to numerous criteria, but did not select a specific characteristic as his basis of a classification.

Unlike Gessner, the Italian botanist Andrea Cesalpino (1525–1603) decided on a classification system based on a single criterion, namely the structure and number of seeds found in fruits. Correspondingly, he classified the flora in trees and shrubs with single or multipart fruits, in herbs with single or many-seeded fruits and also in seedless plants such as ferns.[14] Thus, Cesalpino became the founder of modern botanical classification. He published his results in 1583 in *De plantis*, but had already formulated this classification system during Gessner's lifetime as was proved by a herbarium that he designed prior to 1563 and subsequently assembled.[15]

In places, Gessner's letters suggest that he was also thinking in this direction, but it is not possible to discern a clear botanical system from the surviving documents. He wrote to Theodor Zwinger in Basel on April 24, 1564 that he usually included the fruit and seeds of plants in his illustrations to make them easier to identify.[16] On November 18, 1564, he let Adolph Occo in Augsburg know that he determined relationships among plants based on their seeds.[17] And shortly before his death on November 26, 1565 he wrote to Zwinger: "If you have any rare foreign plants or plant parts stored away, please write to me at your earliest convenience. Along with the main plants, their leaves and their names, I would like you to let me know if you also have the flowers, fruits and roots, or only one of these, or all of them. For all plants, I am very keen to create lifelike representations with all three. From them, in fact, the properties and relationships are much better to recognize than from the leaves of plants. From these characteristics (the fruit, the seed and the flower) it can be easily determined that the plants commonly known as larkspur [*Delphinium staphisagria*] and comfrey [*Symphytum officinale*] are related to wolfsbane [*Aconitum sp.*]."[18]

Today, *Delphinium* and *Aconitum* are still seen as belonging to the same family of the *Ranunculaceae*, while *Symphytum* nowadays belongs to the *Boraginaceae*. Because Gessner studied the relationships among plants based

12 Fischer (1966), p. 95.
13 Gessner (1587a).
14 Mägdefrau (1992), pp. 43f.
15 Taffetani (2012), pp. 18f.
16 Gessner (1577), f. 107v–108v.
17 Gessner (1577), f. 65r–66v.
18 Gessner (1577), f. 113r/v.

on the structure of their flowers and seeds and not other organs such as their leaves, he was headed toward the same conclusions as Carl Linnaeus two hundred years later. It is highly likely that he might have anticipated Linnaeus's system had he lived long enough, even if he made a mistake here and there, such as with the lady slipper [*Cypripedium calceolus*], which he associated with the water plantain [*Alisma*] instead of orchids.[19]

15.3 Friendly Favors

After David Kyber died from the plague in January 1553, Gessner took over the publication of his *Lexicon rei herbariae trilingue* as a friendly favor. The work contains not only the nomenclatural identifications of the Greek, Latin and German names of the plants, but also botanical and etymological information. Gessner's contribution to the volume was in the form of summary tables (*Tabulae collectionum*) which listed the times during which plants bloomed. He dedicated the work to the pharmacist Nikolaus Speicher, who run the pharmacy at Langstrasse 149 in Strasbourg.[20] Gessner stressed that he did not dedicate the tables to him, because they might be of professional use to himself but rather out of their common friendship with Kyber. Another mutual friend was Nikolaus Gerbel, a humanist from Strasbourg. Together, he and Gessner wrote an extensive 91-verse Greek dialog in hexameter to honor the deceased, which followed Gessner's preface to the *Lexicon rei herbariae trilingue*. Gerbel also expressed his grief anew to Gessner in a previously unknown letter of May 1, 1553.[21] Gessner's dedication epistle was dated March 31, 1553 and was devoted to David's father, the theologian Lucius Kyber († 1554). Gessner wrote that he felt especially connected to David, not only because of his good heart, his moral values and his erudition, but also because of a special kinship. He could not remember when the death of a beloved friend had been so devastating to him. Additionally, Gessner mentioned that David Kyber had overseen the printing of the first ten sheets up to page 160, and that Gessner had corrected various minor errors and supervised the printing of the remaining work up to page 548. The book appeared in September 1553.[22]

---

19 Letter from Gessner to Johannes Funck, November 5, 1564, in: Gessner (1577), f. 96v. On Gessner and orchids cf. Wartmann (2019).

20 Kettler (2008), p. 653.

21 Letter from Gerbel to Gessner: StAGR, D V/37 C 36.06.40.

22 Kyber (1553), f. a_4r–a_5r.

Despite being permanently overworked, this was not the only occasion on which Gessner had done such favors for his friends and published partial manuscripts that would have otherwise come to a halt. Such selflessness may have deprived him of his own notoriety in the history of botany. If the time had been used to work on his own *Historia plantarum*, he would not only have had the creation of the *Historia animalium* to his credit but perhaps also its botanical counterpart. On March 31, 1565, he wrote to Achilles Pirmin Gasser about the translation of *Euporista* by Dioscorides that was prepared by the Augsburg physician Johannes Moibanus. Gessner was handling the publication after the doctor's death: "This book has certainly been a lot of work for me. It is easy to translate impeccable books and those in which every sentence can be found nearly word-for-word by other authors. With such references, one can compare, correct and substantiate them, what I already had to do for several translation jobs. This verification work I also had to do here, not only in the pieces I have translated, but also in many that Moibanus had already translated. I wanted to proof most of them just as he would have, had he not passed away. Due to this and other obstacles, I have interrupted my history of plants for many months, which has been progressing very slowly and may never be finished. I have nonetheless gladly done this and do not regret it – nor will I."[23]

Gessner sets a good example for how ambition, fame and a long list of publications are not the only things that count in the life of a researcher. As he stated elsewhere, he would have enjoyed more time to cultivate his friendships and contacts.[24]

Another wide-ranging project that he undertook in the service of friendship was the publication of a not yet released plant book by the doctor and botanist Valerius Cordus, who unexpectedly fell ill and died prematurely during a research trip to Rome. He had studied among others together with Crato von Krafftheim in Wittenberg, who also wished to assist with the publication of his remaining works and who managed only with the help of an unknown German doctor to locate Cordus's grave during a trip to Rome in 1548. One of their teachers had been none other than Luther's close associate Philipp Melanchthon, who introduced them to the *Alexipharmaca* by Nicander. Cordus had also taught in Wittenberg for a short time and had there delivered three lectures on the writings of the Greek physician Dioscorides. Crato sent Gessner two other unedited treatments by Cordus regarding the distillation of oils and the production of sulpheric acid.[25] In early 1559, Gessner went to work on his own

23 Burmeister (1975), pp. 350f.

24 Gessner (1577), f. 139v.

25 Cordus (1561), f. b_2r–b_3r.

plant book, but wrote to Kentmann on February 27 that he would take some free time in the summer to publish Cordus's works.[26] The project was apparently prolonged since it was not until November 25 that he was able to send Cordus's four plant books (*Historia stirpium*) back to the Nuremberg physician Hieronymus Herold (ca. 1515–1566), who had sent them to Gessner in 1558 and had prompted their publication. The physician and pharmacist from Gdansk, Johannes Placotomus (1514–1577), who also belonged to Melanchthon's circle of friends, was particularly helpful in finding further writings by Cordus. Because he had not left behind any drawings of plants, they relied on the woodcuts produced for the book of herbs by Hieronymus Bock, which had also been published in Strasbourg by Wendelin Rihel. Wherever that was the case, a corresponding illustration acknowledgement was also printed. Otherwise, the illustrative material came from Gessner's personal collection or the album that Kentmann had previously made available to him.[27] For the illustration of the sugar cane, Gessner consulted Sebastian Münster's *Cosmographia*.[28] Along with Herold and Placotomus, Cordus's uncle Johannes Ralla († 1560)[29] and the Basel professor of logic Philipp Bächi (1521–1560)[30] helped with the publication of Cordus's edition, which also included a new edition of Gessner's *Annotationes in Dioscoridis de medica materia* that he had already published in 1549.[31]

The impressive volume was released in the summer of 1561 with more than 600 folio pages, but Gessner was anything but inspired by the printer's work. On August 16, he wrote to Crato von Krafftheim that the book had been carelessly and poorly printed.[32] A few days later, on August 29, he wrote to John Caius that he was ashamed to have his name associated with it. The printer, Josias Rihel, had promised to personally oversee the printing preparations, but had failed to do so,[33] leaving out numerous additions and corrections.[34] The polyhistor from Zürich also wrote to Kentmann on March 23, 1563 about his frustration that the Wittenberg Academy, to which he had dedicated the Cordus edition in print, had shown no acknowledgement of it during the previous two

26 Gessner (1584), f. B_3v–C_1r.

27 The volume is now kept in the Herzogin Anna Amalia Library in Weimar (Fol. 323).

28 Cordus (1561), f. 85r/v.

29 Cordus (1561), f. 213r.

30 Cordus (1561), f. 225r.

31 Wellisch (1984), pp. 39 and 90.

32 Letter from Gessner to Crato von Krafftheim, August 16, 1561. Gessner (1577), f. 1r–2v.

33 Gessner (1577), f. 133v–136v.

34 Letter from Gessner to Crato von Krafftheim, February 15, 1562. Gessner (1577), f. 10r/v.

years.[35] Despite these bitter disappointments, he had copies sent to various colleagues as is recorded in letters to Jacques Daléchamps of September 6, 1561, to Johannes Kentmann on March 4, 1562,[36] to Wolfgang Meurer on August 13, 1562[37] and to Johannes Bauhin on August 24, 1562.[38]

Two years later, in 1563, a slender fascicle of 13 folio pages was released, containing part 5 of Cordus's plant history, which also included a letter of December 1, 1544 by the Nuremberg doctor Hieronymus Schreiber to his Leipzig colleague Wolfgang Meurer. Schreiber had been with Cordus in Rome when he died, and he gave an account of Cordus's illness and death.[39] Gessner mentioned in the foreward, which he dedicated to Meurer and which was written on August 13, 1562 while he was a guest at the thermal baths in Baden, that not everything by Cordus had been published. A friend of his was in possession of yet another text, entitled *De Halosantho seu spermate ceti vulgo dicto* (*About Sea Salt or What is Commonly Known as Whale Semen*), which had not yet been sent to Gessner.[40] Crato von Kraftheim was unhappy with certain passages by Hieronymus Schreiber regarding Cordus's demise and wrote to Gessner on June 29, 1563 with a few corrections from an account by Moibanus that Gessner had not been aware of.[41]

15.4 Botanical Gardens

At the request of the printer, Gessner added his famous *De hortis Germaniae* to the Cordus edition. This was to present newer work and not just the results of a colleague who had died 20 years before.[42] In this treatise, Gessner provides an overview of private botanical gardens in Switzerland, Germany, France and Italy that he had either visited or about which he was able to obtain information. The inventory was also intended to help interested readers find out which plants grew where and from whom one might obtain seeds. Thus, Gessner advised Kentmann on March 4, 1562 that he could find out from the

35 Gessner (1584), f. C_4r/v.
36 Gessner (1584), f. C_3v–C_4r.
37 Schmiedel (1754), p. 3.
38 Bauhin (1591), pp. 107f.
39 A German translation is published in: Hanhart (1824), pp. 232–240.
40 Gessner published this work later, cf. p. 370.
41 StAGR, D V/37 C 36.06.27.
42 Cordus (1561), f. 236v.

Horti Germaniae which plants he, Gessner, owned from which he could then request seeds.[43]

Gessner dedicated the work to the Emperor's personal physician, Stephanus Lauraeus (of Amersfoort), whom he had met during a dinner at Johann Jakob Fugger's house in 1559.[44] It is likely that around 1561, Lauraeus became a convert to Paracelsianism,[45] something that Gessner was probably not aware of. In the dedication, Gessner refers to him as a friend whom he wished to win over as an advocate in the dispute about the proper interpretation of ancient plant names. Here, Gessner was alluding to a conflict he had with Pietro Andrea Mattioli,[46] who also worked as a personal physician at the Habsburg court until 1568. Additionally, Gessner explained the concept and purpose of the comprehensive 131 folio-paged addition.[47] He undertook "in a single book, to alphabetically list all the plants that I believe I now have and have ever had in my garden, as well as those in the gardens of my friends. Some of these friends – whose names I mention at the beginning of the book – sent me their lists. However, I have not only included just the names as received, but also further information for most of the plants, either something pertaining to their names or something about their origins and care. If you are thinking of creating and maintaining a garden, and not just common greenery, but one that is adorned by an abundance and variety of plant species, this provides an example and a method for getting started. In the future, posterity may also learn from my book which plant species in this century were common in German lands and were kept in gardens, which ones grew well in them and for how long. Finally, the French and Italians should be impressed that there are German-speaking people under cloudy skies, who are vigorous gardners who have advanced so far as to be experts in the oft-repeated plant nomenclature of antiquity."[48]

Gessner wrote in the introduction to the work: "I have only included actual garden plants in this book that are sown or planted, and in particular only those that Germans are familiar with. Then, I describe a goodly number of foreign plants, citing those that were first planted in gardens in earlier times as

43 Gessner (1584), f. C_3v–C_4r.

44 Gessner writes in Cordus (1561), f. 236r, that he had met Lauraeus in Augsburg in earlier years. Since Lauraeus studied in 1543 in Bologna, he could hardly have made his acquaintance in 1545 as imperial personal physician, but only in 1559 in Augsburg. Gaila Jehoel: Het culturele netwerk van Jan van Scorel, Academisch proefschrift University of Amsterdam 2019, p. 430 (chrome-extension://efaidnbmnnnibpcajpcglclefindmkaj/https://pure.uva.nl/ws/files/32520737/Nawerk.pdf).

45 Kühlmann/Telle (2001/2004), vol. 2, p. 544.

46 Cf. p. 301.

47 Cordus (1561), f. 236r–301r.

48 Cordus (1561), f. 236r.

well as in my own time – which are most of them. Thirdly, I address wild plants that are sometimes found in gardens, the type, for how long, and those that do not fare well, each on a case-by-case basis. Nevertheless, I did not want to make special lists in these three categories; therefore, they are all included with the others alphabetically, so that they can be found as easily as possible."[49]

Impressed by the abundance in the plant world that he had encountered during years of study, and also fascinated by the many new revelations made in his century, he concluded: "As a young man, I thought there might be only a very few plants in nature that could elude my knowledge. As an old man, however, I have travelled across many regions and came across all kinds of plants that I have transplanted to my garden and those of my friends. And my friends have sent me a very great number of rare species. I have read all the literature I could find in Greek, Latin, Italian, French, German, English and Spanish. Now it is completely clear to me that my knowledge makes up only the very tiniest part of that which is left to learn."[50]

As the basis for his compilation of plants in Central European gardens, he not only had his own observations and notes from travels and excursions, but also indices of plant stocks received from various correspondents. He named them in the following list and assigned each a letter abbreviation to specify the source locations of specific plant names:

Benedictus Aretius (Marti), professor at the *gymnasium* in Bern (abbreviation: A)
Georg Aemylius, pastor in Stolberg (AE)
Petrus Coldenbergius, owner of the pharmacy "*The Old Bell*" in Antwerp (C)
Didymus Obrecht, physician in Strasbourg (D)
Franciscus Folliet, pharmacist in the Savoy (F)
Joachim Kreig, pharmacist in Torgau (K)
Georg Ollinger, tradesman in remedies, collector of plant illustrations (O)
Johannes Ralla, pharmacist in Leipzig (R)
Woyssel, citizen of Breslau (Vu)

In addition, he also received details about garden plants from the physician Rurpodius and the pharmacist Jakob Oppenheim, both from Frankfurt, as well as the pastor Christoph Pfäfferlin (Piperinus) from Sigriswil in the canton of Bern and the Meissen physician Christoph Leuschner.[51] He also mentioned

49 Cordus (1561), f. 236v.
50 Cordus (1561), f. 236v.
51 Cordus (1561), f. 243r/v.

more than a dozen other garden owners. The lists from the Esslingen pharmacist Johannes Volmar and from Johannes Schmidlapp, a well-known businessman from Schondorf in Württemberg, arrived too late, so that the former was included in an appendix, and the rare plants in the latter's garden were printed separately in the same book.[52] The catalogue of 144 plants from Piero Micheli a citizen of Venice who lived in the San Gervasio quarter, was also printed separately. Because Gessner concentrated on plant life cultivated north of the Alps, he did not integrate this list despite having received it four years earlier. Nonetheless, it was important enough for him to publish. It was followed on the same page by the ten rarest plants in the garden of the Duke of Florence.

Gessner designated the plants in his own garden with the letter "G."[53] They were followed by those in other Zürich gardens, for example, from the surgeon Peter Hafner, the pharmacist Johann Jakob Clauser and the guild master Funck, who kept a terraced garden.[54] Gessner's prized plants included newcomers from the New World[55] such as the prickly pear, which was called the "Indian fig," and tomatoes (ill. 55), which he described as follows: "Golden apple (*Pomum aureum*) or love apple (*Pomum amoris*), or also New World apple (*Pomum de altero mundo*). It grows on a foreign, perennial shrub and appears to be related to the nightshade plants. One could designate the plant as a second type of apple-bearing nightshade (*Solanum pomiferum alterum*) as opposed to the unhealthy apple (*Malum insanum*). It emits a strong, rank smell. The fruit is nearly odor-free, not unpleasant to eat, and not harmful. It is the size of a spherical or otherwise small round apple. It is frequently larger with an uneven bumpy shape. The color is golden yellow, while another kind is red and yet another is white. The golden apple is easy to grow here, and the fruit ripens early. It thrives when well watered, prefers a dense soil and can be planted in beds or pots."[56]

He frequently invited physician friends to come see his local, alpine and exotic plants, as shown by corresponding letters sent to Cosmas Holzach in Schaffhausen on September 6, 1554 and to Didymus Obrecht in Strasbourg

52 Cordus (1561), f. 287v.

53 Gessner laid out three gardens in Zürich until 1560 and also used the medical gardens of his colleagues, the surgeon Peter Hafner and the pharmacist Johann Jakob Clauser. Cf. Fretz (1948), pp. 72–76.

54 Fretz (1948), pp. 105 and 134.

55 Leu (1992).

56 Cordus (1561), f. 273v.

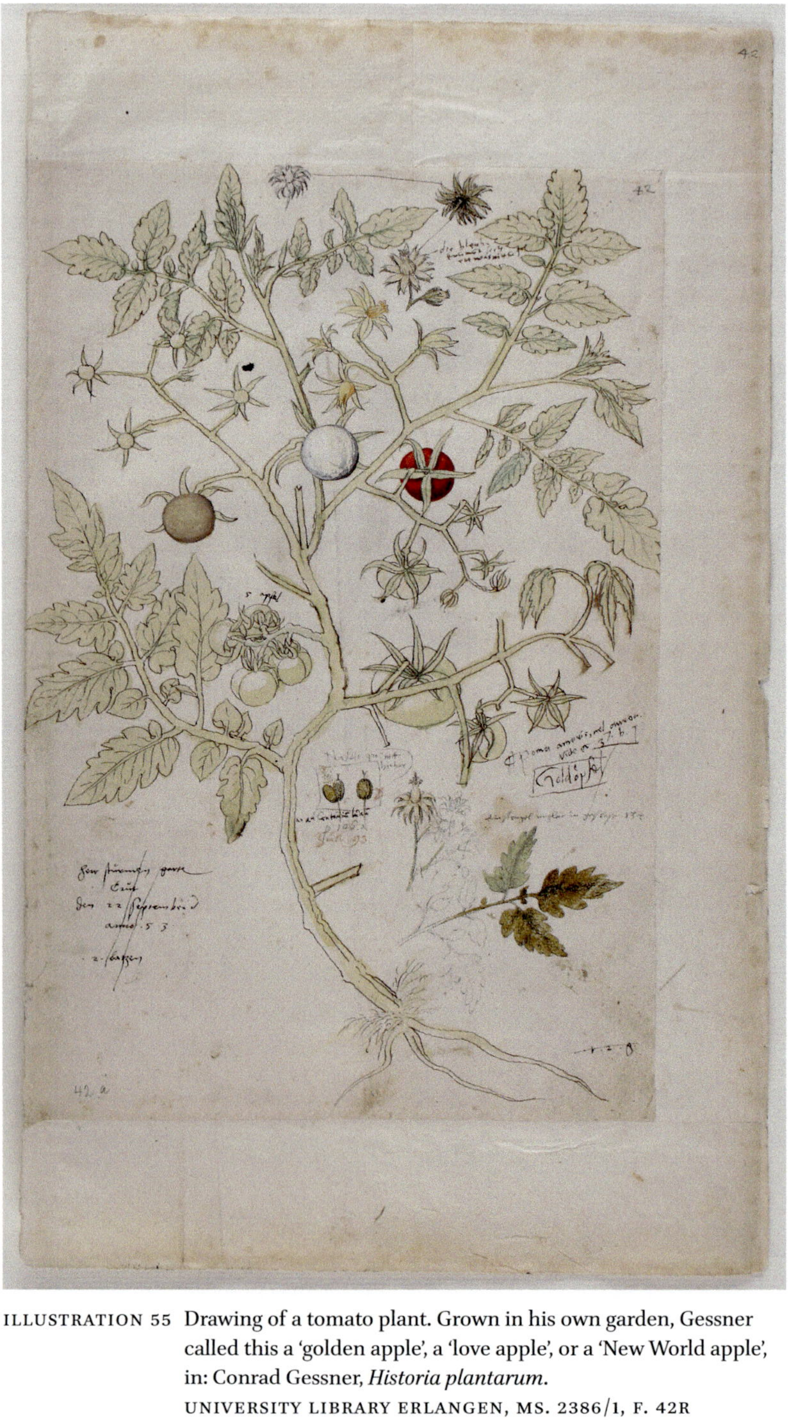

ILLUSTRATION 55 Drawing of a tomato plant. Grown in his own garden, Gessner called this a 'golden apple', a 'love apple', or a 'New World apple', in: Conrad Gessner, *Historia plantarum*.
UNIVERSITY LIBRARY ERLANGEN, MS. 2386/1, F. 42R

on March 18, 1560.[57] He had visited the latter in 1559 on his return trip from Augsburg[58] by way of Basel.[59]

With so much enthusiasm for botany, it is not surprising that Gessner tried to win over the Zürich authorities for the founding of a botanical garden similar to those established in Pisa in 1544 and Florence and Padua in 1545.[60] His colleague Melchior Wieland took charge of the one in Padua in 1561. Gessner sent the following request to the city council in 1558: "Mr. Mayor! Honorable, wise, beneficent, gracious and kind gentlemen! Your municipal physician and other doctors of medicine, whom you have appointed, request of your wellborn to be allowed to find a suitable location in your city in which to plant many sorts of trees and plants (especially foreign ones and those not found in abundance here). They should be planted and grown in a pleasing and fertile manner, so that this site will fully and permanently serve your gracious lords to the honor and satisfaction of the city. Over time it will grow and attract outsiders as well yourselves, benevolent gentlemen, to stroll through and enjoy. And for illnesses and other incidents, there will be refreshments and restoratives, fresh and green, that one cannot find elsewhere, or even in a pharmacy. This will be a boon to those who are otherwise averse to medicines. In addition, this should offer an attractive, well-situated, enclosed and safe location with a great deal of sunshine that is protected from chill northwest winds. Sunlight should reflect off its walls, adding extra warmth to preserve the many exotic plants. Furthermore, in diligent service to your gracious gentlemen, your municipal physician offers to gradually equip, supplement and contribute more to be grown there, for as long as God grants him life and good health. For he knows clients in various locations from whom he can acquire uncommon branches and plants. The same service is offered by the other two doctors of medicine appointed by my lords. In future, it will be recommended to any municipal physician to maintain and promote the gracious gentlemen's arboretum and to make sure that it does not gradually deteriorate, which is pleasurable work and comes at a low cost. It is also a fact that in Italy it is considered praiseworthy that such community gardens are maintained, not just for pleasure and decoration but also to be of use in healing the sick. At any time, many kinds of fruits, herbs, seeds, flowers and roots, which would otherwise not be available,

---

57 Gessner (1577), f. 81v and f. 114v/115r.

58 Gessner (1584), f. C_1r/v.

59 Cf. AK 11, no. 4693, pp. 974–994.

60 Italy is not only the home of botanical gardens, but also of the private garden created for medical, agricultural and botanical studies. Cf. the comments of Pietro de Crescenzi (1233–1320) or Francesco Petrarch (1304–1374): Cardini/Miglio (2002), pp. 92–114.

are good and fresh at hand in such convenient sites. With this, we are completely at the command of, and in service to, your honors' wisdom."[61]

Unfortunately, the request was not granted. It was only in 1837, in the wake of the establishment of the University of Zürich, that the old botanical garden began to be set up. Today, you will find there also a small Gessner Garden where more than fifty species of medicinal plants and a bust of Gessner, from the end of the nineteenth century.

15.5 Excursions

Gessner was involved in botany throughout his youth, collecting and describing plants during his studies in Basel and France and also during his leisure time in Lausanne. Later he undertook various botanical excursions. Leafing through his surviving three-volume handwritten *Historia plantarum*, one can find a great many documented locations where he either observed plants himself or had samples sent to him by others. Both near and far, he also researched the local surroundings of his home town on botanical outings: "Gessner was well acquainted with the still botanically interesting area around Katzensee, as well as Sihl Valley, the Albis Hills, the area of Andelfingen, Greifensee, the Glatt River, the Eglisau district, the suburb of Dübendorf and Kyburg area. He was especially familiar with the vicinity of Lake Zürich and its upper lake. He must have wandered through these stock areas countless times."[62]

There, he encountered plants that have since died out and today no longer grow in the Canton of Zürich, such as the wild form of the broadleaf wild leek (*Allium porrum*).[63]

The universal scholar from Zürich was among the few men in the sixteenth century who had learned to swim. On June 26, 1563, he wrote to Gasser in Augsburg: "Also, I am adding to my herbarium with more of an effort than before. I am having pictures made, and for each illustration (whether painted or a woodcut) providing live plants from the neighboring land, from meadows, fields, pathways, vineyards, gardens, hills, mountains, forests, rivers, streams, ponds, lakes, along banks and every other place that plants grow. To get certain plants, I have gone swimming recently, although I have not done so for years."[64]

61 ZBZ, MS. S 93, no. 181. Cf. Hanhart (1824), pp. 212f.
62 Fischer (1966), p. 120.
63 Landolt (2013), p. 220.
64 Burmeister (1975), p. 238.

Gessner, the overworked scholar, also had paid helpers whom he sent on journeys to collect plants. He even sent one such unnamed person to the Tessin with a letter from September 22, 1559 for Johannes Fabricius Montanus in Chur. He wrote: "I'm sending you the printed story by David Georgius [the Anabaptist David Joris] entitled *Heresiarcha*. I also recommend to you the courier, a poor young man whom I have sent at my own expense to the Alps and high mountains to collect genuine black *Helleborus* plants [Christmas rose][65] and other rare plants with which to adorn my garden and add to my general history of plants. After travelling through your Alps, he will continue on to Bellinzona, then to Tesserete near Lugano, where the excellent doctor Aloysio Quadri and his son might be able to provide him with the *Helleborus* and other plants. For I have heard that the Monte Ceneri near Lugano has a wealth of the rarest plants. Please pass this along to your skillful doctor, my friend, Hieronymus Brixius, whose letter I received a few days ago. Perhaps he knows Quadri and can recommend this good young man to him and also give him letters of recommendation for Lugano, which he may eventually also reach. He does not understand Italian. If possible, provide him with a travelling companion, or give him written instructions for reference as to which route to take. He can read and does know the plants very well. I think he will also make it to the Valle Mesolcina, where he can find the musk yarrow [*Achillea moschata*], one of which I would like him to bring to me. I also want ripe seeds from the herb I mentioned, or perhaps a *Polygalon* [milkwort], which has similar healing qualities. He will very much want to have some roots from the *Gladiolus* [iris], and I ask that you show him the meadow where they frequently grow. He will be returning to us by a different route, namely, through the Gotthard Pass."[66]

This was possibly the same courier about whom Fabricius Montanus wrote to Bullinger on October 9, 1559, remarking that he had arrived too late, and that the mountains were already covered with snow, so he was being sent back to Zürich with letters and some plant roots.[67]

Among Gessner's famous mountain excursions were his trip to the Glarus Alps in 1541, his scaling of Mount Pilatus in 1555 and his first balneologically motivated research trip to the Valtellina Valley in 1561. Additionally, he conducted many more alpine tours, for some of which there is no surviving documentation. Likewise, in the fall of 1562, he wished to climb various high peaks,[68] but his plans and what was ultimately accomplished are not known.

65 *Helleborus niger* is found in Switzerland only in a few locations in Ticino.
66 UB Basel, G2 I 22:1:76. Cf. Hanhart (1824), pp. 244–246.
67 Letter from Fabricius Montanus to Bullinger, October 9, 1559, cf. StAZH: E II 375, 575.
68 Gessner (1584), f. C_3v–C_4r.

While the Glarus trip was briefly mentioned above, and the outing to Valtellina will be covered in the section on balneology, the exploration of Mount Pilatus in the Lucerne region will be taken up here.

Gessner was not the first to take an interest in what was known as *Mons fractus* and in the little Pilatus Lake.[69] The oldest, precise description of the now largely silted-up little lake and its surroundings comes from the Zürich lawyer Felix Hämmerli (1388/89–1458/61). He visited the area in 1447 and commented on the medieval legend about how the evil spirit of Pontius Pilate inhabited the lake and could stir up severe storms.[70] In his *Tractatus de nobilitate*, Hämmerli wrote: "The ability to cause storms has been attributed to it from time immemorial. Those who approach the lake in silence and do not agitate the water with words or deeds can move on unscathed. But if words are uttered, especially if Pilate is summoned, or if even the slightest thing is thrown into the water, or it is touched or caused to move, the greatest of storms will immediately break out of completely clear skies. Unbearable hailstorms and squalls will set in and devastate the surrounding lands. The neighboring areas, mountain meadows and fields and many parts of the city of Lucerne have been repeatedly ravaged. But if a child, a horse or any other animal enters the lake, no harm will come of it."[71]

Hämmerli was the first to consider that these storms might not be related to the spirit of Pilate, but possibly to natural causes. Another famous Swiss visitor to Mount Pilatus was the poet, physician and later reformer from St Gallen, Joachim Vadian. Guided by the Lucerne canon Johannes Xyloctetus (1490–1526), he climbed the mountain up to the so-called *Mittaggüpfi* (1,917 meters above sea level) in August 1518.[72] Also along were Oswald Myconius and Vadian's brother-in-law, Conrad Grebel (ca. 1498–1526), who would later become a leader of the Anabaptist movement.[73] Vadian published his notes about the ascent in his second edition of the Roman geographer Pomponius

69 On oronymy, see Heini (2014), pp. 39–42; Frohne (2010), p. 71.

70 A similar legend was created in the case of Lago di Pilato near Foce in the Italian region of Marche. Pontius Pilate, after his execution in Rome, was supposedly immersed in this lake, which subsequently turned blood-red. The authorities even tried to prevent people from visiting the lake by means of a wall at the entrance to the valley. Cf. Ciabochi (2013), pp. 86f.

71 Cf. Heini (2014), pp. 86f. Gessner also had knowledge of Hämmerli's report, cf. Gessner (1555), pp. 60–62.

72 Weber (1913), p. 251.

73 Vadian very likely did not climb to the top of the mountain. What he described as "ad summum montis verticem" seems to denote a mountain height or a pass. I would like to thank Dr. Rudolf Gamper (Winterthur) for this information. Cf. Suter-Meyer (2020), pp. 204–220.

Mela (1st century), which was published in Basel in 1522. He gave the impression that, at first, he was not sure what he should make of the whole Pilate legend. He did, however, dismiss as nonsensical the fable that Pilate could be seen on the lake in the robes of a judge on every Good Friday, and that those who saw him would die within a year.[74] He later asserted that the entire matter was superstitious humbug.[75]

Conrad Gessner climbed Mount Pilatus on August 20, 1555 together with the Zürich surgeon Peter Hafner, the pharmacist Pierre Boutin from Avignon and the painter Johann Thoma, a relative of Gessner.[76] Lucerne's municipal physician, Johannes Chrysostomus Huber, had made efforts to ensure that the group was allowed to explore this storied region, and that same year, Gessner dedicated his published description of the excursion to him. Permission was granted by the mayor Niklaus von Meggen (before 1497–1564). In the dedication, Gessner asked Huber to send him other texts about mountain climbing so that he could combine them with his own observations to be published in a booklet.[77] At the end of this little book, he included his colleague Rhellicanus's 1536 poetic description of the Stockhorn mountain near Bern. The later material about the plant world of the Stockhorn and Mount Niesen by Benedictus Aretius was published by Gessner in his Cordus edition in 1561.[78]

Gessner's description of Mount Pilatus contains the first published observations of alpine altitude levels, which is why he is also called the founder of phytogeography (plant geography). With his frequent scaling of the North Italian Monte Baldo, the Veronese pharmacist Francesco Calzolari (1522–1609) had possibly made the same discoveries before the universal scholar from Zürich, but he did not publish his findings until 1566.[79] Gessner noticed that the vegetation changed as the altitude increased: "Notably, one can say that there is eternal winter on the peaks of the highest mountains. Somewhat lower it is still spring during the middle of summer or later. At this level, one can see flowers blossoming that bloom in springtime although it is actually mid-summer or the fall, such as *violae* [violet], *flores bechii* [cowslip] and *flores petasitidis* [butterbur = *Petasites albus* or *Petasites hybridus*]. However, there were no fruits except perhaps *fraga* [strawberries] and *vitis idaea* [cranberries; *Vaccinium vitis-idaea*]. Further down, however, there is a level for autumn, with certain

74 Mela (1522), p. 34; Frohne (2010), p. 73.

75 Gessner (1555), p. 59; Frohne (2010), p. 163.

76 Weber (1913), p. 255 assumes that Gessner, unlike Vadian, was at the summit.

77 Gessner (1555), p. 44.

78 Cf. Cordus (1561), f. 232r.

79 Salzmann (1959), pp. 81–89.

trees bearing fruit, especially *cerasa* [cherries], which ripen late, since the sun shines more as if it were springtime than summer. At the bottom now, the sun is warmer and its reflecting rays create a veritable summer. This is why I say it is winter at the highest level of such mountains, since winter and snow dominate there, and if the snow melts further down (as it does on the mountain we write about) then cold and wind are prevalent. The second zone, which spreads out below the summit is springlike and the winter is not extended, but the spring is short. The third is fall-like because beyond spring or winter, it also has the quality of autumn. The lowest level is like summer. Therefore, at the top, one recognizes a single season, below the summit two, at the third level three, and at the bottom, four seasons."[80]

Gessner thoroughly enjoyed hiking and effused about it: "Starting from the middle of the mountain range and the Eigenthal Valley, the ascent is steeper and more difficult up to the highest hut or 'alpine dairy', as the locals call it. A little below this place is a spring on the right side of the slope, hidden in a small cave. Thanks to its crystal clear and ice-cold water, we recovered wonderfully from our fatigue, thirst and the heat. We drank a great deal and ate bread dipped in it. I do not know if the human senses can experience any greater Epicurean enjoyment than this (although it is exceptionally sober and simple)."[81]

He described how the mountains bring pleasure to the senses of touch, sight, hearing and taste, and praised the excellent fruit and dairy products. He recounted how they all blew on an alphorn, and stated that plants that grow in the mountains are better for use in remedies than those from the flatlands: "One also encounters pleasant smells from herbs, flowers and plants in the mountains. The alpine plants smell better than the same ones growing in the plains and are more effective when used to make remedies. The air here is much freer and healthier, unlike the foul-smelling and infectious air in cities and other human settlements. When moving from the nose to the brain and from the windpipe to the lungs and heart, it is not only harmless but even delights and invigorates."[82]

From this statement, not only can one surmise that due to insufficient sewage disposal, even small cities like Zürich smelled bad, but that it was also thought that foul-smelling air carried various diseases such as the plague.

When the group reached Pilatus Lake – more swamp than lake – which Gessner thought was about the size of 12 beds, he expressed his thoughts about the legend in an almost enlightenment manner: "It is said that Pilate

80 Gessner (1555), pp. 45f. Cf. Deubelbeiss (1991), pp. 36f.
81 Gessner (1555), p. 47. Cf. Deubelbeiss (1991), p. 38.
82 Gessner (1555), pp. 48f. Cf. Deubelbeiss (1991), p. 40.

is submerged in this lake, and his wife is in a smaller one nearby. The area is swampy. If something is purposefully thrown into it – it is said – the entire surroundings will be endangered by storms and floods. Because this belief by the locals has no basis in nature, I can give it no credence. Indeed, while superstition and false beliefs are often the result of unexplained and curious occurrences, this is not cause for reasonable men to simply believe them. I tend to think that Pilate was never present in this place, and had he been, his death would not have lent him the power to take either positive or negative action against people. However, if one were to say that evil spirits are permitted by divine guidance to accomplish many such things, and that Pilate has also assumed the nature of such a spirit, my answer would be that it cannot be called 'power' when evil is done by evil spirits, but that the Lord God permits this only to deceive the faith of the godless and superstitious, to strengthen their delusion and thereby make them suffer punishment for their godlessness and superstition."[83]

Finally, Gessner cites the historians Eusebius of Caesarea (ca. 260–340) and Johannes Nauclerus (ca. 1425–1510), who state that no one knows whether Pilate committed suicide in Italy or France, but neither of them brings up the Swiss Confederation.[84] Gessner explained the floods as resulting from streams that lead rainwater from the mountains down to flatland, which can have unpleasant consequences. In this short text, by offering rational argument and explanations, he not only debunked the myths surrounding Pilatus Lake, but he also relieved people of their fear of the inhospitable mountains and certain natural phenomena.[85]

Gessner ended with an account of the plant life on Pilatus and also addressed the superstitious pharmaceutical uses of the so-called moonmilk. It is a moist lime deposit which he analyzed rationally: "As we said, in a cave, high on the mountain there is a loose, white, very fragile and crumbly substance on the vault that one might refer to as a rocklike sponge or a stony larch fungus (*agaricon*). It is called 'monmilch', which means moonmilk (*lac lunae*) and describes the white, foamy substance that hardens to become this stone, if one can call it a stone. … Out of ignorance and superstition, some people believe it will heal the illness of any afflicted person, which is why it is given this name and gathered from this cave."[86]

83 Gessner (1555), pp. 42f. Cf. Deubelbeiss (1991), pp. 53f.
84 Gessner (1555), pp. 53f.
85 Gessner (1555), p. 59.
86 Gessner (1555), p. 66. Cf. Deubelbeiss (1991), p. 50.

It appears that Gessner's observations regarding altitude levels soon became a topic of discussion among scholars. Johannes Fabricius Montanus sent a letter to him on June 26, 1559 in which he listed 19 plants that he found on Mount Calanda together with the physician Zacharias Beeli[87] and the school director Pontisella. He specified that 18 of these were from the middle, and one was from the peak of the mountain.[88]

Gessner's small anthology containing the Pilatus description began with the essay *De raris et admirandis herbis, quae ... Lunariae nominantur* (*About Rare and Remarkable Plants that are Called Moon Flowers*), which was the first work in the history of science to discuss the topic of bioluminescence, which today's science attributes mostly to deep sea creatures.[89] Gessner dedicated the work to the Landshut physician Lorenz Grill (Laurentius Gryllus, 1484–1560), who visited him in Zürich in 1555 and who also maintained a relationship to Ulrich Fugger.[90] Gessner related that he was often asked about flowers that glow at night (*Lunaria*), which led him to write this piece based on the literature as well as his own observations. His assertion was that he knew of no plants that give off light in any form. The longest chapter, which makes up half the publication, is titled: "*De aliis quibusdam lunariis quas novimus, etsi nullam earum noctu lucere observatum nobis sit*" (*About Certain other Moon Flowers* [*Lunaria*] *We Are Familiar with, of which, However, None Has Been Observed by Us to Glow at Night*).[91] Gessner, the empiricist, formulated his argument based on his own observations and discussed a number of plants that had been categorized by scholars as *Lunaria,* despite the fact that they emit no light. This included, for example, the *Tora venenata* (tortoiseshell buttercup = *Ranunculus thora*). The mountain cowslip (*Primula auricula*) was also among these *Pseudo-Lunaria.* Gessner printed an outstanding woodcut of it in his publication (ill. 56).[92] While he critically probed this myth-laden chapter of plant history and found no bioluminescence in plants,[93] he did admit that certain creatures such as the firefly do have it.[94] With the idea that animal eyes could emit light, as he

87 I thank Prof. Dr. Jan-Andrea Bernhard (Ilanz) for identifying the physician Belinus as Zacharias Beeli of the family "Beeli von Belfort."

88 Cordus (1561), f. 235v.

89 Harvey (1957), p. 66.

90 Durling (1965), p. 140.

91 Gessner (1555), pp. 23–42.

92 He had also received a specimen from Ulisse Aldrovandi, which, according to the statements of the Bolognese botanist, was called *Lunaria* by certain scholars. Cf. Gessner (1555), p. 29. Cf. also Zuliani (2017).

93 Meanwhile, bioluminescence has been detected in 71 species of fungi.

94 Gessner (1555), p. 5. Cf. Harvey (1957), p. 67.

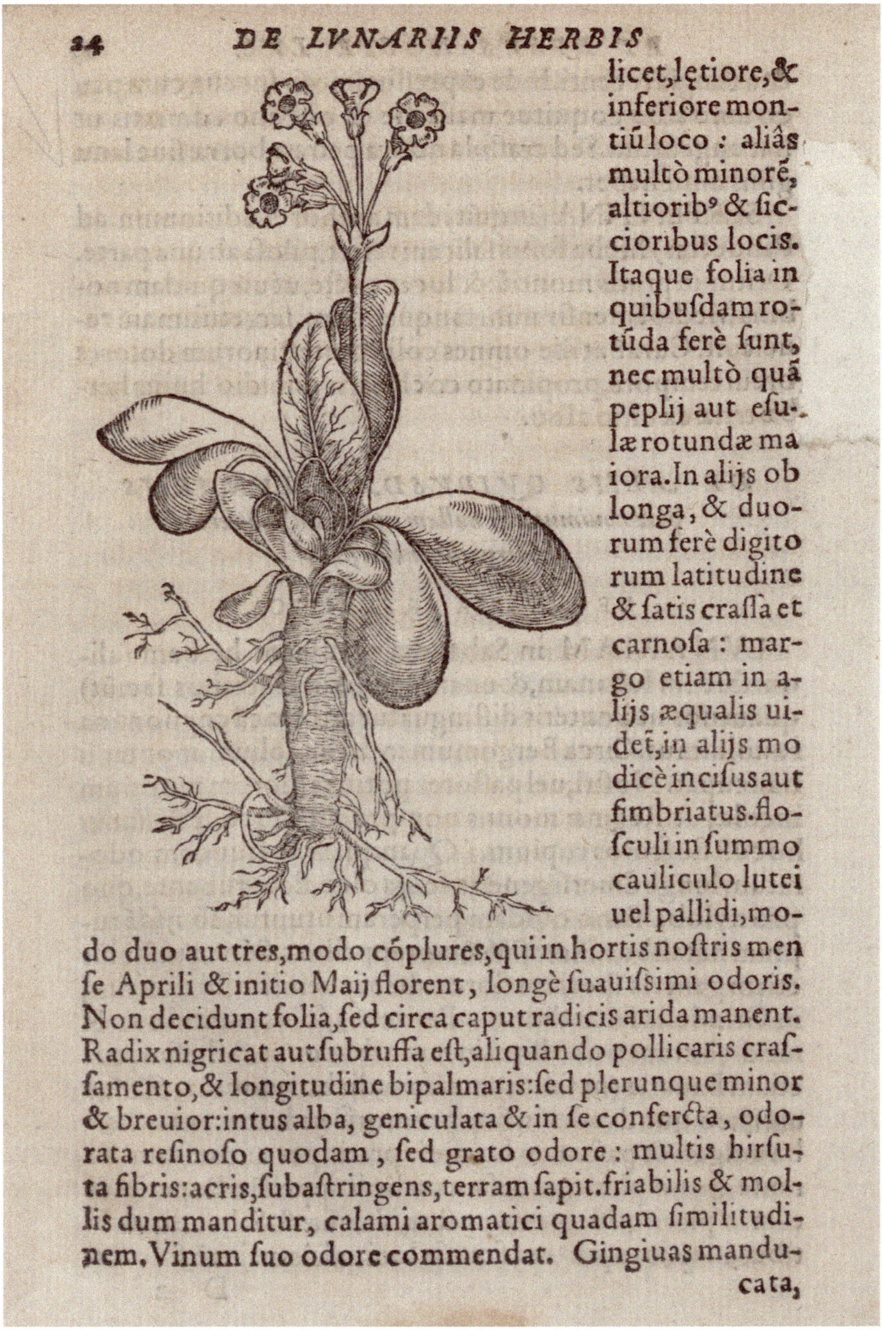

24 DE LVNARIIS HERBIS

licet, lętiore, & inferiore montiũ loco: aliâs multò minorẽ, altioribꝰ & ſiccioribus locis. Itaque folia in quibuſdam rotũda ferè ſunt, nec multò quã peplij aut eſulæ rotundæ maiora. In alijs oblonga, & duorum ferè digitorum latitudine & ſatis craſſa et carnoſa: margo etiam in alijs æqualis uidẽt, in alijs modicè inciſus aut fimbriatus. floſculi in ſummo cauliculo lutei uel pallidi, modo duo aut tres, modo cõplures, qui in hortis noſtris menſe Aprili & initio Maij florent, longè ſuauiſsimi odoris. Non decidunt folia, ſed circa caput radicis arida manent. Radix nigricat aut ſubruffa eſt, aliquando pollicaris craſſamento, & longitudine bipalmaris: ſed plerunque minor & breuior: intus alba, geniculata & in ſe confercta, odorata reſinoſo quodam, ſed grato odore: multis hirſuta fibris: acris, ſubaſtringens, terram ſapit. friabilis & mollis dum manditur, calami aromatici quadam ſimilitudinem. Vinum ſuo odore commendat. Gingiuas manducata,

ILLUSTRATION 56 Mountain cowslip (*Primula auricula*), in: *Gessner* (1555), p. 24.
ZBZ, ABTEILUNG ALTE DRUCKE UND RARA: 5.200

believed he had observed from the *Lucerna* fish, he was in agreement with the conventional wisdom of the time.[95] Here, he also dedicated a few pages to inorganic light emissions, such as electrical sparks given off by hair, glowing gemstones, and atmospheric phenomena such as St Elmo's fire and auroras.[96] The last of these was a topic that he dealt with in a separate 1561 publication under the pseudonym Conradus Bolovesus.[97]

15.6 History of Plants

15.6.1 *Correspondents*

For much of the information gathered for the *Historia animalium*, Gessner was indebted to a wide-ranging circle of correspondents stretching all across Europe. Preparations were similar for the *Historia plantarum*, which he wanted to work on after finishing the books on animals. He mentioned this in a letter to Johann Jakob Fugger on July 19,[98] and to Georg Fabricius in Saxony on the following day.[99] As in his animal history, he thanked 42 correspondents and helpers, who had provided him with illustrations, specimens and information. They were listed by name at the beginning of the newly identified third volume of the surviving *Historia plantarum* manuscript that is now located in the University of Tartu Library.[100] After the alphabetical list, Gessner gave a short statement about how he had acquired plants from the environs of Montpellier. He had collected some himself in 1541, and he had received others from Isaak Keller, Balthasar Hummel, Conrad Forrer, Caspar Wolf, Felix Platter, Valerand Dourez and Johannes Bowhius, a little-known pharmacist.

Just as with Gessner's network of correspondents, with the contributors to his *Historia animalium* and with the guest names in his *Liber amicorum*, there was also a wide-ranging European network of contacts that he utilized very well for his work on plants, and that likewise benefitted from him. However, we tend to forget just how difficult and unreliable the means of communication were at the time. Aside from the fact that privately paid couriers had to deal with war, plague and bandits, it was often difficult just to mail something directly from point A to point B. Detours were frequently taken and middlemen employed, as Gessner recounted on December 5, 1562, when he

95 Gessner (1555), p. 3.

96 Gessner (1555), pp. 6–12 and 83. Cf. Harvey (1957), pp. 68 and 70.

97 Cf. p. 364.

98 ZBZ, MS. C 50a, no. 49.

99 Gessner (1584), f. B_2r/v.

100 Rand (2014) and Leu (2016).

wished to send a letter to Bauhin, who was then in Padua. At the time, there were few couriers in Zürich who were headed to Padua, so he initially sent the letter to Fabricius Montanus in Chur, where there were better connections to the North Italian cities. Gessner recommended to Bauhin to send his reply to the physician Domenico Montesauro in Verona or to send it via the medical student Giovanni Battista Mundella to Aloysio Mundella in Brescia. It was then up to Gessner's colleagues in either Verona or Brescia to send it on to him in Zürich.[101] When Bauhin was in Lyon, Gessner suggested that he send the requested package of books to Zürich by way of Johannes Fries Jr. in Geneva, who had been staying there for some time.[102] Letters and packages were frequently held up somewhere, so that it would take months before being received by the addressee. While not a unique case, this is illustrated in a letter that Gessner sent to Kentmann in Torgau on February 16, 1564 in which he noted that on the previous day he had received Kentmann's letter, sent nearly a half a year earlier, at the end of July.[103] There were any number of reasons that mail was sometimes never received,[104] as Achilles Pirmin Gasser experienced in January 1564, when the courier to whom the letter was entrusted ripped it up and threw it away.[105]

Not all of the letters from Gessner's important correspondents have been preserved. The number of unknown pieces must be significant, as in the case of the Antwerp pharmacist Peeter van Coudenberghe (Petrus Coldenbergius, ca. 1517–ca. 1599) or the Valaisan doctor and pharmacist Caspar Collin (1520–1561), whom Gessner thanked for his support in a printed dedication.[106] The former was mentioned by Gessner in his personal copies of plant books and on the drawings in the *Historia plantarum* as a provider of plants and objects for botanical research. Van Coudenberghe maintained a beautiful private garden, an inventory of which he sent to Gessner in 1557, which the latter also published in the *Horti Germaniae*, as previously mentioned. However, it was only the plants unfamiliar to Gessner that were listed. The Antwerp pharmacist owned a total of 600 exotic plants, which were somewhat challenging for him to care for. During winter, he brought them to an underground room to protect them from the cold. Sadly, his garden was destroyed by Alessandro Farnese (1545–1592) during the Spanish siege of Antwerp in 1584/85.[107]

101 Bauhin (1591), pp. 112–114.
102 Bauhin (1591), pp. 126f.
103 Gessner (1584), f. D_2r.
104 Cf. Burmeister (1975), pp. 296f.
105 Burmeister (1975), p. 289.
106 Cordus (1561), f. 213r.
107 Vandewiele (1993), p. 24.

Another important contributor of plants from whom not a single letter has survived was the Veronese pharmacist Francesco Calzolari. He was a student of the famous botanist from Padua, Luca Ghini da Imola (1500–1556) and was the second university instructor in Italy to hold a chair for the study of medicinal plants (*Lectura practicae medicinae*).[108] Gessner often had Calzolari send him plants from Monte Baldo, which continues to be of interest today due to its many endemic plant species.[109] He placed a high value on this contact and dedicated to him the appendix to the *Horti Germaniae*.[110] Gessner expressed his gratitude for the welcome botanical exchange and pointed out that ancient authors had focused on Mediterranean fauna and flora while barely researching Nordic countries and alpine regions. For this reason, there remained a great deal to be done specifically in these areas.[111] He passed this conclusion on to Johannes Fabricius Montanus as well: "We do not know many of the earlier names of the fish in the ocean, because the ancients did not describe them – only those found in the Mediterranean Sea, on whose shores the Greeks and Latins lived. Therefore, it should not be surprising if the same thing should also occur to us with respect what grows on land. For the authors name only a few alpine plants, although it is clear that nearly every specific region has its own specific plants."[112]

The flora of Italy is a good example of this. The Romans were familiar with 408 species, and during the Middle Ages another 189 were discovered, while in the sixteenth century yet another 1,171 were added.[113]

The following are Gessner's most important botanical correspondents, or those from whom the most relevant letters have been preserved (in alphabetical order): Benedictus Aretius from Bern, Johannes Bauhin the Younger from Basel, Johannes Fabricius Montanus from Chur, Johannes Kentmann from Torgau as well as Felix Platter and Theodor Zwinger from Basel. Aretius and Fabricius Montanus were especially important to Gessner as sources of alpine plants. The former mentions around 40 alpine species in his descriptions of the Stockhorn and Niesen mountains.[114] The latter was not only frequently engaged in collecting plants for Gessner,[115] but also sent him the first

108 Taffetani (2012), p. 12.
109 Salzmann (1959), pp. 92f.
110 Cordus (1561), f. 288r–289r.
111 Cordus (1561), f. 288r–289r.
112 Gessner (1577), f. 93v.
113 Mariotti (2004), p. 150.
114 Hanhart (1824), p. 247; published by Gessner in: Cordus (1561), f. 232r–235v.
115 Fabricius Montanus wrote to Bullinger on March 2, 1562, that he had spent the whole day collecting plants for Gessner. On October 5, 1562, he reported the same to Bullinger in

catalogue of plants from the Calanda mountain.[116] Among others, *Centaurea rhaponticum* and *Lunaria rediviva* are mentioned in it for the first time.[117] Over the years, Gessner grew to become perhaps the most knowledgeable person regarding alpine flora. For example, while the ancient authors knew of only one species of gentian plant, Gessner was familiar with at least 10.[118] Today, we possess descriptions of 40 varieties of gentian (*Gentiana sp.*).[119] In this vein, Gessner let Johannes Kentmann know on March 16, 1555 that he had discovered many rare alpine plants.[120]

There were also many botanical giants who belonged to Gessner's circle of correspondents. These included: Ulisse Aldrovandi and Giacomo Antonio Cortusi[121] from Italy, Leonhart Fuchs from Germany, Jacques Daléchamps and Guillaume Rondelet from France and Charles de l'Écluse from the Netherlands.[122] However, only a few letters from Cortusi have survived and none from de l'Écluse.

15.6.2 *Drawings*

It is not known exactly when Gessner began to collect drawings of plants, but by May 22, 1553, in a letter to Benedictus Aretius, we learn that he was already in possession of more than 100 plant pictures.[123] Clearly, plants unfamiliar to Gessner were of the greatest interest to him, and it was from these that he wished most to have drawings. Therefore, on June 22, 1554, he let Kentmann know that he would appreciate being notified of illustrations that had not yet been published in the relevant works of Hieronymus Bock, Leonhart Fuchs or Pietro Andrea Mattioli.[124] By the spring of 1555 his collection had grown to about 1,000 images.[125] He was aware that the edition of Fuchs's *Historia stirpium* would include more woodcuts, but this did not deter him from his plan.

Zürich. On September 6, 1563, he was again in the mountains all day for the same reason. Cf. Schiess (1905), pp. 369, 415 and 459.

116 Fischer (1940), p. 326. The plant catalogue is published in: Cordus (1561), f. 235v.

117 Braun-Blanquet/Rübel (1932), p. 1.

118 Fischer (1940a), p. 331.

119 Aeschimann et al. (2004), vol. 2, pp. 10–31.

120 Gessner (1584), f. A_3v–A_4r.

121 Cortusi, successor to Melchior Wieland as director of the botanical garden in Padua, sent Gessner a so-called rose of Jericho (*Anastatica hierochuntica*) and other plants. Cf. Gessner (1577), f. 113v.

122 Gessner wrote to Bauhin on October 11, 1565, that Carolus Clusius, who was studying the plants of Spain, had recently written to him from Antwerp. Cf. Bauhin (1591), pp. 157–159.

123 Gessner (1577), f. 116v.

124 Gessner (1584), f. A_2v–A_3v.

125 Letter from Gessner to Kentmann, March 15, 1555, cf. Gessner (1584), f. A_3v.

As he had been able to gather many new alpine plants as well as new findings from his own medicinal garden, he intended to write about all plants and combine all published and as yet unpublished illustrations in a single work that would have a less detailed text compared to the *Historia animalium*.[126] It seems that he changed his plan during the following year and a half, and on October 18, 1556, he wrote to Leonhart Fuchs, who regarded Gessner as somewhat of a competitor. In this important and informative letter, Gessner says that he would only illustrate those plants that were not to be found in Fuchs's works. This is the letter in its entirety: "Learned friend, it gave me great pleasure to see from your letter that you are healthy and working with great diligence to complete your three-part plant history. Less agreeable is the fact that you wish to prevent me from working on the same content. Here, we should be thinking more about the general good than about ourselves. '*Ein Mann, kein Mann*' (a man alone is not a man); this saying applies especially well here. For the number of plants is endless, and due to the variety of locations and regions in which they grow, one person by himself cannot be fully informed about them all. However if, for the common good, each makes his observations known, a complete and comprehensive work can one day be made from our preparatory efforts. I dare not hope that this might be accomplished already in our century, regardless of how much I yearn for it. Your wish is for others to send their observations to you; I am pleased if indeed many plant lovers do so and thereby support your great and exceptional enterprise. I would like to do this as well – for I have the greatest confidence in your knowledge and good judgment – if only I had not already gathered such a great mass of information myself, or if my annotations had been written down clearly enough to be of service to you. But most of my observations are on countless little slips of paper, more jotted down than written out, and not of service to anyone but myself. At present, I do not have the time to sort them all out, and my head contains even more information than my papers. This is because I found my greatest pleasure in botany as a youth, and until I draw my last breath, the engagement with it will be my fondest inclination. So, do not take away my freedom and the joy of my life. I cannot say when I will publish something about plants, for I have not yet completed my large work about the history of animals. I can promise you that when I do write it, your name will be mentioned with the highest of honors. And when my opinion diverges from yours, I will give a modest explanation of my differing view. If I were able to see your illustrations and descriptions of plants before publication, I could give you my honest opinion, and perhaps suggest some improvements with regard to plant names. Were you to look over

126 Gessner (1584), f. A_3v–A_4r.

my collection, you could surely perform the same service for me, for with a science such as botany, there is invariably more to learn. It has always been my viewpoint and wish that many scholars should work in the same areas as myself. At the same time that I was writing about four-legged animals, birds and fish, Belon and Rondelet were also doing so in France. However, they made honorable mention of my writings, and Rondelt even sent important contributions to my work. That is how removed we were from jealousy and envy. It is true that Rondelet sometimes criticizes Belon and Ippolito Salviani, but he does so with justification and without indicating their names. For he has sent them a great many contributions, about which they have not uttered a single word, but have attributed his results to themselves. This reveals the highest level of ingratitude and invidious ambitions. I, on the other hand, have always honorably named those who have sent me a contribution – and not just once, but whenever there is a reason in my works to thank other scholars. And hence, my friend, I will reveal all of my plans to you, so that you can see that I speak to you with only the greatest sincerity, and with a completely open heart, and that I care about your fame and the benefit of Isengrin the bookseller (to whom I owe much). The late Wendelin Rihel, a bookseller in Strasbourg, a childhood friend, and someone to whom I taught the elements of the Greek language, gave me many books that he had printed as well as many other fine gifts. Since he knew of my love for botany, he wanted me to correct the Latin translation of Bock's book of herbs and publish a new edition that is augmented with annotations. He wrote to me that this was also the wish of the elderly Bock himself, whose written work is hampered by old age and illness. It was also the wish of my trusted friend, the young Kyber, who translated the work into Latin. He also asked me to include an appendix about Swiss and alpine plants, to which I all too willingly consented, but without setting a time. So far, under the burden of other work, I have not been able to carry out this plan. However, because I made a promise, and because I received money from Rihel for a trip to the Alps and for illustrations of rare plants, it is only correct that I follow through with the son on that which I promised to the father. Indeed, this will be a modest book and will in no way hinder the publication of your larger work. I will not increase the size of Bock's work, but instead will shorten his often idle ramblings, remove or cite errors and append a short text about Swiss and alpine plants with a few illustrations. As previously mentioned, I have a very large number of notes about plants and plant illustrations, and every day I make new personal observations and receive new contributions from friends in Germany, France and Italy. Therefore, I decided to collect all the useful and accurate writings by authors from ancient and recent times in a single volume, and I have already prepared some of the pages for this book. However, if you or

Isengrin are unhappy with this approach, I will change my plans. I will wait until your entire work has been published, then after I have read it, if still alive, I will write some additions to my book and give them to Isengrin for printing. In these additions, however, there should not be any repetition of information already found in your work. However, if I were to do this for you and give up on my original plan, I would also expect a favor in return from you that would be advantageous to the bookseller and also welcomed by botany enthusiasts. Specifically, I would wish for Isengrin to give me access to all the illustrations in your book, so that I could combine them with mine and publish a single volume with all of them ordered and named as I see fit. It would be similar to that which I have done with four-legged animals and birds, and that I am now doing with fishes. When labeling, I would note which illustrations came from your work and where the plant descriptions may be found. Likewise, I would say which illustrations came from my collection and where the plants are described in my additions. If you like this idea, I would be delighted, and if not, then I will return to my original plan, not consider myself bound by your work, and proceed as I please."[127]

Of special interest in this letter is the importance Gessner placed on making his own decisions regarding the order and names used in his planned book of botanical plates. He emphasized this wish again in a similar letter to Fuchs on February 11, 1557.[128] From this we can clearly surmise that Gessner had developed his own order (*ordo*), a botanical system, about which he unfortunately gave no further details.

On October 1, 1558, Fuchs informed his colleague in Zürich that his three-volume *Historia plantarum* with 1,200 woodcuts was finished and that Isengrin would publish it the following winter in Basel.[129] However, due to Isengrin's death on March 3, 1557, neither he nor his widow, who subsequently ran the print shop, followed through. Additionally, the project failed financially and was ultimately terminated, because Fuchs was not in a position to cover the immense printing costs of 3,000 guilders. He died with the manuscript remaining unpublished, which has been preserved in the Austrian National Library.[130]

Fuchs made clear in a letter written to Joachim Camerarius the Elder (1500–1574) much later, on November 24, 1565, that he rejected any possible collaboration with Gessner: "I'm quite amazed that some people are so overly excited about Gessner's enormous work. It was nearly a year ago that he wrote

127 Gessner (1577), f. 137v–138v, cf. Hanhart (1824), pp. 218–223.
128 Gessner (1577), f. 138v–139r.
129 Letter from Fuchs to Gessner: StAZH E II 356, 169 (copy in: ZBZ, MS. V 318, no. 17a).
130 Baumann et al. (2001), p. 112.

to me about not yet having found enough good new material for this book. Even if the work comes out before mine does, I do not care, since I have seen through Gessner's intentions. Of course, this surely best friend of mine, who writes to me so frequently, flits through every genre of author, taking pleasure in undoing their garlands into bit and pieces out of with which to weave himself a new wreath. This is how he recently appropriated Rondelet's entire book for his commentary about fish. In the same vein, he asked me to allow him discretion in how certain things are presented and in writing addenda after my commentary is published. He wishes to put forth his work as he sees fit, but I will not permit him to change that which I have prepared. He even calls for me to inform him about all of my work. I can see what he is up to, however, and therefore will not be sending him any of my plants any time soon."[131]

Gessner forged ahead with his large work on plants, which was not as comprehensive as the *Historia animalium* but was better than the equivalent work by Fuchs. A few days before his death, Gessner wrote to Adolph Occo: "I am not surprised that Fuchs has completed his botanical work after such a long time. I believe mine will be much better, although I dare not say when it will be finished."[132] Over the years, his correspondence was often about botanical matters, frequently dealing with the ordering and shipping of plants and plant illustrations.[133] Gessner let the Zürich surgeon Johannes Muralt know on August 27, 1565 that he would not shy away from the amount of necessary work and the associated costs of the *Historia plantarum*.[134] For instance, he offered to compensate the Montpellier physician François Fontanon[135] and Joachim Camerarius the Younger in Nuremberg[136] for drawings of plants.

Gessner placed a high value on natural colors and botanical details in the illustrations.[137] For this reason, he wrote to Joachim Camerarius the Younger: "I urgently wish to have drawings of an acacia and a Celtic nard [*Valeriana*

131 Baumann et al. (2001), p. 85.

132 Tilemann (1939), p. 90; letter to Occo written by Gessner in Greek: Gessner (1577), f. 58r.

133 Aretius sent him a good illustration of a mountain crocus (Gessner [1577], f. 94r), Sebald Hauenreuter (1508–1589) of a *serpentaria* (Gessner, *Historia plantarum*, UB Erlangen, MS. 2386, vol. 2, f. 233a r/v), Theodor Zwinger of the blossom of a white poplar (Gessner [1577], f. 111r), Adolph Occo of the blossoms of a *cassia* (Bauhin [1591], pp. 141f.), Crato von Krafftheim of a cinnamon tree (Gessner [1577], f. 21r), etc.

134 Rath (1951), p. 208.

135 Gessner (1577), f. 122r/v.

136 Rath (1950), p. 164.

137 Cf. the letter from Gessner to Theodor Zwinger of March 22, 1563, in which he requests the drawing of a poplar blossom in natural colors, cf. Gessner (1577), f. 111r. Cf. also his instructions for the coloring of the printed colored specimens on various sheets of his handwritten *Historia plantarum* in Erlangen and Tartu.

celtica], and that they be very carefully rendered, especially the Celtic nard blossoms. Every petal should be painted as exactly as possible. Thus, it can be shown whether this plant is structurally related to the common valerian [*Valeriana officinalis*] as I have claimed. If the seeds are available, have some of them drawn as well."[138]

A good illustration not only clarified relationships but also eliminated the necessity for long descriptions: "Usually, I make an effort to include the fruit and seeds in my illustrations. This way, a large number of plants are easier to recognize, and the drawing can substitute for a written description."[139]

15.6.3 *Herbaria*

Dried specimens from various herbaria were often used as references for Gessner's plant pictures. The drying and affixing of plants likely has its roots in the Middle Ages and was practiced by the previously mentioned Luca Ghini and his students. By the second half of the sixteenth century it had become a widely used technique for the documentation and study of local and regional flora. The oldest extant herbarium goes back to the natural scientist Gherardo Cibo (1512–1600), contains 516 plants, and originated in 1532. The second oldest was assembled in 1545 by Michele Merini, the abbot in Lucca, and contains 202 plants. The most comprehensive of these from the sixteenth century was by Ulisse Aldrovandi in Bologna, who collected five thousand dried specimens in fifteen volumes. Today, it is kept in the University of Bologna's botanical garden.[140] The two oldest Swiss herbaria came from Felix Platter, which was started in 1552 and contained 1,800 specimens,[141] and from his student Caspar Bauhin (1560–1624), with 2,500 samples.[142] A letter to Aretius from November 24, 1565 reveals that Gessner also had a collection of dried plants. He mentioned that a bookseller from Aarau had sent him a book as a gift, which held many beautiful and rare pressed plants.[143] In a letter that he sent to Cosmas Holzach in Schaffhausen sometime in 1553, Gessner mentioned receiving roughly three hundred dried plants from Bologna. In the previous

138 Rath (1950), p. 167.

139 Gessner (1577), f. 107v; cf. Peine (1941), p. 75.

140 Taffetani (2012), pp. 17f.

141 Platter's herbarium is kept in the Burgerbibliothek Bern and can be viewed online: https://www.burgerbib.ch/de/bestaende/privatarchive/einzelstuecke/platter-herbarium.

142 Bauhin's herbarium is kept at the Botanical Institute of the University of Basel, cf. https://botges.unibas.ch/herbar/herbar.htm.

143 Gessner (1577), f. 121v.

year there had been another 100, which were possibly a gift from Aldrovandi.[144] Today, no trace remains of Gessner's extensive herbaria.

Gessner repeatedly used dried plants as references for illustrations. In a January 29, 1552 letter to Aretius, he confirmed that he had received the herbarium belonging to a certain Ludovicus.[145] This was very likely Ludovicus Carinus (ca. 1496–1569), a Swiss physician and humanist working in Strasbourg. On October 17, 1556, Gessner had Caspar Wolf, who was studying in Montpellier, ask François Fontanon (who lived there) if he might make his herbarium available.[146] Furthermore Gessner wrote to Felix Platter in Basel on January 16, 1559 that he wanted to look over his collection of dried plants or at least have an index of them.[147] Apparently, Platter complied, for four years later, on March 22, 1563, he asked Gessner to send back his herbarium.[148] In the following autumn, Gessner apologized for not having returned it yet.[149] In the meantime, it seems that Platter had sent him even more specimens, for which Gessner thanked him on May 5, 1565 with the following: "My dear Mr Platter, your dried plants have been received, and I thank you. I will use them as soon as possible and return them per your wishes. My sister's son, who delivered them, says that you have received some from Lyon that you would make available to me. I am very beholden to you for your generosity to me. If they are rare (you could give me the names of some of the rare ones), and complete enough that the artist can use them, it would be a favor to me if you would send them. Indeed, I have a large number of dried plants and am also in possession of Mr Bauhin's collection. I especially like having a selection of two or more of the same plant species because you can find in the one that which is missing in the other."[150]

As Gessner mentioned, the Basel physician and botanist Johannes Bauhin the Younger had made his herbarium available. Gessner thanked him on June 30, 1564 for permission to keep the collection for as long as he needed

---

144 Gessner (1577), f. 82v. It seems to me that much of the correspondence between Gessner and Aldrovandi is lost. Although Gessner complains to Bauhin on December 5, 1562, that he had already written to him three times and still had not received an answer (Bauhin [1591], pp. 112–114), Aldrovandi not infrequently refers to Gessner in his herbarium, which suggests a more than casual exchange. Cf. De Toni (1911/12), pp. 66, 82, and 113. Gessner was also a topic in the correspondence between Calzolari and Aldrovandi, as evidenced by a letter to Aldrovandi dated November 2, 1558. Cf. Cermenati (1908), pp. 110f.

145 Gessner (1577), f. 116r.

146 Gessner (1577), f. 122r.

147 Gessner (1577), f. 97v.

148 Gessner (1577), f. 98r.

149 Gessner (1577), f. 98v.

150 Gessner (1577), f. 101r/v; cf. Peine (1941), p. 57.

it.[151] One year later, on July 9, 1565, he let his colleague in Basel know that he was currently working with the dried plants and would soon return them.[152] In the meantime, his conscience began to bother him, which is why he wrote to Theodor Zwinger on November 26 that he must finally make drawings from Bauhin's specimens or use them to improve existing drawings. Following this, he wanted to use the interesting pieces from Zwinger's herbarium, a few of which were unfortunately damaged,[153] and also work on Aretius's herbarium.[154] It is no wonder that he wrote to Bauhin on March 16, 1565 that at present he was drowning in an ocean of plants,[155] especially since he not only had the herbaria from Aretius, Bauhin, Platter and Zwinger, but also various dried samples from Joachim Camerarius the Younger from Nuremberg that he wished to return soon.[156] Gessner was especially eager to send back Bauhin's herbarium, because Bauhin was also working on his own plant book, which he had not mentioned to Gessner.[157] Gessner did not begrudge him, but instead offered support to his younger colleague.[158]

Neither Platter nor Bauhin had their plant collections returned to them until after Gessner had died, as evidenced in a letter of February 7, 1566 in which Platter requested that the estate executor Caspar Wolf send back both collections.[159] Even when Gessner had asked for patience from one or the other of his colleagues because of his permanent work overload, he was always eager to show his gratitude to them. He frequently asked them[160] to show him a plant from the many undescribed and unnamed specimens so that he could name it after them.

15.6.4 *Painters*

Because dried plants lose some of their color over time, Gessner would paint from nature whenever possible, or he would use freshly plucked specimens for his illustrations. Many plants grew right in his garden, and he collected others

151 Bauhin (1591), pp. 141f.
152 Bauhin (1591), pp. 154f.
153 Gessner (1577), f. 113r.
154 Gessner (1577), f. 121v.
155 Bauhin (1591), pp. 152–154.
156 Rath (1951), p. 195.
157 Gessner (1577), f. 121v.
158 Bauhin (1591), pp. 157–159.
159 Letter from Platter to Wolf, February 7, 1566, in: StAGR, D V/37 C 36.06.55. Occo (Burmeister [1975], pp. 347–349).
160 Such requests by Gessner went, for example, to Johannes Bauhin (Bauhin [1591], pp. 145–147), Benedictus Aretius (Gessner [1577], f. 120r–121r), Achilles Primin Gasser, and Adolph Occo (Burmeister [1975], pp. 347–349).

nearby or went on excursions to the mountains, as he wrote to Camerarius the Younger on July 17, 1564: "It is night now, and I am only writing a quick note since tomorrow morning, because of my history of plants, we are leaving quite early for the alpine peaks."[161] The immense project could not be executed without help from others. He drew several plants himself, such as the grass lily (*Anthericum ramosum*)[162] and the strawberry.[163] The latter is illustrated in the manuscript torso of *Historia plantarum*, which is preserved in Erlangen and proves what an excellent illustrator Gessner was (ill. 57).[164] He received many illustrations from colleagues, several of which were by unknown artists, whom he frequently referred to in his correspondence.[165] The high standards Gessner expected of painters is clear from a letter of July 12, 1560 to Fabricius Montanus in Chur: "In future, I do not wish for something to be painted that I have not specifically requested. This is especially so since some of the plants you sent me grow in my own garden, and they are also better represented by my painter in their dried state than the fresh ones by your painter."[166] A week later, he doubled down: "I thank you for the fact that you had so many drawings made for me, at least with regard to your very devoted affection for me. But I cannot praise the painter's skill. I would prefer that when you think that a plant is rare, you directly send it to me, either fresh if possible, or dried with flowers and leaves pressed into a book. Our painters render the dried plants better and more vividly than yours do the fresh ones. This will also avoid that new ones are painted for me that I already possess in good depictions, as those that you recently sent me."[167]

Nearly a year later, on March 20, 1561, he apologized for maligning the Grisons painter: "If I have openly confessed that I do not care much for your painter's illustrations, and that one should not pay too much for them, and that I would prefer to have the plants themselves, either fresh or dried, I do not wish that in so doing I restrained you from writing and sending them to me. Therefore, I ask for your forgiveness, if I have caused any offence."[168]

---

161 Rath (1950), p. 153.

162 Letter from Gessner to Fabricius Montanus, July 16, 1563, in: Fischer (1946), p. 130.

163 Gessner (1577), f. 84v.

164 Conrad Gessner, *Historia plantarum*, UB Erlangen, MS. 2386, f. 400v. Cf. reproduced in facsimile in: Zoller/Steinmann (1987–1991), vol. 2, p. 85. The facsimile edition contains various references to other Gessner drawings, for example in vol. 1, p. 28 (f. 53r) or pp. 72f. (f. 185v).

165 Kusukawa (2012), pp. 145–147.

166 Gessner (1577), f. 89r.

167 Gessner (1577), f. 89r.

168 Gessner (1577), f. 89v.

ILLUSTRATION 57 Strawberries in Gessner's garden that he drew himself, from: Conrad Gessner, *Historia plantarum*.
UNIVERSITY LIBRARY ERLANGEN, MS. 2386/2, F. 388R

Starting in the summer of 1560, Gessner's letters speak repeatedly of a painter who worked for Gessner. It may have been his relative Johann Thoma,[169] who was better known as Grosshans Thomann,[170] or possibly Jos Murer, who did Gessner's portrait for the *Historia avium*.[171] There is also evidence that Murer contributed various drawings of plants to the *Historia plantarum*, which he signed with his initials "IM."[172] During the summer season, the unknown artist drew plants and flowers out in nature, and in winter he worked from dried specimens. For each illustration, Gessner paid between 4 plappart (about ¼ guilder)[173] and one guilder.[174] Occasionally, Gessner's work progressed more slowly than he wished because his painter was either working on other projects,[175] or was ill,[176] and a replacement was not easy to find. Even the well-to-do naturalist from Bologna, Ulisse Aldrovandi, complained to Gessner on October 9, 1564 that there were barely any good painters available who could master natural-scientific illustration. He was urgently seeking two or three good artists for himself and other scholars. In case Gessner knew of any, he should convince them to go to Bologna, where Aldrovandi and the others could offer a great deal of work with good compensation. Up to that point, due to the dearth of painters, only 300 of Aldrovandi's 3,600 dried plants had been illustrated, not to mention a further 500 birds, insects and other animals that he wished to have rendered (ill. 58a, 58b and 59a, 59b).[177]

Starting in May of 1565, Adrian Chortander from Ijsselstein,[178] was staying with Gessner to help him complete the *Historia plantarum*. Despite all his efforts, however, Gessner was sometimes despondent about the large amount of material, and was prone to self-critical remarks. For example, he wrote to Adolph Occo on December 12, 1564: "On the one hand, I am hindered by the course of events. Almost the entire city and the surrounding areas are in mourning for the dead [because of the plague], and we are overwhelmed from all sides with thoughts of death. On the other hand, my long hours of work at

169 Cf. p. 267.
170 Meyer (1884), p. 233.
171 Gessner (1555a), reverse of title page.
172 Leemann-van Elck (1935), p. 24; Gessner, *Historia plantarum*, UB Erlangen, MS. 2386, vol. 1, f. 50v, 58r, 103r, 137v and 145r. About Gessner and other artists in Zürich cf. Ruoss (2019).
173 Bauhin (1591), pp. 122–124.
174 Kusukawa (2012), p. 145.
175 Bauhin (1591), pp. 141f.
176 Bauhin (1591), pp. 157–159.
177 Cf. Frati (1907), p. 75; Olmi (1992), p. 70. In Frati, the volume number is incorrect. The letter from Aldrovandi to Gessner is preserved in the Bibliotheca universitaria di Bologna, MS. 77, vol. II, f. 23r–24r. The mentioned letter passage is found on f. 23r.
178 Bauhin (1591), pp. 154f.

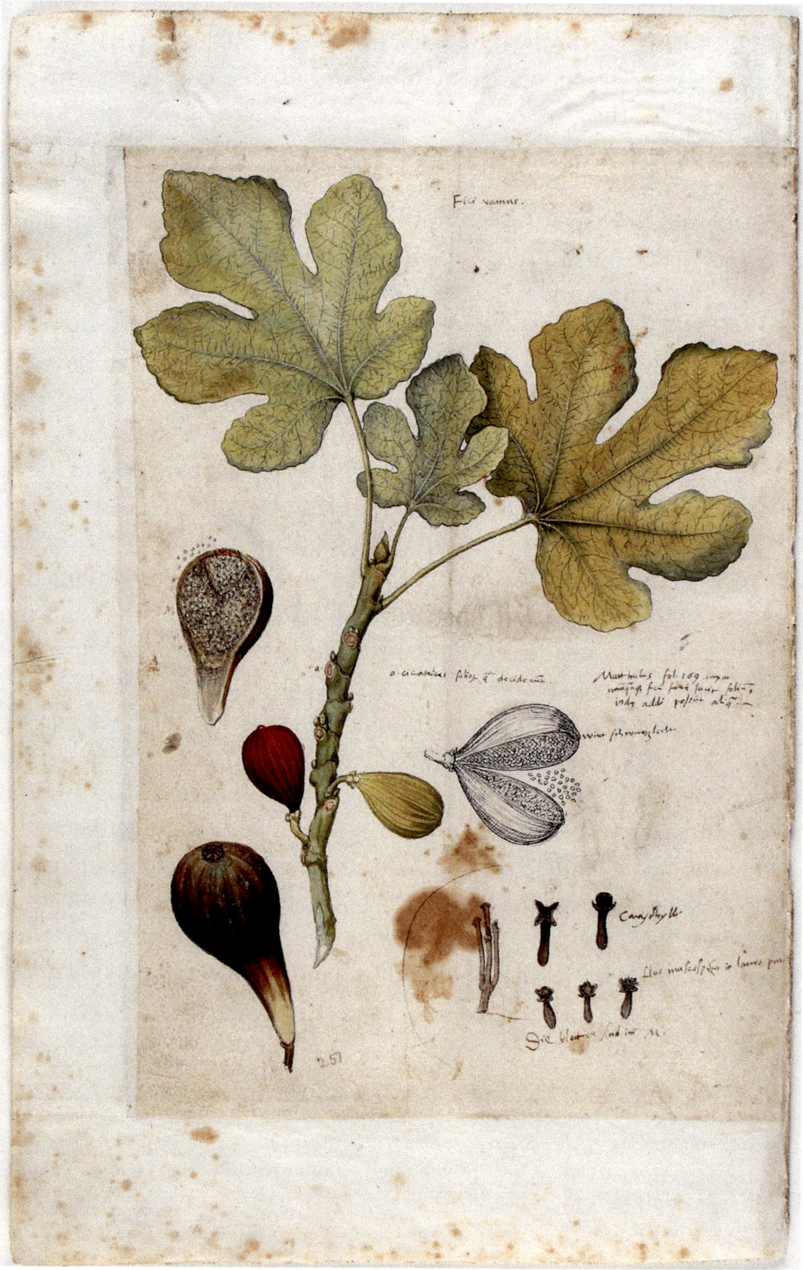

ILLUSTRATION 58A Fig (*Ficus carica*) by Gessner with detailed studies. He examined the individual organs of a plant and attempted to understand them, while contemporaries such as Ulisse Aldrovandi were satisfied with just documenting the flora. Conrad Gessner, *Historia plantarum*.

UNIVERSITY LIBRARY ERLANGEN, MS. 2386/2 F. 346 V

ILLUSTRATION 58B Figs from the illustrated herbarium of Ulisse Aldrovandi, *Tavole di piante*, vol. III, p. 193.
BIBLIOTECA UNIVERSITARIA DI BOLOGNA

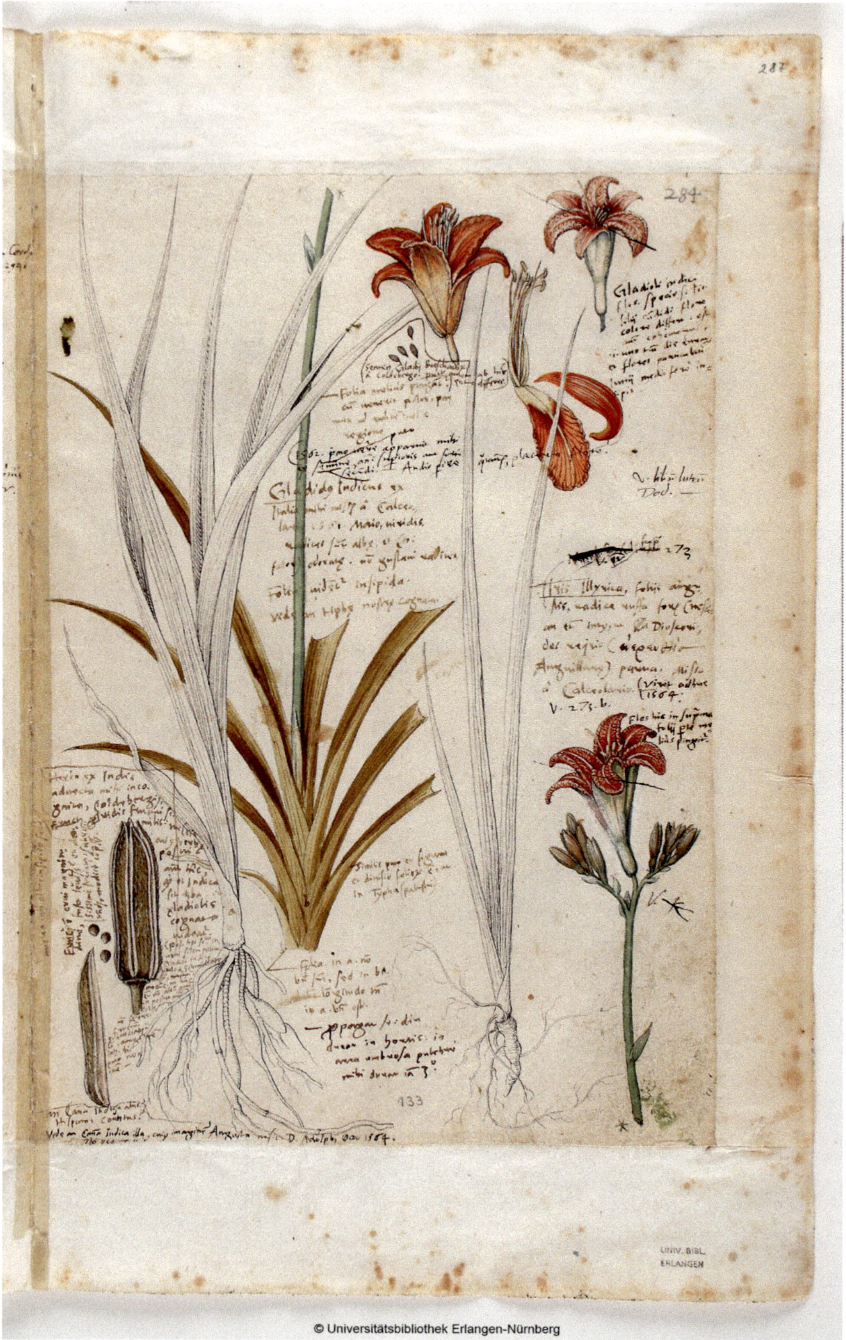

© Universitätsbibliothek Erlangen-Nürnberg

ILLUSTRATION 59A Yellow-red daylilies (*Hemerocallis fulva*) by Gessner with detailed studies. Conrad Gessner, *Historia plantarum*.
UNIVERSITY LIBRARY ERLANGEN, MS. 2386/S, F. 284R

ILLUSTRATION 59B Yellow-red daylilies, in: Ulisse Aldrovandi, *Tavole di piante*, vol. I, p. 87.
BIBLIOTECA UNIVERSITARIA DI BOLOGNA

night, which I consider to be disastrous, hamper me from doing or describing anything else with serious conviction. By the way, I am more suited to be a philologist than a philosopher, and I am not as thorough as many others. I cannot promise myself, let alone anyone else, and give assurances as to when I might be through with my work on plants. So far, my plan is to do a lot of reading and make extractions from it, so that the work I publish will be as complete as possible."[179]

15.6.5 *Drawings in Erlangen and Tartu*

Unfortunately, Gessner died on December 13, 1565 with three volumes of plant illustrations and notes still unfinished. In 1929, two of the codices were rediscovered at the University Library Erlangen with no trace of the third volume. It was recently identified along with a few more illustrations at the University Library in Tartu (Estonia).[180] Gessner's handwritten notes are found on many of the often very artistic renderings. While the entire volume Ms 55 was surely created in the study of the Zürich polyhistor, in Volumes 56 and 57, there are only some individual drawings and hand-written notes that may be clearly identified as belonging to him.[181]

As a rule, Gessner used a single page for each plant and wrote down everything about it.[182] He would also credit those who might have provided the drawing and also write down various instructions for the artists who were creating the woodcuts or colorizing them. On the back side of page 146 on the Ms 55 codex, Gessner wrote: "Hactenus ex libro Io. Kentmanni" (up to this point, from Johannes Kentmann's book). If one looks for the meaning of this in Gessner's correspondence, his March 15, 1555 letter to the Torgau doctor Johannes Kentmann offers an explanation. Gessner wrote that he would return the beautiful volume of plant illustrations to him.[183] The *Codex Kentmannus* is still in existence and is preserved at the Duchess Anna Amalia Library in Weimar, Germany.[184] When leafing through, it is immediately apparent that Kentmann's plant pictures correspond to those in Tartu. This must be the book that Kentmann sent to Gessner in Zürich so that he could copy the plant images

179 Gessner (1577), f. 53r.

180 Rand (2014); Leu (2016). The missing third volume bears the call number MS. 55 and contains 147 leaves. Individual drawings can also be found in MS. 56 with a volume of 49 leaves and MS. 57 with 45 leaves.

181 University Library Tartu, MS. 56, f. 2 and 34; MS. 57, f. 4, 5, 14, 23, 25, 27, 41, 61 and 70.

182 More or less every sheet contained a picture and text about a plant, making each sheet the carrier of a specific botanical *loci* collection. Cf. Kusukawa (2012), p. 160f: "The sheets of drawings functioned as slips of commonplaces that were the basis of compiling a book."

183 Gessner (1584), f. A_3v–A_4r.

184 Herzogin Anna Amalia Bibliothek Weimar, call number: Fol. 323. Cf. Kusukawa (2009); Olariu (2022).

or have them copied, and that was returned in March 1555. Unlike Kentmann, Gessner wrote down further information from his personal research on many of the illustration pages (ill. 60a, 60b and ill. 61a, 61b).

Gessner did not have all the illustrations copied, but only of those plants that interested him and for which he did not yet have an image. In the volume in Tartu, he occasionally made reference to the two volumes that are kept in Erlangen. For example, he noted on page one of the Tartu codex, which shows a palm tree, that there was another illustration of this tree, designated "G155b." This means "*Gesneri historia plantarum*, page 155, back side." Likewise, on page 7 of the Tartu codex he said that he was in possession of a better illustration of this plant, a chrysanthemum, in "alio libro 194 a in the other." By "alio libro" he meant the other two-volume plant book in Erlangen. "The other" book was a reference to the second volume, in fact the front side of page 194. When one follows up on these notes, everything clearly corresponds. Gessner had not only prepared two volumes of plant illustrations for publication, as one had assumed from those in Erlangen, but actually three.

Gessner's pages in Tartu are unique in another regard, in that they are designated with a Roman numeral I or II, followed by an Arabic number. Johannes Kenntmann's previously mentioned codex contained animal illustrations in addition to the plant renderings as well as plant drawings that had been collected by Theophil Kentmann. In 1549, the plant images that were of interest to Gessner were divided by Johannes Kentmann into a *Centuria prima* and a *Centuria secunda*. The pictures from the first group of 100 that Gessner or his illustrator copied were marked with a Roman numeral I and those from the second group were labelled with a II. The additional Arabic number referred to the page number in Kentmann's book.

Further evidence that the two volumes in Erlangen and the one in Tartu belong together comes from illustrations that the Nuremberg physician Joachim Camerarius the Younger published in 1586. They appeared in his reworking of Pietro Andrea Mattioli's *Kreutterbuch* that was partly based on Gessner's preliminary work, which Camerarius had acquired from Caspar Wolf a few years earlier. Camerarius wrote in his *Foreword to the benevolent reader*: "Several years ago, due to our special friendship and as a favor, the honored and highly erudite Zürich physician Mr Caspar Wolf sold me the wide-ranging and laborious botanical work of the very famous and learned late Mr Conrad Gessner. This was in case I might be willing to organize this important work together with my own and other relevant observations and translate it into Latin. Although indexed and collated by him, in part it was still just a pile of papers. Because of a great deal of work and for other reasons, the undertaking was to be drawn out and difficult. In any case, I chose (by God's will) not to reject it,

ILLUSTRATION 60A Tulip from Johannes Kentmann's album of plants: *Herbarum fruticum, arborum nondum ab aliquo depictarum*, Centuria prima.

DUCHESS ANNA AMALIA LIBRARY, WEIMAR: FOL. 323, P. 16

ILLUSTRATION 60B Gessner's copy of Kentmann's tulip with detailed studies, in: Conrad Gessner, *Historia plantarum.*
UNIVERSITY LIBRARY TARTU: MS. 56, F. 3R

ILLUSTRATION 61A Cedar from Johannes Kentmann's album of plants: *Herbarum fruticum, arborum nondum ab aliquo depictarum*, Centuria prima.
DUCHESS ANNA AMALIA LIBRARY, WEIMAR: FOL. 323, P. 89

ILLUSTRATION 61B
Gessner's copy of Kentmann's cedar with detailed studies, in: Conrad Gessner, *Historia plantarum.*
UNIVERSITY LIBRARY TARTU: MS. 56, F. 41R

but instead to diligently finish the project as quickly as possible. Within the late Mr Gessner's library of herbaria, there were many plant illustrations, some of which were already in the form of woodcuts, others only drawn. Many good gentlemen and friends of mine as well as some very noble people suggested and encouraged me to finish the remainder. It is a substantial amount, which should be finished and then, for as a good start, I should republish the German *Kreuterbuch* by the most erudite late Mr Mattioli, for which there is a great demand, but no no more copies of it to be found any longer, with the addition of many fine items."[185]

It is therefore not surprising to find that Camerarius's *Kreutterbuch* contains woodcuts that were based on illustrations from the Tartu codex and the Erlangen codices, for example, the cedar from Tartu[186] and the maple leaves from Erlangen[187] (ill. 62a and 62b). This is because all three plant books were in

185 Camerarius (1586), f.):(8r.
186 University Library Tartu, MS. 55, f. 41r; Camerarius (1586), f. 33v.
187 University Library Erlangen, MS. 2386, vol. 1, f. 7v; Camerarius (1586), f. 35v.

ILLUSTRATION 62A Maple leaves in: Conrad Gessner, *Historia plantarum*.
UNIVERSITY LIBRARY ERLANGEN, MS. 2386/1, F. 7V

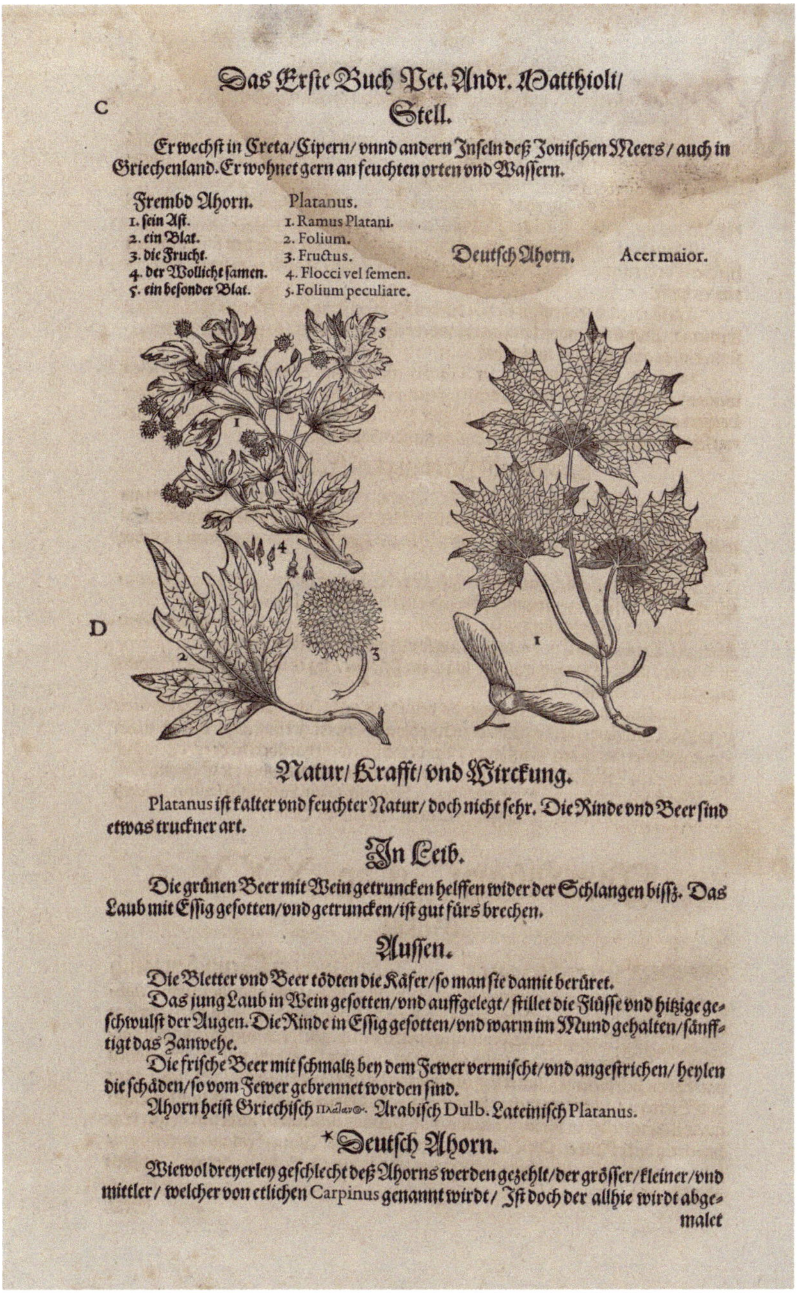

Das Erste Buch Pet. Andr. Matthioli/

C

Stell.

Er wechst in Creta/ Cipern/ vnnd andern Inseln deß Jonischen Meers/ auch in Griechenland. Er wohnet gern an feuchten orten vnd Wassern.

Frembd Ahorn.	Platanus.
1. sein Ast.	1. Ramus Platani.
2. ein Blat.	2. Folium.
3. die Frucht.	3. Fructus.
4. der Wollicht samen.	4. Flocci vel semen.
5. ein besonder Blat.	5. Folium peculiare.

Deutsch Ahorn. Acer maior.

D

Natur/ Krafft/ vnd Wirckung.

Platanus ist kalter vnd feuchter Natur/ doch nicht sehr. Die Rinde vnd Beer sind etwas truckner art.

In Leib.

Die grünen Beer mit Wein getruncken helffen wider der Schlangen bißz. Das Laub mit Essig gesotten/ vnd getruncken/ ist gut fürs brechen.

Aussen.

Die Bletter vnd Beer tödten die Käfer/ so man sie damit berüret.

Das jung Laub in Wein gesotten/ vnd auffgelegt/ stillet die Flüsse vnd hitzige geschwulst der Augen. Die Rinde in Essig gesotten/ vnd warm im Mund gehalten/ sänfftigt das Zanwehe.

Die frische Beer mit schmaltz bey dem Fewer vermischt/ vnd angestrichen/ heylen die schäden/ so vom Fewer gebrennet worden sind.

Ahorn heist Griechisch πλάτανος. Arabisch Dulb. Lateinisch Platanus.

*Deutsch Ahorn.

Wiewol dreyerley geschlecht deß Ahorns werden gezehlt/ der grösser/ kleiner/ vnd mittler/ welcher von etlichen Carpinus genannt wirdt/ Ist doch der allhie wirdt abgemalet

ILLUSTRATION 62B Camerarius published a few plants from Gessner's *Historia plantarum*, among which were these maple leaves: Joachim Camerarius, *Kreutterbuch*, Frankfurt/M. 1586, f. 35v.
ZBZ, ABTEILUNG ALTE DRUCKE UND RARA: RX 12:B

Gessner's estate, which was bought by the Nuremberg physician. Camerarius's work enjoyed broad popularity and was even used by Carl Linnaeus.[188]

Gessner's botanical corpus contained roughly 1,500 plant drawings, of which about 150 were of his own making.[189] They offer many detailed studies of plant organs, especially flowers, fruits and seeds, since Gessner examined all parts for hints of gender differentiation.[190] When it came to such details, even the otherwise lifelike self-print from nature was inadequate in which plant parts were blackened and pressed onto white paper in a technique dating back to the thirteenth century.[191] Gessner was in possession of two such examples.[192] The morphology was true to the original, but this method did not suffice as a representative illustration or for the fine details that become visible after flowers, fruits and seeds have been dissected.

While Leonardo da Vinci is considered the founder of detailed botanical illustration, Gessner was much more systematic in illustrating the relationships among plants. Qualitatively, Gessner was by no means inferior to the Renaissance genius, Leonardo: "Gessner consistently made use of the secondary detail sketch that first appeared 55–60 years earlier in Leonardo's works. It was used to enlarge physical features that would not fit within a holistic study. Alternatively, it did something entirely novel in supplementing the primary drawing with representations of various stages of development. In several instances, Gessner drew spurge plants in the same fashion as Leonardo, so that a direct comparison can be made. With regard to accuracy, Gessner's illustration leaves nothing to be desired when compared to Leonardo's. On the contrary, by observing that there is one seed in each of the three chambers of the ovary, which he has extracted and drawn individually, he has made significant progress toward scientific plant morphology."[193]

Furthermore, when Leonardo represented plants within a painting, he often emphasized aesthetic aspects more than the accuracy of the rendering.[194] Like Dürer, he created great naturalistic artworks, but it was not their intention to scientifically document plants and animals.

Hans Weiditz, who was possibly a student of Albrecht Dürer, and who did the illustrations for Otto Brunfels' book of plants, (ill. 63a and 63b), sometimes added secondary sketches to his plant drawings. However, these have a more

188 Jarvis (2007), p. 141.
189 Wolf (1566), f. 48r.
190 Wolf (1566), f. 49r.
191 Hausinger (2009a), p. 119; cf. Reeds (2006), pp. 211f.
192 University Library Erlangen, MS. 2386, f. 326r and 327r.
193 Zoller (1989), p. 116.
194 Capra (2008).

ILLUSTRATION 63A Helmet iris (*Orchis militaris*), which Hans Weiditz very accurately illustrated, even showing signs of the plant's decay.
BURGERBIBLIOTHEK BERN: BBB ES 71 (3)

ILLUSTRATION 63B Helmet iris by Hans Weiditz, published in: Otto Brunfels, *Herbarum vivae eicones*, Strasbourg 1532, p. 103.

ZBZ, ABTEILUNG ALTE DRUCKE UND RARA: I R 3:A

ornamental quality.[195] Gessner owned 94 renderings by Weiditz that were later acquired by Felix Platter,[196] and which are today kept in the *Burgerbibliothek* in Bern.[197] After Gessner's death, Caspar Wolf mentioned on March 14, 1566 that these detail drawings should also be published along with the full-scale plant images: "Up until now, some who have ordered plant illustrations to be made have not sufficiently taken size and proportions into consideration, neither to keep costs down nor to avoid possible misconceptions by the reader, when both the smallest and largest plants are drawn to the same format. In his plant book, however, Gessner, who was circumspect and thoughtful in all his undertakings, had planned to have the largest plants – some of which measure 4 cubits (two meters) or more – drawn in large format, namely six Roman thumbs high. The smaller ones, on the other hand, were to be executed at scaled-down, proper proportions: those two cubits high at three thumbs, and those at three cubits at four and a half, etc. The very small plants were to be drawn life size, although he wished to avoid illustrations that would be too small and hard to discern. It might be too difficult for the artist to render a complete plant and still have some of the details fine enough or sharp enough for the viewer to properly make them out. Therefore, it was Gessner's intention to add enlarged sketches of constituent parts next to the plant drawings that would depict flowers, seeds, fruits and leaves. There were sometimes even cutaway section drawings of the insides of fruits and seeds."[198]

15.7 The Mattioli Affair

We have already repeatedly mentioned how the identification of ancient plant names was not always easy or clear-cut. As for Gessner, he strove to resolve the resulting chaos of nomenclature in various of his publications. In one instance, however, the disparate interpretations given by the scientific community led to a conflict that lasted for years and also involved Gessner. At issue was a plant that Dioscorides had designated as *Aconitum pardalianches* and described as follows: "*Aconitum*, which is also called *Pardalianches, Kammaros, Thelyphonon, Kynoktonon* and *Myoktonon*, has either three or four leaves that are similar to those of the violet or cucumber. They are, however, smaller and

195 Zoller (1989), p. 114.
196 Lienhard (2014), p. 76.
197 Cf. online: https://www.burgerbib.ch/de/bestaende/privatarchive/einzelstuecke/platter-herbarium.
198 Zoller et al. (1972–1980), vol. 1, pp. 11f. Latin text see: Wolf (1566), f. 48v.

a bit rough. The stem is one hand span in length, and the roots shine like alabaster and resemble a scorpion's tail. If these roots are brought close to a scorpion, it will become lame. It will revive, however, if it comes into contact with a hellebore. It is applied as an analgesic eye remedy, but is otherwise deadly when wrapped up in meat as bait for panthers, wild pigs, wolves and other animals."[199]

The difficulties of identifying this plant persist even today, and it has been designated in the recent literature as both *Aconitum anthora* (healing wolfsbane)[200] and *Doronicum pardalianches* (creeping chamois).[201]

Gessner had already made mention of this plant in his *Catalogus plantarum* of 1542, and equated Dioscorides' *Aconitum* with the plant named *Realgal* in Arabic, or commonly known as *Tora* (today, the *Ranunculus thora*).[202] Since the plant could even kill wolves, he referred to it in detail in the first volume of the *Historia animalium* under the term *lupus* (wolf) and came to the same conclusion. They are, so he explained, abundant in the Northern Italian valleys of the Waldensians, persecuted heretics who had emigrated from France.[203] In 1554, the Italian Pietro Andrea Mattioli published an illustration of the plant and called it *Aconitum primum*.[204] However, the woodcut only portrays a well-developed taproot connected to four large leaves. The stem has been cut and all the flowers are missing. Mattioli claimed that this was how it had appeared to him, and he seems to have owned a dried specimen.[205] Gessner, however, did not believe him and published in 1555 an accurate image of a *Tora*[206] as Dioscorides's *Aconitum*. He had received the plant from his student Alexander Peyer from Schaffhausen, who in turn had his professor of medicine in Padua, Gabriele Falloppio (1523–1562) to thank for it. Gessner also discovered a further *Tora* illustration in Johannes Kentmann's previously mentioned album of plants, which had been provided to him for making copies.[207] Kentmann had received it from Luigi Anguillara (1512–1570),[208] the first director of Padua's

199 Dioscurides (2002), pp. 245f.

200 Dioscurides (2002), p. 245.

201 Gunther (1959), p. 475. Cf. also Dioscorides (1902), p. 411: "All the fathers of botany have struggled with the interpretation of this plant and have given it the most diverse names."

202 Gessner (1542), f. 2v.

203 Gessner (1551), p. 748. Cf. on Gessner and the Waldensians also Zuliani (2017).

204 Cf. Delisle (2004), p. 163. The same illustration is found in Mattioli (1558), p. 537.

205 Ferri (2001), p. 46, footnote 120.

206 Palmer (1985), p. 155; Gessner (1555), p. 39.

207 Cf. p. ... The illustration is in the third volume of Gessner's *Historia plantarum*, which is kept in the University of Tartu Library (MS. 55, f. 96v).

208 Gessner (1555), p. 40, writes that Kentmann received it from Aloisius Romanus, referring to Luigi Anguillara. Cf. Greene (1983), vol. 2, pp. 729f.

botanical gardens, who had also designated it as Dioscorides's *Aconitum*. It had come from the Monte Baldo region.[209] The entire discussion not only had a scientific-theoretical aspect but also a practical one, since the *Acontium* was a common component in ancient collections of remedies. During the Renaissance, these were largely adopted and further developed. Were the wrong ingredients to be mixed into a medicine, it could have had drastic consequences.[210]

Melchior Wieland, who had taken up residence with Falloppio in 1556[211] and to whom Gessner had dedicated a short work in May of the same year,[212] sharply attacked Mattioli in a letter of October 20, 1556, accusing him of insufficient knowledge of the classical languages and of having falsely identified many plants. On March 10, 1557, Gessner wrote to Wieland, asking him to soften his tone; he tried to smooth the waters by pointing out that everyone was after all only human and capable of mistakes.[213] The correspondence between Gessner and Wieland was published by Nicolaus Philesius in 1557. In the foreword, Philesius wrote that when visiting Gessner at home, he had noticed a letter from Wieland and requested permission to publish it along with Gessner's answer. Gessner declined, not wishing to make public what was private and written in confidence. Using his close friendship, however, Philesius had been able to persuade Gessner until the latter approved the publication.[214]

It is possible that Gessner agreed because he felt it was high time that someone challenged the arrogant and dogmatic Mattioli. The Italian had not limited his criticisms to Gessner, but had challenged many other famous scientists as well. He had also angered Leonhart Fuchs, who called him a haughty and self-important Italian.[215] Mattioli's lack of character was apparent in his dealings with Amatus Lusitanus, a Portuguese Jew who dared to criticize him. He responded by accusing Lusitanus of apostacy in his *Apologia adversus Amathum* (Venice 1558), which called him to the attention of the inquisition.[216] Despite Mattioli's disagreeable qualities, his book about plants, titled *Commentarii in libros sex Pedacii Dioscoridis*, was published in 60 editions, and tens of thousands of copies were sold during his lifetime.[217] Likewise, his *Epistolae*

---

209 Gessner (1555), pp. 39f.

210 Cf. on the medical side of the discussion: Kusukawa (2012), pp. 163–170.

211 Trevisan (1995), p. 60.

212 Conrad Gessner: *De fictis in germanica lingua aquatilium nominibus* …, in: Gessner (1556), pp. 245–280.

213 Gessner/Wieland (1557).

214 Gessner/Wieland (1577), p. 3.

215 Baumann et al. (2001), p. 90. On Mattioli's bad character, see also Palmer (1985), pp. 152f.

216 Arieti (2014), p. 53.

217 Palmer (1985), p. 152, speaks of 32,000 copies sold by 1561.

medicinales was widely read, finding 400 purchasers alone at the Frankfurt book fair of spring 1561.[218] Because of Mattioli's renown and his activity as personal physician to the emperor, expressing any hasty or rash criticism of him was ill advised.

It was to be expected that Mattioli would soon answer Wieland's accusation. In January of 1558, he published an open letter to Falloppio that contained a diatribe aimed at Wieland. In the same year, another edition of Mattioli's plant book was printed (*Commentarii in libros sex Pedacii Dioscoridis*), this time with Gessner in his sights. The keywords *Gesneri lapsus* (Gessner's error) could even be found in the index, although it appears in only two places. In his personal copy of the work, Gessner actually added it four more times where he had been criticized – probably with a smile on his lips.[219] Mattioli not only criticized Gessner's identification of *Aconitum*, but also his *Historia animalium*, which according to him contained a series of errors.[220]

The dispute between the two camps concerning the attribution of *Aconitum* expanded to include ever more scholars, who became involved or felt it necessary to make a statement. On October 1, 1558, Leonhart Fuchs wrote to Gessner that Wieland had written harshly about Mattioli, alluding to the letter published in 1557 that Wieland had sent to Gessner on October 20, 1556. On the other hand, Fuchs showed understanding for Wieland's comments, stating: "So far, Mattioli has bitten everyone else; so it is no wonder that he too has now been bitten."[221]

In a letter of November 1, 1558 the doctors Girolamo Donzellini (1513–1587) and Johannes Hess from Nuremberg admonished the Zürich polyhistor: they would never have thought that two such erudite figures as Gessner and Mattioli could argue in this manner. Gessner had attacked Mattioli's illustration of *Aconitum*, although Dioscorides had not described any more than what Mattioli had drawn. They felt Gessner to be more at fault and called on him to end the squabble. Mattioli was well disposed toward Gessner, and now Gessner should take the first step toward reconciliation.[222] Donzellini also sent Gessner a 13-page statement, a summary of which Gessner wrote out by hand: "*H. Donzellinus defendit Mattioli aconitum et meam de eo sententiam impugnat nulla solida ratione*" [Donzellini defends Mattioli's *Aconitum* and, without a

218 Raimondi (1906), p. 173.

219 Mattioli (1558), f. β_1r. Gessner's own copy: ZBZ, call number: Dr M 438.

220 Mattioli (1558), p. 538.

221 Letter from Fuchs to Gessner, October 1, 1558, in: StAZH E II 356, 169: "Momordit hactenus omnes Matthiolus, quid mirum si rursus mordeatur."

222 Donzellini and Hess to Gessner: ZBZ, MS. C 50a, no. 31.

good reason, disagrees with me on this matter].[223] It seems that in general Gessner did not care much for Donzellini, who was a Paracelsian.[224] He wrote to Crato von Krafftheim on March 17, 1561: "I believe this man is very learned, but also ambitious and self-important. That is how we humans are!"[225]

Mattioli was actually behind the letter to Gessner from Donzellini and Hess.[226] He wrote to Aldrovandi on November 26, 1558 that he had won over Donzellini and two German physicians as referees in this dispute, and with their help, he had hoped to make Gessner back down and that he would respond.[227] The two German doctors were likely Hess and Hieronymus Herold.[228] In a letter of October 6, 1569 to Crato von Krafftheim, it is clear that Gessner saw through it all: "Regarding *Aconitum*, I am unable to sort things out with Mattioli. If you have the time, read the remarks by Mattioli in the chapter about the *Aconitum* in his latest edition. Donzellini is not fit to mediate since he's in collusion with Mattioli."[229]

By January 1 of the following year, the situation had still not been settled, as was apparent in the next letter to Crato von Krafftheim: "Meanwhile, he accepts my friendship, as long as I choose not to disagree with him. I cannot make him any promises."[230] Crato von Krafftheim also tried to mediate,[231] which prompted Gessner to add on March 17, 1561 that Mattioli had tried to convince him of the existence of this unique plant as he had described it (ill. 64): "I sincerely thank you for what you so thoroughly and kindly wrote regarding my reconciliation with Mattioli. I do not think it is necessary to write privately to him any further about my reasons for taking a position against his *Aconitum* and for defending my own. I have written twice, although since then I have received several items from him that I believe support my position. He can toss out all of my reasoning, and I would be his best friend if he only sent me the weed that he believes to be *Aconitum*, or any part of it (leaf or root). This is all that I have publicly and privately asked of him. As long as he finds

223 Undated letter from Donzellini to Gessner. He wrote his brief summary below the address on the letter. ZBZ, MS. C 50a, no. 35.

224 Celati (2014).

225 Gessner (1577), f. 8v.

226 Cf. also the letter from Mattioli to Gessner of March 18, 1559, as well as his reply of August 8, 1559, which are preserved in the ZBZ, MS. C 50a, nos. 36 and 37/1.

227 Raimondi (1906), p. 162.

228 Cf. letter from Gessner to Kentmann, February 27, 1559, in: Gessner (1584), f. B_3v–C_1r.

229 Gessner (1577), f. 7r.

230 Gessner (1577), f. 7r.

231 Cf. also the letters from: Gessner to Crato von Krafftheim of August 16, 1561 [Gessner (1577), f. 1r–2v] and from Crato to Gessner of March 25, 1563, in: StAGR D V/37 C 36.06.41.

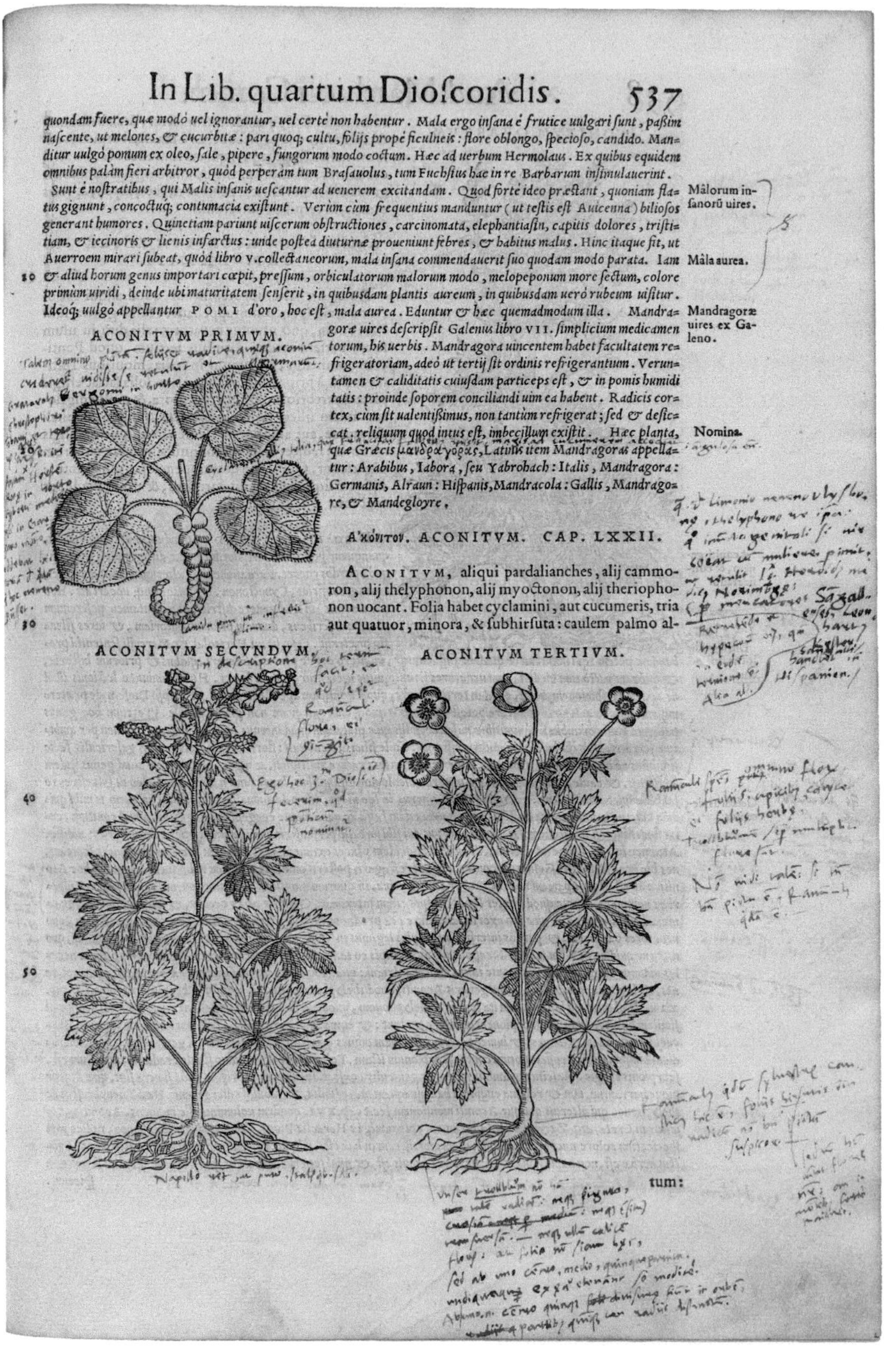

In Lib. quartum Dioscoridis. 537

quondam fuere, quæ modò uel ignorantur, uel certè non habentur. Mala ergo insana è frutice uulgari sunt, passim nascente, ut melones, & cucurbitæ: pari quoq; cultu, folijs propè ficulneis: flore oblongo, specioso, candido. Manditur uulgò pomum ex oleo, sale, pipere, fungorum modo coctum. Hæc ad uerbum Hermolaus. Ex quibus equidem omnibus palàm fieri arbitror, quòd perperàm tum Brasauolus, tum Fuchsius hac in re Barbarum insimulauerint.

Sunt è nostratibus, qui Malis insanis uescantur ad uenerem excitandam. Quod fortè ideo præstant, quoniam flatus gignunt, concoctuq; contumacia existunt. Verùm cùm frequentius manduntur (ut testis est Auicenna) biliosos generant humores. Quinetiam pariunt uiscerum obstructiones, carcinomata, elephantiasin, capitis dolores, tristitiam, & iecinoris & lienis infarctus: unde postea diuturnæ proueniunt febres, & habitus malus. Hinc itaque fit, ut Auerroem mirari subeat, quòd libro V. collectaneorum, mala insana commendauerit suo quodam modo parata. Iam & aliud horum genus importari cœpit, pressum, orbiculatorum malorum modo, melopeponum more sectum, colore primùm uiridi, deinde ubi maturitatem senserit, in quibusdam plantis aureum, in quibusdam uerò rubeum uisitur. Ideoq; uulgò appellantur POMI d'oro, hoc est, mala aurea. Eduntur & hæc quemadmodum illa. Mandragoræ uires descripsit Galenus libro VII. simplicium medicamentorum, his uerbis. Mandragora uincentem habet facultatem refrigeratoriam, adeò ut tertij sit ordinis refrigerantium. Veruntamen & caliditatis cuiusdam particeps est, & in pomis humiditatis: proinde soporem conciliandi uim ea habent. Radicis cortex, cùm sit ualentissimus, non tantùm refrigerat; sed & desiccat. reliquum quod intus est, imbecillum existit. Hæc planta, quæ Græcis μανδραγόρας, Latinis item Mandragoras appellatur: Arabibus, Iabora, seu Yabrohach: Italis, Mandragora: Germanis, Alraun: Hispanis, Mandracola: Gallis, Mandragore, & Mandegloyre.

Malorum insanorũ uires.

Mala aurea.

Mandragoræ uires ex Galeno.

Nomina.

ACONITVM PRIMVM.

Ἀκόνιτον. ACONITVM. CAP. LXXII.

ACONITVM, aliqui pardalianches, alij cammoron, alij thelyphonon, alij myoctonon, alij theriophonon uocant. Folia habet cyclamini, aut cucumeris, tria aut quatuor, minora, & subhirsuta: caulem palmo al-

ACONITVM SECVNDVM.

ACONITVM TERTIVM.

tum:

ILLUSTRATION 64 Mattioli's *Aconitum primum*, in: Pietro Andrea Mattioli, *Commentarii secundo aucti, in libros sex Pedacii Dioscoridis Anazarbei de medica materia*, Venice 1558, p. 537.

ZBZ, ABTEILUNG ALTE DRUCKE UND RARA: DR M 438

me unworthy of this favor, I must do without his friendship, which I otherwise greatly value."[232]

Crato von Krafftheim continued to attempt mediation between the two, but was unsuccessful. Nonetheless, out of consideration to him and other friends, Gessner was prepared to halt his criticisms of Mattioli, as he wrote to Adolph Occo: "I could easily give answers to all that Mattioli has written against me, and I would have done so, had not Mr Crato and others interfered. This is why I wish to defer it to my plant history, and despite everything, I will simply write about the things themselves and will not argue by name with anyone."[233] But even in the last letter in which he mentions Mattioli, he has nothing nice to say about him: "In his latest edition, he shamelessly writes that a great deal in my animal books is misleading and in error, and that the animal illustrations are fabricated. However, I wish to ignore this. If he were as erudite as he is vain, we would all rightfully admire him. But his scholarship is derived from others."[234]

Next to Gessner's epistolary dispute with Mattioli, a smaller publication-related disagreement developed in which Gessner did not take part. In Mattioli's above-mentioned 1558 public letter that attacked Falloppio and Wieland, the latter replied with an *Apologiae adversus Petrum Andream Matthiolum liber* (Padua 1558). To end the confrontation, Wieland received 70 gold ducats from Falloppio with which to remove himself from the scene and travel to the Orient. Unfortunately, he was captured by Algerian pirates off of Sardinia, and Falloppio had to pay 200 gold *scudi* to set him free. On the return trip, he was shipwrecked off the coast of North Africa but finally made it back to port in Genoa.[235] In Padua once again, in 1561, he continued, now as the new director of the botanical garden, the argument with Mattioli. His friend, the German physician Paul Hess (1536–1603), wrote a *Defensio XX. problematum Melchioris Guilandini* (Padua 1562), to which Mattioli answered with his *Adversus viginti problemata Melchioris Guilandini disputatio* (Padua 1562). Even ten years later, Wieland found it necessary to deliver another blow to his opponent and published his text *Papyrus: hoc est commentaria in tria C. Plinii maioris capita de papyro: ubi Mattheoli errores non pauci deteguntur* (Venice 1572).[236]

In Zürich as well, this topic had not yet been lain to rest. In the fall of 1577, Caspar Wolf published the three books that contained Gessner's medical letters (*Epistolarum medicinalium libri III*), which he had collected from the correspondents of the polyhistor. Two smaller treatments by Gessner were

232 Gessner (1577), f. 8v.
233 Gessner (1577), f. 47r.
234 Letter from Gessner to Joachim Camerarius, August 27, 1565, in: Rath (1950), p. 166.
235 Trevisan (1995), p. 60.
236 Ferri (2001), pp. 33 and 46.

appended: one about the *Aconitum* and the other about sour honey. Wolf stated that this was just a small taste of Gessner's *Historia plantarum* that he, Wolf, had promised to publish but had not yet had time to produce.[237] Presumably, the booklet he intended to publish held Gessner's comments on *Aconitum*, according to a letter to Crato von Krafftheim of October 6, 1560.[238] As editor, Wolf had made some additions to the text.[239] In that same year, 1560, the tora plants (tortoiseshell buttercups = *Ranunculus thora*) were flourishing in Gessner's garden, which enabled him to make personal observations.[240] In his remarks, Gessner covered the interpretations of *Aconitum* by Mattioli, Barolomeo Maranta (1500–1571) and Leonhart Fuchs, whereby his attitude toward Mattioli remained unchanged.[241] Gessner's estate contained two good drawings of the plant, from which Wolf had woodcuts made (ill. 65).[242] The original images were from the third volume of the *Historia plantarum*, which is kept in Tartu.[243] The larger of the two versions was acquired from Aloysio Quadri from Monte Generoso in Ticino, and the smaller of the two was from Caspar Collin in the Valais.[244]

Perhaps it was the tiresome quarrel with Mattioli that elicited the following bit of wisdom from Gessner in April 1564: "The more arrogant someone is and the more removed from piety and humanity, the more impatiently he will reject those who think differently (also in those things that are not part of the essence and basis of our salvation), and the more easily he will judge them. The more ambitious someone is, the more he dares to comment on all things and to arrive at conclusions, for fear of appearing ignorant. The best people, by contrast, distinguish themselves through great modesty, so that they overlook many things but remain sincere, or when required pretend to be unknowing and turn a blind eye on this, so that they win over weak hearts through Christ and their humility."[245]

---

237 Gessner (1577a), f. 2r.

238 Gessner (1577), f. 7r.

239 Cf. for example: Gessner (1577a), f. 11r.

240 Gessner (1577a), f. 5r.

241 Cf. Gessner (1577a), f. 9v–11r.

242 Gessner (1577a), f. 4r.

243 University of Tartu Library, MS. 55, f. 97r.

244 Gessner (1577a), f. 5r.

245 In the private library of Johannes Gessner (1709–1790) there was a note containing this piece of wisdom of Conrad Gessner. The document is lost, but a copy was made in the eighteenth century by Johann Jakob Simmler (1716–1788), which is preserved in: ZBZ, MS. S 109, no. 148, pp. 3f.

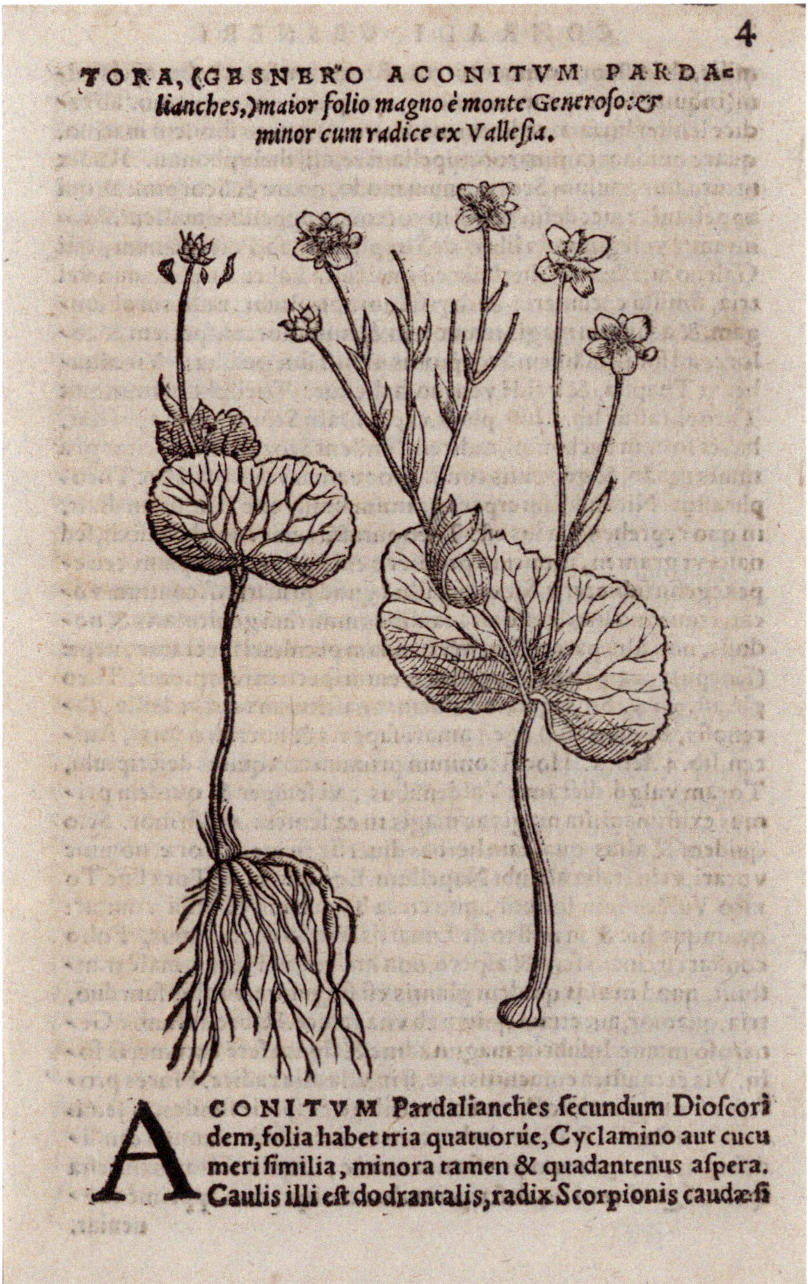

4

TORA, (GESNERO ACONITVM PARDA-lianches,) *maior folio magno è monte Generoso: & minor cum radice ex Vallesia.*

ACONITVM Pardalianches secundum Dioscoridem, folia habet tria quatuorúe, Cyclamino aut cucumeri similia, minora tamen & quadantenus aspera. Caulis illi est dodrantalis, radix Scorpionis caudæ si-

ILLUSTRATION 65 Gessner's *Aconitum*, in: Gessner (1577a), p. 4.
ZBZ, ABTEILUNG ALTE DRUCKE UND RARA: 5.124

In this regard, there is an interesting passage among the works of the Italian physician Gerolamo Cardano (1501–1576). More than a decade before the passage just cited, in the fall of 1552, he described Gessner, whom he had visited when travelling through Zürich, as a modest and reserved figure: "I was travelling home from Scotland when I arrived in Zürich, a day's journey from Sion. I visited a friendly, modest man – as those who are engaged in the high arts must be – highly learned and diligent, and a friend not only of wisdom but also of those who seek wisdom. He had anonymously written *Euonymus* [*Thesaurus Euonymi Philiatri de remediis secretis*, Zürich 1552], and gave it to me as a gift. I did not ask the reason why his name was left off, but I was especially impressed that he had written a dedication to me, a stranger. I hold the honesty of this man in high regard. I found much in the book that I liked, and as is my habit, I added notes in certain places, just a few in the chapter about distillation. Indeed, Gessner wrote perhaps more and better than any previous scholar. However, I admire him even more for his sincerity and integrity, and for that I wish to praise him (not for his scholarship and skills, not for his reputation, his writings or for other things of questionable value)."[246]

246 Salmann (1956), pp. 54f.

CHAPTER 16

Medicine and Pharmacology

16.1 A Compendium of Remedies

The book that we have just heard Cardano mention, entitled *Thesaurus Euonymi Philiatri de remediis secretis, liber physicus, medicus, et partim etiam chymicus, et oeconomicus in vinorum diversi saporis apparatu, medicis et pharmacopolis omnibus praecipue necessarius*, was a *pharmacopoeia* (book of medicinal remedies) that Gessner published in 1552 under the pseudonym Euonymus Philiater (the doctor's blameless friend). The title of the English edition of 1559 reads: "The treasure of Euonymus, conteyninge the wonderfull hid secretes of nature, touchinge the most apte formes to prepare and destyl medicines, for the conservation of helth; as quintessence, aurum potabile, hippocras, aromatical wynes, balmes, oyles, perfumes, garnishyng waters, and other manifold excellent confections. Wherunto are joyned the formes of sondry apt fornaces, and vessels, required in this art."

A decade later, Gessner explained that he had turned the incomplete work over to his cousin Andreas Gessner the Younger, who was a partner in the printing firm of Rudolf Wyssenbach[1] since 1551 and was looking for suitable titles. He decided against publishing the unfinished work under his own name. It may be suspected that Gessner also chose a pseudonym because the title was characterized as *chymicus* which at that time also included the alchemists, and because he did not want to be counted among certain quacks or the Paracelsians who were impopular in Zürich. Such people often worked from a "chemical" basis, although it should be noted that there was not yet a differentiation between chemistry and alchemy in the sixteenth century.[2] Alchemy and chemistry did not begin to diverge until the late seventeenth century. When Gessner realized that his book had become very popular, he found the courage to add his name to the title page of the 1554 Latin edition that was published in Zürich.[3] In any case, no other work from Gessner's pen turned out to be this popular, and for decades there were various editions published in German, English, French, Italian and Latin. In Lyon alone, the French capital of medical books, there were twelve editions printed between 1554 and 1559 – four in

1 Leemann-van Elck (1939), p. 1.
2 Cf. Webster (1990), p. 22.
3 Gessner (1562), f. B_1r.

© KONINKLIJKE BRILL NV, LEIDEN, 2023 | DOI:10.1163/9789004541696_017

French and eight in Latin. Unlike the Zürich editions, the Lyon ones contained a great number of explanatory woodcuts.[4] Gessner had planned to follow up with a more complete edition in 1562[5] but never got to it. Caspar Wolf, who had been attracted to Paracelsianism in his youth,[6] published a posthumous version of the second part in 1569 as prepared by Gessner. It was translated into German in 1608 by the physician Hans Jakob Nüscheler (1551–1620).

On the first page of his text, Gessner explained that this book was about medicines that were not found anywhere as natural substances, but that needed be produced through a process of distillation. In his dedication of the 1552 *Thesaurus* to Nikolaus Zurkinden (1506–1588), the Bern Commissioner General for Vaud, who was no stranger to the distillation of plants, Gessner explained that he had gathered together information from books and manuscripts in Latin, German, French and Italian, then ordered the material in a reasonable manner, adding some of his own observations. On the following pages, he offered an overview of the history of distillation, which, Gessner explains, belongs to the science that is referred to variously as *Ars chymistica, Chymia, Alchymia, Alkimia, Chemia* or *Alchemia*. Gessner mentioned as an important practitioners in Germany Hieronymus Brunschwig (before 1450–before 1512), who had introduced this art to Strasbourg 70 years earlier. It did however not originate with him, but had been developed in the North African and Arabic regions, but had subsequently been adopted by the Greek and Byzantine physicians Aëtius of Amida (5th/6th century), Oribasius (ca. 320–403), Johannes Zacharias Actuarius (1275–1328) and Michael Psellos (1017–1078). The Arabic scholars who mastered this art included Avicenna/Abū Alī al-Husain ibn Abdullāh ibn Sīnā (ca. 980–1037), Geber Hispalensis/Abu Muhammad Dschabir ibn Aflah al-Ischbili (ca. 1100–1160) and Rhazes/Abū Bakr Muhammad ibn Zakarīyā ar-Rāzī (ca. 864–925). In closing, Gessner mentioned that he was not particularly well versed in either chymistry or the associated apparatuses. He had only occasionally tried his hand at them or learned a little from friends. Nonetheless, he wished to write something about it and offer an introduction to the topic while also giving a warning about charlatans to the unsuspecting.[7]

4 Régnier-Roux (2014), pp. 12 and 17.

5 Gessner (1562), f. B_1r.

6 Milt (1953), p. 194. Cf. Gessner's remark to Crato von Krafftheim of October 18, 1562 (instead of 1561) that especially younger people tended toward Paracelsianism (Gessner [1577], f. 9v). For the dating of the letter, see Kühlmann/Telle (2001/2004), vol. 2, p. 762.

7 Gessner (1552), pp. 5–20.

Pastor Johann Rudolf Landenberg (ordained in 1553, † 1590) wrote in the foreword of the first German edition that a physician must be familiar with medicines, make the correct prescription, and take into consideration the constitution and physical condition of the patient. Additionally, the prescribed medicines must be produced in a professional and clean way, which certainly did not happen in every case. Too many quacks earned their living with secret recipes and dubious machinations, which Gessner also greatly opposed with this publication. There was even talk of untrustworthiness among professional pharmacists: "One often hears that sick people get even bigger and more severe diseases because of their repulsion to the medicines that have been prepared crudely and uncleanly by the pharmacists (who would be better working as cooks for the pigs)."[8] Distillation was the focus of these observations, for through distillation, Landenberg explained, one could arrive at the purest form of a substance, the *quinta essentia*, which "is isolated and extracted from the impure, gross, earthy, useless and unsuitable substance. Therefore, in this work you will find the excellent and virtuous medicines that artisans, doctors and alchemists have kept hidden as special secrets of nature. ... Therefore, we can well understand that the medicines, which have been pure essences by distillation, will be much more useful, better and more powerful because of their pure, subtle and penetrating substance and essence than the medicines made by the pharmacists and wound doctors in the traditional, careless way."[9]

Gessner was the first doctor in Zürich to write a pharmacopoeia, let alone a chymical one, and to publish the corresponding prescriptions as previously well-kept medical secrets.[10] This would explain the extraordinary demand for the work. It amounted to a textbook for the production of medicines through distillation (ill. 66). The second volume, printed in 1569, "partially repeats the procedures and preparations from the first volume, but the addition of new procedures for many medicines clearly makes it an expansion of volume I. In these volumes one can find various medicines such as *Decocta, Fomenta, Succi, Vina medicinalia* and the preparation of metal precipitates and sublimates."[11]

With his 600-page comprehensive *Thesaurus*, Gessner made an attempt to give a complete overview of ancient and new medicines. Significant progress in this regard was made by the addition of a number of new chemical compounds and materials. These were based on combinations containing

---

8 Gessner (1555e), f. †$_{ii}$r. Cf. Leu (2019).

9 Gessner (1555e), f. †$_{ii}$v.

10 Milt (1953), pp. 178–184.

11 Dobler (1955), pp. XIIf.

440 EVONYMI

fingeretur, oleum & operam perderes: licet minima tantum rima aut fiſſura ei accideret. Itaque ut facilius igni reſiſtat, loricanda eſt luto, eoq́; duplici aut triplici, hoc eſt poſt primum ſiccatum alterum inducendum lutum, aut etiam tertium. Fornacis figura huiuſmodi fermè erit.

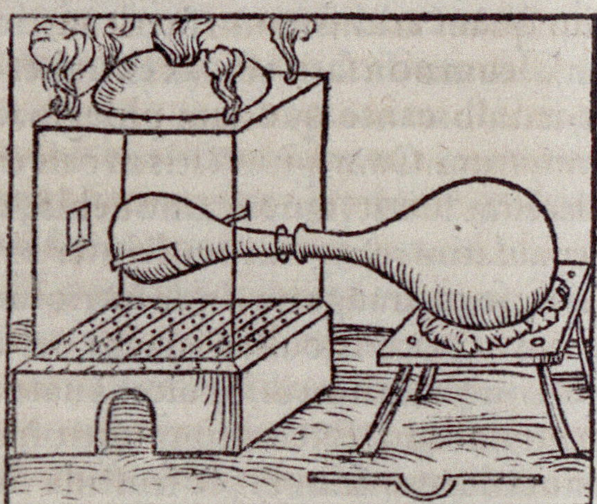

Hæc fornax cõſtruenda eſt quadrilatera, & in ſummo caminanda ſeu cõcamerãda modicè. In medio ſummæ fornicis faciendũ eſt foramẽ ſeu infumibulũ, ea amplitudine ut ſingulis horis uel

quoties

ILLUSTRATION 66 A complex distillation apparatus, in: Gessner (1552), p. 440.
ZBZ, ABTEILUNG ALTE DRUCKE UND RARA: MD E 377

antimony, arsenic, lead, iron, gold, silver, mercury, copper and other elements.[12] Distillation was the most important chemical process for the production of medicines, and at least a third of the medicines introduced by Gessner were based on this procedure.[13] Of these, there were three categories: *Destillatio per ascensum* (distillation by ascent; a heat source is placed beneath the ingredients), *Destillatio per inclinationem ad latus vel obliqua* (distillation occurs at an angle with vapors flowing to nearby containers) and the *Destillatio per descensum* (vapors are forced downward). An actual temperature measurement was not yet possible, so one got by with four very rough levels of heat (the warmth of a warm bath, warmer than a warm bath, fire, and glowing embers). Achieving good seals at the connections between vessels posed a special challenge, since neither rubber gaskets nor milled connectors were available. This was resolved with so-called "vessel glue" which mainly consisted of clay. Horse manure, sunlight focused through a magnifying glass, warm baths, lamps, candles and ovens were all used as heat sources. For chilling, cooling hoods made of cow's bladders and glass were used, as well as pipes leading through cold water.[14]

In 1955, the pharmacist Friedrich Dobler tested a large number of Gessner's recipes and came to the conclusion that: "At the time, most of the preparations were indeed effective and were shown to have active ingredients in greater amounts than has been generally believed."[15] In places, he described Gessner's procedures as ingenious and certified the *Thesaurus* as being much more than just a literary compilation. Additionally, he found there to be no tendency toward alchemy nor a commitment to any alchemical or gnostic natural philosophy.[16] This was also apparently understood by Gessner's contemporaries, which is why there were no such accusations made against him. Not only did Gessner send a complimentary copy to Heinrich Bullinger,[17] head of the church in Zürich, but also the physicians Johannes Kentmann in Saxony and Cosmas Holzach in Schaffhausen, from both of whom he asked for their judgement. These have not been preserved, but were unlikely to have been negative.[18]

12 Dobler (1955), p. 3.
13 Dobler (1955), p. 53.
14 Dobler (1955), pp. 18–33.
15 Dobler (1955), p. 114.
16 Dobler (1955), pp. XIII, 94 and 113.
17 The copy is kept in the ZBZ, call number: Md E 377.
18 Letter from Gessner to Kentmann, 22 June 1554 (Gessner [1584], f. A_2v–A_3v) and to Holzach, 6 September 1554 (Gessner [1577], f. 81v).

16.2 Paracelsus

While we cannot make a connection between Gessner and alchemy in the modern sense, the facts are quite different for his 20-year-older contemporary Philippus Aureolus Theophrastus Bombastus von Hohenheim (1493–1541), known as Paracelsus, who was considered *persona non grata* in Zürich. Admittedly, Leo Jud (1482–1542), an ally of Zwingli, saw to it that Paracelsus's interpretation of Halley's comet in August 1531 was published by Froschauer in the same month. The thin, cryptically formulated booklet was interpreted to the effect that Zwingli would triumph over his Catholic opponents.[19] But the attitude toward the dazzling natural philosopher changed when the people of Zürich experienced just the opposite with the Second War of Kappel in which both Zwingli and Gessner's father, Urs Gessner, were killed. It is surely no coincidence that among 20 or so astrological tracts that had been published by Paracelsus by 1539, Gessner listed only this one in his *Bibliotheca universalis*. He described Paracelsus as uneducated and professionally not worth mentioning. At Christoph Clauser's home, Gessner had seen handwritten texts by Paracelsus with obscure, barbaric, self-important and ridiculous statements and sayings.[20] The later Clauser had by then also expressed criticism of Paracelsus.[21]

Aside from this, Paracelsus's concept of natural philosophy and medicine, which included stars and spirits having influence over humans, was contrary to Gessner's Galenic-Hippocratic approach. Influenced by reformed theology, Gessner distanced himself from astrology, the mystification of nature, and esoteric elements to the extent that he, as a sixteenth-century man, could recognize them.[22] He made remarks about Paracelsus in various letters. He described him to Didymus Obrecht as a sorcerer and spagyrist,[23] and on August 16, 1561 he wrote to Crato von Krafftheim: "In Basel, there lives a physician, the son of the late theologian Karlstadt, who is obviously a follower of Theophrastus [Paracelsus]. A year ago, he published a booklet about his long life. I see that most people of this sort are Arians and deny the divinity of our Christ. … Oporinus from Basel has once been a student and friend of Theophrastus, and he tells of Theophrastus's miraculous encounters with demons. They traffic in vain astrology, geomancy, necromancy, and similar arts, which are forbidden. I harbor a heavy suspicion that they are remnants of the

19 Gantenbein/Weidmann (2006), pp. 24f.

20 Gessner (1545), f. 614v. Cf. also Webster (1990), p. 15.

21 Kühlmann/Telle (2001/2004), vol. 1, p. 93.

22 Cf. for example on the reformers Bullinger and Vadian: Leu/Weidmann (2004), pp. 174f.; Gamper (2006) and Gamper (2017), pp. 144–146.

23 Letter from Gessner to Obrecht, March 18, 1560, cf. Gessner (1577), f. 114v/115r.

druids, who lived among the Celts in underground caves and were instructed by demons for many years. Of course, it is known that even today some of them still exist in Salamanca, Spain. From that school, there are those whom people call 'travelling students', which included the much celebrated and recently deceased Faust."[24]

Similar criticisms are found in a letter Gessner probably wrote in March of 1565 to Paracelsus's publisher, Michael Toxites (1514–1581).[25]

In December 1562, Gessner encouraged his colleague Crato von Krafftheim to publish something in opposition to Paracelsus, with assurances of his support: "You will certainly have various opportunities to write about Paracelsus and his deceptions, dealing specifically it in either as a booklet or letter, given how important the matter is. Once you have written something along these lines, but cannot find a printer, send it to me if you like! I will do as much as I can to see that it is published."[26]

In the second half of 1563, a medical-therapeutic work by Crato von Krafftheim was printed by Pietro Perna in Basel. In the dedication foreword of July 1, 1563, he unequivocally expressed opposition to Arians and alchemists, by whom he meant Paracelsus and his followers.[27] Gessner congratulated him[28] and brought the new publication to the attention of Achilles Pirmin Gasser.[29]

Despite his rejection of Paracelsus, the occultist, Gessner gave him credit for various medicines that were shown to be effective, and therefore of interest to him. On February 8, 1564, Gessner wrote to the Gdańsk Paracelcist Alexander von Suchten (1520–1575) – whom he elsewhere called a "good-for-nothing" (*nebulo*)[30] – that he did not deny that Parcelsus had noteworthy accomplishments in the healing arts.[31] Gessner also favored his remedy against the plague, which included a portion of chicken flesh.[32] Likewise, on December 12, 1564 he related to Adolph Occo that he did not doubt the healing power of various inorganic compounds: "I am grateful for this book by Paracelsus the fanatic. Dr Acronius in Basel has prepared his [Paracelsus's] theriac, made from pissasphalt. I understand, however, that it was of no benefit to him or others. I have

24 Gessner (1577), f. 1v.

25 Letter from Gessner to Toxites 1565, in: ZBZ, MS. C 50a, no. 22 (copy made by a *famulus*). Cf. also Milt (1929), p. 508.

26 Letter from Gessner to Crato von Krafftheim, December 28, 1562, in: Gessner (1577), f. 11r.

27 Crato von Krafftheim (1563), f. $^{*}{}_{6}$r–$^{*}{}_{7}$v.

28 Letter from Gessner to Crato von Krafftheim, March 15, 1563. Gessner saw the manuscript ready for printing, cf. Gessner (1577), f. 12r.

29 Letter from Gessner to Gasser, February 20, 1564, in: Burmeister (1975), pp. 296f.

30 Burmeister (1975), p. 301.

31 Hubicki (1960), p. 58.

32 Letter from Gessner to Fabricius Montanus, July 21, 1564, cf. Gessner (1577), f. 93r.

no doubt that there is a great deal of power in sulphur, gold and other properly prepared materials. But I dislike Paracelsus's other misjudgments: his lack of religious piety, which denies the Son of God, etc., his condemnations of other doctors, including the leading men in our field, and finally, his ignorance of the method and its entire subtlety."[33]

Despite Gessner's respect for specific pharmacological achievements, he felt compelled to issue warnings about some of Paracelsus's medicines, for instance, the pain relief remedy *laudanum*, which contained opium and from which people had died by the dozens, probably including the book printer Johann Froben.[34]

Against this backdrop, it is not surprising that Gessner repeatedly tried to procure works by Paracelsus. On October 31, 1558, he asked Felix Platter in Basel to send him all the publications by Paracelsus that he could find.[35] On January 16, 1559 he sent his thanks for two books, but the titles are not specified in the letter.[36] He asked his colleague in Basel once more on October 17, 1563 to send him a work by Paracelsus that had recently been published by Adam von Bodenstein[37] and that was available through the bookseller Peter von Mecheln.[38] Again, it was not clear which title Gessner had in mind, since that year there were many printed works to choose from. From the previously mentioned letter to Adolph Occo of December 12, 1564 we learn that Gessner owned the titles *Über die Pest* (*About the Plague*) and *Kleine Chirurgie* (*Minor Surgery*).[39] Gessner's personal copy of the booklet *Von der Pestilentz und ihren Zuofällen* (*About the Pestilence and Its Consequences*) (Strasbourg 1564) has been preserved, as has the title *Drey Bücher des theuren, hocherfarnen, von Gott hochgelehrten, weisen Theophrasti Paracelsi* (*Three Books by the Illustrious, Highly Experienced, with God-given Scholarship, Wise Theophrast Paracelsus*).[40] He received the latter as a gift from the editor Michael Toxites. He also owned an edition of *Grossen Wund Arzney* (*Compendium of Wound Medicine*) that was printed in 1536 and is now kept at the National Library of Medicine in Washington, DC.[41]

33 Gessner (1577), f. 54r.
34 Kühlmann/Telle (2001/2004), vol. 2, p. 480.
35 Gessner (1577), f. 97r.
36 Gessner (1577), f. 97v.
37 Gessner accused him of Arianism, cf. Kühlmann/Telle (2001/2004), vol. 1, pp. 473f.
38 Gessner (1577), f. 98v.
39 Gessner (1577), f. 54r.
40 ZBZ, call numbers: 17.662_5 and 18.2028. The latter contains *Das Holtzbüchlin*, *Von dem Vitriol* and die *Kleyne Chyrurgy*. Cf. Leu et al. (2008), pp. 189f.
41 The copy is now held at the National Library of Medicine in Washington, call number: WZ 240 f C39dm 1952.

Gessner received a noteworthy visit on April 23, 1563 from the English astrologer and alchemist John Dee (1527–1608), who was also active at the court of Queen Elizabeth I, where he influenced several of her decisions in a significant way.[42] According to his entry in Gessner's guest book, Dee was travelling on his way to Venice.[43] His stop in Zürich is testimony to the high regard in which Gessner was held as a natural scientist. Dee had great admiration for Paracelsus, as Gessner briefly remarked: "He paused here for a few hours on a spring day in April of 1563. Incidentally, he was exceptionally well educated in Latin as well as marvellously well-versed in the areas of mathematics and philosophy and a foremost admirer of Theophrastus Paracelsus, remarking that, of course, he did not approve of his errors. [Johannes Scultetus] Montanus and Carolostadius [Adam von Bodenstein] had not fully grasped Theophrasus' teachings. He was on his way to Venice, but might also go to Vienna, where there is a physician Bartholomaeus, a particularly ardent follower of Theophrastus; he [Dee] defended the view that according to Theophrastus, it is possible to heal injuries *in absentia*."[44]

Dee probably did not appeal much to Gessner. Not only did he not share Dee's admiration for Paracelsus, but he categorically rejected such magical healing practices by which a doctor who was absent could heal a patient through occult powers.

16.3 Self-experiments and Animal Experiments

Gessner did not shy away from testing various active ingredients from medicinal plants on either himself or animals. His experiments with sometimes very toxic plants, such as deadly nightshade (*Atropa belladonna*), hellebore (*Helleborus niger*), tobacco (*Nicotiana tabacum*), chamois root (*Doronicum sp.*), common water buckthorn (*Eupatorium cannabium*), Lord's wort (*Gratiola officinalis*), common nux vomica (*Strychnos nux vomica*) and the true lover's knot (*Paris quadrifolia*) are well known.[45] Various colleagues of Gessner, including Achilles Pirmin Gasser and Adolph Occo, found such self-experiments to be reckless. Gessner wrote to the latter on April 3, 1565: "You two, meaning you and Mr [Achilles Pirmin] Gasser, do not approve of me testing certain strong

42 Doran (2015), pp. 230 and 234.

43 Cf. Gessners *Liber amicorum*, entry no. 175 (National Library of Medicine, Washington, MS. E 77).

44 Cf. Kühlmann/Telle (2001/2004), vol. 2, p. 337.

45 Büchi (1984), p. 131.

remedies such as hellebore on myself. Nonetheless, I have many reasons for doing so. First, my sanguine and phlegmatic physical constitution demands some sort of medical remedy, and the common sorts are not effective enough. Secondly, I wish to be more sure of my knowledge of simple remedies, so that under other circumstances, I and others can better prescribe them. Not only am I passing along knowledge from others in my writings, but also my own observations. We can actually administer better and more safely to our own body than to others. I am not haphazardly testing just anything, but rather that which I know to be very strong but certainly not fatal remedies, such as both kinds of hellebore and grace of God [*Gratiola officinalis*]. I take others in the smallest dosage at first, then increase it by a speck at a time. If it is a liquid, I gradually add drop-by-drop until the proper dosage for a purgative or the effective potency has been reached. I do not wish to try remedies on myself that are cold, since I am cold by nature. Of the warm medicines, I only wish to try those that I know to have been successfully tried by others, such as hellebore, which was tested by the ancient physicians, especially Hippocrates. I find no shame in learning something from the common folk and then replicating it. Trusting in the judgment of my own senses, there are a few things I have wished to become familiar with, that as far as I know, no one has tried before. After long usage, I am practiced enough that I am usually able to immediately notice through smell and taste, not only changes in initial properties, but also the degree of change. From smell alone, I can usually discern the purgative powers, and I usually apply the substance with more confidence in my judgment that it will be of particular benefit to me."[46]

Medicinal plants were not spared from Gessner's frequently described debunking of myths about nature. He was able to precisely describe their effects based on his own experiments, even if the medicinal applications seem questionable from today's perspective. He related the following about chamois root (*Doronicum*) to Crato von Krafftheim on April 17, 1564: "I find it laughable that *Doronicum*, which the quacks use, is considered poisonous for people and animals. I have often found it in the mountains and I eat it with pleasure in large amounts, both the plant and its roots. I have often grown them in my garden – an especially beautiful, fragrant flower with the sweetest of scents. We name this herb 'chamois root' after the chamois goat because they love to graze on it. The roots are used to counter dizziness and epilepsy, among other things."[47]

46 Gessner (1577), f. 71v.

47 Gessner (1577), f. 18v.

Also with regard to the active ingredient in hellebore, which had already been described by the Greeks, Gessner wanted to form his own opinion. On August 27, he wrote to the Zürich pharmacist Muralt, with barbs aimed at Mattioli: "I have decided to find out what the purgative effects of the grape-shaped black hellebore are by testing it on myself tomorrow (to date, Mattioli and others have had nothing to say about this). Gradually, and at personal risk, I am becoming familiar with a great many things, so that the readers will be presented, as far as I can, with facts and not with the speculations offered by that fool [Mattioli]."[48]

While the effects of European plants were partially known, one was mostly in the dark when it came to the medical applications of those from the New World. Thanks to Thevet's *Singularitez de la France antarctique*, Gessner had some knowledge of tobacco plants, which he tested on a dog as well as himself.[49] He wrote about it to Adoph Occo on November 5, 1565: "When I tasted even the smallest part of the leaf I have received from Mr Funck and you had sent to him – I think it was sent to you without mentioning its name from France – not by swallowing it, but by chewing it, it immediately had a wondrous effect on me. I felt completely drunk and overcome with dizziness, as when one rides down a river in a boat. I had the same experience when I tried it three or four more times, which I wrote about in several letters to Mr Funck with the request that he pass them on to you. After I wrote about this to him, I gave a small piece of the ground-up leaf together with some meat to a dog. After some hours, he vomited a great deal, but as far as I can tell he did not otherwise suffer. Then, I let him go. In the French monk [André] Thevet's book describing America, I read that the Americans' *petum* herb – like the bugloss – has similar effects. However, the people there smoke it on a daily basis, so I threw some of the partially ground up leaf into the coals and breathed in the smoke through a funnel over my mouth and nose. Except for a sharpness, I noticed no ill effects. The next day, I tried a larger amount and experienced some dizziness. This was less, however, than from chewing the leaf. The power of this leaf is astonishing in the speed with which is brings about dizziness as well as causing a kind of drunkenness. This is the reason that those people breathe in this smoke before going into battle or other dangerous situations."[50]

While the experiments described were relatively harmless, others yielded more dire results, for instance the jatropha nut. It contains different kinds of

48 Rath (1951), p. 212.

49 Gessner was also interested in veterinary topics, cf. Gessner's letter to Adolph Occo of December 12, 1564, in: Gessner (1577), f. 53r.

50 Gessner (1577), f. 79v.

nerve toxins whose effects are dependent on the quality of the drug and how full one's stomach is. Gessner wrote to Achilles Pirmin Gasser on February 12, 1564: "Although I had doubts as to whether the jatropha nut is a narcotic, I have now confirmed it. Recently, I gave 10 grains of it with some meat to a dog, who swallowed it. After a half hour, it fell asleep; then after four hours, while still asleep, it suffered head and leg cramps and then it died. Subsequently, we gave the same amount of this nut to another dog in the same manner. After nearly three hours, during which it was in the grip of severe symptoms, we gave it a Saxon antidote made from wolfberries, etc., with which you are surely familiar (I first heard of it from Mr Moibanus). This was a half *drachma*, given together with meat. After swallowing it (although this did not occur right away, but only after I offered it to him several times), he recovered with no lasting ill effects, even though he did not vomit, maybe because the antidote was administered so late."

Gessner also had negative experiences with the highly poisonous, so-called white hellebore (*Veratrum album*), which can cause death within three to twelve hours. When taking it himself, he experienced right away a burning sensation on the tongue and gums. On February 18, 1565, he continued his report to Adolph Occo as follows: "I felt hot, burning sensations in my shoulders, and on my face and head. I continued to eat. And I also drank without concern. I already knew the nature of the hellebore from experience with other substances. Soon came belching and a swelling of the esophagus, as if I were suffocating, but not as severe. I walked back and forth with neither trepidation nor weakness. After the belching lasted for about a half hour, I used a feather and a finger down my throat to cause vomiting three or four times. Then, I quickly drank a lukewarm extract of mastic and, thank God, felt fine."[51]

Presumably, the *Thesaurus Euonymi Philiatri* and its rich treasure of pharmacological information was responsible for the Zürich city council entrusting Gessner and his surgeon colleague, Jakob Ruf (who possessed the necessary expertise),[52] with drawing up Zürich's first pharmacists' ordinance of March 6, 1553. It was intended to remedy deficiencies in Zürich's pharmacy system.[53] The entire document reads: "Rules and regulations that pharmacists here in the city – those currently present, or who will be in the future – must solemnly swear to abide by, 1553.

First, they must never give anyone poison.

Second, give nothing harmful to women and girls to force childbirth.

---

51 Gessner (1577), f. 69r.

52 Müller (2008).

53 Büchi (1983), pp. 123–127.

Third, do not to use cures for anything other than what doctors allow and agree to.

Fourth, do no sell counterfeited medicines.

Fifth, do not give out that which is too old or expired.

Sixth, do not allow any unfamiliar travelling salesmen or those with insufficient experience to prescribe or produce medicines without permission from our municipal or rural physicians. However those outside the jurisdiction of our municipal and rural physicians may purchase from and submit prescriptions to our pharmacists for that which they wish and need. This is with the clear requirement, that if anything suspicious is noted about these unfamiliar doctors, whether in or outside of the city, then the pharmacists are obligated to notify the municipal physician.

Seventh, they must adequately prepare for and see to it that in the future, if the municipal physician might want them to mix ingredients to make spirits, to be used as internal medicine, and which is otherwise known as golden water, then they are obligated to do so. It is to be cooked and provided at a good price. Those spirits that are used as medicine externally, should be cooked in the usual manner.

To date, pharmacists have not had general instructions for how to produce individual medicines, nor regulations to determine the price for selling each remedy. Therefore, Dr Gessner and Master Jakob Ruf, as doctors assigned by the municipality and familiar with such issues, are to provide consistent corresponding instructions for this purpose. This includes drawing up a price list that is acceptable to them and the common man, by which to live and act accordingly.

When travelling tradesmen and rootdiggers put their various wares on offer and sell their roots and other medicines to many people, they can do more harm than good. To prevent this, foreign traders and peddlers should only be allowed to offer their goods at the two fairs and on one day between the two fairs as well as on one day in May. And the municipal physician, together with a pharmacist, can look over all of their wares and determine what is unsuitable and not to be sold here. These things should be thrown away or they will be punished. This is what they are supposed to abide by.

Furthermore barbers and those who administer internal medicines without experience or approval should be closed down and permission no longer granted. This way, all the less people will get sick or die.

And to ensure that all act correctly and are properly compliant, the municipal physicians, as well as Mr Hans Wegmann [guild master for Saffron], Mr Felixen Peyer [guild master for the Meisen-guild] should initially oversee the pharmacists and all their goods with diligence, and also every year thereafter. The outdated herbs, stale waters and other things that the pharmacists have neglected should be replaced with fresh wares. And everything that might be harmful to people should be changed and improved to the best of their abilities. Said ordinance has been authorized on Monday, March 6, 1553, in the attendance of Mayor Lavater and both councils."[54]

Apparently, shortly after the death of Christoph Clauser, Gessner took over his duties as municipal physician together with the surgeon Jakob Ruf († 1558), although he did not officially receive the appointment as *Archiater* (city doctor) until May 2, 1554.[55] Ruf was an exceptionally gifted surgeon, or "municipal cutting doctor" who, like his colleagues in other cities, had no academic training. He had taken over responsibilities, "which were listed in the job specifications of the municipal physician, that the ill Clauser had no longer been able to carry out. He inspected the water fountains, examined the sick, and also oversaw midwife training and testing. Without a title as medical doctor, which was formally required of a municipal physician, Ruf nonetheless fulfilled the associated responsibilities well beyond Clauser's death in 1552."[56]

His *Trostbuechle* (*Little Book of Comfort*) of 1554 became famous as a textbook for midwives and doctors, with illustrations in the field of embryonic development that surpassed even Vesalius.[57] Gessner contributed a dedication poem in which he praised Ruf as "an outstanding surgeon, very famous everywhere for his practical experience."[58] By 1553 at the latest, Gessner was not only responsible for quality control over both of Zürichs pharmacies,[59] but he was also invited by the city council of Schaffhausen to visit theirs as well.[60] This is possibly the reason for his acquaintance with Tobias Stimmer, who painted the famous March 7, 1564 portrait of Gessner (ill. 67) that hangs in Schaffhausen's *Museum Allerheiligen*. This valuable documentary image was almost destroyed by an American air raid on April 1, 1944. However, the air pressure from an explosion threw it clear of the museum building, and it was retrieved from the street in a nearly undamaged condition.[61]

54 StAZH A 77.9, 133; published in: Keller (1893), pp. 18f.
55 StAZH A 77.9, 5.
56 Steinke (2008), p. 96.
57 Steinke (2008a), p. 112.
58 The poem can be found edited and translated in: Keller (2008), vol. 4, p. 341.
59 Gessner (1577), f. 88v. By comparison, Venice, with 172,000 inhabitants, had 71 pharmacies in 1565. Palmer (1985a), p. 103.
60 Keller (1979), p. 37.
61 Fischer (1966), p. 94.

ILLUSTRATION 67 Portrait of Conrad Gessner, March 7, 1564, oil, by Tobias Stimmer.
MUSEUM ALLERHEILIGEN SCHAFFHAUSEN

16.4 Balneology

Gessner added a work by the French mathematician Jacques Besson († 1573) to the 1559 Zürich edition of the first volume of the *Thesaurus Euonymi Philiatri* that also dealt with distillation techniques.[62] The second, posthumously

62 Régnier-Roux (2014), p. 12. Gessner donated a copy of Besson's writing, printed in Zürich, to Crato von Krafftheim. The book is kept in the library of Nostitz and bears the handwritten dedication: "D. D. Io. Cratoni Con. Gesnerus" (call number: cg 60).

published volume of the *Thesaurus* was supplemented by an excerpt from Gabriele Falloppio's writing, *De medicatis aquis et de fossilibus*, in which methods for inspecting the water from baths and hot springs were described.[63] This was also a topic of interest to Gessner. because he and his wife had experienced relief from physical ailments through repeated visits to various baths. From Zürich, they were likely to have visited the baths most often in nearby Urdorf and Baden.[64] The latter was likewise visited by Reformed Protestants, Lutherans and Catholics, which occasionally led to rather interesting discussions on religion. While in Baden, Gerhard thom Camph (from Emden) wrote about this on May 15, 1545 to Bullinger and Pellikan. He described how he and a Catholic bailiff from Uri had tried to convert each other, and in this effort, had constantly done good to each other. On the other hand, a Lutheran denied Camph salvation due to his reformed ways and said he would rather be an adulterer and a fornicator than a Reformed Protestant. He nonetheless arranged for two Catholic clerics, from Freiburg in Breisgau and from Bavaria, to visit Zürich. The people of Zürich should welcome them kindly.[65]

On a return trip from Basel to Zürich in March of 1535, Gessner visited the hot springs in Baden for the first time together with his newlywed bride.[66] Subsequent visits to these baths, which had been highly prized by the Romans, were booked for the years 1547[67] and 1559. However, because the distance between Zürich and Baden posed no great challenge, Gessner surely spent some additional time there. During the *Hirsebreifahrten* (porridge races) of 1456 and 1576, athletic rowers from Zürich were able to cover the distance to Strasbourg within a day's time.[68] It is therefore not surprising that Gessner went to Baden at the end of August 1553 to give medical attention to the French secretary of state, Sébastien de L'Aubespine, abbot of Bassefontaine (1518–1582).[69] This mission was particularly sensitive, because Bullinger had intervened with him on June 1st on behalf of five Lyon students who had been executed for their Protestant beliefs a short while before.[70]

On August 25, 1559, Gessner wrote to Kentmann that he had spent a good portion of that spring and summer with trips to Augsburg and Strasbourg.

63 Milt (1945), pp. 2f. and 14–16.

64 Cf. Heinrich Bullinger's stays at the Urdorf baths in 1534, 1547 and 1552: Bullinger (1904), pp. 24, 34 and 42.

65 HBBW 15, pp. 327–332.

66 Cf. p. 31.

67 Gessner (1553c), f. 293r.

68 Cf. Vischer (2015), pp. 58–62.

69 Letter from Bullinger to Calvin, August 26, 1553, in: Calvin (1875a), pp. 597f.

70 Büsser (2005), vol. 2, p. 192.

Additionally, he had spent some time "in our hot springs" (*in thermis nostris*), by which he would have meant those in Baden.[71] One year later, the pain from his sciatica was severe enough to rob him of sleep,[72] and in August 1560 he spent 10 days undergoing a cure in Baden.[73] In 1560 and 1562, at the least, Gessner had resided in the *Gasthaus Hinterhof* (Rear Courtyard Guest House) which had its own hot bath facilities.[74] It seems he visited his wife in Baden in 1562, as he wrote to Gasser on June 28: "I was also recently there [Baden] for four days to visit my wife, who was bathing there. During those four days, I also sat in the baths for an hour in the morning and again in the evening with a large sponge covering my entire head, upon which warm water from the springs was poured. I found this method to be more effective than the repeated shower out of a container from above, which I tried out two years ago."[75]

A short time later, on August 13, he travelled once again to the hot baths in Baden with plans to return home six days later. He had hoped to visit Johannes Bauhin there, and to return with him to Zürich on foot, or alternately by horse if Bauhin did not accompany him.[76] Gessner also took a cure in Baden in the summer of 1563, which he mentioned to Kentmann on August 25,[77] after receiving a letter from him on August 21.[78] On Whitsunday 1564, he explained his hot drink cure in Baden (*thermae nostrae*)[79] to Crato von Krafftheim. From this, one can conclude that at least once in every summer season from 1559 to 1564 he visited the baths in Baden, with the exception of 1561, because his wife was at the baths in Aeugst,[80] and Gessner spent a month travelling in the Grison Alps and Valtellina. This trip was for botanical research, but it was also an opportunity for him to personally look over various hot springs and spas. He wrote to Holzach in Schaffhausen on July 18, 1561: "During an entire month, I was away on alpine peaks and in the Rhaetian valleys, and advanced as far as the extremely high Valtellina Valley. I researched plants and hot springs and

---

71 Letter from Gessner to Kentmann, August 25, 1559, in: Gessner (1584), f. C_1r.

72 Letter from Gessner to Bauhin, in: Bauhin (1591), p. 97.

73 Cf. the letter from Gessner to Bauhin of August 21, 1560 (Bauhin [1591], pp. 97f.) and Gessner to Wolf of September 1, 1560 (Gessner [1577], f. 123r.

74 Gessner (1577), f. 88r; Bauhin (1591), pp. 106f.

75 Burmeister (1975), p. 202.

76 Bauhin (1591), pp. 108f.

77 Gessner (1584), f. D_1r–D_2r.

78 Letter from Kentmann to Gessner dated March 5, 1563, which Gessner received in Baden on August 21, according to the note on the envelope (StAGR D v/3.271, no. 13).

79 Gessner (1577), f. 18r.

80 Letter from Gessner to Cosmas Holzach, July 18, 1561: Gessner (1577), f. 85r.

got some exercise. There were two locations where I took baths for a few days, and once I also took a drip bath."[81]

By December 1557, Gessner followed his plan to visit the famous Bormio baths in Bormio, the main village in Upper Valtellina, in today's northern Italian province of Sondrio. He enquired with Johannes Fabricius Montanus about a text about the local baths that was owned by Friedrich von Salis.[82] Actually, the excursion was to take place in 1560, but had to be postponed due to the ill health of Gessner's wife. Gessner finally left Zürich on June 14, 1561 and arrived in Chur on the evening of June 15, where he was joined by Johannes Fabricius Montanus[83] and the mayor of Chur, Johannes Tscharner. It is unclear when Johannes Bauhin the Younger joined them, but he was present for the crossing of the Septimer Pass. Their route went through Chiavenna, where they paused to visit the garden of the religious refugee Franciscus Portus, then on to Teglio. They lodged there with the reformed physician Nicola Guicciardi, who, in the summer of 1564 together with Tscharner, travelled to Zürich to pay Gessner a reciprocal visit.[84] The route continued through Tirano and on to Bormio, where they arrived on either June 23 or 24. They stayed for a week to visit the baths, and on the Umbrail Pass they collected plants that both Gessner and Bauhin pressed and added to their herbaria at home. Their next planned stop was Scuol, where they stayed for two days, researching plant life and the mineral content of the hot springs as well as boiling the water to examine the resultant distillates. Susch (Süs) was the next rest stop; it was the home of Ulrich Campbell, to whom Gessner sent plants that same year. It is likely that they then headed for Samaden, where the statesman Friedrich von Salis (1512–1570) and the jurist Jakob Bifrun (1506–1572) lived. The latter praised Gessner for his philological contributions in *Mithridates*. The journey continued on over the Albula Pass to Alvaneu and its sulphur hot springs. Bauhin and Gessner probably rode through the Prättigau district to Weesen, then back to Zürich, where they arrived on July 13.[85]

81 Gessner (1577), f. 85r.

82 Letter from Gessner to Fabricius Montanus dated December 3, 1557 (ZBZ, MS. C 50a 25/2). The work that Gessner would have liked to have seen from Friedrich von Salis probably had the following title: Paul. P. Paravicini, *De Masiniensium, et Burmiensium, Thermarum hactenus incognitarum situ, natura, et miraculis ad Io. Ambrosium Cavenagum Archiatron epistola desideratissima*, Milan 1545. A copy of this rare writing is kept in the Kantonsbibliothek Chur, call number: Bo 96 (1) (I thank Prof. Dr. Jan-Andrea Bernhard, Ilanz, for this kind information).

83 Johannes Fabricius Montanus wrote a poem about this journey, which he dedicated to Gessner, cf. Döpp (2012).

84 Letter from Gessner to Bauhin, June 30, 1564, cf. Bauhin (1591), pp. 145–147.

85 Steiger (1978).

Despite Gessner's extensive collection of materials as well as his research and use of baths, there is only one published work of his on this topic. It was only a short treatment of 21 folios, *De Germaniae et Helvetiae thermis* (*About the Thermal Baths of Germany and Switzerland*), which appeared in the anthology *De balneis* (*About Baths*), published by the Venetian book printer Tommaso Giunta in 1553. Especially after his excursion to the Valtellina, Gessner had intended to publish an extended edition of this work, but it never came to pass. He wrote to John Caius in England on August 29, 1561 with the request that Caius publish his book on the thermal baths of England before anything untoward might happen to him. Gessner explained that he was satisfied to write only about the Swiss thermal baths, although he wished to surpass previous publications by writing on the topic with more care, and obviously, he would make mention of Caius's book about English baths. He continued by narrating that he had recently spent a month in the Grison Alps, which are shared by Italy and Switzerland, and the journey from Zürich to this area had lasted nearly eight days. The purpose had been to enjoy the thermal baths for his own health, to learn more about them, to search for plants, visit old friends, meet new ones, to recuperate from his studies and to firm up physically through exercise. On this journey, he even encountered a spring with sodium bicarbonate, a rarity in Europe.[86]

Tommaso Giunta's published anthology contained almost 1,000 folio pages, giving an overview of thermal baths that had been mentioned by Greek, Latin and Arabic authors and describing the effects of their metallic and mineral content. Gessner dedicated his text to the publisher of the volume and subsequently listed more than 20 authors as his sources, some of which were extensively quoted. To explain the warmth of the water, he referred to the Italian physician Gerolamo Cardano, who attributed the cause to subterranean fire and heat. This was then a fairly modern theory, since the sun or hot winds in the earth's interior were otherwise cited as the source.[87] Gessner followed this with various literary quotes about the thermal baths in Baden and Pfäfers. Regarding the latter, he even referred once to Paracelsus and printed a shortened version of his list of illnesses that would benefit from these hot springs.[88] Although Gessner did not use the baths in Pfäfers when he was there, he was familiar with Bad Fideris from personal observation. There, he had been

86 Gessner (1577), f. 133v–136v.

87 Hardmeyer (1980), pp. 32f.

88 Gessner (1553), f. 294r/v; Kaufmann (2009), pp. 119 and 132; Danzi (2012), p. 185.

DE THERMIS HELVETICIS 295

A	Locus in quo aqua Balnei hauritur è puteo.	H	Orbis ligneus.
B	Rota quam riuus montanus impellit.	I	Corius circumaffixus.
C	Canalis ligneus.	K	Canalis ex quo aqua effunditur.
D	Aqua defluens è riuo montano impellens Rotam.	L	Lebes in quo calefit aqua.
E	Aqua riui montani seorsum defluens quando non impellit rotam.	M	Lignum quod est operculum impediens riuum montanum.
F	Hastæ immissæ quæ implent tubos.	N	Baculum præcipuũ quod sustinet Cyconiam.
G	Tubi in puteum demissi.	O	Tabula quæ lebetem operit.
		P	Baculum paruum infixum operculo Lebetis.

ILLUSTRATION 68 The pump works at Bad Fideris, in: Gessner (1553c), f. 295r.
ZBZ, ABTEILUNG ALTE DRUCKE UND RARA: MD U 2

impressed by the pump works, of which he published a detailed illustration in his essay (ill. 68).[89]

He also mentioned the baths at Brig, Leuk, Urdorf, the *Wengibad* at Aeugst, Grüningen, Hinteregg on the Pfannenstil, the *Nidelbad* at Rüschlikon, Lustdorf, Nuolen, the *Rietbad* in the Toggenburg, Alvaneu, Schamps and Andeer, Osterfingen in the canton of Schaffhausen, a small hot springs near Basel and one in Knutwil in the canton of Lucerne. Finally, he cited various German baths with which he presumably had little personal experience.[90]

16.5 Medical Text Editions

Besides the *Thesaurus Euonymi Philiatri*, Gessner did not publish any other medical texts that essentially advanced or revolutionized the field in a similar fashion to his bibliographical, botanical and zoological works. The main benefit from his medical publications came from text editions of ancient and humanist physicians. In so doing, he contributed to the dismissal of the old Arabo-Latin tradition, to which for example Vadian, his professional colleague in St Gallen, had maintained a lifelong commitment.[91] Gessner wrote to Crato von Krafftheim on August 16, 1561 that the new, humanist medicine was based on Greek sources, given that the medical foundations had been laid down by Galen and Hippocrates.[92] Naturally, one was also aware that there were errors in ancient knowledge and that it needed to be reworked and supplemented. In this vein, Gessner wrote to Michael Toxites in 1565: "I realize that Galen was only human and that he knew nothing of the true religion. I also know that he was mistaken about many things, and that there was much that he was not aware of, as has been shown for anatomy by Vesalius and Fernel and for other fields by other researchers. Should I live long enough, I myself will show that his observations about the qualities of certain plants and their ordering in the plant world were erroneous. As you know, among the philosophers, there was a 'school of eclectics', who picked out the ideas they liked best and that corresponded best to their own research and reasoning from various, often mutually conflicting, teachings. If the majority of people did likewise with regard to religion, philosophy and the other arts, especially in the healing arts, and did

89 Gessner (1553), f. 295r.
90 Milt (1945), pp. 10f.
91 Müller (2011).
92 Gessner (1577), f. 2r.

not decide dogmatically for only one master or school of thought, then many of passionate disputes would soon end."[93]

The new medicine appeared to contemporaries almost like the 'new' Gospel preached by Luther and Zwingli since 1517 and 1519, respectively. The early 1530s witnessed German doctors mounting resistance in their publications to Arab medicine.[94] To this context belongs a dialogue between a midwife and a doctor, written by Christoph Clauser and printed in 1531, *Das die Betrachtung des Menschenn Harns on anderen Bericht unnütz, und wie doch der Harn zuo empfachen und zuo urteylen am gschicktisten syge* (*That the Observation of Human Urine is Useless without Further Instruction, and how Urine is Obtained and Best Evaluated*). At one point, the midwife says: "I almost get the impression as if you wished to make a new medicine, like a new Gospel."[95] And somewhat later she asks: "What novelty do I hear now? Is that a special Gospel of yours?" To which the doctor answers: "If you have not forgotten my earlier words, you will remember that the doctor is an assistant of nature, because in medicine nature stands much more as the representative of God than the Pope sitting in Rome, and the evangelists that I recite are Hippocrates, Aristotle, Dioscorides and Galen and what more there are of authors. These four teachers have bequeathed us the proper foundation for medicine. This requires little more than modesty and a wise judgment of the brain on every experience. Therefore, in sum, there are three spiritual gifts or virtues that sprout and bear fruit like no others, namely the natural causes, the four above-mentioned commanding authorities, and day-to-day experience, which is the master of everything and is the greatest of these."[96]

In the dialogue, the doctor also does not hold back in his attack on traditional medicine, for instance when criticizing Avicenna[97] or the popular book on remedial plants *Gart der Gesundheit* (*Garden of Health*), calling it *Gart der Krankheit* (*Garden of Sickness*).[98] It is exactly the four "medical evangelists" mentioned by Clauser, Hippocrates, Aristotle, Dioscorides and Galen, who

93 Gessner to Toxites in 1565 (not dated more precisely), in: Milt (1929), p. 508.

94 Siraisi (1987), p. 71. Various authors, among them recently Poskett (2022), draw a false picture of the influence of the Arab sciences on early modern Europe. While Poskett's statements are correct for astronomy and physics, the situation is different for medicine, botany, zoology, etc. (life sciences). The influence of Greek models was greater than that of Arab researchers, even if the texts came to Europe in part through Islamic intermediaries. Moreover, Poskett overuses the term Renaissance and almost completely ignores the effect of the Reformation on the natural sciences. Cf. Leu (2023).

95 Clauser (1531), f. B_{ii}v.

96 Clauser (1531), f. C_{i}v–C_{ii}r.

97 Clauser (1531), f. C_{ii}r/v.

98 Clauser (1531), f. C_{i}r.

were the subjects of a major portion of Gessner's work and of various of his published texts.

The editions that Gessner produced were often due to his efforts to choose appropriate texts for his relatives' printing shop that would also generate a certain amount of sales. To a degree, he obviously felt responsible for family members, as is expressed in September 1558 in his request to Bullinger for a raise in pay.[99] On at least one occasion, Gessner took some questionable steps in this regard. It concerned a book of fishes written by the Constance patrician Gregor Mangolt (1498–ca. 1577). When Constance was recatholicized, Mangolt was driven from the city and settled in Zürich, where he worked as manager, bookkeeper and bookbinder at Froschauer's printing house. There, among other projects, he was at work on a treatise on the fish of Lake Constance, which also contained dietetic information. In 1557, the work was published without Mangolt's permission by the print shop of Andreas and Jakob Gessner in a much abridged form. Their famous cousin had provided them with the text, as Mangolt explained in his notes: "When I had composed and finished this fish booklet, as it now reads, Dr Conrad Gessner came to me in 1557. He asked for permission to take it home and read it over, to which I had no objection. He gave the book to his cousin Mr Andreas Gessner, the printer, who completely reorganized it and added illustrations of some fish, such as those recently printed next to the bloodletting letter, along with the sayings which Jacob Ruoff contributed but which are not relevant to my book. When the booklet was printed, he sent me several complimentary copies, and I could not have been more stunned. For I had never intended that it be published, let alone that I approved or permitted this."[100]

After being named Zürich's municipal physician, the first publication by Gessner was probably *Enchiridion medicae triplicis* in 1555. In the foreword to Achilles Pirmin Gasser of August 5, 1554, Gessner explains that the work had been inspired by Gasser. It contained commentaries on Galen and Hippocrates by the French doctors Jean Vassès (1486–1550) and Jacques Dubois († 1576), the Italian Marco Gattinara (1442–1496) and the Portuguese Jewish physician Manuel Brudo (ca. 1500–1585), the latter being discussed most extensively in Gessner's foreword: "The books by the Portuguese Brudo that were printed in Venice ten years ago were mishandled and ruined by the most inexperienced printers. There is not a single page that is not teeming with errors, and not just countless printing mistakes. Because of how sentences have been separated with periods and commas, one frequently cannot find their beginning or end.

99 Cf. p. 348.

100 Keller (2008), vol. 4, p. 972.

Nevertheless, it is said that a well-structured book is a kind of commentary. Some years ago, when I began to read these books to expand my knowledge as well as for practical purposes, I wanted to correct the errors and give them to a more meticulous printer. My desire was even greater because I met Brudo himself in Venice, where in personal conversation and on the basis of the testimony of doctors, I found him to be a very experienced physician. Moreover, I myself was the instigator and initiator of the publication of these books, as Brudo himself testified to by prefacing the Venetian editions with my letter to him."[101]

Gessner elaborated that Brudo not only reiterated the ancients' opinions about fevers and their treatments, but he also corrected errors and drew his own conclusions. He had practiced in Portugal, England and Italy, and his father and grandfather had been physicians before him.

In the same year, 1555, an anthology supervised by Gessner was published under the title *Chirurgia*. The dedicatory foreword to the Augsburg municipal physician, Gereon Seiler († 1562) of January 5, 1555 states that the initiative was taken by the Gessner print shop. Gessner's cousins had wanted to publish a volume with works by famous surgeons. Gessner had made the effort to gather together texts on the topic that had not yet been printed in German-speaking countries. The submissions, featuring interesting works about surgery, were put together – so the foreword explained – in a single volume that was convenient for students. Aside from this, printers and booksellers preferred thick books because they received more attention, and students would take better care of such comprehensive books.[102]

The topics ranged from ulcers, stone diseases, head injuries, and wound surgery, to syphilis and the application of dressings. There was also an alphabetically arranged encyclopedia of surgeons through the centuries that Gessner wrote for it. Except for two excerpts from the ancient authors Galen and Oribasius and the *Themata chirurgica XI* by the German physician Johannes Lange (1485–1565), the texts were by Renaissance doctors from France and Italy.[103]

Froben's extensive edition on Galen was published in 1562, for which Gessner wrote an introduction and contributed a bio-bibliography about Galen.[104] The *Naturales et medicinales quaestiones* by the Greek physician Cassius Iatrosophista (2nd/3rd century) was printed in the same year. Gessner's cousin,

---

101 Burmeister (1975), p. 136f. Cf. Gessner's letter to Brudo in: Brudo (1544), f. $*_8$r/v.

102 Gessner (1555d), f. $\dagger_2$r/v.

103 Cf. Nutton (1985), pp. 90f.

104 Cf. p. 151.

the printer Jakob Gessner, had pressured him to publish it because he needed the business. In the foreword to Kentmann, Gessner mentioned that he had improved the Greek text in several places, in part because of a codex he had once received from Arlenius in Venice, and in part on the basis of Alexander of Aphrodisias and Aristotle's *Problemata*. In this way, the accompanying Latin translation could be checked against the original. Gessner also wrote a 38-page commentary on this work. The Greek portion of the text was dedicated to his former student Anton Schneeberger, whose as-yet-unpublished catalogue of plague remedies was added to the work.

Released separately in 1562, *De venenis* (about poisons and drugs) by the Italian doctor Sante Arduino (15th century), was not published by Gessner, but he did provide the manuscript, which was edited by Theodor Zwinger.[105] Gessner's relatives were also the driving force behind the publication of *Ars magirica* (*Art of Cooking*) by the philologist and physician Jodocus Willich (1501–1552), who taught at the university at Frankfurt (Oder). Gessner dedicated the work to the Chur school rector Johannes Pontisella, who had supported his botanical work in various ways. Gessner thanked him on August 30, 1563.[106] Gessner wrote in the foreword of August 5, 1563 that his cousin had sought an appropriate text for publication. Because Gessner did not have anything on hand, he acquired this work by Willich for him. Several years before, Oporinus had received it as a torso from a relative of Willich, who had died while working on it. Three years prior, when Gessner was a guest of Oporinus, he had seen the work in the latter's extensive library and had asked for permission for his cousin to publish it. Oporinus agreed. Gessner gave a detailed description of how he corrected and finished it, whereby some passages remained unclear or, at the least, difficult. The work contained much of practical use, especially for doctors who needed to know which foods were appropriate for either ill or healthy people. To Willich's work, he also added the letter about foods that Jakob Bifrun wrote to Gessner at Pontisella's request about the milk products of the Italian-speaking Raetians (as opposed to those of the German-speaking Raetians).

By the middle of the 1550s, Oporinus still wanted to publish the title.[107] Gessner thanked him on June 25, 1556 for sending him a work written by Willich,[108] which will have been a collection of rhetorical pieces that Oporinus

---

105 Hieronymus (1997), p. 1212.
106 StAGR D V/37 C 36.06.29.
107 Simmler (1555a), f. 73r. Cf. also Gilly (2001), pp. 77f.
108 Letter from Gessner to Oporinus, in: ZBZ, MS. C 50a, no. 29.

had printed.[109] He apparently refrained from publishing the incomplete *Ars magirica* because of the effort involved.

Bifrun's treatise added by Gessner mentioned the fact that milk products yielded two sorts of cheese in Engadin, which were the so-called "house cheese" and "fat cheese," the latter was imported from Italy 30 years ago. Bifrun described the production of both cheeses and remarked that cheese and butter from the Grisons were so famous that large quantities were exported to Italy and Germany with demand for both young and aged cheese. While a pound of young cheese cost two *kreuzer*, aged cheese cost twice as much. Bifrun calculated that exports from about 1,000 households in Upper Engadin earned 15,000 guilders. This was equivalent to 225 tons of young cheese or 112.5 tons of aged cheese.[110] The high production of butter and cheese out of Graubünden continued for centuries. In the eighteenth century, Davos, Prättigau and Avers were still exporting "a great deal of butter and cheese, expecially to Montafon, Vorarlberg and Chiavenna. Just from the Sertig Valley alone, there were more than 6,000 small *Krinnen* (6,750 pounds) during the appropriate season."[111]

Gessner made corrective suggestions concerning the medical encyclopaedia by Erotianus (1st century AD, which was contained in Henri Estienne's *Dictionarium medicum*, Geneva 1564).[112] He also wrote a preface to the *Viaticum novum* (Zürich 1565), which was in truth a work by the French physician Jacques Houiller († 1562), but which Caspar Wolf had plagiarized, without Gessner being aware of it.[113] The Dioscorides edition of 1565 was however the last genuine medical text edition that Gessner published during his lifetime. As mentioned above,[114] Gessner took on the edition planned by the Augsburg physician Johannes Moibanus († 1562), whom he had met and come to appreciate while in Augsburg. Moibanus was able to complete the first and biggest part of Dioscorides's second book, *Euporista*.[115] It was his wish for Gessner to bring the work to a satisfactory completion. Gessner wrote in the foreword to the Augsburg city council that doing this work was done as a favor to his friend, despite all his other projects: "I wished to hold onto the memory of our true friendship and loyalty even after his death."[116]

---

109 Jodocus Willich: *Erotemata in rhetoricen ad Alexandrum* … Basel (VD 16 W 3239).

110 Bifrons (1563), pp. 220 and 226. Cf. 60 kreuzer = 1 guilder. For 2 kreuzer, one received 1 pound of young cheese.

111 Sprecher (1976), p. 60.

112 They are listed in: Estienne (1564), pp. 179f.

113 Houiller (1565), f. a_6r/v; cf. Wellisch (1984), p. 97.

114 Cf. p. 256.

115 The title *Euporista* is actually wrong. Correctly the work is called Περὶ ἁπλῶν φαρμάκων.

116 Dioskorides (1565), f. a_{iiii}v.

Several pages long, Gessner's dedication to Crato von Krafftheim (Moibanus's teacher) was written in Greek on July 10, 1564 with a thorough discussion about the character of true friendship. To a degree, a genuine friend is an *alter ego*, who is prepared to give everything – even his life – for the other. The great role model in this case is Jesus Christ, who out of love for humankind became human himself and accepted death, thus making brotherly love an objective. True friendship lasts beyond death and continues on through one's descendents. This was why Gessner passed on his affections to Moibanus's widow and orphans after his death, and why he made representations on their behalf to the Augsburg city council. Due to the triangle of friendship between Moibanus, Crato and Gessner, the latter fulfilled Moibanus's wish by completing his translation as well as his commentary to *Euporista*. Gessner transcribed the Greek version the Augsburg city council had sent to him years before from the city library. It had been full of orthographic and other errors, so he amended it and proofed it against Moibanus's text. He dedicated the work to Crato von Krafftheim since, as a foremost scholar of Greek, he would be able to appreciate it. Gessner emphasized that Crato's qualities were wide ranging, which is why he had been called to the court of Ferdinand I in Vienna to care for the ill ruler together with the court physician Julius Alexandrinus. Furthermore, because of Crato von Krafftheim's charm, his friendly personality, and his sincere faith in God, he was also popular in non-intellectual circles. Gessner ended his praise of Crato at this point so as not be taken for a flatterer, for flattery was against his nature, he explained, and he wished to stay true to himself for the short remaining duration of his life.[117]

In the spring of 1561, Moibanus had enquired of Gessner whether the Greek text should be added to the Latin translation and whether magical and superstitious assertions should be left out.[118] A letter from Gasser to Gessner in July 1562 illustrates the value Moibanus placed on Gessner's opinion, which nearly led him to abandon the project: "Along with other honorable intentions, he [Moibanus] wishes that the books *About Simple Remedies* (the title of our manuscript) reach our contemporaries. Within are cures from head to toe that to date have not been published either in Greek or translated by anyone into Latin. He had begun to translate it with the greatest care, but from the beginning, 'from the threshold', as the saying goes, your thesis that it would be wrong to attribute these books to Dioscorides has made it difficult for him. He subsequently also considered Cornarius's suspicion that they had been compiled by a younger Alexandrian author, also called Dioscorides,

117 Dioskorides (1565), f. b_8r–c_3r.

118 Letter from Moibanus to Gessner dated April 30, 1563 (StAGR D V/37 C 36.06.05).

from titles about sicknesses in Dioscorides' five older books on medical *materia* for Arius. Cornarius's theory had made a strong impression on him. Since it is in Moibanus's nature to waver, I saw that he had become inhibited and was completely undecided as to whether he should invest any more time and money. As is frequently the case, in the end, he followed my advice not to give up the entire work already begun and to press on as best he could. We know from experience that there is hardly a book in which we cannot find something of value. Then, over time, he became more familiar with this still unknown Greek author, about whom we had doubts, investing himself more earnestly, verifying and considering the uses and categorizations of each remedy. Finally, with a sharpened judgment, he tried to track him down by researching the collected works of Galen in particular as well as other authors, and on the basis of daily proof, he reached the solid conviction, of which he could assure me without doubt, that these two books addressed to Andromachus about healing illnesses of the human body with easy-to-produce remedies must be attributed to none other than Dioscorides of Anazarbos himself."[119] This view is still held by modern classicists.

On December 28, 1562, Gessner was able to write to Crato von Krafftheim that the volume was finished, and he asked Crato to send him a short biography of Moibanus,[120] which was to be prefixed to the edition.[121] The Strasbourg printer Josias Rihel agreed on January 2, 1563 to print the work, of which Gessner immediately informed Gasser: "Today, I received a letter from the book printer Josias Rihel, who promises to print Moibanus's *Discorides* in Greek and Latin. He is also willing to pay an honorarium, even right away, once he is sure that Moibanus's children are in dire need. I have requested an honorarium of one thaler for four sheets once they have been printed. If they are indeed desperate, please admonish him to pay a portion now, assuming – as I hope – that there are no unforeseen events.[122]

Gessner had hoped that Rihel would begin printing after the start of the spring fair of 1563,[123] but preparations for printing dragged on until the spring of 1565, a delay that was mostly due to the long wait for imperial printing privileges.[124] On March 31, 1565, Gessner wrote to Gasser that he would remind Rihel to send the ten sample copies to Augsburg that Gasser had asked

119 Burmeister (1975), pp. 207f.

120 Letter from Gessner to Crato von Krafftheim, December 28, 1562. Cf. Gessner (1577), f. 11r.

121 Dioskorides (1565), f. b_1r–b_7v.

122 Burmeister (1975), p. 213.

123 Letter from Gessner to Crato von Krafftheim, February 15, 1563. Cf. Gessner (1577), f. 11v.

124 Cf. the letters from Gessner to Gasser of April 7, 1563 (Burmeister [1975], p. 226) and from Gessner to Crato von Krafftheim of January 8, 1564 (Gessner [1577], f. 15v.

for, which was actually only half of what he had requested for him. Gessner explained that he was in need of 30 author copies, 20 of which he had to pay for. In addition, he had incurred expenses for the printer's privilege. In case the Augsburg council decided to make a monetary donation to Gessner for his dedication to them, he would pass it on to Moibanus's orphans, who had also lost their mother shortly before their father's death. Should however Moibanus's children not be in dire straits, he would accept the monetary give, since he himself had a nephew in poverty and others whom he had to support financially.[125] From another letter to Gasser on July 29, 1565, we learn that he had given up the dedication money to the orphans but was able to keep the author's honorarium from the printer than had originally been intended for the children: "Yesterday I received your letter of July 20, from which I learned to my great joy, firstly, that your High Council has been very generous to the orphans of Moibanus, secondly, that you are asking me to keep the money from the printer that was meant for them. Both circumstances make me very happy. I had already planned in the coming eight days to enclose ten thalers in a barrel of books to be sent to you by Froschauer. Now, thank God, this is an unexpectedly good outcome, which makes me feel more than a little beholden to you. Nonetheless, I will always be there for the orphans, and for as long as I live, whenever they become desperate, I will feel responsible for providing them with this amount or even more."[126]

Three further medical text editions were arranged to be published by Caspar Wolf after Gessner's death. The gynecological writings by the African physician Mustio (6th century), which were based on the works of Soranos of Ephesus (2nd century) and were released in 1566 by Thomas Guarinus in Basel. The humanists had confused Mustio with the Greek doctor Moschion (1st century BC). Efforts to produce a Moschion (or "Mustio") edition appear as a topic in Gessner's correspondence starting in 1556.[127] The project was left to his successor Wolf, however, because Gessner never found the time for it. The foundational manuscript from the library in Augsburg had been received by Gessner from Gasser, as Wolf explained in his dedicatory foreword to Johannes Kentmann. In 1584, Leonhard Straub in St Gallen printed Hippocrates's *Aphorisms* that had been reworked by Gessner. In the same year, a comparative work by the Italian physician Domenico Massaria about medical weights and measures was published by Froschauer's printing house in Zürich, which included a commentary by Gessner. From the foreword, we learn that this

125 Burmeister (1975), pp. 350f.
126 Burmeister (1975), pp. 358f.
127 Letter from Gratarolus to Gessner, June 29, 1556 (ZBZ, MS. F 60, f. 87r–88v).

edition was based on a manuscript that Gessner had received from Petrus Coldenbergius from Antwerp.

16.6 Medical Practice

Because Gessner was frequently in ill health – he suffered from sciatica and his wife from painful varicose veins – he knew well what it might mean to be sick or near death. There were also eye problems, which Gessner called "weak-sightedness," by which he likely meant severe nearsightedness, especially affecting one eye.[128] It may have made him unable to deal with certain issues among his patients, as he admitted in a letter to Cosmas Holzach on January 31, 1562: "I have never observed genital warts and similar ailments, especially of the anus, as well as many other less unattractive illnesses. This is partly due to my poor eyesight and partly because I do not like performing surgery."[129]

The following lines are from a letter to the Danzig physician Johannes Placotomus, who apparently also suffered from nearsightedness. They testify to the explanation, then current, of the process of seeing, which was based on the existence of visual rays (*spiritus visivi*), which emerge from the eye like feelers and establish the connection between the sense organ and the object: "Obviously, your visual rays dissipate too rapidly, as you cannot recognize distant objects, but only those that are close by. It has been the same with me since my youth, as I am weak-sighted and myopic. In order to see a little into the distance, I wear a kind of spectacles that gather the *spiritus* [visual rays] and make the objects appear smaller. I would like to know if you use spectacles, and if so, what kind. I actually believe that you too are nearsighted. This malady is incurable, but has as an advantage that the eyes do not get worse with age."[130]

In addition to the aforementioned ailments, Gessner repeatedly described himself as forgetful, which he believed was due to his many labors. On January 31, 1562, he wrote to Holzach in Schaffhausen: "I was sure that I answered your last letter, which shows how all of these many preoccupations have left my memory confused. While recovering a little today and straightening up the mess on my desk, I found your letter (to my astonishment) without the mark I use to

128 Cf. the letters from Gessner to Cosmas Holzach of November 15, 1560 and July 18, 1561, in: Gessner (1577), f. 84r and 85r.

129 Gessner (1577), f. 85v. For eye ailments he recommended the stonemason Master Petrus, cf. Gessner (1577), f. 84r.

130 Quoted from Koelbing (1961), p. 19.

indicate that I have answered it. Therefore, please forgive one who is as busy as he is forgetful, but who is nevertheless sincerely devoted to you."[131] And to Adolph Occo, on November 18, 1564, he wrote concisely: "Too much work usually ruins my memory."[132]

As mentioned, Gessner was reluctant to do surgical procedures and left those to surgeons such as his colleague Jakob Ruf. One should not conclude, however, that his anatomical knowledge was in any way deficient.[133] He was not only familiar with Vesalius's anatomy, but also dissected animal and human corpses (ill. 69a and 69b). There are three skeletons from his period that were still being shown to students in the seventeenth century.[134] Gessner's comparative anatomical knowledge is made clear in a letter to Gasser of March 19, 1564: "I laughed at your joke about the pigs and your neighbors. Surely, there is a lot that will help both men and animals with the same ailments in a similar way. I could document this from elsewhere, but also from Greek and Roman veterinary books. As you know, in particular, the internal body parts of pigs correspond very closely to those of humans."[135]

At the time, Gessner's reluctance when it came to surgery was also in line with the separation of duties between the academically trained doctors and the "craftsmen," meaning the surgeons and wound doctors who had no academic training.

Gessner's circle of patients was primarily made up of people from the city and canton of Zürich. It seems that he also played a major medical role for the city of Winterthur, with its continual requests for consultation. This is supported by the correspondence with his doctor colleague, the physician and theologian Conrad Forrer, who was especially interested in the interpretation of urine with regard to epilepsy and melancholy.[136] For the treatment of the latter psychological disorder in a depressed woman, Gessner prescribed, as it were, a Christian therapy, consisting of encouragement and prayer: "Give her consolation through your wisdom and piety, and give her positive hope that she will undoubtedly (with God's help) get better in a short time, both

131 Gessner (1577), f. 85v.

132 Gessner (1577), f. 65v.

133 Gessner (1545), f. 42r/v, praises Vesalius's good illustrations and quotes from the preface to Charles V. He thus had the book in his own hand. Gessner's colleague Rudolf Gwalther owned a copy of Vesalius's *De humani corporis fabrica* (Basel 1543), which is now kept in the ZBZ, call number: Md P 2a. Cf. Leu/Weidmann (2020), pp. 71f. and 296f.

134 Gessner's dissection of animal carcasses is evident from various passages in the bird book, cf. above p. 173. Cf. Bucher (1945), p. 131.

135 Burmeister (1975), p. 302.

136 Gessner (1577), f. 125r–128r.

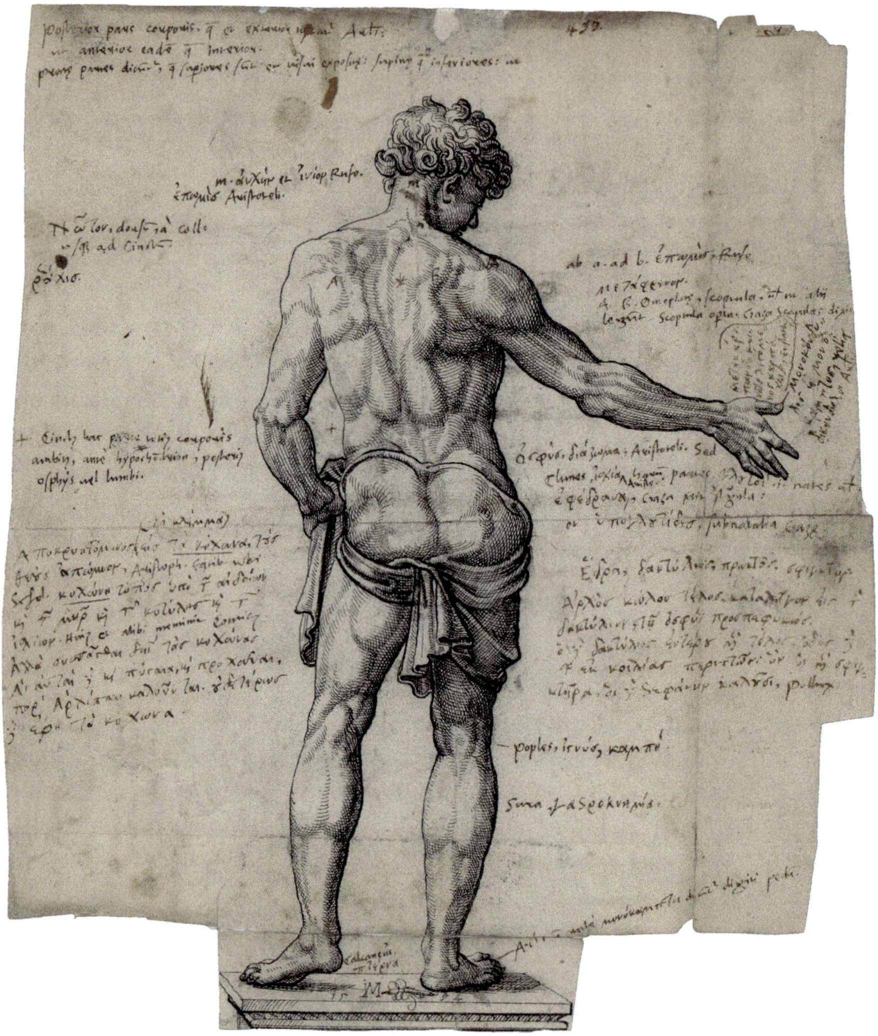

ILLUSTRATION 69A Rear view of nude, drawn by Jos Murer with Gessner's Latin and Greek handwritten labels of individual body parts from ancient authors. Although Gessner's interest here was primarily onomatological, this kind of illustration was otherwise used for learning scientific anatomical terms, as shown by 69b.

ZBZ, HANDSCHRIFTENABTEILUNG: MS. Z VII 118

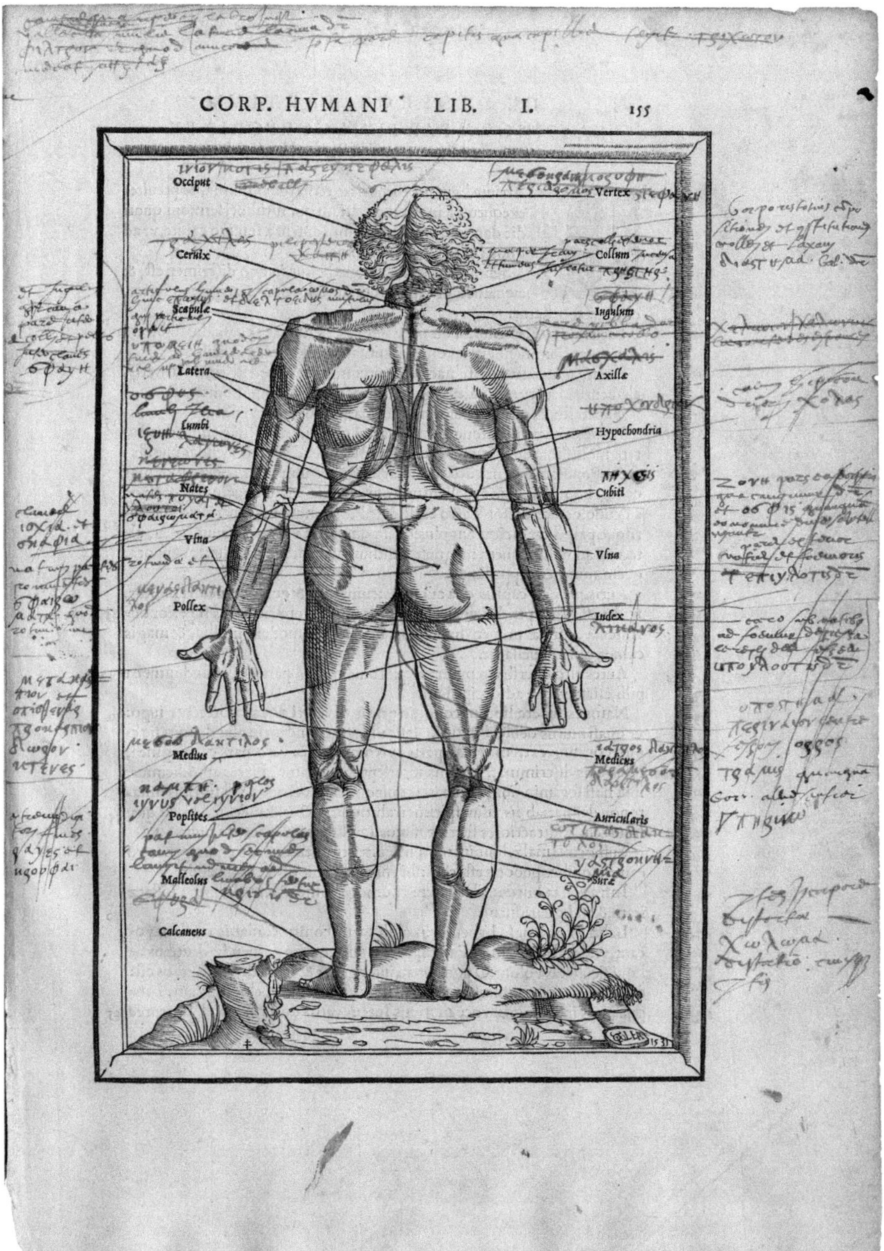

ILLUSTRATION 69B Similar rear-view nude with somewhat comparable anatomical labeling by an unknown scholar. This was probably meant to emphasize the illustration of specific body parts. Charles Estienne, *De dissectione partium corporis humani libri*, Paris 1545, p. 155.

ZBZ, ABTEILUNG ALTE DRUCKE UND RARA: AW 63

according to your own conviction and my prediction. In general, it is necessary to help melancholic people to the greatest and certain hope. In the meantime, you will also see to it that public intercession is made for the sick person, and you will command her to place the greatest hope and confidence in it."[137]

Due to permanent headaches, Samuel Kessler († 1590), pastor in the Winterthur countryside, also sought help from Gessner in Zürich. Gessner's letters to Kessler are available in his 1577 printing of his *Epistolae medicinales*,[138] while the letters from Kessler to Gessner are found in an extensive three-volume collection of remedies that Wolf posthumously organized and to which he personally contributed.[139] Gessner tried to relieve Pastor Kessler's pain with various medicines, including juniper juice. On January 18, 1562, Kessler wrote that the drink had made the headache worse, so he had tried some aged white wine, which did help. Kessler's is one of countless cases that show how helplessly many ailments were dealt with at the time.

It was not only sufferers in the canton of Zürich who sought help from Dr Gessner, but also patients who were well beyond the limits of the canton as well as confessional affiliation. This included various high-level politicians of the Confederation such as Aegidius Tschudi from Glarus and Nikolaus Zurkinden from Bern. Gessner remarked to Gasser that he had cured Tschudi of quartan ague and from asthma that always befell him whenever he became excited.[140] Zurkinden, on the other hand, sought advice for an old chest ailment and stomach complaints.[141] Gessner also came to the aid of his friend Fabricius Montanus in Chur,[142] from which city also the town clerk Daniel Gugelberg von Moos turned to the Zürich physician. The local physician, Dr Beli, could not help any further, and Gessner had such a good reputation. Gugelberg von Moos wrote that a relative of his had severe headaches every three to four weeks that lasted for three or four days. During this time, she had neither thirst nor hunger. At the same time, saliva would run from her mouth, and the pain on the right side of her head was so severe, it felt as if her right eye was being squeezed out of her head.[143]

137 Quoted from Baumann (1953), p. 139.

138 Gessner (1577), f. 128r–129v.

139 *Thesaurus medicinae practicae e Conradi Gesneri schedis autographis et celeberrimorum sui aevi medicorum epistulis, scriptis propriaque experientia et observatione collectus a Casparo Wolphio ... anno 1596*, ZBZ, MS. S 204 a–c, f. 42r/v.

140 Burmeister (1975), pp. 256f.

141 Letter from Zurkinden to Gessner dated September 20, 1557 (ZBZ, MS. S 204 b, f. 93r).

142 For example, in the case of a stone ailment, cf. the letter from Gessner to Fabricius Montanus of July 21, 1564, in: Gessner (1577), f. 92v.

143 Letter from Gugelberg von Moos to Gessner, May 12, 1562, in: ZBZ, MS. S 204a, f. 51v.

The flow of saliva suggested that she may have suffered from epilepsy, which Gessner explained to a student in Tübingen from the background of the four-fluid humoral pathology:

"The ailment causing you concern originates from too much moisture in the brain, which is increased by rising vapors. In turn, the vapors are caused by too much moisture in general. To inhibit this, it is therefore especially necessary to dry out the body – even more so, the head. Of course, this also applies to the stomach, from where the vapors arise. Thus everything must be avoided that ascends as vapors to the head, such as onions, mustard, radish, garlic, nuts, fatty foods, pork, and foods cooked in butter. But also anger. Avoid wine, since its vapors are very strong. One should not drink strong wine, but only in small amounts, well watered down and only at mealtimes. Milk and milk products are not good for the head. Avoid everything that makes one dry and thirsty so as to require less to drink. Large meals and many courses are extremely harmful. Three good swallows at a time are enough; in summer half wine and half water, unboiled if you like; in winter, a little water, boiled in an earthen container and either plain or with a little sugar and cinnamon. So, if you have a headache or otherwise do not feel well, whatever the cause, avoid wine altogether and drink only this kind of boiled water. For this ailment comes on suddenly and often together with other illnesses. Therefore, do not overlook remedies, and be sure to quickly seek consultation from a good physician, such as the excellent doctors Schegk and Fuchs in Tübingen. Please pass along my greetings to both. Should you need to receive treatment from them, let them know which medicines and prescriptions you have received from us. I also advise you to be bled two or three times a year, under the tongue or somewhere else on the head in early March or in the autumn when the grape harvest begins."

Further, Gessner advised him: "Run from love and run from women for they weaken the head and the brain." He introduced the final section with these words: "Live austerely in all that you do and get used to getting by with less."[144]

The above-mentioned collection of remedies with the title *Thesaurus medicinae practicae* contained 1,600 folio pages put together from cut out and glued together snippets of paper about illnesses and their therapies. They were organized according to a medical system of *loci* (*loci medicinales*), which followed the time-honored principle of *de capite ad calcem* (from head to toe) (see

144 Quoted from Milt (1928), p. 892.

ill. 16). It not only included the revealing document about epilepsy just cited, but also some rarities such as the only medical *consilium* by Vadian for kidney stones, the Basel lecture by Paracelsus, *Über offene Schäden und Geschwüre* (*About Open Wounds and Ulcers*) and an astonishing remedy for goiter.[145] The most frequent cause of goiter growth (*struma*) is iodine deficiency. Without knowing its true cause and effect, Gessner's therapy prescribed that the patient consumes sea creatures that contained iodine in the form of powdered sponges, starfish and algae. This would lead to a reduction in swelling of the thyroid and elimination of the goiter.[146]

One medicine that was frequently mentioned in Gessner's correspondence, which he was proud of and that he recommended frequently was the so-called *Niesswurz-Sauerhonig* (*Oxymel*). This was a honey mixture containing hellebore (*Helleborus*) and other ingredients of which there was a weaker and stronger version. He revealed the recipe on separate sheets of paper only to his closest colleagues, but none of these formulas have survived. He prescribed the remedy for the widest variety of ailments, such as asthma, shortness of breath, missed periods and epilepsy.[147]

Gessner's medical practice was probably situated in his home. There is no information about his office hours, patient management, the price for treatments or other questions. In a letter to Cosmas Holzach from the 1560s, Gessner mentioned that his practice was small, and that there were many doctors in Zürich, which allowed him to concentrate on his scientific projects.[148] This meant that the situation had eased a great deal, compared to the circumstances in 1549, when Bullinger pessimistically described the conditions for medical care in the city.[149] Meanwhile, along with Gessner, his two colleagues and successors as municipal physicians Caspar Wolf[150] and Georg Keller were now practicing in Zürich, as was the religious refugee from Locarno, Taddeo Duno (1523–1613).[151] For a time, they were also accompanied and advised by the Memmingen physician Johannes Funck (1536–after 1588).[152] Records show that the following surgeons were also active: Peter Hafner,[153] Felix Kerer,[154]

145 Cf. Büchi (1984), pp. 148–152.

146 Milt (1931).

147 Letter from Gessner to Crato von Krafftheim, February 15, 1562. Cf. Gessner (1577), f. 10r.

148 Gessner (1577), f. 87r.

149 Cf. p. 69.

150 Cf. the letter from Gessner to Wolf of September 1, 1560, in which he delegates the treatment of the pastor of Pfungen to Wolf: Gessner (1577), f. 123r.

151 Salzmann (1940).

152 Salzmann (1940), p. 341; Burmeister (2015), p. 234.

153 Cf. p. 261 and 267.

154 Burmeister (2015), p. 345.

Johannes Muralt,[155] Jakob Ruf, Felix Wyss,[156] Erhard Heidegger and Heinrich Schwarzenbach.[157] One may presume that there were also others: "There were about 15 surgeons – or rather, master cutters – who were surgically active in a wide variety of ways in the city. There were two, who were hired by the municipality, of whom one was a surgeon for stones and fractures, or city surgeon, which was the office that Ruf held; the other was the hospital doctor, who was responsible for treating wounds and the more uncouth operations. Furthermore, there were five or six salaried midwives and, on average, about five bathing facilities, but only two pharmacies. This list covers the municipally approved and recognized medical personnel, to which we must also add the hospital caregivers."[158] Compared to other years, in the 1560s there was an abundance of medical professionals available to the 7,000 residents of Zürich.

16.7 Municipal Physician

After the death of Christoph Clauser in 1552, the surgeon Jakob Ruf took over many of his responsibilities, which came with a raise in pay.[159] In 1553, Conrad Gessner nominally appeared as the town physician as a signatory of the Apothecary Ordinance, but his official appointment did not occur until 1554. As a special encouragement to him in his new office, Ruf dedicated his *Libellus de tumoribus* (*A Booklet about Tumors*) to him in 1556 (ill. 70).[160] In their "Advise on what Dr Conrad Gessner is expected to do" of May 2, 1554, the authorities stipulated the following tasks for him: "First, as a lecturer in physics at the *Hohe Schule*, he should continue to give his lectures as before, with the usual hours and days that such lecturers are assiduously expected to perform, and to maintain the same salary. In addition, my gracious lords want the said doctor to be appointed and work as the municipal physician. And he is to wait upon and serve the whole city and their inhabitants, rich and poor, young and old, in their physical ailments and illnesses that manifest themselves in time, and to help and advise them diligently and to the best of his abilities with his art, his remedies and by means of uroscopy, and to receive for this a fair and adequate reimbursement; but if the sick are poor, he should not ask a fee and do the best for them according to God's will. One counts on

155 Rath (1950), pp. 152f.
156 Letter from Gessner to Johannes Funck, January 21, 1564 (ZBZ, MS. S 109, no. 18).
157 Wehrli (1934), p. 83.
158 Steinke (2008), p. 99.
159 Steinke (2008), p. 96; Wehrli (1934), p. 50f.
160 Steinke/Müller (2008), pp. 801–803.

his modesty and trusts him in this matter. Furthermore, Dr Gessner will not be allowed to leave the city on horse or foot for other places without the consent and permission of the mayor. Dr Gessner is further required to be present in the examination those suspected of suffering from leprosy and to provide help and advise, as is custom. Likewise, he is required to advise and provide help to pregnant women and midwives in various ways and as emergencies may demand. To the best of his ability, he must take heed of, review and instruct midwives in all professional areas at each of four fasting holidays and when called upon. In addition, he is to visit the sick in the hospital in the sick rooms, examine them, and give them comfort and help, whether at usual times or when he is called. He is also ordered to oversee pharmacies together with the authorized personnel and doctors. Deficiencies and errors are to be corrected and improved, in accordance with the contents the new pharmacy regulations. All of this, Dr Gessner should swear to uphold, carry out and faithfully execute all the above. Because Dr Gessner has readily agreed to do his best under the auspices of the honored gentlemen, he burdens himself with a greater amount of responsibilities and work with respect to before, when he earned his living only from his lectureship at the school, so that he got first of all 80 guilders, then also 10 bushels of seed, 10 buckets of wine, 2 *Malter* of oats, and finally 10 bushels of seed and 10 buckets of wine for his animal book. Per year, this amounts to 122 guilders. Therefore, the good gentlemen have increased his compensation as municipal physician by 20 guilders. Going forward, this is to be alloted and paid from the city treasury – 5 guilders at each of the four fasting holidays. Enacted in the presence of Mayor Haab, the municipal governor, and both city councils on Wednesday, May 2, 1554."[161]

With this, Gessner's income reached a total of 142 guilders. Additionally, there were profits from his publishing activities, which were sometimes substantial amounts,[162] and payments from patients. The council decision of March 5, 1554 prohibited doctors (but not pharmacists) from selling medicines[163] used in the city hospital, which was housed in a former Dominican monastery, or in other official facilities. Nonetheless, Gessner was clearly not among Zürich's financially disadvantaged residents. All the same, it seems that the money was still not enough to cover his many obligations, especially those regarding his relatives. He complained about this to Bullinger in mid-September 1558 in a long and famous letter: "Now that by God's mercy, I have finally finished my book on aquatic creatures, I must, tormented me, assume once more new and

161 StAZH, Shelf mark: H I 321, pp. 60f.

162 Cf. p. 82.

163 StAZH, Shelf mark: H I 321, p. 61.

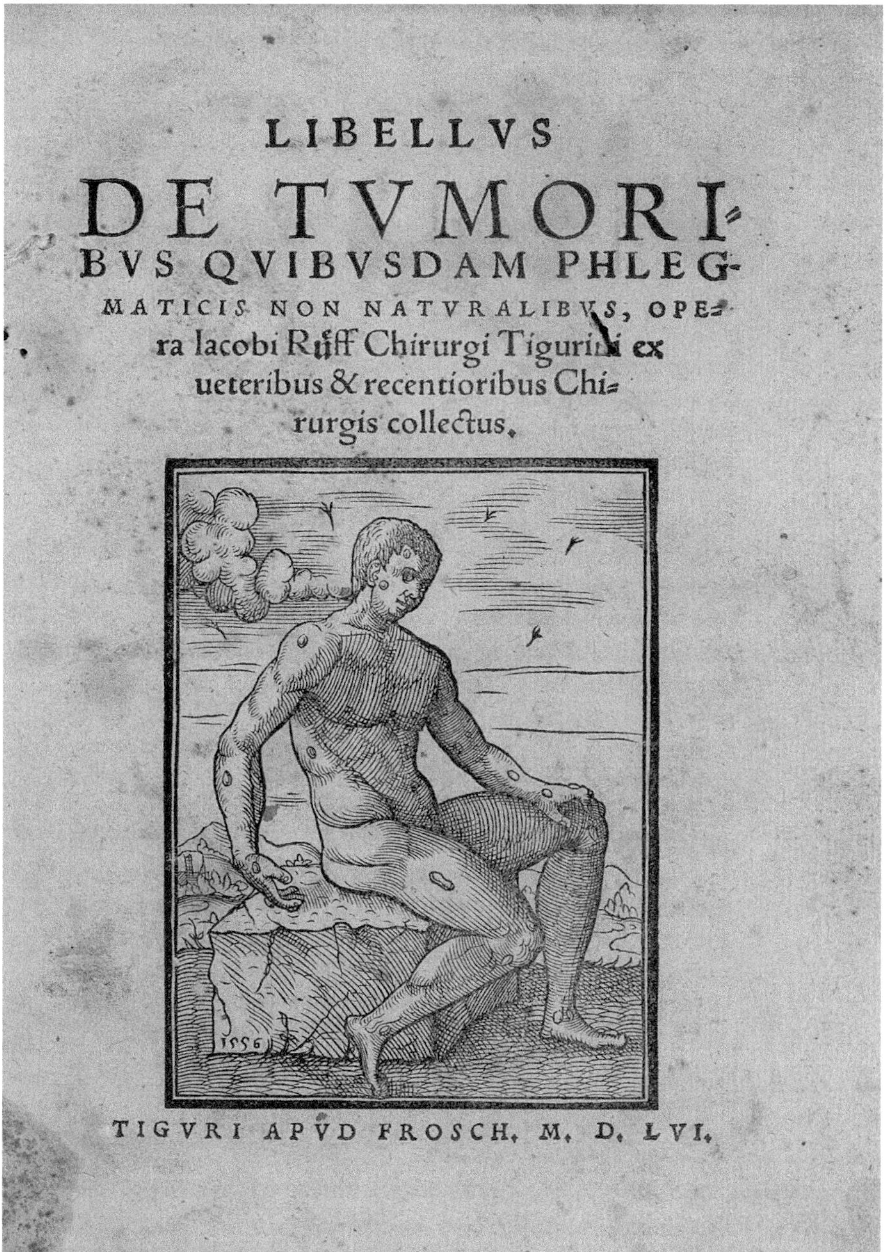

LIBELLVS
DE TVMORI-
BVS QVIBVSDAM PHLEG-
MATICIS NON NATVRALIBVS, OPE-
ra Iacobi Rüff Chirurgi Tigurini ex
ueteribus & recentioribus Chi-
rurgis collectus.

1556

TIGVRI APVD FROSCH. M. D. LVI.

ILLUSTRATION 70 Cover page of *Libellus de tumoribus* …, by Jakob Ruf, Zürich 1556.
ZBZ, ABTEILUNG ALTE DRUCKE UND RARA: 5.166

large projects. For twenty years now, I have never been fortunate enough to recover even once from my uninterrupted and stressful nighttime work. I do in no way desire an inert or unproductive rest, but just some more leisure time and a lifestyle that is more fitting to my position as a doctor and public teacher. So far, I could dedicate myself to this position only in passing, as I have constantly been busy writing books and dealing with their time-consuming publication. Were I able to find more spare time, I could also enjoy better health, the decline of which is currently indicated by my deathly pallor and emaciated body. I could revive my spirit, which at present is nearly always gloomy and downbeaten. My duties as a physician would be more happily executed, and my teaching assignments as a professor would be more fruitful. But why do you not set aside the book writing and just be content with your salary, someone might object. But when I was a young man, who would have satisfied my creditors had I not worked so diligently after returning from France, where I was supported by only a meager stipend? Who would have sustained me and my family, since for such a long time the gentlemen in Zürich's *Grossmünster Stift* granted no more than 30 guilders per year? How could I have purchased a house, when previously I had to exchange one apartment for another on a moment's notice? How could I have supported my nieces and nephews, most of whom are poor – and my dear mother as well? Should I also mention that, due to my poor health, I required somewhat better nutrition, and that many expensive books were indespensable to the advancement of my studies? Currently, these conditions and the associated expenditures are continuing, while the great number of my poor relatives adds still more to the costs. Therefore, I am forced to live not as a physician nor as a friend and devotee of the sciences but like the most indigent and lowly day laborer. Perhaps some believe that I must be wealthy, because I have owned a house for quite some time, subsequently bought a garden outside the city and have recently been building at great expense. But I had to buy a house, if I wished to avoid always having to move from one apartment to another. I purchased the little garden in order to plant herbs and gain knowledge about them that is essential for a physician. Due to financial difficulties, however, I have now sold it again to use the money for having some comfortable rooms installed in my house that I have done without for a long time. Please excuse me, my dear Bullinger, if I am repeating what you have mostly heard before. I do so to seek your advice concerning my situation.

Froben, the Basel bookseller, wants me to proof the Latin translation of Galen's complete works against the original Greek text (an enormous project) within a month. Froschauer, the famous book printer in Zürich, wants me to prepare an excerpt from the three large volumes[164] of the natural history of animals. But I am exhausted, emaciated, debilitated, half-blind and sometimes not myself (it is no wonder since I am pressured into so many huge projects). So, should I strap into the same old yoke and take on work that will leave me no room to breathe for the next two or three years? Would you recommend this, dear Bullinger? So that I do not fall into despair and for the health of both mind and body, please give me some compassionate advice, so that, finally, I must not spend the greater part of a long night awake. I am robbed of my sleep by a mind that is overstimulated by work activity lasting deep into the night and the misery of no outlook or glimmer of hope for a better and more peaceful situation. Can you not decide to provide me with a better salary for my double effort as municipal physician and publicly hired teacher who has served his home town for twenty years already? With this I could enjoy a bit more tranquility as I get older. Indeed, I can no longer endure such strenuous nighttime work. The book printers insist on large books only and will not accept the small ones, even if nothing is asked for in return. By God's grace, I am able to perform as a doctor, as long as I can spend time on it, the sick adhere to my prescriptions and do not call me too late for help. The most famous doctors at royal and princely courts are friends of mine who often share rare remedies or important guidance with me in their confidential letters, and I return the favor, so some of these men call me even their teacher. This is not written for you, dear Bullinger, but for when it is necessary – and it is quite necessary – for you to recommend me to our councilmen for these and other reasons.

One might argue that other, younger doctors also require support. To be sure, this could be done from the same source as where I obtain my salary, in part from the state and in part from the *Grossmünster Stift*. By the way, they are young, possess a certain amount of wealth and will surely earn more from their skills than I will. For, by my nature, I am inept at chasing profits as it is contrary to my entire being. What's more, soon I will also be making room for them since my body feels the premature aging, and this shows on my face. My friend, if you do not stand by me, I do not know what to do. If I can find no encouraging help from you, I will be torn up inside by the ingratitude of my fatherland. If you are determined to apply yourself on my behalf, you will accomplish a great deal. You are my only protector, and I will gladly owe my freedom to you. Whatever you promise in my name and however you might

164 Gessner's *Historia animalium* actually consists of four volumes, but the second is so thin that it was often bound together with the first.

praise me, that is what I will try to accomplish with conscientious faithfulness. My wish is to have my case brought before the council, because I deserve to receive the same salary as the other teachers for the sake of my lectures alone. By this I do not wish to compare myself to them, but I believe it is acceptable for you to mention this about me. At the same time, my book on aquatic creatures should be handed over to the council so that the councillors can see how much work and effort it must have cost me to write twelve similarly thick volumes over a 20-year period. It was for this reason that my work as a physician was completely diminished. It is appropriate for me to give them an account of how I have used my time, especially now that I have completed such a large work and should begin a new one. If the council members wish for me to continue, then I will. If they want to help me with their generosity, so that I might dedicate myself more to my medical practice, then this too will happen. If, as a physician, I carefully practice this profession, with God's help I would be capable of saving great men from the most dangerous of ailments and even – if I may say so – from the jaws of death. God forbid, should the plague strike again, I know of some effective and unique remedies that require adequate preparation time, which I do not have at the moment. If our councilmen wish for me to be a good and capable doctor, they must show themselves to be kind and generous patrons. None of today's doctors apply the opening of veins at the temple and few understand the use of uterine flushing in difficult cases. I have often successfully used these and other healing methods, which were known to the most ancient physicians since Hippocrates. When not called upon too late, and when my prescriptions have been followed, I have been able to heal severe cases of edema, stroke and epilepsy. I have kept some frail old men alive who seemed close to their demise. With regard to the knowledge of animals, herbs and remedial ingredients with which to prepare medicines, the most famous doctors of our time acknowledge me to be in first place above all past and present physicians. My facility with the Greek language is indispensible for understanding ancient medical writings – of which there are many, and I myself possess several manuscripts that have not yet been published – it would not be easy to find anyone who surpasses me in this language, for I write and speak it as fluently as my mother tongue. I do not like to tout myself this way, and wish to do so with no one besides yourself and a few others who will not suspect me of immodesty.

Excuse me for openly pouring out my heart and please advocate for me in the way that I trust you will. If your recommendations are effective, and more leisure time is afforded – not so much for my benefit as for the community – I will be thankful to you and yours. If my situation does not change, then I will pass my days as before, in excessive toil and sorrow, and I will either waste away entirely, or else, I do not know what other decision to take. Farewell!"[165]

Before sending off this letter, Gessner had the opportunity to verbally present his situation to Bullinger. He still sent the letter, adding the following note: "I had written this before revealing my situation and my wishes to you personally. All the same, I am sending you the letter because you have promised to advocate for me, so that you will be more completely informed and able to remember everything. I fear that concern for the younger doctors [Georg Keller and Caspar Wolf] may do great harm to my own efforts. Some of the supporters of these young fellows, whom I myself find dear, good and skilful, could wish to secretly or openly hinder me, believing that whatever is awarded to me would be withdrawn from them. Should you notice that they are actually feeding such suspicions, please let them know that were I receive a larger salary, this would also be advantageous to the younger doctors. This is not just for when they return from their journey abroad, but even more so for when I leave this earthly existence and pass on to a better and immortal one beyond, which is likely to be soon. Therefore, that which you do for me, will also have been done for them. I do not expect an answer from you, since you have already given it to me orally. If you think it is a good idea, then please present the main points of my Latin exposition to the mayor and most distinguished councilmen. Or do you think it would be better that I elaborate a German version of my request? If so, I will send a copy for you to read. I will remember with gratitude everything you have done for me in this matter. Your Conrad Gessner."[166]

Bullinger, who was a member of the guild *zur Meise*, had good contacts to the ruling class in Zürich.[167] He found it reasonable for Gessner to present his case in personally to the council, and he himself must have given his support to this request, otherwise it was not likely to have been approved. A plea penned in German for a higher salary was submitted to the council on September 14, 1558.[168] It was approved on the same day, which is confirmed on a parchment document that bears the city seal.[169] Gessner was awarded the title of canon and with it an income that relieved his financial bottleneck. This salary raise was a veritable act of liberation, for Gessner was now free of the printers and

165 Hanhart (1824), pp. 139–147.

166 Hanhart (1824), pp. 148–150.

167 Leu (2001), p. 144f.

168 The draft of the letter is preserved in the ZBZ, MS. C 50a, no. 33, f. 225–226 . Original to the Council, StAZH, E II 338, 1564f. A modern transcription can be found in: Fischer et al. (1967), pp. 224–226.

169 The original document is kept with a seal in: StAZH, C II 1, no. 968.

their badgering, as he reported to the Torgau physician Johannes Kentmann on February 27, 1559.[170]

Starting on May 2, 1554, Gessner became the municipal physician and therefore the highest ranking medical person in Zürich. He was also the chairman or member of various special medical commissions, was responsible for oversight of medical personnel, led the inspection of pharmacies, oversaw doctors and treatments in the hospitals, maintained his own medical practice and took care that midwives and surgeons received training. Additionally, it is known that he taught an introduction to the natural sciences at the *Hohen Schule*.[171] The education of his pupils, Caspar Wolf and Georg Keller was dear to his heart too, judging by the study rules he set down for them on May 22, 1555: "Order for the two medical students who are sent to Padua in 1555 at the city's expense. Georg Keller in Padua and Caspar Wolf in Montpellier should apply themselves diligently to the medical arts and their studies as well as whatever else is necessary to complete their courses, especially pertaining to the art itself and the method of healing, the nomenclature of simple and mixed medicines, diet, surgery or treatment of injuries.

First, they should attend three lessons each day, two concerning the healing of physical diseases and one about the treatment of wounds.

Secondly, they should carefully write down valid exegeses and comments concerning the books on which they follow lectures, so that they are able to reproduce them at home if asked to do so.

Third, they should exercise themselves in anatomy so diligently that they can show to the surgeons and scholars, if they so desire, by opening and demonstrating all external and internal organs on a human body, along with all blood vessels.

Fourth, each day they go along with general practitioners and wound doctors on their visit, diligently observe and note down all the things in a book that they can show [upon their return] as a proof of their diligence, and reporting on what happened to many patients and what advice or medicines were applied to them with either good effect or harmful results.

Fifth, they should send an annual report of their studies to noble gentlemen [of the city council], from which one can gather their diligence in lectures and visits to the sick.

Sixth, they should obtain their doctorate in medicine and surgery at a veritable school of higher learning on the basis of a test by disputation, and they

170 Gessner (1584), f. B_3v–C_1r.

171 Cf. Büchi (1983), pp. 121f.

should report of this sincerely and with the concomitant diploma, unless religion [i.e., anti-Protestant prohibitions] provides an obstacle to this.

Seventh, once they become physicians, they should devote themselves completely to the practice and observe all manner of cases and ailments at large, well-known hospitals or those of other doctors. This should be done until called upon by our gracious lords.

Lastly, with regard to clothing, food and drink and all things that come at a price, they should be proper, measured and mannerly, as well as honorable and pious in every regard.

The aforementioned two, Georg Keller and Caspar Wolf, have accepted, agreed and promised to faithfully carry out all of the above.

Enacted in Zürich on May 23, 1555 in the presence of both mayors, the financial accountable and other scholars and the foremost and most distinguished friends of the two young men."[172]

Various letters to Keller and Wolf show that Gessner took an interest in their studies and that their professional progress was a concern of his. On March 6, 1554, he had to inform Keller, who was studying in Paris at the time, that his father had died from an intestinal ailment. Along with words of solace, he also expressed his wish that the two students from Zürich make progress in their medical studies: "I also want very much for you and your friend Caspar Wolf to make quick and great strides in your medical studies and that you will surpass me by far – which should not be difficult for you – and be a joy and an asset to your hometown. I would like you to send me a sample of your progress, for example, some dubious suppositions expounded clearly, briefly and in good order, or maybe a report from one of your erudite professors about your studies and hard work so that we may praise you and demonstrate that the expenditures on your behalf have not been wasted."[173]

On October 17, 1556 Gessner also let Caspar Wolf in Montpellier know that the progress report that he had brought to Bullinger's attention was well received.[174]

Among the greatest challenges Gessner faced as municipal physician was dealing with epidemics such as leprosy, bubonic plague and syphilis. The latter caused the death of Zürich's famous imperial knight and humanist Ulrich von Hutten (1488–1523), from whose private library Gessner owned two books.[175]

172 ZBZ, MS. L 39, p. 155f.; second version in: ZBZ, MS. S 85, no. 11. Not quite completely reproduced in Meyer-Ahrens (1860), pp. 3f.

173 Gessner (1577), f. 123v.

174 Gessner (1577), f. 122r.

175 Leu et al. (2008), pp. 167 and 224.

Hutten had tried to cure the disease with the help of guaiac wood from the New World, which already the Mayas in South America had boiled for treating syphilis. Gessner was in possession of Hutten's *De Guaiaci medicina et morbo Gallico* (*About the Guaiac Medicament and Syphilis*)[176] and knew about the plant's medical application.[177]

Syphilis was not uncommon in the sixteenth century, which is why a special hospital was installed with a doctor in charge in the former women's convent at Ötenbach.[178] It was named *Blaternhaus* (pox house), because syphilis was identified by its skin rashes and sores. The oldest surviving regulations for *Blaternhaus* are from August 28, 1555. They stipulated that because of misuse, no one would be further admitted without first being evaluated and approved by Conrad Gessner, Jakob Ruf, Lux Esslinger and councilman Heinrich Trüb (1497–1587). If the person in question was not suffering from syphilis or something similar, then they would be sent to the hospital located in the former convent of the Dominicans. The standard tariff for medical treatment was 4 pounds and 5 shillings per person, and the head doctor was expected to look after the patients in the morning and evening.[179] As delegates to the *Blaternhaus*, Esslinger and Trüb received roughly 1000 liters of grain per year for their services.[180] Thanks to the isolation of syphilis patients in the Ötenbach hospital, it was possible to reduce the epidemic.[181] Oddly enough, unlike the approach in Italy, this step was not taken for those who were sick with the plague.

At the time, along with the Ötenbach delegates, Zürich had at least two other medical authorities. One was responsible for the general testing of local and foreign patients whose medical treatments were to be paid for by the authority. Besides being responsible for the *Wundgschau* (examination of injuries), *Grosse Gschau* (general examination) and *Malezeygschau* (disease examination), it also had advisory functions. The *Sondersiechenschau* (special examination for leprosy) and the *Kleine Gschau* (minor examination) were performed by a smaller commission, whose primary activity was the evaluation of lepers.[182] The *Gschau-Protokoll* (examination record) confirmed that the very professionally executed examinations of patients were often sought by others, especially from the Catholic areas of the Confederation: "In the years 1554,

176 Leu et al. (2008), p. 147.

177 Cf. letter from Gessner to Holzach of February 23, 1560, in: Gessner (1577), f. 83r/v.

178 Wehrli (1934), pp. 30f. and 61.

179 StAZH H II 31.

180 StAZH H I 321, p. 70.

181 Dobay et al. (2013).

182 Wehrli (1934), pp. 76–85.

'58, '59, '66, '67, '69 and '70, for example, suspected lepers came with letters of recommendation from the canton magistrate and council, also from Uri, Unterwalden and Glarus. At the same time, some of them were recommended for admittance to the leprosarium if leprosy was verified by the officials."[183]

Gessner took his responsibility and execution of duties as municipal physician very seriously. Ambrosius Blarer from Winterthur wrote to Bullinger on July 15, 1560, that a guest had told him how grateful he was that his wife had been admitted to hospital in Zürich with daily visits by Dr Gessner.[184] Gessner's friend, Johannes Fabricius Montanus in Chur, comforted Bullinger on September 16, 1560 that the Zürich municipal physician would do his utmost to restore his son-in-law Ludwig Lavater to good health.[185] With the outbreak of the plague, in contrast to Vadian, Gessner went so far as to remain in the city to care for the sick in an act of self-sacrifice. As he wrote to Holzach in Schaffhausen on September 24, 1564, he was well aware of the danger: "If you have found, or eventually find, anything that is effective in treating the plague, let me know. It has struck my surrounding area from all directions, and since then, I have visited many of the sick, and if God does not protect me, I am in mortal danger."[186]

The plague returned to the city and canton of Zürich time and again, taking thousands to their death. The Zürich canon Wolfgang Haller (1525–1601) made a daily handwritten record of the weather on printed calendars from 1544 to 1576.[187] Unsystematically and rather haphazardly, he also remarked on noteworthy events in the surrounding area.[188] This included lists of plague victims' names as well as the summary counts of plague deaths for 1564 and 1565. They communicate a profound image of suffering and helplessness that befell the population. Haller noted on the cover of the 1564 calendar[189] that about 8,000 people had died from the plague in Basel,[190] which was exaggerated since this was the estimated count for the entire population.[191] The remark nonetheless gives an indication that, at the time, people must have had the

183 Wehrli (1934), p. 83; StAZH, H I 321, pp. 57–82.
184 StAZH, E II 357a, pp. 583f.
185 StAZH, E II 376, p. 21.
186 Gessner (1577), f. 88r.
187 Cf. on this type of literature: Keller (2008a). Cf. on the plague waves from 1562 to 1566: Gunnoe (2019).
188 ZBZ, MS. D 269–271. Cf. Flohn (1949) and Mauelshagen (2010), pp. 46f.
189 *Lassbüchlin samt der Schrybtafel, Mässen und Jarmarckten, uffs Jar M.D.LXIIII* [gestellt durch Caspar Wolffen ...] Zürich 1563. ZBZ, MS. D 2706.
190 In his letter to Gasser of September 16, 1564, Gessner speaks of 5,000 dead, cf. Burmeister (1975), pp. 323f.
191 Cf. for Zürich p. 1.

perception of great numbers of deaths. A few pages later, there is a credible note regarding the deaths of 300 people per week in July. From these mostly unnumbered and interleaved pages, we learn that in Bern from October 1 to November 5, 1564, there were 180 plague deaths, a further 300 in November and 200 more in December. Within ten weeks, Bern had suffered 1,100 deaths. For Zürich, Haller listed the names of more than 150 people who had died in the first half of 1564, including the book printer Christoph Froschauer on April 1 and Gessner's mother on April 2. Both died on a beautiful, warm day.[192] A profound loss of life in Zürich began on August 1, taking the lives of prominent victims such as Samuel Pellikan, Theodor Bibliander, Bullinger's wife Anna Adlischwyler, Lavater's wife Margarita Bullinger and Ambrosius Blarer in Winterthur. The total number of dead in the canton reached 3,700. Bullinger himself fell ill from the plague on September 15, was then treated by Gessner, and on December 15 was able to resume his preaching activities.[193] It is suspected that either his infection was mild or his immune system was stronger than that of his contemporaries.[194] Gessner informed Gasser on October 20, 1564 that he had burned out a bubo on Bullinger with a hot iron,[195] which would have otherwise been performed by a wound doctor. Afterward, he wrote to Culmann on January 25, 1565 about the treatment of the church leader: "First, I gave Mr Bullinger, who sends his greetings, a mild solvent solution because he had been sweating a bit prior to summoning me. Then I opened one of two carbuncles with a glowing iron, and the other with a caustic agent. He felt better afterwards."[196] At the time, cutting buboes open was the most effective form of treatment.[197]

Haller continued his notes in the calendar for 1565.[198] There was an outbreak in Wädenswil and Horgen in May, with 200 deaths in Wädenswil alone. In early July there were 18 deaths in Adliswil within 5 days, and more than 500 died in August. In October, the small community of Hausen reported 140 victims, and in Kappel there were 64. Even the mountainous regions of Bern, Freiberg, Glarus and Wallis were not spared from the plague. The city of

192 In his letter to Theodor Zwinger of April 8, 1564, Gessner gives no reason for the death of his 80-year-old mother. In Froschauer's letter, he mentions malignant pleurisy (Gessner [1577], f. 107r), which probably means primary pneumonic plague. Cf. Waldis (1983).

193 Cf. also Bullinger (1904), p. 78.

194 I owe this information to PD Dr. Albert Wettstein of the Zürich City Medical Service.

195 Burmeister (1975), pp. 326f.

196 Gessner (1577), f. 46v.

197 Koelbing (1984), p. 10.

198 *Lassbüchlin samt der Schrybtafel, Mässen und Jarmarckten, uffs Jar M.D.LXV* [gestellt durch Caspar Wolffen ...] Zürich [1564]. ZBZ, MS. D 270_7.

Zürich was beleaguered once again, with Gessner's colleagues Josias Simmler and Rudolph Gwalther both losing their wives in November. In the night of Thursday, December 13 to Friday, December 14, Gessner himself also succumbed to the plague.

Against a background of the plague's latent threat to break out at any time and send people to their graves, it is no wonder that the pestilence and its possible antidote were a frequent subject in Gessner's correspondence. Furthermore, it was his duty as municipal physician to think about prevention and how to fight the disease. In 1564, a booklet was published by Froschauer with the title *Ein kurtzer bericht wie man sich mit der hilff Gottes vor dem praesten der Pestilentz hueten unnd bewaren/unnd so der mensch damit angriffen/was mittel er darfür bruchen solle: mengklichem zuo guotem/durch die S[t]attartzet der S[t]att Zürych gestellt [A Short Report on How, with God's Help, to Protect and Defend Oneself against an Outbreak of the Plague and in Case a Person Has Been Infected, What Remedies Should Be Used against it: For the Good of the People Set down by the Municipal Physicians of the City of Zürich*].[199] A letter of November 5, 1564 from Gessner to Bauhin states that the text had been written by Gessner, Keller and Wolf together.[200] However, a letter to Gasser on December 17, 1564 makes clear that Zürich's booklet about the plague was mostly the work of Georg Keller.[201] In the introduction, the plague was regarded as a punishment from God – a notion that corresponded to the general thinking of the times. This was followed by a discussion of preventive measures, which included dietary recommendations, taking special pills and making sure the air was pure: "The air, without which man cannot live, should also be kept clean, pure and pleasant, whether in chambers, parlors or other places. Therefore, it is beneficial in the morning to breathe in sweet-scented wood whenever there is a foul-smelling haze and evenings when the sun goes down."[202] Behind this was the ancient conviction that unclean, pestilential air was a significant factor in an outbreak of the plague. Should someone be afflicted with the disease, it was advisable to confess one's sins to God and seek a doctor within 12 to 20 hours at the latest. The patient should be given a purgative as well as medicines for inducing perspiration and strengthening of the heart. Bleeding was also recommended, but not for the weak or those who had become weakened. Any buboes should be cut open and treated with a salve.

199 I know of only one copy of the print, which is kept in the University Library Basel, call number: Lm VIII 1:8.

200 Bauhin (1591), pp. 148f.

201 Burmeister (1975), pp. 335f.

202 Keller et al. (1564), f. A_{iii}r.

It was clear from Keller's explanations that people were aware that infection could also result from contact with those who were diseased.[203] This had first been pointed out by the Italian physician Girolamo Fracastoro (ca. 1478–1553): "He taught that the plague can be transferred by contagion, not only by direct contact (*per contactum*), but also through an intermediary carrier (*per fomites*). The third possibility was across some distance through the air (*ad distans*)."[204] It is all the more astounding that the plague patients in Zürich were not kept isolated, as had been done with syphilis patients or that the residents did not leave the city. Such action had already been described in Giovanni Boccaccio's (1313–1375) *Decamerone*, and it was also the reaction by Vadian during the plague of 1519, when 1,500 people fell victim in St Gallen.[205] It was not until the 1585 treaty between Milan and the Swiss that trade and travel restrictions were introduced to keep the plague from travelling across the Gotthard Pass, with measures for quarantining both people and goods.[206]

---

203 Keller et al. (1564), f. A_vv.
204 Waldis (1979), p. 209.
205 Müller (2011), p. 39.
206 Koelbing (1986).

CHAPTER 17

Freaks, Monsters and Northern Lights

According to Gessner and his contemporaries, it was not just through the plague but also through other catastrophes and wonders that God communicated with the human race. As it is the Creator who has established the forms and laws of nature, whenever He repeals them partially so as to let a wonder or an "abnormality" occur, then there must be good reasons for this.[1] It is thus no wonder that various scholars in Zürich collected news, flyers and broadsheets about cosmic phenomena, natural disasters, freaks and monsters. In this context, *monstra* meant everything that "God presented to man as a sign of his decrees – but to which our seeing eyes would prefer to be blind."[2] Conrad Lycosthenes's *Prodigiorum ac ostentorum chronicon* (Basel 1557) is a well-known example, as is Johann Jakob Wick's (1522–1588) assembled collection of broadsheets about all sorts of spectacular occurrences, ranging from comets to the birth of conjoined twins.[3] Even Zürich's church leader Bullinger kept relevant broadsheets in his private library,[4] and demonstrably, four examples were found in Gessner's study, of which three were about celestial and cloud phenomena.[5] The view was prevalent that unusual events were connected to divine reasoning and man's moral failings, as can also be seen from Wolfgang Haller's previously cited notes about the weather and the plague, which also mentioned the murder of Prince Mustafa by his father Sultan Suleiman I († 1566) in November of 1533.[6] Haller recorded that in the first night, when the news reached Venice, 18 houses were destroyed by an earthquake.[7] As is often the case, here too, the causal nexus between the events recorded remain unexamined and unexplained.

It is possible that Gessner found puzzling cosmic and atmospheric phenomena to be the most interesting, which was why he attempted to analyze the northern lights that were visible on December 28, 1560. He published his thoughts anonymously under the title *Historia et interpretatio prodigii* in 1561.[8]

1 Cf. Leu (1990), pp. 91f.
2 Holenstein Weidmann (2013), p. 218.
3 Cf. Mahlmann-Bauer (2002) and Mauelshagen (2011).
4 Leu/Weidmann (2004), pp. 174–191.
5 Leu/Weidmann (2004), p. 181; Leu et al. (2008), pp. 113 and 175f.
6 Suleiman I blamed the crime on the grand vizier Rüstem Pasha. Cf. Clot (2005), p. 158.
7 *Lassbüchlin sampt der Schrybtafel uff das Jar M.D.LIII*, Zürich [1552]: ZBZ, MS. D 2698.
8 Cf. Mauelshagen (2011), pp. 79–82.

© KONINKLIJKE BRILL NV, LEIDEN, 2023 DOI:10.1163/9789004541696_018

The flyers in Wick's collection document 26 occurrences of the northern lights during the years 1547 to 1587. From this we can discern that they appeared at this latitude much more frequently in those times than they do today.[9] In the century from 1550 to 1650, northern lights were somewhat common in Central Europe. From the middle of the 17th century until the beginning of the 18th century, they were hardly seen anymore due to reduced solar activity (so-called Maunder Minimum).[10] To explain the phenomenon, Gessner referred back to Aristotle who believed that the sublunar zone was filled "with terrestrial evaporation of either a dry or humid nature, the former of which (fire) accumulates in an upper layer, while the latter (air) is found in the layer beneath. Luminous phenomena in the sky are caused by ignition of the fire element on the celestial sphere, or it is the result of contraction processes of the air element."[11]

The colors of the northern lights on December 28, 1560 were extremely intense, which was reported in corresponding broadsheets (ill. 71). For this reason, many thought it to be a fire or associated it with a fire nearby.[12] Bullinger made note of this in his diary: "The sky was burning at several places in the Swiss Confederation toward the end of the year on December 28."[13] Wolfgang Haller also recorded it in his personal copy of the printed calendar for 1560: "*Verum incendium coeli. Prognosticon motuum Gallicorum, Germanicorum, Turcicorum*" (The sky is truly on fire; a sign that the French, Germans and Turks will be on the move). Then he wrote in German: "On the same morning there was a meteor in the sky coming from the west and heading to the east. It was seen far and wide, and people believed it to be fire."[14]

Contrary to Haller, Gessner reported that it appeared from out of the southern sky. His interpretation reflected a compilation of anonymous letters and notes, as well as printed contemporary and ancient literature.[15] It led to the blood-red sign of God across the firmament being seen as a harbinger of war and persecution. He believed that because the phenomenon was of short duration, the coming catastrophe would be short as well. And since the red in the sky was neither homogenous nor dense, and the stars shone through it, it followed that the faithful and the Gospel would be preserved. The blood-red threat came from the south, which Gessner equated with Rome and its military allies, but the Babylonian beast (Rome) would ultimately be defeated. Thus, Gessner's tract belonged to the anti-Catholic polemics that compared

9 Schlegel/Schlegel (2011), pp. 68.
10 Eddy (1976).
11 Gutwald (2002), p. 254.
12 Schlegel/Schlegel (2011), p. 64.
13 Bullinger (1904), p. 65.
14 *Lassbüchlin sampt der Schrybtafel uff das Jar* M.D.LX, *Zürich* [1559]: ZBZ, MS. D 2702.
15 Gutwald (2002), p. 252; Leu (1990), p. 94.

ILLUSTRATION 71 Depiction of the northern lights on December 28, 1560 on a broadsheet printed in Nuremberg from the news collection of Johann Jakob Wick (Wickiana).
ZBZ, GRAPHISCHE SAMMLUNG UND FOTOARCHIV: PAS II 1/14

Rome to the Antichrist and the Whore of Babylon, a comparison that had previously been made in Heinrich Bullinger's *Vom Antichrist und seinem Reich* (Frankfurt 1541), Theodor Bibliander's *Relatio fidelis* (Basel 1545) and Rudolf Gwalther's *Endtchrist* (Zürich 1546). Gessner's *Historia et interpretatio prodigii* was published under the impression of a strengthening papacy with the Peace of Cateau-Cambrésis (spring 1559) between France, Spain and England affirming the Catholic Habsburgs as the preeminent power in Europe. In parallel, the six Catholic cantons of the Swiss Confederation struck an alliance with pro-Spanish Savoy in 1560. This forced the Protestants to re-focus on God, as Gessner wrote to Crato von Kraffheim on January 1, 1561: "On this day (early morning on the Saturday after Christmas), a fiery, colorful and marvelous sign appeared in the sky that was seen near and far. With this, God has wondrously called for repentance."[16]

16 Gessner (1577), f. 7v.

Gessner's brief text is considered the first scientific description of the northern lights to be taken seriously. Because the booklet was clearly outside his main areas of competence, its author published it under the pseudonym Conradus Bolovesus von Fridemontanus. In the printed list of his published books in the form of a letter to William Turner, Gessner mentions that the Englishman liked the booklet and had published an English translation. In addition, a friend of Gessner's had translated it into German, but never published it.[17]

Indeed, the interpretation of wondrous phenomena and astrological topics in general were outside of Gessner's field of activities, and he generally took a critical view of astrology. At the time, it was barely distinguished from astronomy, but Gessner very clearly expressed his scepticism, if not rejection of it. Bullinger did likewise when, for example, in 1556 he forcefully rejected his son Heinrich's wish to draw up a horoscope.[18] In his short treatment *Sanitatis tuendae praecepta* (*Prescription for Preserving Health*), Gessner passed along advice for a healthy lifestyle with a rebuke of gluttony, binge drinking[19] and the astrological calendar for bloodletting, which he a sham. He warned against the foolish draining of man's life-giving fluids merely because it was recommended by a certain astrological constellation. A weakened body could be caused to suffer even more from this, and instead of protecting against sickness, it would lead to even more serious ailments. Gessner could not fathom how a "favorable" alignment of the stars on an astrologer's calendar could make promises that would urge people to commit such destructive acts. He described how nonetheless, these calendars were full of "favorable" predictions that required bleeding, and when such signs appeared, the surgeons were then called upon to open up veins. Subsequently, people would flock to the barber shops; some had one vein bled and others several veins opened to pointlessly drain away their life-giving blood and weaken the body. Thus, not a few ruined their health or contracted new illnesses. Gessner found these calendars to be senseless and dangerous. He saw the observation of heavenly bodies in this context to be erroneous and superstitious. He acknowledged that his text would win him no friends, but such long and dangerous abuse had led him to take this position.[20]

In the following year, Jakob Ruf also joined Gessner in his criticism, castigating astrological *prognostica* with the following words: "It is all ungodly and unsubstantiated by the Holy Scriptures, [to pretend] that God's future decrees as hidden in his providence is revealed in the upper elemental bodies

17 Gessner (1562), f. B_vv, no. 60.
18 Leu/Weidmann (2004), p. 174.
19 On Gessner and alcoholism cf: Baumann (1953), pp. 143f.
20 Gessner (1556a), pp. 21–23.

and stars. No reasonable, living Christian person can find, seek, proclaim or hope to prophesy about this world without God's grace and the righteous, holy and pure Scriptures. Regardless how highly knowledgeable one might be on this earth, all the elements and stars were not created for such human prognostications."[21]

Despite these outright enlightened insights by Gessner and Ruf, making calculations with such calendars continued in Zürich into the eighteenth century.

In the relevant broadsheets about astrological occurrences, miraculous signs and monsters, freaks were a frequent topic, such as conjoined twins and man-beast creatures. In both instances, people in Zürich, especially Jakob Ruf, attempted to offer reasonable explanations for such unique individuals. In the case of the so-called Monster from Krakow, a child born with an elephantine trunk and extra faces on its body, the origin was regarded to be demonic. Ruf avoided speculation and calmly stated that he would leave the interpretation of this creature up to God, who was all-knowing. Additionally, he referred to Pliny's *Natural History*, who had described that various strange animal hybrids were to be found in Africa.[22] Likewise, when observing the conjoined twins of Schaffhausen in 1543, he again avoided folkloristic explanations and suggested natural causes, namely, a fragmentation of the male semen within the mother's body. He further explained that this was caused by certain alignments of the planets, a view that he probably would not have taken 20 years later. In the end, however, he also acknowledged that God's mercy was behind it all: "In any case, the almighty and benevolent God allows such things to occur in his mercy, and not without reason. For just as we can only correctly interpret extraordinary occurrences, all sorts of signs and wonders, the comets and similar light and atmospheric phenomena as nothing else than a fatherly admonition out of God's pure mercy to do penance for our earlier life (otherwise, our eyes will be opened in our envisaged punishment), so this is likewise the meaning of birth defects, as each can judge for himself."[23]

How exactly Gessner thought about these things hasn't been reported, but we may presume that he will have seen things similarly or the same as Ruf.

21 Steinke (2008b), p. 951.
22 Spinks (2008), p. 56.
23 Steinke (2008b), p. 425.

CHAPTER 18

Earth Sciences

When Gessner was awarded the sinecure of a canon on September 14, 1558, which assured him of a good income, the Zürich city council set down conditions. It was decreed that he had to content himself with this benefice and dedicate himself solely to his task as municipal physician, to the cure of the sick in the city and the countryside and to his lectures, but that he should no longer write books or enter into any contracts with printers.[1] As mentioned above, Gessner remarked to Cosmas Holzach in Schaffhausen that he had only a small medical practice, that Zürich had sufficient doctors at the time,[2] and that he could reduce his activities as a physician and continue with his scientific projects.[3] He might have made the case to the authorities that his medical and pharmacological publications were in the spirit of his activities as municipal physician. However, such an explanation was certainly not possible with his work on the earth sciences that was published in the summer of 1565. The anthology was titled *De omni rerum fossilium genere, gemmis, lapidibus, metallis et huiusmodi, libri aliquot* (*Several Books about the Types of Fossils, Gems, Stones, Metals and the Like*). In the sixteenth century, the term "fossil" did not just refer to that which was petrified, but to everything that came out of the ground, including minerals, prehistoric implements and archaeological finds. As Gessner explained,[4] the main title page of the volume (ill. 72) shows a diamond ring and a somewhat larger ring made of amber with a bug enclosed in it,[5] which his student Anton Schneeberger from Poland, had given him as a gift. The two pieces of jewelry are encircled by twelve gems, which, Gessner explains, had decorated the breastplate of the Old Testament High Priest (Exodus 28).

In this volume, Gessner brought together eight related contributions about fossils, minerals, stones and objects that were relevant to the earth sciences. The first, with the title *Nomenclaturae rerum fossilium, quae in Misnia praecipue, et in aliis quoque regionibus inveniuntur*, contained a list of the minerals and fossils in the collection of his Torgau colleague Johannes Kentmann. There

1 Council decision of September 14, 1558, in: StAZH C II 1, no. 968.

2 Cf. p. 346.

3 Gessner (1577), f. 87r.

4 Gessner (1565), f. 98v.

5 Rudwick (1976), p. 21, has wrongly interpreted this amber encluse as a scarab. Bugs are not rare in Baltic amber, cf. Weitschat/Wichard (2002), pp. 120–127.

© KONINKLIJKE BRILL NV, LEIDEN, 2023 | DOI:10.1163/9789004541696_019

DE OMNI RERVM FOSSILIVM GENERE, GEMMIS, LAPIDIBVS, METALLIS, ET HVIVSMODI, LIBRI ALIQVOT, PLERIQVE NVNC PRIMVM EDITI.

Operâ Conradi Gesneri: Quorum Catalogum sequens folium continet.

Tiguri, excudebat Iacobus Gesnerus: Anno M. D. LXV.

ILLUSTRATION 72 Title page of Gessner (1565a) depicting an amber ring with an encased insect, a diamond ring and 12 gems from the Old Testament breastplate of the High Priest.

ZBZ, ABTEILUNG ALTE DRUCKE UND RARA: NG 1910

was also an illustration of the collection's cabinet in which the latter stored 1,608 pieces from 135 locations and organized into 26 groups. The transcribed list reads:[6]

1. Soils
2. Natural mixtures
3. Efflorescent mixtures
4. Grease mixtures
5. Stones
6. Stones found in organisms
7. Fluorites
8. Silicates
9. Gems
10. Marbles
11. Rocks
12. Petrified wood
13. Sands
14. Gold
15. Silver
16. Mercury
17. Ore or Copper
18. Metal, calamine, galena
19. Gravel
20. Lead
21. Bismuth
22. Tin
23. Antimony
24. Iron
25. Steel
26. Marine minerals, miscellaneous

Gessner hoped that the work, which he had initiated once again to generate business for his relatives in the printing trade, could be presented at the 1564 spring book fair. He wrote to Crato von Krafftheim on January 8, 1564: "My cousin, the printer, is pressuring me to publish my little work *De lapidibus* [*About Stones*], despite the fact that I am so very busy with other things. Because he spent a lot on the illustrations and his finances are tight, I will try to accommodate him. God willing, I will somehow release it at the next fair

6 Prescher et al. (1980), pp. 25 and 30.

in Frankfurt."[7] Unfortunately, the publishing date was delayed, which is why he asked Kentmann on June 24, 1564 to send a revised copy of the inventory. Since the last list was sent in 1559,[8] the collection was sure to have grown.[9] On January 27, 1565, Gessner expressed his gratitude for receiving a piece of basalt and some leaves from the extended *Catalogus fossilium*.[10] Kentmann wrote a dedication foreword to Gessner in which he cited him as a celebrated pioneer in the natural sciences: "For you have so clearly, deftly and thoroughly examined the field of natural science that includes living creatures in the air, on the earth and in the seas, like no one before. And in all other areas, you have made the utmost effort to always achieve your goals with diligence, not only fulfilling the initial expectations you have engendered in so many, but far surpassing them. I am busy every day with the results of your studies ever more avidly, and I try to assess all that you have so beautifully accomplished and that you are currently working on, to the benefit of all of posterity, and I can't help but admire your spiritual strength and your physical stamina which is undiminished despite your constant toils. I have reached the conviction that you are not only driven by nature as your guide, but also by divine influence, to bring so many things into the the light of day and to explain correctly that have so far been mostly in the dark and unknown. How significant and welcome to posterity will therefore also be your research of that part of nature that you have included in your study of minerals."[11]

The second text, *Calculorum qui in corpore ac membris hominum innascuntur genera XII*, was also by Kentmann and dealt with 12 kinds of stones that can be found in the human body. The third, about metals, with the title *De metallicis rebus ac nominibus*, was written by Kentmann's colleague Georg Fabricius. It addresses 11 kinds of metals and where they can be located and covers, among other things, remarks about them found in ancient classics. The fourth work, *De succino* (*About Amber*), was written by Severin Göbel (1530–1612), a doctor from Königsberg, who had sent his treatise to Gessner, who subsequently lost it. Gessner then asked Achilles Pirmin Gasser whether he could send him his own copy, which he did, and for which Gessner thanked him on April 17, 1565.[12] In the first part, Göbel compares those who seek amber allegorically with Jesus Christ. Amber represents mankind, and those who look for it are often kept like slaves, and they can only find the stone with great effort. Similarly, Christ also sought out mankind under great hardship. As with amber, there are also

7 Gessner (1577), f. 15v.
8 Prescher et al. (1980), p. 29.
9 Gessner (1584), f. D_3r–D_4v.
10 Gessner (1584), f. E_1r/v.
11 Prescher et al. (1980), p. 29.
12 Burmeister (1975), pp. 352f.

different qualities among people. White amber can be compared to genuine members of Christ's church, whereas the darker version – the most abundant and least useful – is like those people who only indulge in carnal pleasures. In part two, Göbel writes about the origins of amber, believing it to be hardened bituminous petroleum, created in the interior of the Earth, then spilled out onto the ocean floor. On June 24, 1563 Gessner remarked to Johannes Culman, the personal physician to the Prince of Württemberg, that amber was related to natural asphalt.[13] A few months later, he told Crato von Krafftheim that he owned many different kinds of Bernstein.[14] Among them was a piece from Ethiopia that was transparent and had a pleasant scent. This was probably not amber, but frankincense resin, which has a similar appearance.[15] Additionally, Gessner received various pieces of amber with encased insects from the Danzig physician Johannes Placotomus.[16]

The fifth piece, *De halosantho seu spermate Ceti* (*About a Type of Sea Salt or Whale Sperm*) was written by Valerius Cordus. According to the dedicatory foreword that Gessner addressed to Andreas Pellizerus, he had received the book two years ago by the pastor Georg Aemylius from Stolberg. On August 1, 1563, Gessner wrote to Crato von Krafftheim: "Today, I received another piece by Cordus, *De halosantho*. He believes this is what common folk call *Cetus* [whale] sperm, which I seriously doubt. If I publish it, I will most certainly present opposing evidence to contradict this view."[17] According to the preface, Gessner also received additional copies of the work from the physicians Wolfgang Meurer from Leipzig and Hieronymus Herold from Nuremberg. Gessner remarked that he wanted to publish the descriptions of 66 kinds of fish by Cordus that were prepared 25 years earlier in Venice as well as other zoological works, but he never got around to it. The text *De halosantho* was about deposits of sea salt that were used to treat skin conditions. While folklore held it to be whale sperm, Gessner, in a progressive manner, interpreted the phenomenon in his commentary to be salt crystals, with no connection to whales.

The sixth contribution, *De duodecim gemmis quae erant in veste Aaronis* (*On the Twelve Precious Stones that Were on Aaron's Vestments*) was by the Church Father Epiphanius of Salamis (ca. 315–403). Epiphanius there discusses the precious stones, which, according to Exodus 28, 15–21, this Old Testament high

13 Gessner (1577), f. 45v.
14 Letter from Gessner, November 12, 1563, cf. Gessner (1577), f. 15r.
15 Letter from Gessner to Johannes Funck, November 19, 1564, cf. Gessner (1577), f. 95v.
16 Cordus (1561), f. 217r.
17 Gessner (1577), f. 12r.

priest wore on his chest. In the translation of the New International Version, the instructions Moses received were as follows: "Fashion a breastpiece for making decisions – the work of skilled hands. Make it like the ephod: of gold, and of blue, purple and scarlet yarn, and of finely twisted linen. It is to be square – a span long and a span wide – and folded double. Then mount four rows of precious stones on it. The first row shall be carnelian, chrysolite and beryl; the second row shall be turquoise, lapis lazuli and emerald; the third row shall be jacinth, agate and amethyst; the fourth row shall be topaz, onyx and jasper. Mount them in gold filigree settings. There are to be twelve stones, one for each of the names of the sons of Israel, each engraved like a seal with the name of one of the twelve tribes."

According to the dedicatory foreword to Adolph Occo of July 7, 1565, Gessner had received the incomplete manuscript from Caspar von Niedbruck many years before.[18] He wrote that he had heard that the work had been published in Paris, but he had never seen a printed copy in either Latin or Greek. This was why he published a bilingual version with the Latin translation by his friend Jolas Hierotarantinus. He added that he owned yet another manuscript with the title *Physiologus*,[19] in which Epiphanius described 39 animals. He asked Occo to confirm whether the two texts about the stones and animals were attributed to Epiphanius in the *Myriobiblon* (notes about authors and their works) by the Byzantine patriarch Photius (ca. 820–891) that was located in the Fugger library in Augsburg.

There is also an Armenian version (*Codex Vaticanus Borgianus Armenus 31*), located in the Vatican Library in Rome that was edited in 2014.[20] By comparison, Gessner's version does not amount to even half of the text, and it contains numerous discrepancies. The second part, which features the allegorical arrangement of the precious stones and the various comments about their magical powers, is missing. Gessner's edition primarily contains descriptions of the gems with information about where they are found as well as their names and medicinal uses at the time. Some aspects of related folklore and superstition were also included, but Gessner did not comment on the latter at this point. He stated his relevant opinions in the subsequent seventh title of the anthology.

This seventh title, *De gemmis aliquot, iis praesertim, quarum D. Ioannes in Apocalypsi meminit, et aliis quarum usus hodie apud omnes percrebruit*

18 Cf. p. 120.

19 Presumably, the fragment preserved in the ZBZ (MS. S 112, no. 132) is a part of this manuscript or copy in Gessner's hand.

20 Epiphanius of Salamis (2014).

(*About Some Precious Stones, Especially Those Mentioned by John in the Book of Revelations, and others that Everyone Uses Today*) was written by the Parisian physician François la Rue. In the first chapter, la Rue wrote about the various powers of the stones. In chapters 2–13 he dealt with each of the 12 stones that adorn the heavenly Jerusalem in Revelation 21. In a postscript, Gessner informed the reader that he had edited out any magical elements regarding the effects of stone amulets and the like. After this, he printed a letter by the canon Paschase Balduin from Phalempin in Northern France, who struck the same note and refuted the stones' supposed healing powers – a position that Gessner had taken for many years. For instance, on December 28, 1562, he wrote to Crato von Krafftheim: "When regular doctors add to the remedy against the plague that you recently communicated also gold, precious stones and the bones of deer hearts, I have found them to be dubious and unconvincing and would never use them. … I have fewer doubts about pearls and corals; in fact, I have used corals with frequent success, but I find it highly questionable that precious stones should possess various powers."[21] Concerning his own contribution to the anthology, Gessner wrote in the foreword that a second extensive work on the topic of stones and metals was planned, which would include explanations about each kind (*genus*), their power (*vis*), their qualities (*natura*) and linguistic designation (*philologia*).[22] With *vis*, however, he was not referring to magical or esoteric powers, but rather the medicinal effects of inorganic substances. For example, in the following passage from a letter to Gasser on December 3, 1565, he expressed his gratitude for receiving an exceptionally large piece of malachite: "And perhaps paint or pigments for artists can be made from it as is done with celadonite and lapis. These are similar, if I am not mistaken, and are produced for metallic objects. Along these lines, I have decided to test its medicinal effects as a purgative. I will not take it myself but give it to some people who are melancholic."[23] Despite Gessner's editorial efforts, some questionable assertions were left in the work, such as that this or that stone provides protection against demons. And given his interest in Epiphanius' explanations on the gems of the priestly ephod of the Old Testament, he also felt he had to include Francois la Rue's thoughts on the gems of the heavenly Jerusalem, which pertains to the New Testament and could prove to be helpful to theologians. It was not until about 200 years later

21 Gessner (1577), f. 11r.

22 Gessner (1565), f. Aa$_2$v. These and other passages were misinterpreted by Rudwick (1976). Gessner was wrongly declared by him to be a Neoplatonic esotericist, as Fischel (2010) has already pointed out.

23 Burmeister (1975), p. 367.

that a first natural scientific commentary on the Holy Scriptures was published in the four-volume *Kupferbibel* or *Physica sacra* by the Zürich physician and universal scholar, Johann Jakob Scheuchzer (1672–1733),[24] whose goals were therefore similar to Gessner's in publishing la Rue's work.[25]

The eighth and last part was written by Gessner himself, titled *De rerum fossilium, lapidum et gemmarum maxime, figuris et similitudinibus liber, non solum medicis, sed omnibus rerum naturae ac philologiae studiosis, utilis et iucundus futurus* (*A Book About the Shapes and Appearances of Fossils, Notably Stones and Gems, Which Is Useful and Enjoyable not Only to Doctors but to All Who Are Dedicated to the Study of Nature and Philology*). As mentioned before, there is a ring on the cover page of the anthology that was given to him by his Polish student and later colleague Anton Schneeberger from Kraków. The dedication of this text went also to Kraków: shortly before printing had finished, on July 28, 1565, Gessner devoted the work to Andrzej Szadkowski (Andreas Schadcovius), *notarius* of the Kraków salt mines. He did not know him personally, but Schneeberger had informed him that Szadkowski thought highly of the Zürich polyhistor's work. He had apparently also made some financial contribution. Gessner explained that the subjects were not presented in alphabetical order. To a degree, he adhered to a natural system that followed the ascending levels on the ladder of nature. Starting with the most simple forms, such as lines, corners, circles and others, which are also found in the elements or the cosmos, Gessner then ascended toward more complex forms as found in stones that appeared similar to plants or animals.[26] His system included the following 15 groups:

1. Stone objects that have lines, points or corners (e.g., asbestos fibers, that run more or less in lines, or pyrite with its regular corners).
2. Stones that have something in common with the stars, the sun, the moon or the Aristotelian elements (e.g., star-shaped fossils like the limbs of the crinoid *Pentacrinus*).
3. Stone objects that derive their names from things in the atmosphere (e.g., prehistoric tools called "thunderbolt stones"; people believed that they came from millstones that had been struck by lightning and eroded by the wind).
4. Stones or metals that were named after inanimate things on Earth (e.g., Chrysolith [= goldstone] or magnetic stone).

24 Cf. Leu (2012) and Leu (2022).
25 Fischer (1946a).
26 Cf. p. 163.

5. Unearthed objects that resemble man-made products (e.g., belemnites that look like little arrows).
6. Objects fashioned from stone, metal, minerals or precious stones (e.g., rings with gems and graphite pencils).[27]
7. Stones that resemble plants (e.g., mushroom-shaped stones and *corallina officinalis*, a calcareous red seaweed).
8. Shrubs and similar objects that have become petrified, or those that look like shrubs or parts of them.
9. Trees or parts of trees that are stonelike (e.g., petrified wood).
10. Corals.
11. Other petrified aquatic plants (e.g., fossilized corals[28] and tube worms).
12. Stones named after persons or animals or that are found inside them.
13. Stones named after birds or that are found inside them.
14. Stones with images of aquatic animals (petrified sea creatures).
15. Stones with snake or insect images (e.g., ammonites or fossilized sea urchins = snake eggs).

Gessner did not yet possess a clear idea about the true nature of fossils. In the case of terrestrial and aquatic plants, and what he identified as such, he speaks of organic substances turning to stone (*lapidescere*).[29] However, in the case of fossilized animals, his formulation was more cautious (ill. 73a, 73b, 73c), comparing stones with images of animals (*effigiem referre*).[30] His uncertainty was a reflection of the times, in that scholars were not in agreement. Was it a *vis plastica* that produced fossils, or were they relics of the great flood? Or must they be interpreted as organic remnants that had become petrified within the context of the earth's history and its local catastrophes and processes – this being a view defended among others by Leonardo da Vinci?[31]

While the facts seemed clear to Gessner regarding petrified wood, he made no comment about the process that produced it. In 1557 in Joachimsthal (today, Jáchymov in the Czech Republic), a three-meter long fossilized tree trunk of the genus *Laurinoxylon* was found in the palaeozoic layer of a mining site. With a diameter of roughly 50 cm, it included both roots and branches and was located in basaltic tuff from the Oligocene period. Pieces of bark and imprints of leaves were also preserved. Today, it is still unclear how this

27 Gessner (1565), f. 104v, shows the oldest printed image of a pencil.
28 The true nature of corals was not recognized until the eighteenth century, until then they were considered plants. Cf. Schouppé (1991), pp. 27–33.
29 Gessner (1565), f. 135r.
30 Gessner (1565), f. 162r. Cf. Schierl (2019); concerning the crab: Etter and Schmidt (2019).
31 Dominici (2010), p. 5; Cioppi (2010), pp. 77f.

ILLUSTRATION 73A Fossilized crab *Palaeocarpilius macrochelus* from Gessner's private collection. It is very likely from the Eocene epoch and came from the Lessin Mountains north of Verona.
MUSEUM OF NATURAL HISTORY, BASEL, GEOLOGICAL SECTION, NO. F 1281

geologically young fossil ended up in these palaeozoic layers of the Earth's history.[32] Johannes Kentmann sent Gessner a piece of the famous find and reported that it had been discovered unaltered in a petrified state (*conversa in lapidem*) with roots and leaves.[33] It seems that Gessner easily accepted the idea of organic substances turning to stone, but was unable to explain the process. This was a question that he would certainly have wanted to pursue in view of the planned, more detailed version of his work on fossils and minerals.

In his dedicatory foreword to Szadkowski, Gessner mentioned that most of the illustrations shown were of objects that he had at home. Of the 200 objects depicted, at least six have been preserved in the Natural History Museum in

32 Dupéron et al. (2008).
33 Gessner (1565), f. 125r.

ILLUSTRATION 73B Drawing of the crab at original size from Felix Platter's fossil and mineral album, which also contains the preliminary drawings for Gessner's *De rerum fossilium … liber*. The fossil was lightly magnified by the artist.

UNIVERSITY LIBRARY BASEL: MS. K I 2, F. 68R

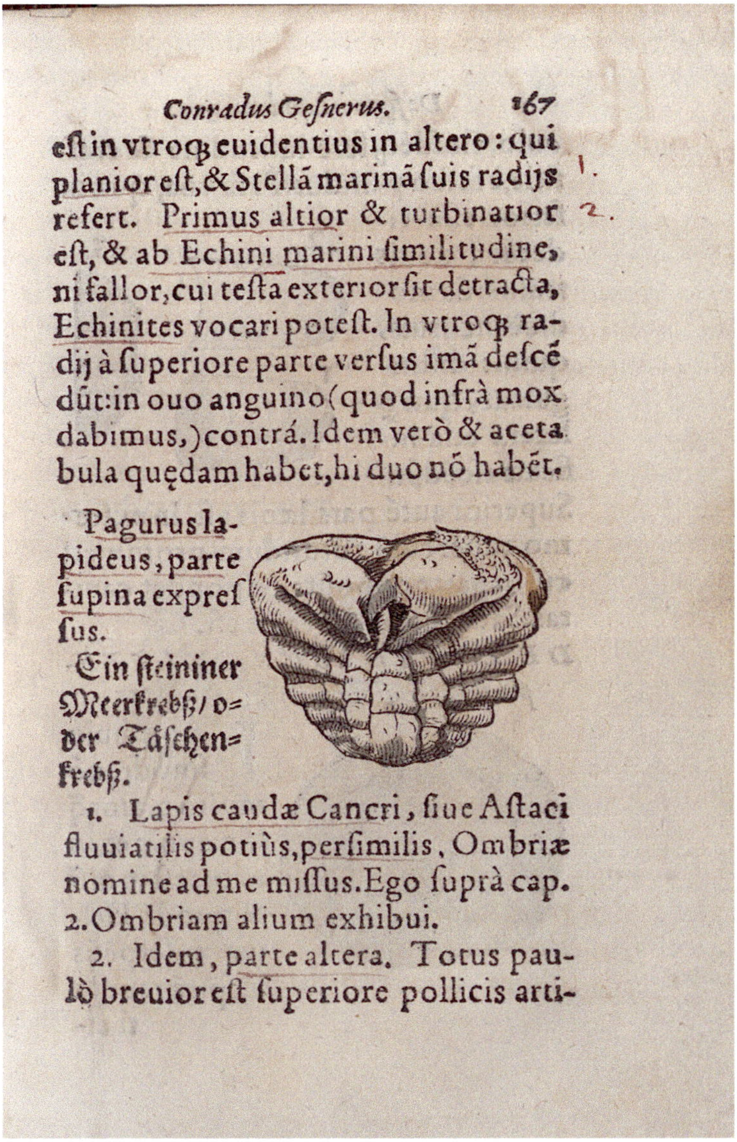

Conradus Geſnerus. 167

eſt in vtroq; euidentius in altero: qui planior eſt, & Stellã marinã ſuis radijs refert. Primus altior & turbinatior eſt, & ab Echini marini ſimilitudine, ni fallor, cui teſta exterior ſit detracta, Echinites vocari poteſt. In vtroq; radij à ſuperiore parte verſus imã deſcẽdũt: in ouo anguino (quod infrà mox dabimus,) contrá. Idem verò & acetabula quędam habet, hi duo nõ habẽt.

Pagurus lapideus, parte ſupina expreſſus.

Ein ſteininer Meerkrebß/ oder Täſchenkrebß.

1. Lapis caudæ Cancri, ſiue Aſtaci fluuiatilis potiùs, perſimilis, Ombriæ nomine ad me miſſus. Ego ſuprà cap. 2. Ombriam alium exhibui.

2. Idem, parte altera. Totus paulò breuior eſt ſuperiore pollicis arti-

ILLUSTRATION 73C Partially colorized woodcut of the crab at a smaller scale, from Gessner (1565), f. 167r.
ZBZ, ABTEILUNG ALTE DRUCKE UND RARA: FF 1264

Basel.[34] Gessner's was not the first to publish on such matters,[35] but his was the sixteenth century's most richly illustrated book on earth sciences. From his correspondence, we learn that in addition to the illustrated objects, he must have had various other fine specimens. For instance, there were the aforementioned amber pieces with encased insects, and a butterfly fossil,[36] which might have actually been an insect fossil from the Solnhofen limestone deposits. It is also likely that there were a few interesting, well-preserved specimens of Pliocene conchylia from Northern Italy that had had obtained from Francesco Calzolari.[37] It seems that Gessner did not have any of the famous fish and plant finds from Monte Bolca (near Verona) in his museum, even though Mattioli, who had been made aware of it by Diego Hurtado de Mendoza, had mentioned the site by 1554 at the latest.[38] Gessner not only received fossils, but also gave them as gifts. For instance, he sent bivalve fossils to both Occo[39] and Calzolari.[40] He also sent his book on fossils and minerals to the latter,[41] as well as to other scholars, including Johannes Bauhin the Younger,[42] Joachim Camerarius the Younger,[43] Wilhelm Stucki[44] and Achilles Pirmin Gasser.[45] The book sold fairly well, and was out of print by 1571.[46]

---

34 Etter and Schmidt (2016) and (2019).

35 Illustrations of fossils can already be found in Sebastian Münster's *Cosmographia* (Basel 1550, p. 432) and Christoph Entzelt's *De re metallica* (Frankfurt 1551, pp. 227 and 229).

36 Letter from Gessner to Theodor Zwinger, December 16, 1561, cf. Gessner (1577), f. 106r. The year of the letter according to AK 11, p. 976.

37 Salzmann (1959), pp. 98f.

38 Mattioli (1554), p. 568.

39 Letter from Gessner to Occo, September 28, 1565, cf. Gessner (1577), f. 78r/v.

40 Cerutus/Chioccus (1622), pp. 396f.

41 Cerutus/Chioccus (1622), p. 373.

42 Bauhin (1596), pp. 154f.

43 Rath (1950), pp. 163f.

44 Gessner (1577), f. 124v/125r.

45 Burmeister (1975), p. 364.

46 Helm (1971), p. 186.

CHAPTER 19

Testament and Estate

In his meteorological description of the weather on Thursday, December 13, 1565, Wolfgang Haller used the words, "desolate weather, wet, snow." He described the next day with "rather good weather." During the night that fell between, Conrad Gessner succumbed to the plague. Haller also makes an interesting note about this, namely, that up until his last minutes, Gessner had apparently not allowed himself any rest: "December 13. At about 11 o'clock at night, the esteemed Dr Gessner died from the plague after an ulcer had emerged over his heart on December 9. During his illness, he did not lie down until he was about to die."[1] It seems that Gessner was busy trying to organize his estate until his strength finally gave out, which was also the impression left by Caspar Wolf, the administrator of Gessner's holdings. During his last four days, Gessner had given him extensive instructions about how to continue with the *Historia plantarum*. Wolf described the condition of the estate literally as chaos.[2]

Another description that fills in the details of Gessner's last hours is found in the biography that Josias Simmler wrote shortly after Gessner's death. He reported that on December 9, Gessner had noticed a plague bubo above his heart and hence close to the axillary lymph nodes. He noticed no headache, no fever and no other symptoms. He would occasionally take a short rest, but was not bedridden. Ultimately, he called his friends together and made his testament with bequests to his wife and his nephews on his sister's side. His only living sister was designated the primary inheritor. In particular, aside from other visitors on the day before his death, Gessner spoke with his highly beloved Bullinger, discussing many family matters. He declined the offers by friends to stay with him during the following night. As had been his way throughout his life, even in his last hours, he did not wish to be a burden to anyone. Finally, with only a servant girl present, he was led to his bedroom where he prayed fervently to God and lay down. At 11 PM, when he sensed that he would soon die, he was helped to his museum, where the day before he had arranged for a bed to be placed. There, in the arms of his wife, he gently passed away. The very

1 *Lassbüchlin sampt der Schrybtafel, Mässen und Jarmarckten, uffs Jar* M.D.LXV, *Zürich* [1564] (ZBZ, Shelf mark: MS. D 270_7).

2 Wolf (1566), f. 47r.

© KONINKLIJKE BRILL NV, LEIDEN, 2023 | DOI:10.1163/9789004541696_020

next day, accompanied by many mourners, he was buried in the cloister garden of the Grossmünster church next to his friend Johannes Fries.[3]

Gessner had written out his testament already on September 18, 1564.[4] By the fall of 1562, he sensed that he probably did not have long to live. He wrote to Fabricius Montanus in Chur: "I, too, feel every day more human, and that my time is near."[5] In his will, Gessner arranged for the passage imperial grant of arms and for the establishment of a family foundation for the needy, for which he and other family members set aside 100 guilders. It was to be supplemented with regular contributions from the family. An annual family meal was another desire of his, whereby a goblet was to be drunk from that came from Kaplan Frick, Gessner's erstwhile foster father on his mother's side of the family.[6] Each of these family gatherings was to close with a reading from the New Testament and viewings of his animal books (*Icones animalium, Icones avium* and *Nomenclator aquatilium animantium*). From another document that is currently kept in the Gessner family archive, it appears that a sum of 2,661 pounds and 8 shillings were owed to him by various people, of which a third was to go to his wife.[7] These figures clearly show that Gessner certainly did not die a poor man.

The death of the universal scholar was a bitter blow for many of his contemporaries. Before he was aware of his illness, on December 4, he consoled himself in his last surviving letter, written in Greek to Adolph Occo, as follows: "If you were to leave this earth before I do, I am encouraged by the hope that I will soon follow you – me, with the long gray hair, who has been pale, gaunt and in poor health since childhood. The greatest and strongest consolation, however, will come from God. I know and trust with firm confidence that all things are for the best of those who trust in God, who are His children through Jesus Christ, that not a hair will fall from their heads without His fatherly will, without His omnipotent providence. ... Therefore, I also nurture the hope that we will soon be free from the bonds of earthly life, and together we will enjoy the one true and blessed life with Christ."[8]

Gessner's death was very difficult for Bullinger and others. The head of the Zürich church expressed his grief on December 14 in a letter to Fabricius

3 Simmler (1566), f. 18r–20r. In the meantime, the tombs in the cloister have been removed. Cf. Fischer (1966), p. 133.

4 The autograph is in the ZBZ, call number: FA Gessner 1516.3 (transcribed in Hanhart [1824], pp. 285–291).

5 Gessner (1577), f. 92r.

6 Cf. p. 1.

7 ZBZ, FA Gessner 1516.2.

8 Gessner (1577), f. 55v–58r; the Greek letter is translated into German in: Hanhart (1824), pp. 277f.

Montanus in Chur. He remarked that Gessner had worked extremely hard all through his last days. A week later, he confessed in a letter to an acquaintance in Graubünden that Gessner's death had left him so depressed that he could barely write.[9] Likewise, on December 14 he informed Johannes Kessler in St Gallen of the sad news: Gessner, the glory of Zürich and Switzerland, had died. Even on the previous Sunday, December 9, as the first symptoms of the illness appeared, Gessner had sat as an attentive listener to Bullinger's sermon. Thereafter, he had never laid down. If he sat for a moment to read, he would quickly stand again and keep moving. He constantly discussed theological and important topics. The day before, he had done a great deal of writing, had said goodbye to Bullinger and confessed his faith in Jesus Christ. One could see, Bullinger wrote, that God called even the best and most brilliant people away from this earth.[10] Four days later, Bullinger sent a letter with similar content to Theodor Beza in Geneva. He was in deep mourning, because after his beloved daughter had died of the plague, Gessner had now also died of it, who was not just a close friend (*amicissimus*) but also an extremely exemplary (*sanctissimus*) and illustrious (*clarissimus*) man.[11]

The lines that Bullinger wrote to his son Hans Rudolf on December 15, 1565 are also of interest: "The late pious doctor [Gessner] said good-bye to me on Thursday afternoon. He expressed thanks for all that is good, confessed his faith and gave assurance that with his trust in Christ and the true Christian faith, he was quite ready to die. For he knew in his heart that God in his mercy would accept him. He also asked me to take care of some things for him and to communicate his parting words and his gratitude for their friendship to his friends, the scholars, whom I also know well and who are our shared friends, at the imperial court, in France, in England, etc. They should all remain honest to the true Christian faith, the only way to be truly blessed. Then, not without tears, I asked him not to let his fine materials for remedies be lost along with him. He replied that he would show Dr Wolf and his brother-in-law Dr Keller all that was necessary, and on that same day, he made many arrangements with Dr Wolf and his brother-in-law Dr Keller. On Thursday at 11 p.m. he was dead. He never laid down for any length of time, but was constantly on the move, eating, working and organizing his herbarium, which he assigned to Dr Wolf. Yesterday at 4 in the afternoon, he was buried with a large presence in church."[12]

---

9 Schiess (1905), pp. 656f.

10 Vadianische Briefsammlung, Kantonsbibliothek Vadiana St Gallen, Shelf mark: MS. 38 (VBS IX), p. 228.

11 CThB 6, p. 224.

12 Fischer et al. (1967), p. 234.

Shortly before his death, Gessner sold his private library – of which there are still around 400 titles in existence today[13] – to Caspar Wolf, along with his collected manuscripts, including the *Historia plantarum*. In Wolf's own words, it was for a not insubstantial price (*non exiguis sumptibus*).[14] On July 10, 1566, Wolf wrote to Theodor Zwinger in Basel that Gessner was still thinking about publishing 24 more titles during his lifetime:

1. About the physician in general.
2. A catalogue of famous physicians.
3. An edition of certain books by Hippocrates.
4. An edition of certain books by Galen.

4a. Physiologica.

5. Anatomica.
6. On health care.
7. On the dialectic of health and disease.
8. On prognostics.
9. On simplicia.
10. Collected consults, arranged according to individual symptoms.
11. On practice.
12. On fevers.
13. On "extreme" diseases.
14. On corpulence, etc.
15. On the dissolution of unity (*De solutione unitatis*).
16. On cosmetics.
17. On diseases of the head.
18. On diseases of the thorax.
19. On diseases of the abdomen and anus.
20. On diseases of the genitals.
21. On diseases of women.
22. On wound medicine.
23. Miscellaneous and doubtful.
24. About veterinary medicine. [15]

13 Leu et al. (2008).

14 Wolf (1581), f. 2r; cf. Leu et al. (2008), pp. 5–9. A substantial part of Gessner's private library was dispersed in the course of the subesquent centuries; for example, a large lot of the formerly indexed Protestant author even made its way to Rome. Cf. Serrai (1990), pp. 357–360.

15 Letter from Wolf to Zwinger, July 10, 1566: UB Basel, Frey-Gryn Mscr II 4, no. 341 (I thank Dr. Walter Tilmann, Würzburg, for this information).

Unfortunately the great Zürich polyhistor never got around to any of this. It appears to be a character trait of great minds that, as long as they are healthy, they never cease to make new plans and contemplate further-reaching projects. In this sense, the death of a scholar comes always too soon.

In 1577, Wolf published the three books of Gessner's *Epistolae medicinales*, the compilation of which had been begun by Josias Simmler, who had however died while working on it.[16] The publication of Gessner's lectures followed in 1586. These were only the two most important works among other texts[17] that Wolf published from Gessner's estate. Despite his published promise of 1566,[18] Wolf found it impossible to edit the extensive and fragmentary *Historia plantarum*, and so he sold the three volumes with its about 1,500 drawings in 1581 to Johannes Camerarius the Younger in Nuremberg[19] for 150 guilders. Camerarius published only a few of the illustrations from it in the above-mentioned revision of Mattioli's *Kreuterbuch*[20] and its corresponding Latin version (*De plantis epitome utilissima*, Frankfurt/M. 1586) as well as in his *Hortus medicus philosophicus* (Frankfurt/M. 1588). He sent a complimentary copy of the revision of Mattioli's *Kreuterbuch* to Wolf, who thanked him on August 29, 1587.[21]

After the death of Johannes Camerarius the Younger, Gessner's botanical estate was passed on to his son, Ludwig Joachim Camerarius (1566–1642), who sold it to Nuremberg's municipal physician, Johann Georg Volkamer (1616–1683), who then bequeathed it to his son who bore the same name and occupation. The latter offered to sell it to the Zürich city library, which could however not afford to pay the 300 guilders that Volkamer requested. Finally, the plant illustrations were bought by Christoph Jakob Trew (1696–1769), who was a municipal physician, botanist and the president of Nuremberg's Natural Historical Society. Trew in turn entrusted its publication to the professor of botany Casimir Christoph Schmiedel (1718–1793). The two volumes were published in 1754 and 1759 in Nuremberg, where most of the German-language publications featuring illustration plates were printed in the eighteenth century. The first volume contained 22 plates with 198 woodcuts originating from Gessner's time, and 20 copperplates with 177 illustrations. The second volume

---

16 Gessner (1577), f. α_2r.

17 Cf. Wellisch (1984), pp. 101–107.

18 Wolf (1566). He published individual drawings in 1566 and 1577, cf. Wolf (1566), f. 53r–56r, and Gessner (1577a), f. 4r, 11v and 13r. Cf. Funk (2018).

19 Cf. Trew Letter Collection, Erlangen University Library, letter from Wolf to Camerarius dated June 26, 1581, available online at: www.trew-letters.com.

20 Cf. p. 291.

21 Cf. Trew Letter Collection, Erlangen University Library, letter from Wolf to Camerarius dated August 29, 1587, available online at: www.trew-letters.com.

held 31 large-format, colorized copper engravings that were executed according to Gessner's templates. Two of the three volumes from Gessner's botanical estate remained in the region and were located in Erlangen in 1929. The third volume was only recently discovered in the Library of Tartu University. The associated drawings had somehow made their way into the possession of the Erlangen physician Johann Georg Schneider (1744–1833), who sold them in May of 1804 to Tartu.[22]

As has been explained before, Gessner's entomological estate travelled by way of Camerarius to England.[23] His earth science collection, together with his sketches for the book on fossils and minerals, were bought by Felix Platter in Basel,[24] who also bought some bird drawings that are currently located in Basel,[25] along with some of fish and other animals that are now kept in Amsterdam.[26] The Zentralbibliothek Zürich holds a collection of Gessner's medical prescriptions with additions by Wolf,[27] as well as some hand-written pieces of correspondence.[28] Because Gessner was in the habit of cutting up the letters he received according to topic, only a modest number of handwritten correspondence has survived. Nonetheless, there are some surprises in this regard. Some years ago, about 50 letters addressed to Gessner were found in the Grisons State Archive in Chur whose provenance is still unclear.[29] Even though much has been lost along the way, Gessner as a person as well as his accomplishments, the extant printed and handwritten work as well as his incompletely examined illustrative material will keep researchers busy for generations.

One of the questions that arise from Gessner's biography is the influence of the Reformation on the non-theological sciences. The famous sociologist Max Weber (1864–1920) found the Protestant work ethic to be a factor critical to the

22 Fischer (1966), pp. 134–138; Hausinger (2009), pp. 86f.; Henglein (2010), pp. 65–80; Rand (2014), pp. 16f.

23 Cf. p. 183.

24 Etter/Schneider (2016). The fossil and mineral drawings are kept at the University Library Basel, MS. K I 2.

25 University Library Basel, MS. K I 1, digitally available at: www.e-manuscripta.ch/doi/10.7891/e-manuscripta-12849. The bird drawings acquired by the New York Historical Society do not represent preparatory drawings for Gessner's bird book. This becomes clear when one compares the body posture of the eagle with Gessner's woodcut (cf. Olson/Mazzitelli [2007], p. 450). Furthermore, the preliminary drawing in question is preserved in the aforementioned volume in the UB Basel (f. 28r).

26 Egmond (2013).

27 ZBZ, MS. S 204 a–c.

28 ZBZ, MS. C 50a. Cf. also the list of letters in the appendix.

29 The letters were handed over to the Grisons State Archives a few years ago by Anna Marie Salome Linder-von Tscharner, owner of Ortenstein Castle. It is possible that they originate from the family Tscharner or Travers von Ortenstein. Cf. Leu (2016b).

prosperity of countries during the early modern period.[30] In Gessner's case, indications are that the Renaissance's new, empirically trained view of nature, along with the Reformation, influenced him in his work, research and conclusions. As a devout Christian, who took the Bible as God's word, it was obvious that he distanced himself from those things that were condemned in the Holy Scriptures, such as astrology, magic and occult practices. He demystified and debunked various areas of medicine and the natural sciences and made corresponding advances in natural knowledge. It would be of interest to examine the influence of the Reformation on other Swiss researchers as well as its reception among the Protestant population, and to assess whether modernity and scientific progress were inhibited in Catholic areas of the Confederation.[31]

30 Cf. on the more recent discussion about Max Weber and the development of the natural sciences: Pollack and Pohlig (2020), especially Eire (2020).

31 Today we are confronted with a counter-movement on the part of esotericism, homeopathy etc. The process described by Max Weber as the disenchantment of the world is thus being reversed, and here and there our society is being transported back into medieval categories of thinking. Cf. Weber (2019), p. 19; Rüegg (2014).

APPENDIX 1

Gessner's Testament

Translation of the German original from the Gessner family archive, which is held at the Zentralbibliothek Zürich (call number: FA *Gessner* 1516_3*).*

I Conrad Gessner, citizen and municipal physician of Zürich in my beloved fatherland and lecturer of philosophy in the *Hohe Schule*, legitimate son of Urs Gessner, a tanner and citizen of Zürich (my dear father, who died in the mountains near Zug or Baar in 1531), born in the year 1516 on Palm Sunday, etc.

Among other books I have written and had printed, there is a very large Latin book about fishes and all sorts of animals that live in the sea and in other waters, which I dedicated to Emperor Ferdinand I and sent him a copy. His Majesty received the book with special favor, and a short time later, during an Imperial Diet in Augsburg, his personal physicians invited me to join them, as the Emperor wished to see me and make my acquaintance. When I arrived, they led me to His Imperial Majesty in his chambers. The Emperor declared that my book on fishes was pleasing him greatly, and we discussed subjects related to nature for more than an hour. His Majesty then promised to take care of me. A short time later, the said personal physicians wrote to me again, that because they had seen no family crest on my letters or seal, his Imperial Majesty had ordered a grant of arms to honor me and my family. If I so pleased, I was also allowed to pass the award along to my blood relatives to honor them. I knew this would please my cousins and relatives. Because I have no children, I have mentioned my cousin Andreas Gessner the elder and his descendants born in wedlock in this coat of arms letter and give and present to him first, but also to all members of the Gessner family and to all who are born of them, this imperial honor, jewel and coat of honor with the appeal and encouragement that they use and present it with great modesty and civility and with great humility and gentility as honorable people. God did not grant us this award for wantonness and arrogance, but for us to lead an upstanding, honorable and God-fearing lifestyle. Thus can God the Almighty give you happiness and honor forever more, and not just during our transient human pilgrimage on this earth, but eternal glory and joy in Heaven. Furthermore, this crest should remind you not to let your children grow up in idleness but in the respectable and fine arts, whether with lessons, work, handicrafts, studies or otherwise. So I, too, was sent by my father at a young age to the German and then to the Latin school. With the help of the Council and the Foundation of the *Stift*, I received a stipend and was able to study in France. When I became an adult, I was hired and earned a good yearly income teaching the Greek language at Lausanne in the Savoy. Afterward, in Basel I obtained a doctorate

© KONINKLIJKE BRILL NV, LEIDEN, 2023 | DOI:10.1163/9789004541696_021

in medicine, and then returned home to Zürich to become a lecturer of philosophy (physics and ethics) and ultimately also municipal physician.

However, it is my strong will and wish that after my death the eldest in our family living here in Zürich (the eldest among successors in charge) should retain the Latin grant of arms, adorned with the emperor's seal, and keep it all his life, conducting himself respectably as an honored citizen. If not, the letter should be amiably passed to another in our family, namely the eldest who is blameless.

Further, in addition to the grant of arms, I have assigned 100 guilders (to which my dear cousin, M. Andreas Gessner the Elder, his son Jacob and the grandsons Anton and Hans the Younger have contributed 10 talers apiece) and also have ordered to the elder who has the letter of arms to bring in the same amount if it should become necessary. He shall give written confirmation to the elder after him that he has a letter of interest, and shall withdraw and pay the annual interest of 5 guilders yearly to clothe two of the most needy children who live in Zürich on St Michael's Day [September 29]. If there are no poor children, the money should be distributed to two or three of the most destitute adults within the Gessner family. The four eldest in the family (fewer or more of them, depending on who is in Zürich) should determine to whom the money is best given among the poorest and most needy. Because at the moment, the children of my sister Elsbet, the wife of Hans Aberli, are the poorest, (despite not being Gessners) the amount should be properly shared among these children until they are grown and come into their own. Thus, if the next pregnancy results in greater poverty than a pregnancy among our blood relatives, it is intended that the four eldest in the family should have the power to provide for these children.

Further, it is my desire that whoever retains the letter should invite the other Gessners to a friendly dinner each year, especially during the cold of winter – if he can afford it. If not, those who are better off should help out, or each should contribute something and together, in the name of love, provide for a fine meal. For in this way, they will be reminded of the affection, friendship and unanimity among them, which should be encouraged by the eldest. Indeed, the eldest should summon the others, and if he knows that there is envy, hate or discord among some of them, he should make an effort beforehand that they accede to agreement and togetherness. If not, they should not even come to the meal and should be regarded by all as stubborn and disobedient. From then on, the family should consider them unworthy of their love for not behaving or living in a Christian manner.

I have also arranged to have a gold-plated goblet with a cover (weighing 15 lot) at this dinner. It was bequeathed to me by my late mother, Agatha Fritz, and she received it from her cousin, the late Hans Fritz, who was a chaplain here in Zürich and who took me in as a child, raised me and provided for my schooling. This drinking vessel should be in the care of the eldest, who also has the grant of arms and the note of interest, and he should also testify to this in the above-mentioned confirmation. He should never use it except at the aforementioned dinner, when all have come together to eat and drink with love, joy and friendship. For we are related and committed to one another not only in flesh and blood, but in the heart of Christ and the true knowledge and love of God. Particularly those should drink together from this vessel who have had disagreements with one another and bear witness to Christian reconciliation from the heart with no duplicity, so as not to incur the wrath of God. Likewise, if conflict should arise during the course of the year and must be dealt with, the eldest should make an effort to resolve it. People should be brought together for a friendly meal and simply hold and drink together from the goblet. Otherwise, it should never be used.

Lastly, I ask and exhort each of those who come together for this meal of love to contribute a little or a lot, to the eldest, who receives the interest from the 5 guilders. In this way, should the 5 guilders not be enough, the poor in our family can be supported with sufficient funds.

When the dinner is finished, the eldest should praise and thank the Lord God, and read aloud two or three short passages from the New Testament (such as Matthew, etc.). This should serve to support joy, love and unanimity (for this, I have also added a small copy of the New Testament to the grant of arms and note of interest), in which they are further urged to live and love as Christians. Thus will God the Almighty grant them His mercy here and beyond. Then, after this, he should present to them my illustrated treatises on animals, the three that I have bound into a single book put together with the grant of arms. They should look at the books and be motivated by the memory of me to educate their children according to solid teaching and else to good professions and activities. Praise be to God the Almighty, with honor and glory forever through Jesus Christ our Lord. Amen.

Dated: Zürich, September 18, 1564.

APPENDIX 2

Additions to Gessner's Private Library

Additions to the 2008 Repertory of Gessner's Private Library: Newly Found Books.[1]

1. Actuarius: *De medicamentorum compositione Ioanne Ruellio interprete. Adiecimus quoque in Medicinae candidatorum gratiam succidaneorum medicaminum tabulam, quorum usus habetur reciprocus, Graece et Latine, per Conradum Gesnerum …*, Basel, Robert Winter, 1540.
With handwritten annotations by Gessner.
Location: Bibliothek Sprecher von Bernegg, Maienfeld, call number: HV 2_2

2. Aphtonius: *Praeludia*, Paris, Christian Wechel, 1531.
With ownership note and handwritten annotations by Gessner.
Location: Trinity College Library, Cambridge, Call number: II.9.9 (4)[2]

3. Apollonius Rhodius: *Argonautica …*, Venice, Aldus Manutius, 1521.
With ownership note and handwritten annotations by Gessner.
Location: Trinity College Library, Cambridge, Call number: M. 12. 74[3]

4. *La Bible qui est toute la Saincte Escripture …*, Neuchâtel, P. de Wingle dict Pirot Picard, 1535.
With ownership note and handwritten annotations by Gessner.
Location: Bibliothèque des Pasteurs, Neuchâtel, Call number: PA P 104.1.7 (new call number: past 8559).[4]

5. Confalonieri, Giovanni Battista: *De vini natura, eiusque alendi et medendi facultate …*, Basel, Johannes Bebel, 1535.
With ownership note and handwritten annotations by Gessner.
Location: National Library of Medicine, Washington Bethesda, Call number: NLM ID: 8601149

1 Leu et al. (2008).
2 Communication from Prof. Dr. David McKitterick (Cambridge).
3 Communication from Prof. Dr. David McKitterick (Cambridge).
4 Communication from PD Dr. Reinhard Bodenmann (Brugg).

© KONINKLIJKE BRILL NV, LEIDEN, 2023 | DOI:10.1163/9789004541696_022

6. Dryander, Johannes: *Anatomia capitis humani* ..., Marburg, Eucharius Cervicornus, 1536.
With handwritten annotations by Gessner.
Location: Bibliothèque publique et universitaire, Lausanne, Call number: No RERO: 0297668

7. Fuchs, Leonhart: *Methodus seu ratio compendiaria per veniendi ad veram solidamque medicinam* ..., Basel, Michael Isengrin, 1541.
With ownership note: Sum Conradi Gesneri Tigurini
Location: Bibliothek Sprecher von Bernegg, Maienfeld, Call number: HV 2_1[5]

8. Gatinaria, Marco: *M. Gatenariae summi medici omnes, quos scripsit, libri, artis opera* ..., Basel, Heinrich Petri, 1537.
With a few handwritten annotations by Gessner.
Location: Zentralbibliothek Zürich, Call number: Md A 36_2

9. Guilandinus, Melchior: *De stirpium aliquot nominibus vetustis ac novis* ..., Basel, Nicolaus Episcopius iunior, 1557.
With many handwritten annotations by Gessner.
Location: National Library of Medicine, Washington Bethesda, Call number: NLM ID: 8601107

10. Paracelsus: *Grosse Wund Artzney von allen Wunden, Stich, Schüssz, Bränd, Bissz, Beynbrüch, und alles was die Wundartzney begreifft* ..., Ulm, Hans Varnier, 1536.
With handwritten annotations by Gessner.
Location: National Library of Medicine, Washington Bethesda, Call number: NLM ID: 2752426R[6]

11. *Porcellus monstruosus*, [Vilnius 1562].
Drawing of a pig with a misshapen head seen in Vilnius in 1562. Gessner noted below that he received the drawing in 1564 from the Pole Stanisław Starzechowski (Stanislaus Starzechovius).
Location: Zentralbibliothek Zürich. Call number: MS. F 27, p. 24a.

12. Schneeberger, Anton: *Medicamentorum simplicium corpus humanum a pestilentiae contagione praeservantium catalogus* ..., Kraków, Lazarus Andreae, 1556.
With numerous handwritten annotations by Gessner.

---

5 Communication from Prof. Dr. Jan-Andrea Bernhard (Ilanz).
6 Communication from Prof. Dr. Massimo Danzi (Geneva).

Location: National Library of Medicine, Washington Bethesda, Call number: NLM ID: 8601155

13. Schnellenberg, Tarquinius: *Experimenta. Von XX. Pestilentz Wurzeln und Kreutern* ..., Frankfurt/M., Hermann Gülfferich, 1553.
With handwritten annotations by Gessner.
Location: National Library of Medicine, Bethesda (MD), Call number: NLM ID: 8601121

14. Sidonius Apollinaris: [*Opera*], Basel, Heinrich Petri, 1542.
Mit handschriftlicher Widmung auf dem Titelblatt: „Conradi Gesneri ex dono typographi Basilee 1546".
Location: The work was sold in 2010 by the antiquarian Maggs Bros LTD in 2010 to a private person.[7]

15. Stobaeus, Johannes: *Sententiae ex thesauris Graecorum delectae ..., nunc primum a Conrado Gesnero ... in Latinum sermonem traductae* ..., Zürich, Christoph Froschauer, 1543.
With numerous handwritten annotations by Gessner.
Location: British Library, London, Call number: C.134.d.3

16. Theocritus: *Idyllia.*
Anonymous handwritten translation into Latin of Idylls 1 to 13 with a few annotations by Gessner.
Location: Zentralbibliothek Zürich, call number: MS. C 86a (no. 4)

17. Theophrastus: *De historia plantarum libri VIIII* ..., Basel, Andreas Cratander, 1534.
With numerous handwritten annotations by Gessner.
Location: Universitätsbibliothek Leipzig, Call number: Phil.gr.16m

7 The buyer is known to me and has sent me copies of the volume. It is copy no. A 59, which we missed in 2008. Cf. Leu et al. (2008), p. 278.

APPENDIX 3

Gessner's Correspondence

In the following table, reference is made primarily to where the individual letters have been published, since it is easier to get hold of the printed editions than the handwritten texts. For those letters that have never been published, the location of the original is indicated.[1]

Name	Date (year, month, day)	Letter to Gessner (location)	Letter from Gessner (location)
Aldrovandi, Ulisse	1564.10.09	Biblioteca Universitaria Bologna, MS. Aldrovandi 77, vol. II, f. 23r–24r	
Alexandrinus, Julius	1563.01.23	StAGR D V/37 C 36.06.20	
Alexandrinus, Julius	1563.02.01		Dedication letter Gessner (1563)
Altieri, Baldassare	1547.01.29	Gessner to Bullinger	HBBW 19, pp. 212f.
Amerbach, Basilius	1562.01.23		AK 11, no. 4693, pp. 974–994
Amerbach, Basilius	1562.03.03		AK 11, no. 4706, pp. 1035–1037
Amerbach, Basilius	1562.07.16		AK 11, no. 1706, p. 1035
Amerbach, Bonifacius	1545.11.18		AK 6, no. 2761
Amerbach, Bonifacius	1554.01.21	AK 9/1, no. 3711, p. 222	
Amerbach, Bonifacius	1557.10.31		AK 10, no. 4233, pp. 533–535

1 I would like to thank Prof. Dr. Ann Blair (Harvard) as well as Dr. Tilman Walter and Anne Rappert-Sälzer (Arbeitsstelle "Frühneuzeitliche Ärztebriefe," Würzburg) for their assistance in compiling the list of correspondents. Cf. Blair (2016) and (2017).

© KONINKLIJKE BRILL NV, LEIDEN, 2023 | DOI:10.1163/9789004541696_023

(*cont.*)

Name	Date (year, month, day)	Letter to Gessner (location)	Letter from Gessner (location)
Ammann, Johann Jakob	1542		Dedication letter Gessner (1542)
Anonym	1551.02.05		ZBZ, MS. C 50a, no. 57
Anonym	156?	Dedication letter Göbel (1565), f. i5r–7r	
Anonym	1563(?).12.19	ZBZ, MS. F 15, f. 344v/345r	
Anonym	1564(?)		ZBZ, MS. F 38, f. 339r
Anonym aus Chiavenna	1564.05.15	ZBZ, MS. F 15, f. 454v	
Aretius, Benedictus	1551.07.01		Gessner (1577), f. 115v–116r
Aretius, Benedictus	1552.01.29		Gessner (1577), f. 116r
Aretius, Benedictus	1553.05.22		Gessner (1577), f. 116v
Aretius, Benedictus	1553.08.27		Gessner (1577), f. 117r
Aretius, Benedictus	1562.03.05		Gessner (1577), f. 117r/v
Aretius, Benedictus	1562.05.27		Gessner (1577), f. 117v–118r
Aretius, Benedictus	1562.06.19		Gessner (1577), f. 118r/v
Aretius, Benedictus	1562.07.03		Gessner (1577), f. 119r/v
Aretius, Benedictus	1564.10.31		Gessner (1577), f. 119r
Aretius, Benedictus	1564.12.06		Gessner (1577), f. 118v
Aretius, Benedictus	1565.01.26		Gessner (1577), f. 119v–120r
Aretius, Benedictus	1565.02.04		Gessner (1577), f. 120r–121r

(*cont.*)

Name	Date (year, month, day)	Letter to Gessner (location)	Letter from Gessner (location)
Aretius, Benedictus	1565.03.02		Gessner (1577), f. 121r
Aretius, Benedictus	1565.08.22		Gessner (1577), f. 121r/v
Aretius, Benedictus	1565.11.24		Gessner (1577), f. 121v–122r
Bächi, Philipp	1560.12.23		Dedication letter Cordus (1561), f. 225r
Baldenberger, Jakob	1561.03.05	Vadiana MS. o-Fragm I.4	
Bale, John	1555.08.01		Dedication letter Gessner (1555c)
Bauhin, Johannes	1560.06.24		Bauhin (1591), pp. 95f.
Bauhin, Johannes	1560.07.14		Gessner (1577), p. 97
Bauhin, Johannes	1560.08.21		Bauhin (1591), pp. 97f.
Bauhin, Johannes	1560.10.21		Bauhin (1591), pp. 98f.
Bauhin, Johannes	1561.01.07		Bauhin (1591), pp. 99f.
Bauhin, Johannes	1561.04.20		Bauhin (1591), pp. 100f.
Bauhin, Johannes	1561.10.24		Bauhin (1591), pp. 102–104
Bauhin, Johannes	1561.11.09		Bauhin (1591), pp. 104–106
Bauhin, Johannes	1562 (vor)		Bauhin (1591), p. 110
Bauhin, Johannes	1562		Bauhin (1591), pp. 106f.

(*cont.*)

Name	Date (year, month, day)	Letter to Gessner (location)	Letter from Gessner (location)
Bauhin, Johannes	1562.08.13		Bauhin (1591), pp. 108f.
Bauhin, Johannes	1562.08.24		Bauhin (1591), pp. 107f.
Bauhin, Johannes	1562.09.30		Bauhin (1591), pp. 109f.
Bauhin, Johannes	1562.10.20	Bauhin (1591), pp. 115f.	
Bauhin, Johannes	1562.11.05		Bauhin (1591), pp. 111f.
Bauhin, Johannes	1562.12.05		Bauhin (1591), pp. 112–114
Bauhin, Johannes	1563.07.11		Bauhin (1591), pp. 116–119
Bauhin, Johannes	1563.07.19		Bauhin (1591), pp. 119f.
Bauhin, Johannes	1563.08.01		Bauhin (1591), pp. 122–124
Bauhin, Johannes	1563.08.09		Bauhin (1591), pp. 120–122
Bauhin, Johannes	1563.10.03		Bauhin (1591), p. 125
Bauhin, Johannes	1563.10.28		Bauhin (1591), pp. 128–133
Bauhin, Johannes	1563.11.14		Bauhin (1591), pp. 126f.
Bauhin, Johannes	1563.12.12		Bauhin (1591), pp. 133–136
Bauhin, Johannes	1564(?)		ZBZ, MS. F 38, f. 338r
Bauhin, Johannes	1564		Bauhin (1591), pp. 141f.
Bauhin, Johannes	1564.01.06		Bauhin (1591), p. 133

(*cont.*)

Name	Date (year, month, day)	Letter to Gessner (location)	Letter from Gessner (location)
Bauhin, Johannes	1564.01.07		Bauhin (1591), pp. 139–141
Bauhin, Johannes	1564.02.11		Bauhin (1591), pp. 143–145
Bauhin, Johannes	1564.02.15		Bauhin (1591), pp. 138f.
Bauhin, Johannes	1564.03.24		Bauhin (1591), pp. 147f.
Bauhin, Johannes	1564.04.05		Bauhin (1591), pp. 136–138
Bauhin, Johannes	1564.06.30		Bauhin (1591), pp. 145–147
Bauhin, Johannes	1564.11.05		Bauhin (1591), pp. 148f.
Bauhin, Johannes	1564.12.17		Bauhin (1591), p. 150
Bauhin, Johannes	1565.02.25		Bauhin (1591), pp. 151f.
Bauhin, Johannes	1565.03.16		Bauhin (1591), pp. 152–154
Bauhin, Johannes	1565.07.09		Bauhin (1591), pp. 154f.
Bauhin, Johannes	1565.08.31		Bauhin (1591), pp. 155f.
Bauhin, Johannes	[1565].10.11		Bauhin (1591), pp. 157–159
Bauhin, Johannes	1565.10.29		Bauhin (1591), pp. 159–161
Bauhin, Johannes senior	1565.10.06		Bauhin (1591), pp. 161f.
Bebel, Johannes and Isengrin, Michael	1548		Dedication letter Gessner (1548), f. 43r/v

(*cont.*)

Name	Date (year, month, day)	Letter to Gessner (location)	Letter from Gessner (location)
Beck von Beckenstein, Leonhard	1545.02.01		Dedication letter Thylesius (1545)
Beck von Beckenstein, Leonhard	1545.07.??		Dedication letter Gessner (1545)
Beck von Beckenstein, Leonhard	1548.08.23		Dedication letter Gessner (1548)
Beck von Beckenstein, Leonhard	1555.08.26		Gessner (1555c), Dedication letter to the plate
Belfort, Albert	1541.06.01		Dedication letter Gessner (1541a), f. 36v–38v
Belon, Cl.	1564.11.16	ZBZ, MS. C 50a, no. 45	
Berauld, François	1558.02.26	ZBZ, MS. C 50a, no. 51	
Berauld, François	1558.03.??		ZBZ, MS. C 50a, no. 48
Berg, Johann vom and Neuber, Ulrich	1548.02.02		Dedication letter Gessner (1548), f. 95r
Beza, Theodor	1549.08.22	CThB, Bd. 1, pp. 49–55	
Bifrun, Jakob	1556.01.27	Dedication letter Willich (1563), pp. 220–227	
Billing, Heinrich	1540.08.09		Dedication letter Gessner (1541)
Blarer, Thomas	1565.02.20	StAGR D V/37 C 36.06.37	
Boner, [Hans]	15??		ZBZ, MS. C 50a, no. 52
Brixius, Hieronymus	1561.12.22	MS. S 204a, f. 273v	
Brothers in Faith in Italy	1561.01.06		ZBZ, MS. S 99, no. 9
Brudo, Manuel	1544		Dedication letter Brudo (1544)

(*cont.*)

Name	Date (year, month, day)	Letter to Gessner (location)	Letter from Gessner (location)
Brunner, Fridolin	1564.07.21		ZBZ, MS. F 42, f. 98r
Brylinger, Nikolaus	1548		Dedication letter Gessner (1548), f. 59r
Bullinger, Heinrich	15??		ZBZ, MS. C 50a, no. 50a
Bullinger, Heinrich	15??	ZBZ, MS. C 50a, no. 50a	
Bullinger, Heinrich	15??		ZBZ, MS. C 50a, no. 61
Bullinger, Heinrich and Bibliander, Theodor	1532.08.10		HBBW 2, pp. 173f.
Bullinger, Heinrich	1532.10.??		HBBW 2, pp. 252f.
Bullinger, Heinrich and Bibliander, Theodor	1533.02.17		ZBZ, MS. E 25, between p. 480 and 481
Bullinger, Heinrich	1533.02.25		Gessner and Joh. Fries: HBBW 3, p. 74
Bullinger, Heinrich	1533.04.14		HBBW 3, pp. 105f.
Bullinger, Heinrich	1534.05.12	Gessner and Fries: HBBW 4, pp. 172f.	
Bullinger, Heinrich	1534.08.26		HBBW 4, pp. 290f.
Bullinger, Heinrich	1534.12.27		HBBW 4, pp. 457–461
Bullinger, Heinrich	1536.11.11		HBBW 6, p. 468
Bullinger, Heinrich	1536.12.30		HBBW 6, pp. 498f.

(*cont.*)

Name	Date (year, month, day)	Letter to Gessner (location)	Letter from Gessner (location)
Bullinger, Heinrich	1537.09.19		HBBW 7, pp. 241–243
Bullinger, Heinrich; Frey, Felix; Megander, Kaspar; Fabricius, Erasmus; Gwalther, Rudolf	1544.03.07		Dedication letter Martial (1544) = HBBW 14, pp. 116–121
Bullinger, Heinrich	1548/49 oder 1553		Calvin (1875), p. 726
Bullinger, Heinrich	1557.02.05		Dedication letter Athenagoras (1557), pp. 79f.
Bullinger, Heinrich	1558.09.??		ZBZ, MS. S 93, pp. 179–183
Bullinger, Heinrich	1564/1565		ZBZ, FA Gessner 1516.1
Bullinger, Heinrich	1565.05.??	StAZH E II 359, f. 3150	
Burcardus, Apollinaris	1560.05.22	ZBZ, MS. C 50a, no. 38	
Caius, John	1556.03.15		Dedication letter Gessner (1556), pp. 12–14
Caius, John	1561.08.29		Gessner (1577), f. 133v–136v
Caius, John	1565	Dedication letter Caius (1570), f. 1v–13r	
Calandrinus, Scipio	1561.05.28	StAGR D V/37 C 36.06.07	
Calvin, Johannes	1544.08.28		Calvin (1873), p. 745
Calzolari, Francesco	1561.06.01		Dedication letter Cordus (1561), f. 288r–289r

(*cont.*)

Name	Date (year, month, day)	Letter to Gessner (location)	Letter from Gessner (location)
Calzolari, Francesco	1561.06.01		Dedication letter Cordus (1561), f. 297v–298r
Camerarius, Joachim	1554.02.26		Freytag (1831), pp. 60–62
Camerarius, Joachim jun.	1558.08.05		Rath (1950), pp. 150f.
Camerarius, Joachim jun.	1564.07.17		Rath (1950), pp. 152f.
Camerarius, Joachim jun.	1564.08.30		Rath (1950), pp. 155f.
Camerarius, Joachim jun.	1565.01.27		Rath (1950), pp. 159f.
Camerarius, Joachim jun.	1565.08.27		Rath (1950), pp. 163f.
Camerarius, Joachim jun.	1565.09.22		Rath (1951), p. 195
Camerarius, Joachim jun.(?)	15??		Rath (1951), p. 197
Cardano, Gerolamo	1555.08.07	Salzmann (1956)	
Cassander, Georg	1562.10.09		ZBZ, MS. S 105, no. 132
Chytraeus, David	1553(?)		Omont (1891); the year 1543 is wrong
City Council Augsburg	1564.06.20		Dedication letter Dioscorides (1565)
City Council Basel	1560.04.30		Dedication letter Gessner (1560a), pp. 279f.
City Council Zürich	15??		StAZH A 77.9, no. 4 (Gessner together with C. Wolf and G. Keller)
City Council Zürich	1551.08.??		Dedication letter Gessner (1551)

(*cont.*)

Name	Date (year, month, day)	Letter to Gessner (location)	Letter from Gessner (location)
City Council Zürich	1558		ZBZ, MS. S 93, no. 181
City Council Zürich	1558.09.14		StAZH, E II 338, 1564f. (draft: ZBZ, MS. C 50a, no. 33)
Cognatus, Gilbertus	1555.02.26	ZBZ, MS. C 50a, no. 27	
Collin, Caspar	1560.12.23		Dedication letter Cordus (1561), f. 213r
Comte, Beat	1542.03.??		Dedication letter Gessner (1542b: Commentarius Porphyrii …)
Constantin, Robert	15??	StAGR D V/37 C 36.06.45	
Constantin, Robert	1562.12.12	StAGR D V/37 C 36.06.19	
Constantin, Robert	1563.03.16	StAGR D V/37 C 36.06.23	
Cozelius	15??		ZBZ, MS. C 50a, no. 70
Crato von Krafftheim, Johannes[2]	1559.10.18	Dedication letter Cordus (1561), f. b2r–b3r	
Crato von Krafftheim, Johannes	1560.10.06		Gessner (1577), f. 6v–7r
Crato von Krafftheim, Johannes	1561.01.01		Gessner (1577), f. 7r–8r
Crato von Krafftheim, Johannes	1561.03.17		Gessner (1577), f. 8r–9v
Crato von Krafftheim, Johannes	1561.08.16		Gessner (1577), f. 1r–2v

2 Concerning the lost letters see Huth/Walter (2019).

(*cont.*)

Name	Date (year, month, day)	Letter to Gessner (location)	Letter from Gessner (location)
Crato von Kraftheim, Johannes	1561/62	Sennert (1676), pp. 526f.	
Crato von Kraftheim, Johannes	1562		Gessner (1577), f. 2v–3v
Crato von Kraftheim, Johannes	1562.02.15		Gessner (1577), f. 10r/v
Crato von Kraftheim, Johannes	1562.03.01	StAGR D V/37 C 36.06.13	
Crato von Kraftheim, Johannes	1562(?).03.25	StAGR D V/37 C 36.06.41	
Crato von Kraftheim, Johannes	[1562].10.18		Gessner (1577), f. 9v–10r
Crato von Kraftheim, Johannes	1562.11.22	Dedication letter Dioscorides (1565), f. b1r–b7v	
Crato von Kraftheim, Johannes	1562.12.28		Gessner (1577), f. 10v–11v
Crato von Kraftheim, Johannes	1563.02.15		Gessner (1577), f. 11v–12r
Crato von Kraftheim, Johannes	1563.03.15		Gessner (1577), f. 12r/v
Crato von Kraftheim, Johannes	1563.04.24		Gessner (1577), f. 3v–6v

(*cont.*)

Name	Date (year, month, day)	Letter to Gessner (location)	Letter from Gessner (location)
Crato von Krafftheim, Johannes	1563.06.29	StAGR D V/37 C 36.06.27	
Crato von Krafftheim, Johannes	1563.08.01		Gessner (1577), f. 12v–13r
Crato von Krafftheim, Johannes	1563.08.09		Gessner (1577), f. 13r–14v
Crato von Krafftheim, Johannes	1563.11.12		Gessner (1577), f. 14v–15v
Crato von Krafftheim, Johannes	1563.12.06	StAGR D V/37 C 36.06.35	
Crato von Krafftheim, Johannes	1563.12.31		Treviranus (1830), S. 61f.
Crato von Krafftheim, Johannes	[1564]		Gessner (1577), f. 18v–19r
Crato von Krafftheim, Johannes	1564.01.08		Gessner (1577), f. 15v–16r
Crato von Krafftheim, Johannes	1564.02.06		Gessner (1577), f. 20r/v
Crato von Krafftheim, Johannes	1564.02.21		Gessner (1577), f. 16r–17r
Crato von Krafftheim, Johannes	1564.02.28	ZBZ, FA Gessner 1516: 2	

(*cont.*)

Name	Date (year, month, day)	Letter to Gessner (location)	Letter from Gessner (location)
Crato von Krafftheim, Johannes	1564.03.05		Gessner (1577), f. 17r–18r
Crato von Krafftheim, Johannes	1564.03.26		Gessner (1577), f. 21v–22v
Crato von Krafftheim, Johannes	1564.04.17		Gessner (1577), f. 19r–20r
Crato von Krafftheim, Johannes	1564.04.27	ZBZ, FA Gessner 1516: 2	
Crato von Krafftheim, Johannes	1564.05.21		Gessner (1577), f. 18r
Crato von Krafftheim, Johannes	1564.07.10		Gessner (1577), f. 19r
Crato von Krafftheim, Johannes	1564.07.10		Dedication letter Dioscorides (1565), f. b8r–c3r
Crato von Krafftheim, Johannes	1564.07.24		Gessner (1577), f. 20v–21v
Crespin, Jean	1562.07.25	StAGR D V/37 C 36.06.16	
Culmann, Johannes	1563.06.25		Gessner (1577), f. 44v–45v
Culmann, Johannes	1564.01.21		Gessner (1577), f. 45v–46r
Culmann, Johannes	1565.01.25		Gessner (1577), f. 46r/v

(*cont.*)

Name	Date (year, month, day)	Letter to Gessner (location)	Letter from Gessner (location)
Curio, Caelius Secundus	1562.02.01		UB Hamburg, Sup. ep. 59, p. 63, cf. Krüger (1978), p. 319
Curio, Caelius Secundus	1562.01.20	StAGR D V/37 C 36.06.10	
Curio, Caelius Secundus	1562.02.08		UB Basel, G I 66, f. 98
Curio, Hieronymus	1548.02.01		Dedication letter Gessner (1548), f. 87r
Daléchamps, Jacques	155?.12.01	StAGR D V/37 C 36.06.48	
Daléchamps, Jacques	1560.01.06		BNF, MS. lat. 13063, p. 264
Daléchamps, Jacques	1560.01.20		BNF, MS. lat. 13063, pp. 261f.
Daléchamps, Jacques	1561.01.24		BNF, MS. lat. 13063, pp. 259f.
Daléchamps, Jacques	1561.02.04	BNF, MS. lat. 13063, pp. 31–34	
Daléchamps, Jacques	1561.03.27	BNF, MS. lat. 13063, pp. 27–30	
Daléchamps, Jacques	1561.09.06		BNF, MS. lat. 13063, pp. 257f.
Daléchamps, Jacques	1562.01.08		Gessner (1746)
Donzellini, Girolamo	15??	ZBZ, MS. C 50a, no. 35	
Donzellini, Girolamo and Hessius, Johannes	1558.11.01	ZBZ, MS. C 50a, no. 31	
Donzellini, Girolamo	1559.08.07		ZBZ, MS. C 50a, no. 37/2
Dourez, Valerand	1563	Göbel (1565), f. i7r/v	

(*cont.*)

Name	Date (year, month, day)	Letter to Gessner (location)	Letter from Gessner (location)
Dourez, Valerand	1565.06.29		Göbel (1565), f. 30v–[31]v
Du Choul, Jean	15??	StAGR D V/37 C 36.06.46	
Du Choul, Jean	1559.01.30		Dedication letter Hanno Carthaginiensis (1559)
Duno, Thaddeo	1560.08.20		Gessner (1577), f. 88r/v
Echtius, Johannes	1559.09.25		ZBZ, MS. S 204, f. 72r
Echtius, Johannes	1559.09.25		Dedication letter Xenokrates, in: Dubravius (1559)
Elisabeth I	1560.06.13		Dedication letter Gessner (1560)
Episcopius, Nicolaus	1553.01.03	ZBZ, MS. C 50a, no. 59	
Episcopius, Nicolaus	1560.09.30	StAGR D V/37 C 36.06.01	
Episcopius, Nicolaus	1563.04.17	StAGR D V/37 C 36.06.25	
Erastus, Thomas	15??.10.13	ZBZ, MS. Z VII 119/2, f. 1r/v	
Erastus, Thomas	[1557]. 04.11	StAZH E II 361, f. 56	
Erastus, Thomas	1564.04.17 [nach]	Erastus (1590), pp. 68–73.	
Erastus, Thomas	1564.09.17	ZBZ, MS. Z VII 119/2, f. 2r–18r	
Erastus, Thomas	1564.11.17		ZBZ, MS. Z VII 119/2, f. 18r/v
Estienne, Henri	[1550].??.??		ZBZ, MS. C 50a, no. 54/1
Estienne, Henri		[1550].??.??	UB Bremen, msa 008, f. 237, 239 (draft or copy: ZBZ, MS. C 50a, no. 53)
Estienne, Henri	[1551].??.??		ZBZ, MS. C 50a, no. 54/2
Estienne, Henri		1551.02.05	ZBZ, MS. C 50a, no. 57
Estienne, Robert	1548.01.23		Dedication letter Gessner (1548), f. 73r

(*cont.*)

Name	Date (year, month, day)	Letter to Gessner (location)	Letter from Gessner (location)
Etschenreuter, Gallus	1561 (or later)	ZBZ, MS. S 204c, f. 92r/v	
Fabricius Montanus, Johannes	15??.06.04		Gessner (1577), f. 93r/v
Fabricius Montanus, Johannes	1553.01.22		ZBZ, MS. C 50a, no. 25/1
Fabricius Montanus, Johannes	1557.12.03	ZBZ, MS. C 50a, no. 25/2	
Fabricius Montanus, Johannes	[1558].08.02		Gessner (1577), f. 93v
Fabricius Montanus, Johannes	1559.06.16		Gessner (1577), f. 93v–94r
Fabricius Montanus, Johannes	1559.06.26	Dedication letter Cordus (1561), f. 235v	
Fabricius Montanus, Johannes	1559.07.18		Fischer (1940), pp. 331f.
Fabricius Montanus, Johannes	1559.09.22		StAZH E II 340, f. 233
Fabricius Montanus, Johannes	1560.07.12		Gessner (1577), f. 88v–89r
Fabricius Montanus, Johannes	1560.07.19		Gessner (1577), f. 89r/v
Fabricius Montanus, Johannes	1560.10.25		Gessner (1577), f. 88r/v

(*cont.*)

Name	Date (year, month, day)	Letter to Gessner (location)	Letter from Gessner (location)
Fabricius Montanus, Johannes	1561.03.07		Gessner (1577), f. 90v
Fabricius Montanus, Johannes	1561.03.20		Gessner (1577), f. 89v–90r
Fabricius Montanus, Johannes	1561.04.13		Gessner (1577), f. 90r
Fabricius Montanus, Johannes	1561.09.03		Gessner (1577), f. 90v–91r
Fabricius Montanus, Johannes	1561.12.21		Gessner (1577), f. 91r/v
Fabricius Montanus, Johannes	1562.02.06		ZBZ, MS. C 50a, no. 41/1
Fabricius Montanus, Johannes	1562.03.20		Gessner (1577), f. 91v–92r
Fabricius Montanus, Johannes	1562.10.12		Gessner (1577), f. 92r
Fabricius Montanus, Johannes	1562.10.23		ZBZ, MS. C 50a, no. 41/2
Fabricius Montanus, Johannes	1562.11.06		ZBZ, MS. C 50a, no. 41/3
Fabricius Montanus, Johannes	1563.07.16		Fischer (1946)

(*cont.*)

Name	Date (year, month, day)	Letter to Gessner (location)	Letter from Gessner (location)
Fabricius Montanus, Johannes	1563.08.06		ZBZ, MS. C 50a, no. 43/2
Fabricius Montanus, Johannes	1564.07.21		Gessner (1577), f. 92v–93r
Fabricius Montanus, Johannes	1564.09.25	ZBZ, MS. F 38, f. 316r–317v	
Fabricius, Georg	1551.03.01		UB Leipzig, MS. 0354 b, f. 68
Fabricius, Georg	1554.06.22		Gessner (1577), f. 131r–132r
Fabricius, Georg	1558.07.11		Gessner (1584), f. B2r/v
Fabricius, Georg	1563.07.25	StAGR D V/37 C 36.06.28	
Ferdinand I	1558.08.05		Dedication letter Gessner (1558)
Ferrerius, Johannes	1561.08.20	StAGR D V/37 C 36.06.09	
Forrer, Konrad	15??		Gessner (1577), f. 125r/v
Forrer, Konrad	15??		Gessner (1577), f. 125v–126r
Forrer, Konrad	15??		Gessner (1577), f. 127v–128r
Forrer, Konrad	15??	MS. S 204a, f. 136v	
Forrer, Konrad	1563.03.13		Gessner (1577), f. 126r/v
Forrer, Konrad	1565.01.05		Gessner (1577), f. 126v–127r
Forrer, Konrad	1565.08.31		Gessner (1577), f. 127r/v
Frellon, Jean	1548.06.16		Dedication letter Gessner (1548), f. 261r
Fricker, Hieronymus	1544.01.??		Dedication letter Herakleides Pontikos (1544)
Fries, Johann Jakob	15??	ZBZ, MS. C 50a, no. 47	
Fries, Johann Jakob	15??		Gessner (1577), f. 130r/v
Fries, Johann Jakob	1563.05.14		Gessner (1577), f. 129v

(*cont.*)

Name	Date (year, month, day)	Letter to Gessner (location)	Letter from Gessner (location)
Fries, Johann Jakob	1563.08.07		Gessner (1577), f. 129v–130r
Fries, Johannes	1558.11.09		Gessner (1577), f. 104r/v
Fries, Johannes	1561.03.26		ZBZ, MS. Briefe Gessner, no. 14
Froben, Hieronymus	1561 (vor)	StAGR D V/37 C 36.06.47	
Froben, Hieronymus	1561.01.15		UB Leiden, VUL 105: 2
Froben, Hieronymus and Episcopius, Nicolaus	1549.02.??		Dedication letter Gessner (1549)
Fröhlich, Georg	1546.02.01		Dedication letter Antonius Melissa (1546), p. 346
Froschauer, Christoph	1548.08.21		Dedication letter Gessner (1548), f. *5v
Fuchs, Leonhart	1556.10.18		Gessner (1577), f. 137v–138v
Fuchs, Leonhart	1557.02.11		Gessner (1577), f. 138v–139r
Fuchs, Leonhart	1558.10.01	StAZH E II 356, f. 169r	
Fugger, Johann Jakob	1546.02.05		Dedication letter Antonius Melissa (1546, latin translation)
Fugger, Johann Jakob	1555.03.15		Dedication letter Gessner (1555a)
Fugger, Johann Jakob	1556.04.26		Dedication letter Aelianus (1556)
Fugger, Johann Jakob	1558.07.10		ZBZ, MS. C 50a, no. 49
Fugger, Ulrich	1555.03.13		Dedication letter Gessner (1555b)
Fulgonus, Jacobus	1561.02.22	ZBZ, MS. F 12, f. 197v	
Funck, Johannes	1564		ZBZ, MS. F 17, f. 55r

(*cont.*)

Name	Date (year, month, day)	Letter to Gessner (location)	Letter from Gessner (location)
Funck, Johannes	1564,01.21		ZBZ, MS. S 109, no. 18
Funck, Johannes	1564.03.26		Gessner (1577), f. 94r/v
Funck, Johannes	1564.05.22		Gessner (1577), f. 94v
Funck, Johannes	1564.11.05		Gessner (1577), f. 96r–97r
Funck, Johannes	1564.11.19		Gessner (1577), f. 95r/v
Funck, Johannes	1564.12.17		Gessner (1577), f. 95v
Funck, Johannes	1565.01.21		Gessner (1577), f. 96r
Funck, Johannes	1565.05.02	MS. S 204a, f. 295v	
Gasser, Achilles Pirmin	1554.08.05		Burmeister (1975), pp. 134–136
Gasser, Achilles Pirmin	1562.06.07		Burmeister (1975), pp. 196f.
Gasser, Achilles Pirmin	1562.06.28		Burmeister (1975), pp. 200f.
Gasser, Achilles Pirmin	1562.07.??	Burmeister (1975), pp. 203–206	
Gasser, Achilles Pirmin	1563		Burmeister (1975), pp. 256f.
Gasser, Achilles Pirmin	1563.01.02		Burmeister (1975), p. 212
Gasser, Achilles Pirmin	1563.02.17	StAGR D V/37 C 36.06.21	
Gasser, Achilles Pirmin	1563.02.27		Burmeister (1975), pp. 214f.
Gasser, Achilles Pirmin	1563.03.??		Burmeister (1975), pp. 219–221
Gasser, Achilles Pirmin	1563.03.17		Burmeister (1975), pp. 223f.
Gasser, Achilles Pirmin	1563.04.??		Burmeister (1975), pp. 228f.
Gasser, Achilles Pirmin	1563.04.07		Burmeister (1975), p. 226
Gasser, Achilles Pirmin	1563.04.22		Burmeister (1975), pp. 230–232

(*cont.*)

Name	Date (year, month, day)	Letter to Gessner (location)	Letter from Gessner (location)
Gasser, Achilles Pirmin	1563.06.26		Burmeister (1975), pp. 236f.
Gasser, Achilles Pirmin	1563.08.11		Burmeister (1975), pp. 240–242
Gasser, Achilles Pirmin	1563.09.26		Burmeister (1975), pp. 245f.
Gasser, Achilles Pirmin	1563.11.11		Burmeister (1975), pp. 247–251
Gasser, Achilles Pirmin	1563.12.20		Burmeister (1975), pp. 258f.
Gasser, Achilles Pirmin	1564.01.08		Burmeister (1975), pp. 284f.
Gasser, Achilles Pirmin	1564.01.21		Burmeister (1975), pp. 287–289
Gasser, Achilles Pirmin	1564.02.12		Burmeister (1975), pp. 291–293
Gasser, Achilles Pirmin	1564.02.20		Burmeister (1975), pp. 296f.
Gasser, Achilles Pirmin	1564.03.19		Burmeister (1975), pp. 299–301
Gasser, Achilles Pirmin	1564.04.28		Burmeister (1975), pp. 304f.
Gasser, Achilles Pirmin	1564.06.24		Burmeister (1975), pp. 306–308
Gasser, Achilles Pirmin	1564.08.29		Burmeister (1975), pp. 319–321
Gasser, Achilles Pirmin	1564.09.16		Burmeister (1975), pp. 323f.
Gasser, Achilles Pirmin	1564.10.20		Burmeister (1975), pp. 326f.
Gasser, Achilles Pirmin	1564.11.06		Burmeister (1975), pp. 332f.
Gasser, Achilles Pirmin	1564.12.17		Burmeister (1975), pp. 335f.

(*cont.*)

Name	Date (year, month, day)	Letter to Gessner (location)	Letter from Gessner (location)
Gasser, Achilles Pirmin	1565.01.15		Burmeister (1975), pp. 338f.
Gasser, Achilles Pirmin	1565.02.25		Burmeister (1975), pp. 343f.
Gasser, Achilles Pirmin	1565.03.04		Burmeister (1975), pp. 345f.
Gasser, Achilles Pirmin	1565.03.31		Burmeister (1975), pp. 347–349
Gasser, Achilles Pirmin	1565.04.04	ZBZ, MS. S 204a, f. 49r	
Gasser, Achilles Pirmin	1565.04.17		Burmeister (1975), pp. 352f.
Gasser, Achilles Pirmin	1565.04.19		Burmeister (1975), pp. 355f.
Gasser, Achilles Pirmin	1565.07.29		Burmeister (1975), pp. 357f.
Gasser, Achilles Pirmin	1565.10.25		Burmeister (1975), pp. 361f.
Gasser, Achilles Pirmin	1565.12.03		Burmeister (1975), pp. 363–365
Gasser, Achilles Pirmin	1565.12.23	Burmeister (1975), p. 369	
Gerbelius, Nicolaus	1551.02.15		Fecht (1684), p. 858
Gerbelius, Nicolaus	1553.05.01	StAGR, D V/37 C 36.06.40	
Gessner, Jakob	1548.07.20		Dedication letter Barbaro (1548)
Giunta, Tommaso	1548.07.30		Dedication letter Gessner (1548), f. 329r
Giunta, Tommaso	1553.03.07		Dedication letter Gessner (1553c)
Göbler, Justin	1562.09.13	StAGR D V/37 C 36.06.18	

(*cont.*)

Name	Date (year, month, day)	Letter to Gessner (location)	Letter from Gessner (location)
Göbler, Justin	1563.09.16	StAGR D V/37 C 36.06.31	
Gratarolus, Wilhelm	15??.01.21		ZBZ, MS. Z VII 118, f. 1v
Gratarolus, Wilhelm	1556.05.31		Dedication letter Gessner (1556), pp. 245–247
Gratarolus, Wilhelm	1556.06.29	ZBZ, MS. F 60, f. 87r–88v	
Gratarolus, Wilhelm	1561.11.06		ZBZ, MS. F 12, f. 252v/253r
Gravius, Valentinus	1554.02.11		Dedication letter Gessner (1554)
Grey, John and Thomas	1553.07.28		Dedication letter Gessner (1553a)
[Gryllus], Laurentius	15??		StAZH E II 340, f. 372b
Gryllus, Laurentius	1555.07.01	Dedication letter Gessner (1555), f. A2r/v	
Gryphius, Antoine	1563.11.17	StAGR D V/37 C 36.06.34	
Gryphius, Sebastian	1548.03.04		Dedication letter Gessner (1548), f. 117r
Gugelberg von Moos, Daniel	1562.05.12	ZBZ, MS. S 204a, f. 51v	
Gwalther, Rudolf	1539.06.14		CdR, Bd. 5, pp. 333–335
Gymnich, Johann	1548.05.28		Dedication letter Gessner (1548), f. 237r
Hartung, Johannes	15??	StAGR D V/37 C 36.06.43	
Hauenreuter, Sebald	1559.09.24		Dedication letter Xenocrates, in: Dubravius (1559)
Hauenreuter, Sebald	1560.12.24		Dedication letter Cordus (1561), f. 229v

(*cont.*)

Name	Date (year, month, day)	Letter to Gessner (location)	Letter from Gessner (location)
Hauenreuter, Sebald	1561.01.15	UB Erlangen, MS. 2386/2, f. 233a r/v	
Hener, Johann	1558.02.04	MS. S 204b, f. 50r/v	
Herold, Hieronymus	1559.11.25		Dedication letter Cordus (1561), f. 85r/v
Herold, Hieronymus	1565.01.27		Rath (1951), pp. 197–199
Herold, Hieronymus	1565.04.04		Rath (1951), p. 205
Herwagen, Johannes	1548.04.24		Dedication letter Gessner (1548), f. 165r
Holzach, Cosmas	15??		Gessner (1577), f. 84r/v
Holzach, Cosmas	1553 (nach)		Gessner (1577), f. 82v–83r
Holzach, Cosmas	1553.06.22		Gessner (1577), f. 82r
Holzach, Cosmas	1554.09.06		Gessner (1577), f. 81v–82r
Holzach, Cosmas	1560.02.23		Gessner (1577), f. 83r/v
Holzach, Cosmas	1560.11.15		Gessner (1577), f. 83v–84r
Holzach, Cosmas	1561.05.30		Gessner (1577), f. 84v–85r
Holzach, Cosmas	1561.07.18		Gessner (1577), f. 85r/v
Holzach, Cosmas	1561.10.10		Gessner (1577), f. 85v
Holzach, Cosmas	1562.01.??	ZBZ, MS. C 50a, no. 24	
Holzach, Cosmas	1562.01.15		Gessner (1577), f. 86r
Holzach, Cosmas	1562.01.31		Gessner (1577), f. 85v–86r

(*cont.*)

Name	Date (year, month, day)	Letter to Gessner (location)	Letter from Gessner (location)
Holzach, Cosmas	1562.05.01		ZBZ, MS. S 104, no. 94
Holzach, Cosmas	1563.04.15		Gessner (1577), f. 86r/v
Holzach, Cosmas	1564.04.02		Gessner (1577), f. 86v–87r
Holzach, Cosmas	1564.04.14		Gessner (1577), f. 87r/v
Holzach, Cosmas	1564.09.24		Gessner (1577), f. 87v–88r
Hospinian, Johannes	1557.10.08		Gessner (1577), f. 102r
Hospinian, Johannes	1558.03.05		Gessner (1577), f. 102r/v
Hospinian, Johannes	1560.04.18	ZBZ, MS. C 50a, no. 39	
Hospinian, Johannes	1561.08.24		Gessner (1577), f. 102v
Hospinian, Johannes	1562.10.13		Gessner (1577), f. 102v–103r
Hospinian, Johannes	1563.04.08		Gessner (1577), f. 103r/v
Hospinian, Johannes	1563.10.03		Gessner (1577), f. 103v–104r
Hospinian, Johannes	1563.10.28	StAGR D V/37 C 36.06.33	
Hospinian, Johannes	1565.11.27		Gessner (1577), f. 104r
Huber, Johannes	1561.02.10	StAGR D V/37 C 36.06.04	
Huber, Johannes Chrysostomus	1555.08.28		Dedication letter Gessner (1555), p. 44
Hug, Jakob	1561.03.30	MS. S 204b, p. 57	
Hug, Jakob	1561.11.04	MS. S 204a, f. 51r	

(*cont.*)

Name	Date (year, month, day)	Letter to Gessner (location)	Letter from Gessner (location)
Hug, Johannes	15??	MS. S 204b, p. 85	
Hurtado de Mendoza, Diego	1545.08.??		Dedication letter Gessner (1545a)
Iacobus, Petrus et Stephanus	1541		Dedication letter Gessner (1541a)
Isengrin, Michael	1554.03.01		ÖNB, Cod. 9737i, f. 100r/v
Jordán, Tamás	1563.03.15	StAGR D V/37 C 36.06.22	
Jud, Leo	1532		ZBZ, MS. A 51, no. 53, f. 96r–97r
Keller, Georg	1554.03.06		Gessner (1577), f. 123r/v
Keller, Isaak	1556.11.14	ZBZ, MS. S 204b, f. 145r	
Kentmann, Johannes	1550.04.22		Gessner (1584), f. A1r/v
Kentmann, Johannes	1550.12.01		Gessner (1584), f. A1v/2r
Kentmann, Johannes	1551.03.03		Gessner (1584), f. A2r/v
Kentmann, Johannes	1554.06.22		Gessner (1584), f. A2v–3v
Kentmann, Johannes	1555.03.16		Gessner (1584), f. A3v–4r
Kentmann, Johannes	1556.10.29		Gessner (1584), f. A4r/v
Kentmann, Johannes	1558.03.17		Gessner (1584), f. B1r–B2r
Kentmann, Johannes	1558.08.25		Gessner (1584), f. B2v–3v
Kentmann, Johannes	1559.02.27		Gessner (1584), f. B3v–C1r
Kentmann, Johannes	1559.08.25		Gessner (1584), f. C1r/v
Kentmann, Johannes	1561.06.16		Gessner (1584), f. C1v–C2r

(*cont.*)

Name	Date (year, month, day)	Letter to Gessner (location)	Letter from Gessner (location)
Kentmann, Johannes	1561.08.28		Gessner (1584), f. B2v–C3r
Kentmann, Johannes	1562.01.15		Dedication letter Cassius (1562), f. A2r–A3v
Kentmann, Johannes	1562.03.04		Gessner (1584), f. C3v–C4r
Kentmann, Johannes	1563.03.05	StAGR D V/3.271, no. 13	
Kentmann, Johannes	1563.03.23		Gessner (1584), f. C4r/v
Kentmann, Johannes	1563.08.25		Gessner (1584), f. D1r–D2r
Kentmann, Johannes	[1564]	Dedication letter Kentmann (1565a)	
Kentmann, Johannes	1564.02.16		Gessner (1584), f. D2r
Kentmann, Johannes	1564.06.24		Gessner (1584), f. D3r–D4v
Kentmann, Johannes	1564.08.27		Gessner (1584), f. D3r
Kentmann, Johannes	1565.01.27		Gessner (1584), f. E1r/v
Kentmann, Johannes	1565.03.23		Gessner (1584), f. E1v–E2r
Kentmann, Johannes	1565.03.23		Gessner (1584), f. E2r–E3v
Kentmann, Johannes	1565.08.27		Gessner (1584), f. E2r–E4r
Kentmann, Johannes	1565.11.08	Dedication letter Kentmann (1565)	
Kessler, Josua	1551.05.05		Vadiana MS. 33,28
Kessler, Samuel	1561(?)		Gessner (1577), f. 128r/v

(*cont.*)

Name	Date (year, month, day)	Letter to Gessner (location)	Letter from Gessner (location)
Kessler, Samuel	1562.01.01	ZBZ, MS. S 204a, f. 42r	
Kessler, Samuel	1562.01.02		Gessner (1577), f. 128v–129v
Kessler, Samuel	1562.01.18	ZBZ, MS. S 204a, f. 42v	
Koch, Huldrych	1562.02.23	StAGR D V/37 C 36.06.12	
Kyber, Lucius	1553.03.31		Dedication letter from Gessner in Kyber (1553)
Lauraeus, Stephanus	1560.01.25		Dedication letter Cordus (1561), f. 236r
Lazius, Wolfgang	1563.09.26	Gessner and Fries: StAGR D V/37 C 36.06.32	
Lemnius, Levinus	1560.06.05		Dedication letter Gessner (1560a), p. 370
Leuschner, Christoph	1563	UB Erlangen, MS. 2386/2, f. 374 r/v	
Lismanino, Francesco	1563.11.23	ZBZ, MS. S 108, no. 128	
Lüpolt, Martin	15??.01.27	ZBZ, MS. S 204a, f. 48v	
Manuzio, Paolo	1548.02.13		Dedication letter Gessner (1548), f. 107r
Martin (Student)	1555	ZBZ, MS. S 204a, f. 211–211a	
Massarius, Hieronymus	1561 (vor)	Cordus (1561), f. 243r	
Mattioli, Pietro Andrea	1559.03.18	ZBZ, MS. C 50a, no. 36	
Mattioli, Pietro Andrea	1559.08.08		ZBZ, MS. C 50a, no. 37/1
Maximilian II	1560.06.13		Dedication letter Gessner (1560a), f. aa2r–aa3v
Medical Faculty Wittenberg	1561.05.30		Dedication letter Cordus (1561), f. a2r–a3r
Meurer, Wolfgang	1562.08.13		Dedication letter Cordus (1563)

(*cont.*)

Name	Date (year, month, day)	Letter to Gessner (location)	Letter from Gessner (location)
Meurer, Wolfgang	1564.01.01	Dedication letter Aristoteles (1564)	
Meurer, Wolfgang	1564.03.09	ZBZ, MS. S 204c, p. 12	
Millet, Claude	1541.08.??		Dedication letter Gessner (1541a)
Moibanus, Johannes	1561.04.30	StAGR D V/37 C 36.06.05	
Mundella, Aloysius	[1551].10.5.	ZBZ, MS. C 50a, no. 34/2	
Mundella, Aloysius	[1551].10.20	ZBZ, MS. C 50a, no. 34/1	
Mundella, Aloysius	1560.10.18	StAGR D V/37 C 36.06.03	
Mundella, Aloysius	1562.04.10	StAGR D V/37 C 36.06.15	
Munzinger, Johann Heinrich	15??		ZBZ, MS. C 50a, no. 70
Muralt, Johannes	1565.08.27		Rath (1951), pp. 206–209
Muralt, Martin von	15??		ZBZ, MS. C 50a, no. 50
Musculus, Wolfgang	1550.02.21 (terminus ante quem)		Stadtbibliothek Zofingen, Pa 14:1, 65 (73)
Musculus, Wolfgang	[1551].05.10	ZBZ, MS. C 50a, no. 34/2	
Myconius, Oswald	1532.12.14		Vadiana, MS. 33 (VBS 4), 97
Myconius, Oswald	1533.01.07		Vadiana, MS. 32 (VBS 3), 147
Myconius, Oswald	1533.04.14		Vadiana, MS. 33 (VBS 4), 22

(*cont.*)

Name	Date (year, month, day)	Letter to Gessner (location)	Letter from Gessner (location)
Myconius, Oswald	1534.12.21		Vadiana, MS. 32 (VBS 3), 272
Myconius, Oswald	1535.06.30		Vadiana, MS. 32 (VBS 3), 343
Myconius, Oswald	1536.04.17		Vadiana, MS. 33 (VBS 4), 24
Myconius, Oswald	1536.06.12		Vadiana, MS. 33 (VBS 4), 34
Myconius, Oswald	1536.07.03		Vadiana, MS. 33 (VBS 4), 37
Myconius, Oswald	1541.11.21		Vadiana, MS. 34 (VBS 5), 76
Mythobius, Burckhardt	155?		MS. C 50a, no. 27, f. 216v
Mythobius, Burckhardt	1556.03.23		Dedication letter Gessner (1556), pp. 94–96
Niedbruck, Caspar von	1555.03.07		Dedication letter Simmler (1555a)
Niedbruck, Caspar von	1555.03.22		ÖNB, Cod. 9737i, f. 199r/v
Niedbruck, Caspar von	1555.06.12	ÖNB, Cod. 9737i, f. 273r/v	
Niedbruck, Caspar von	1555.12.12	ÖNB, Cod. 9737i, f. 435r/v	
Niedbruck, Caspar von	1556.02.02		ÖNB, Cod. 9737k, f. 23r/25r
Niedbruck, Caspar von	1556.04.20		ÖNB, Cod. 10364, f. 9r/v
Niedbruck, Caspar von	1557.01.12		ÖNB, Cod. 9737k, f. 183r
Nüscheler, Heinrich	1535.03.24		ZBZ, MS. L 530, f. 608v
Obrecht, Didymus	1560.03.18		Gessner (1577), f. 114v/115r
Occo, Adolph	15??		Gessner (1577), f. 47r–48v

(*cont.*)

Name	Date (year, month, day)	Letter to Gessner (location)	Letter from Gessner (location)
Occo, Adolph	1555.07.11	ZBZ, MS. C 50a, no. 58	
Occo, Adolph	1563.12.26		Gessner (1577), f. 48v–50r
Occo, Adolph	1564(?).08.28		Gessner (1577), f. 66v–67r
Occo, Adolph	1564.01.22		Gessner (1577), f. 50r–51v
Occo, Adolph	1564.06.19		Gessner (1577), f. 51r–52r
Occo, Adolph	1564.10.05		Gessner (1577), f. 52r–53r
Occo, Adolph	1564.11.18		Gessner (1577), f. 65r–66v
Occo, Adolph	1564.12.12		Gessner (1577), f. 53r–55r
Occo, Adolph	1565.01.07		Gessner (1577), f. 59v–61v
Occo, Adolph	1565.01.15		Gessner (1577), f. 67v
Occo, Adolph	1565.02.03	ZBZ, MS. C 50a, no. 69	
Occo, Adolph	1565.02.04		Gessner (1577), f. 61v–62r
Occo, Adolph	1565.02.18		Gessner (1577), f. 67v–69v
Occo, Adolph	1565.04.03		Gessner (1577), f. 69v–73v
Occo, Adolph	1565.04.18		Gessner (1577), f. 73v–75r
Occo, Adolph	1565.05.05		Gessner (1577), f. 62r–63r
Occo, Adolph	1565.06.16		Gessner (1577), f. 63v
Occo, Adolph	1565.07.07		Dedication letter Epiphanius (1565)
Occo, Adolph	1565.07.08		Gessner (1577), f. 63v–64v
Occo, Adolph	1565.08.27		Gessner (1577), f. 75r–76r
Occo, Adolph	1565.09.02		Gessner (1577), f. 76v
Occo, Adolph	1565.09.26		Gessner (1577), f. 64v–65r
Occo, Adolph	1565.09.28		Gessner (1577), f. 76v–79r
Occo, Adolph	1565.10.25		Gessner (1577), f. 76r
Occo, Adolph	1565.11.05		Gessner (1577), f. 79r–81r
Occo, Adolph	1565.12.04		Gessner (1577), f. 55v–58r
Occo, Adolph	1565.12.30	Gessner (1577), f. 58v–59v	
Oporinus, Johannes	1548.01.05		Dedication letter Gessner (1548), f. 49r/v
Oporinus, Johannes	1556.06.25	ZBZ, MS. C 50a, no. 29	
Ostrorog, Wenzel von	1563.08.14	StAGR D v/37 C 36.06.30	

(*cont.*)

Name	Date (year, month, day)	Letter to Gessner (location)	Letter from Gessner (location)
Parkhurst, John	1559.05.21	Robinson (1842), pp. 31f.	
Parkhurst, John	1560.02.26		Dedication letter Gessner (1560b), p. 128
Pellizerus, Andreas	1565.06.30		Dedication letter Cordus (1565)
Petreius, Johannes	1548.01.23		Dedication letter Gessner (1548), f. 77r
Petri, Heinrich	1548.01.30		Dedication letter Gessner (1548), f. 81r
Peyer, Alexander	1554.05.16	ZBZ, MS. S 204b, p. 42	
Pfäfferlin, Christoph	1560.12.22		Dedication letter Cordus (1561), f. 232r
Phrygius, Paulus Constantinus	1565.06.16	StAGR D V/37 C 36.06.39	
Pictorius, Georg	1550.10.10	ZBZ, MS. S 204a, f. 150r–151v	
Placotomus, Johannes	15??.03.26	StAGR D V/37 C 36.06.42	
Placotomus, Johannes	1557.03.27		Gessner (1577), f. 136v–137v
Placotomus, Johannes	1559.12.13		Dedication letter Cordus (1561), f. 217r
Placotomus, Johannes	1560.12.22		Dedication letter Cordus (1561), f. 232r
Platter, Felix	15??		Landesbibliothek Hannover, LBr. 809, f. 80
Platter, Felix	15??		Gessner (1577), f. 101v–102r
Platter, Felix	1558.10.31		Gessner (1577), f. 97r/v
Platter, Felix	1559.01.16		Gessner (1577), f. 97v
Platter, Felix	1561.05.16		Gessner (1577), f. 97v/98r
Platter, Felix	1562		Gessner (1577), f. 98r
Platter, Felix	1563.03.22		Gessner (1577), f. 98r
Platter, Felix	1563.10.17		Gessner (1577), f. 98v/99r
Platter, Felix	1564.08.23		Gessner (1577), f. 99r–100r

(*cont.*)

Name	Date (year, month, day)	Letter to Gessner (location)	Letter from Gessner (location)
Platter, Felix	1564.11.06		Gessner (1577), f. 100r/v
Platter, Felix	1564.12.20		Gessner (1577), f. 100v–101r
Platter, Felix	1565.01.28		Gessner (1577), f. 101r
Platter, Felix	1565.05.05		Gessner (1577), f. 101v
Pontisella, Johannes	1562.08.07		ZBZ, MS. C 50a, no. 57a; Bernhard/Müller (2021)
Pontisella, Johannes	1563.08.05		Dedication letter Willich (1563), f. *1v–*7r
Pontisella, Johannes	1563.08.30	StAGR D V/37 C 36.06.29	
Portus, Franciscus	1561.04.20	StAGR D V/37 C 36.06.06	
Portus, Franciscus	1562.07.20	StAZH E II 368, f. 262	
Portus, Franciscus	1563.02.06		Gessner (1577), f. 133r/v
Querculus, Thomas	15??.11.30	UB Bremen, msa 0008, f. 241	
Ralla, Johannes	1561.08.01		Dedication letter Cordus (1561), f. a3v–a4r
Raphael, Antoine de	1565.08.29		Bauhin (1591), pp. 162f.
Ribit, Jean	1542.03.??		Dedication letter Gessner (1542b: Commentarius Procli …)
Ribit, Jean	1547.10.01	BNF MS. Latin 8641, f. 9v	
Ribit, Jean	1548.04.17	BNF MS. Latin 8641, f. 17r	
Ribit, Jean	1550.04.14	BNF MS. Latin 8641, f. 37v	
Ribit, Jean	1552.06.13	BNF MS. Latin 8641, f. 60r	
Ribit, Jean	1554.01.18	BNF MS. Latin 8641, f. 69v–70r	
Rihel, Wendelin	1548.02.05		Dedication letter Gessner (1548), f. 99r
Ritter, Matthias	1548.07.10	ZBZ, MS. C 50a, no. 55	
Ritter, Matthias	1549.03.31		Gessner (1577), f. 123v–124v

(*cont.*)

Name	Date (year, month, day)	Letter to Gessner (location)	Letter from Gessner (location)
Röist, Jakob and Markus; Stapfer, Jakob and Markus; Haab, Jakob; Grebel, Georg; Escher, Georg; Engelhart, Felix; Reublin, Heinrich and Georg; Meier, Wilhelm and Gerold	1549.01.01		Dedication letter Gessner (1549a)
Rondelet, Guillaume	1562.06.30	StAZH E II 368, f. 550	
Ruf, Jakob	1556.03.01	Dedication letter Ruf (1556)	
Sambucus, Johannes	1560.01.18	ZBZ, MS. F 61, f. 99r/v	
Sambucus, Johannes	1562.02.13	StAGR D V/37 C 36.06.11	
Scaliger, Sylvius Caesar	[1560].12.28	Gessner (1577), f. 132r/v	
Scaliger, Sylvius Caesar	1561		Gessner (1577), f. 132v–133r
Schegk, Jakob	1561.06.05	StAZH E II 361, f 287	
Schneeberger, Anton	1562.01.23		Dedication letter Cassius (1562), f. 29v–30v
Scoto, Girolamo	1548.07.15		Dedication letter Gessner (1548), f. 311r
Seiler, Gereon	1555.01.05		Dedication letter Gessner (1555d)
Seman, Maximilian	1556.07.13	ZBZ, MS. C 50a, no. 28	

(*cont.*)

Name	Date (year, month, day)	Letter to Gessner (location)	Letter from Gessner (location)
Siegmund Freiherr von Heberstein	1560.05.23		Dedication letter Gessner (1560a), p. 338
Sinckeler, Sebastian	1541.08.??		Dedication letter Gessner (1541b: De iuventute ...)
Skalic, Paul	1560.10.09	StAGR D v/37 C 36.06.02	
Skalic, Paul	1561.03.04		Glesinger (1950), pp. 40f.
Speicher, Nikolaus	1553.04.04		Dedication letter Gessner (1553b)
Steiger, Johann	1554.02.18		Gessner (1554), Appendix, p. 1
Stenglin, Lukas	1562.08.01	StAGR D v/37 C 36.06.17	
Stucki, Johann Wilhelm	1560.02.23	StAGR D v/37 C 36.06.50	
Stucki, Johann Wilhelm	1563		Rath (1951), pp. 213f.
Stucki, Johann Wilhelm	1564.10.02		Gessner (1577), f. 124v
Stucki, Johann Wilhelm	1565.08.29		Gessner (1577), f. 124v–125r
Suchten, Alexander von	1564.02.08		Stadtbibliothek Danzig, Hs. no. 2318
Sulzer, Simon	1561.08.13	StAGR D v/37 C 36.06.08	
Sulzer, Simon	1563.04.30	StAGR D v/37 C 36.06.26	
Szadkowski, Andreas	1565.07.28		Dedication letter Gessner (1565)
Toxites, Michael	1565.03.??	ZBZ, MS. C 50a, no. 22	
Tschudi, Aegidius	1541.08.??		Dedication letter Gessner (1541b: De longitudine ...)
Turner, William	1557.11.01	Gessner (1558), pp. 1294–1296	
Turner, William	1562.08.28		Dedication letter Gessner (1562)
Ursinus, Zacharias	1561.09.15	Benrath (1964), pp. 99f.	

(*cont.*)

Name	Date (year, month, day)	Letter to Gessner (location)	Letter from Gessner (location)
Ursinus, Zacharias	1561.12.10	ZBZ, MS. F 12, f. 292r	
Ursinus, Zacharias	1562.03.22	Benrath (1964), p. 100	
Vadian, Joachim	1543.04.07		Arbenz (1908), vol. 6, pp. 211f.
Vadian, Joachim	1545.03.18		Arbenz (1908), p. 405
Vadian, Joachim	1551.01.13		Arbenz (1908), p. 904f.
Valgrisi, Vincenzo	1548.07.10		Dedication letter Gessner (1548), f. 303r
Viret, Pierre	1542.03.??		Dedication letter Gessner (1542b)
Vogel, Jakob	1541.06.??		Dedication letter Gessner (1541c)
Vulteius, Justus	1563.04.03	StAZH E II 367, f. 195	
Wattenwyl, Hans Jakob and Nägeli, Hans Franz	1543.06.13		Dedication letter Stobaeus (1543)
Wechel, Christian	1548.04.05		Dedication letter Gessner (1548), f. 165r
Wegmann, Johann and Peier, Felix	1556.12.28		Dedication letter Gessner (1556a)
Welser, Johannes and Herbrot, Jakob	1546.02.01		Dedication letter Antonius Melissa (1546)
Werdmüller, Abel	1555.02.20		Dedication letter Fabricius Montanus (1555), pp. 121–126
Werdmüller, Otto	1542.08.??		Dedication letter Gessner (1542c)
Werthern, Anton von	1559.02.??		Dedication letter Marcus Aurelius (1559)

(*cont.*)

Name	Date (year, month, day)	Letter to Gessner (location)	Letter from Gessner (location)
Wieland, Melchior	1556.05.31		Dedication letter Gessner (1556), pp. 245ff.
Wieland, Melchior	1556.10.20	Gessner/Guilandinus (1557), pp. 7–23	
Wieland, Melchior	1557.03.10		Gessner/Guilandinus (1557), pp. 24–45
Wieland, Melchior	1562.05.27		ZBZ, MS. C 50a, no. 42
Wirz, Melchior	1541.08.??		Dedication letter Gessner (1541b: De divinatione …)
Wolf, Caspar	1543(?)	Clauser and Gessner: Wehrli (1924), pp. 107f.	
Wolf, Caspar	[1558].10.17		Gessner (1577), f. 122r
Wolf, Caspar	1557.04.19		Gessner (1577), f. 122r/v
Wolf, Caspar	1560.09.01		Gessner (1577), f. 123r
Wolf, Johannes	1543	Gessner and Christoph Clauser: ZBZ, MS. S 53, no. 220	
Wolf, Johannes	1564		ZBZ, MS. F 38, f. 370r/v
Xylander, Wilhelm	1558.07.30	Gessner and Fries, StAZH E II 356, 1033	
Zanchi, Girolamo	[1556]	Zanchi (1619), p. 173	
Zanchi, Girolamo	1556.06.11		Zanchi (1619), p. 173
Zanchi, Girolamo	1562.12.15		Zanchi (1619), p. 132
Zanchi, Girolamo	1563.04.01	StAGR D V/37 C 36.06.24	
Zurkinden, Nikolaus	1552		Dedication letter Gessner (1552)
Zurkinden, Nikolaus	1557.09.20	ZBZ, MS. S 204b, f. 93r	

(*cont.*)

Name	Date (year, month, day)	Letter to Gessner (location)	Letter from Gessner (location)
Zurkinden, Nikolaus	1558		Gessner (1577), f. 130v–131r
Zwinger, Theodor	15??.12.14		Gessner (1577), f. 105v–106r
Zwinger, Theodor	15??.12.16		Gessner (1577), f. 106r/v
Zwinger, Theodor	1560.02.30 [sic!]		Gessner (1577), f. 106v
Zwinger, Theodor	1560.06.07		Gessner (1577), f. 104v–105r
Zwinger, Theodor	1560.12.07		Gessner (1577), f. 105r–106v
Zwinger, Theodor	1562.10.21		Gessner (1577), f. 106v–107r
Zwinger, Theodor	1563.03.22		Gessner (1577), f. 110v–111r
Zwinger, Theodor	1564.02.05	StAGR D v/37 C 36.06.36	
Zwinger, Theodor	1564.02.15		Gessner (1577), f. 109v–110r
Zwinger, Theodor	1564.04.07		Gessner (1577), f. 107r/v
Zwinger, Theodor	1564.04.24		Gessner (1577), f. 107v–108v
Zwinger, Theodor	1564.07.14		Gessner (1577), f. 108v–109r
Zwinger, Theodor	1564.07.22		Gessner (1577), f. 109r
Zwinger, Theodor	1564.12.13		Gessner (1577), f. 109r/v
Zwinger, Theodor	1565	StAGR D v/37 C 36.06.49	

(*cont.*)

Name	Date (year, month, day)	Letter to Gessner (location)	Letter from Gessner (location)
Zwinger, Theodor	1565.02.11	StAGR D V/37 C 36.06.38	
Zwinger, Theodor	1565.02.23		Gessner (1577), f. 110r/v
Zwinger, Theodor	1565.04.08		Gessner (1577), f. 111r–112r
Zwinger, Theodor	1565.06.14		Gessner (1577), f. 112r/v
Zwinger, Theodor	1565.08.09		Gessner (1577), f. 112v
Zwinger, Theodor	1565.08.28		Gessner (1577), f. 113r
Zwinger, Theodor	1565.11.26		Gessner (1577), f. 113r/v
Zwinger, Theodor	1565.11.30		Gessner (1577), f. 114r/v
Zwinger, Theodor(?)	1565(?)		UB Basel, Frey-Gryn Mscr I 12: no. 170
Zwinger, Theodor(?)	1565		UB Basel, Frey-Gryn Mscr I 13: no. 83
Zwingli, Huldrych	[1530].10.27		Zwingli (1929), pp. 324f.

Literature

Aelianus, Claudius (1556): Opera ..., ed. by Conrad Gessner, Zürich.

Aeschimann, David et al. (2004): Flora alpina. Ein Atlas sämtlicher 4500 Gefässpflanzen der Alpen, vol. 2, Bern etc.

Alt, Karin (1998): Homers Nymphengrotte in der Deutung des Porphyrios, in: Hermes 126, pp. 466–487.

Ammann, Jost (1960): Das Ständebuch, Insel-Bücherei, no. 133, Frankfurt/M.

Andrés de Uztarroz, Juan Francisco (1878): Progresos de la historia en Aragon y vidas de sus cronistas [...], Zaragoza.

Arbenz, Emil (1908): Die Vadianische Briefsammlung der Stadtbibliothek St Gallen, vol. 6, Mitteilungen zur vaterländischen Geschichte 30, Dritte Folge, St Gallen.

Arieti, Stefano (2014): Amato Lusitano: un grande clinico del XVI secolo, in: Medici ebrei e la cultura ebraica a Ferrara, Ferrara, pp. 52–61.

Aristophanes (1910): Werke, übersetzt von Ludwig Seeger, vol. 3, Stuttgart.

Aristoteles (1564): ΠΕΡΙ ΨΥΧΗΣ ΒΙΒΛΙΑ ΤΡΙΑ. Praefixa est praefatio ... autore Wolfgango Meurero, Leipzig.

Aristoteles (1957): Tierkunde, herausgegeben von Paul Gohlke, Paderborn.

Athenagoras von Athen (1557): Apologia pro Christianis ... eiusdem, De resurrectione mortuorum ..., Zürich.

Aubert, Hippolyte; Aubert, Fernand et Meylan, Henri (1960): Correspondance de Théodore de Bèze, Tome I (1539–1555), Travaux d'humanisme et renaissance 40, Geneva.

Backus, Irena (1990/1991): Randbemerkungen Zwinglis in den Werken von Giovanni Pico della Mirandola, in: Zwingliana 18, pp. 291–309.

Bächtold, Hans Ulrich (1982): Heinrich Bullinger vor dem Rat. Zur Gestaltung und Verwaltung des Zürcher Staatswesens in den Jahren 1531 bis 1575, Zürcher Beiträge zur Reformationsgeschichte 12, Bern und Frankfurt/M.

Bächtold, Hans Ulrich et al. (1999): Schola Tigurina. Die Zürcher Hohe Schule und ihre Gelehrten um 1550. Katalog zur Ausstellung vom 25. Mai bis 10. Juli 1999 in der Zentralbibliothek Zürich, Zürich und Freiburg i. Br.

Barbieri, Edoardo and Irace, Erminia (2010): L'Italie degli incunaboli, in: Sergio Luzzatto and Gabriele Pedullà (eds.): Atlante della letteratura italiana, vol. 1, Turin, pp. 525–530.

Barbaro, Ermolao (1548): Naturalis scientiae totius compendium ..., ed. by Conrad Gessner, Basel.

Barbero, Alessandro (2022): Inventare i libri. L'avventura di Filippo e Lucantonio Giunta, pionieri dell'editoria moderna, Florence.

Bath, Michael (2008): Emblems for a Queen. The Needlework of Mary Queen of Scots, London.

Baudrier, Henri et Julien (1964/65): Bibliographie Lyonnaise. Recherches sur les imprimeurs, libraires, relieurs et fondeurs de lettres de Lyon, 13 tomes, Paris.

Bauhin, Johannes (1591): De plantis a divis sanctisve nomen habentibus … additae sunt Conradi Gesneri Medici Clarissimi Epistolae hactenus non editae, Basel.

Baumann, Brigitte et al. (2001): Die Kräuterbuchhandschrift des Leonhart Fuchs, Stuttgart.

Baumann, Dieter (1953): Psychiatrisches bei Conrad Gessner, in: Gesnerus 10, pp. 123–150.

Bäumer, Änne (1990): De piscibus et aquatilibus libelli III novi (Zürich 1556). Ein bisher unbeachtetes zoologisches Werk von Conrad Gesner, in: Berichte zur Wissenschaftsgeschichte 13, pp. 177–181.

Bay, J. Christian (1916): Gesner, the Father of Bibliography, Papers of the Bibliographical Society of America 10, pp. 53–88.

Becker, W. (1892): Zacharias Ursins Briefe an Crato von Crafftheim, in: Theologische Arbeiten aus dem rheinischen wissenschaftlichen Prediger-Verein 12, pp. 41–107.

Benrath, Gustav Adolf (1964): Briefe des Heidelberger Theologen Zacharias Ursinus (1534–1583), in: Heidelberger Jahrbücher 8, pp. 93–141.

Bernhard, Jan-Andrea (2009): Conrad Gessner Ungarn: Kommunikations- und bibliotheksgeschichtliche Erkenntnisse, in: Christian Moser and Peter Opitz (eds.): Bewegung und Beharrung. Aspekte des reformierten Protestantismus, 1520–1650, Festschrift für Emidio Campi, Studies in the History of Christian Traditions 144, Leiden and Boston, pp. 159–180.

Bernhard, Jan-Andrea (2015): Konsolidierung des reformierten Bekenntnisses im Reich der Stephanskrone. Ein Beitrag zur Kommunikationsgeschichte zwischen Ungarn und der Schweiz in der frühen Neuzeit (1500–1700), Refo 500 Academic Studies 19, Göttingen.

Bernhard, Jan-Andrea (2019): Wasser, Kräuter, Käse und eine Prise Rätoromanisch: Bemerkungen und Erkenntnisse zu Gessners Kontakten mit den Drei Bünden, in: Leu and Opitz (2019), pp. 337–354.

Bernhard, Jan-Andrea and Müller, Clemens (2021): Der Gelehrte, der Schüler und ein Gimpelpaar: Ein vergessener griechischer Brief von Conrad Gessner an Johannes Pontisella III., in: Zwingliana 48, S. 129–175.

Bernstein, Lawrence F. (1973): The Bibliography of Music in Conrad Gesner's Pandectae (1548), in: Acta Musicologica 45, pp. 119–163.

Biaudet, Jean Charles (1982): Histoire de Lausanne, Lausanne.

Bibliander, Theodor (1548): De ratione communi linguarum et literarum commentarius, Zürich.

Bibliander, Theodor (1551): Amplior consideratio decreti synodalis Tridentini, Zürich.

Bifrons, Jakob (1563): De operibus lactariis, in: Iodocus Willich: Ars magirica …, Zürich, pp. 220–227.

Binder, Georg (1535): Acolastus. Ein Comoedia von dem Verlornen Sun / Luc. am 15. vertütscht unnd gehalten zuo Zürich im jar 1535, Zürich.

Blair, Ann (2010): Too Much to Know. Managing Scholarly Information before the Modern Age, New Haven, CT.

Blair, Ann (2016): Conrad Gessner's Paratexts, in: Gesnerus 73 (2016), pp. 73–122.

Blair, Ann (2017): The Dedication Strategies of Conrad Gessner, in: Gideon Manning and Cynthia Klestinec (eds.): Professors, physicians and practices in the history of medicine. Essays in honor of Nancy Siraisi, Cham, pp. 197–231.

Blair, Ann (2019): A new quire for fish: Gessner's *Halieuticon* and *De piscibus* (1556), in: Leu/Opitz (2019), pp. 535–555.

Blair, Ann (2021): L'entour du texte. La publication du livre savant à la Renaissance, Paris.

Bock, Hieronymus (1552): De stirpium, maxime earum, quae in Germania nostra nascuntur … libri tres, Strasbourg.

Bodenmann, Reinhard (2000): Wolfgang Musculus (1497–1563). Destin d'un autodidacte lorrain au siècle des Réformes, Travaux d'Humanisme et Renaissance 343, Geneva.

Bodenmann, Reinhard (2016): Les perdants. Pierre Caroli et les débuts de la Réforme en Romandie, Turnhout.

Boscani Leoni, Simona (2019): Conrad Gessner and a Newly Discovered Enthusiasm for Mountains in the Renaissance, in: Leu/Opitz (2019), pp. 119–128.

Boucher, Jacqueline (2001): Vivre à Lyon au XVIe siècle, Lyon.

Braun-Blanquet, Josias and Rübel, Eduard (1932): Flora von Graubünden. Vorkommen, Verbreitung und ökologisch-soziologisches Verhalten der wildwachsenden Gefässpflanzen Graubündens und seiner Grenzgebiete, Veröffentlichungen des Geobotanischen Institutes Rübel in Zürich, 7. Heft, Bern and Berlin.

Brésard, Marc (1914): Les foires de Lyon aux XVe et XVIe siècles, Paris.

Brudo, Manuel (1544): Liber de ratione victus in singulis febribus …, Venice.

Brunnschweiler, Thomas (1989): Johann Jakob Breitingers „Bedencken von Comoedien oder Spilen". Die Theaterfeindlichkeit im Alten Zürich, Edition – Kommentar – Monographie, Zürcher Germanistische Studien 17, Bern etc.

Bucher, Otto (1945): Die Anfänge der wissenschaftlichen Anatomie in Zürich, in: Gesnerus 2, pp. 131–141.

Büchi, Albert (1928): Glareans Schüler in Paris (1517–1522) nebst 15 ungedruckten Briefen, in: Emmanuel Scherer et al. (eds.): Aus Geschichte und Kunst. Robert Durrer. Zur Vollendung seines sechzigsten Lebensjahres dargeboten, Stans, pp. 372–431.

Büchi, Jakob (1983): Conrad Gessner (1516–1565) als Zürcher Stadtarzt, in: Fehlmann, Hans-Rudolf and Ledermann, François (eds.): Festschrift für A. Lutz und J. Büchi, Veröffentlichungen der Schweizerischen Gesellschaft für Geschichte der Pharmazie 2, Zürich, pp. 113–138.

Büchi, Jakob (1984): Die Entwicklung der Rezept- und Arzneibuchliteratur. 2. Teil: Die Autoren, ihre Werke und die Fortschritte im 16. Jahrhundert, Veröffentlichungen der Schweizerischen Gesellschaft für Geschichte der Pharmazie 4, Zürich.

Büeler, Gustav (1920): Petrus Dayspodius (Peter Hasenfratz). Ein Frauenfelder Humanist des XVI. Jahrhunderts, Beilage zum Programm der thurgauischen Kantonsschule, Schuljahr 1919/20, Frauenfeld.

Bührer, Peter (2001): Johannes Fries (1505–1565). Pädagoge, Philologe, Musiker. Leben und Werk, in: Zürcher Taschenbuch 2002, Neue Folge, Hundertundzweiundzwanzigster Jahrgang, Zürich, pp. 151–231.

Buess, Heinrich (1948): Conrad Gessners Beziehungen zu Basel, in: Gesnerus 5, pp. 1–29.

Büsser, Fritz (1960): Das Bild der Natur bei Zwingli, in: Zwingliana 11, pp. 241–256.

Büsser, Fritz (1971): Zwingli und Laktanz (Beobachtungen bei der Lektüre von Zwinglis „De providentia Dei"), in: Zwingliana 13, pp. 375–399.

Büsser, Fritz (1999): „Prophezei" – „Schola Tigurina". Prototyp, Ideal und Wirklichkeit, in: Hans Ulrich Bächtold (Red.): Schola Tigurina. Die Zürcher Hohe Schule und ihre Gelehrten um 1550. Katalog zur Ausstellung vom 25. Mai bis 10. Juli 1999 in der Zentralbibliothek Zürich, Zürich und Freiburg i. Br., S. 18–21.

Büsser, Fritz (2005): Heinrich Bullinger. Leben, Werk und Wirkung, 2 vols., Zürich 2005.

Buffière, Felix (1973): Les mythes d'Homère et la pensée grecque, Paris.

Bullinger, Heinrich (1574): De origine erroris et de conciliis. Das ist: Vom Ursprung, Aufkommung und Fuergang aller Irrthumben, so je bey den Heiden, Juden unnd Christen gewesen und noch sind ..., Heidelberg.

Bullinger, Heinrich (1904): Heinrich Bullingers Diarium der Jahre 1504–1574, ed. by Emil Egli, Basel.

Bullinger, Heinrich (1987): Studiorum ratio – Studienanleitung, ed. by Peter Stotz, Heinrich Bullinger Werke, Sonderband, In zwei Teilbänden, Zürich.

Burckhardt, Andreas (1967): Johannes Basilius Herold. Kaiser und Reich im protestantischen Schrifttum des Basler Buchdrucks um die Mitte des 16. Jahrhunderts, Basel and Stuttgart.

Burmeister, Karl Heinz (1964): Briefe Sebastian Münsters lateinisch und deutsch, Ingelheim am Rhein.

Burmeister, Karl Heinz (1975): Achilles Primin Gasser 1505–1577. Arzt, Naturforscher, Historiker und Humanist, III. Briefwechsel, Wiesbaden.

Burmeister, Karl Heinz (2015): Magister Rheticus und seine Schulgesellen. Das Ringen um Kenntnis und Durchsetzung des heliozentrischen Weltsystems des Kopernikus um 1540/50, Forschungen zur Geschichte Vorarlbergs 11 (N.F), Constance and Munich.

Caius, John (1570): De canibus britannicis libellus, London.

Calvin, Jean (1873): Corpus Reformatorum 39 = Calvini Opera 11, Braunschweig.

Calvin, Jean (1875): Corpus Reformatorum 41 = Calvini Opera 13, Braunschweig.

Calvin, Jean (1875a): Corpus Reformatorum 42 = Calvini Opera 14, Braunschweig.

Camerarius, Joachim d.J. (1586): Kreutterbuch Deß hochgelehrten unnd weitberuehmten Herrn D. Petri Andreae Mattioli, Jetzt widerumb mit viel schoenen neuwen Figuren …, Frankfurt/M.

Campi, Emidio (2004): Bullingers Rechts- und Staatsdenken, in: Evangelische Theologie 64/2, pp. 116–126.

Campi, Emidio (2022): Bullinger versus Ochino. Anatomie eines Zerwürfnisses, in: Gergely Csukás and Ariane Albisser (eds.): Wirkungen und Wurzeln der Schweizer Reformation. Festschrift für Peter Opitz, Zürcher Beiträge zur Reformationsgeschichte 30, Zürich.

Capra, Fritjof (2008): La botanica di Leonardo. Un discorso sulla scienza delle qualità, International Lectures on Nature and Human Ecology 2, Sansepolcro.

Cardini, Franco and Miglio, Massimo (2002): Nostalgia del paradiso. Il giardino medievale, Rome and Bari.

Cassius Iatrosophista (1562): Naturales et medicinales quaestiones LXXXIIII … accedit catalogus medicamentorum simplicium et parabilium quae pestilentiae veneno adversantur …, Zürich.

Celati, Alessandra (2014): Heresy, Medicine and Paracelsianism in Sixteenth-Century Italy: The Case of Girolamo Donzellini (1513–1587), in: Gesnerus 71, pp. 5–37.

Cermenati, Mario (1908): Francesco Calzolari e le sue lettere all'Aldrovandi, in: Annali di Botanica 6, pp. 83–138.

Cerutus, Benedictus and Chioccus, Andrea (1622): Musaeum Francisci Calceolari iunioris Veronensis …, Verona.

Choptiany, Michał (2013): Conrad Gesner jako czytelnik Carmen de Bisonte Mikołaja Hussowskiego perspektywy badawcze, in: Kwartalnik Historii Nauki i Techniki 58, pp. 111–137.

Christ, Georg (2005): Das Fremde verstehen. Biblianders Apologie zur Koranausgabe im Spiegel des Basler Koranstreites von 1542, in: Christ-von Wedel, Christine (ed.): Theodor Bibliander (1505–1564). Ein Thurgauer im gelehrten Zürich der Reformationszeit, Zürich, pp. 107–124.

Christ, Hermann (1917): Jacques Dalechamp. Un pionnier de la flore des Alpes occidentales au XVIme siècle, in: Bulletin de la Société Botanique de Genève, 2me série, 9, pp. 137–164.

Christ, Karl (1922): Ein bibliothekarischer Briefwechsel des 16. Jahrhunderts. De ordinanda bibliotheca, in: Hermann Degering et al. (eds.): Aus der Handschriften-Abteilung der Preussischen Staatsbibliothek. Abhandlungen und Nachbildungen von Autographen, Ludwig Darmstaedter zum 75. Geburtstag, Berlin, pp. 64–77.

Christ-von Wedel, Christine (2005): Theodor Bibliander in seiner Zeit, in: dies. (ed.): Theodor Bibliander (1505–1564). Ein Thurgauer im gelehrten Zürich der Reformationszeit, Zürich, pp. 19–60.

Christ-von Wedel, Christine and Leu, Urs B. (2007): Erasmus in Zürich. Eine verschwiegene Autorität, Zürich.

Christoffel, Hans (1953): Grundzüge der Uroskopie, in: Gesnerus 10, pp. 89–121.

Ciabochi, Claudio et al. (2013): Marches: Off the Beaten Path, Fabriano.

Ciccolella, Federica and Speranzi, David (2010): Le lingue orientali e la cultura greca nel Rinascimento, in: Sergio Luzzatto and Gabriele Pedullà (eds.): Atlante della letteratura italiana, vol. 1, Turin, pp. 438–448.

Cioppi, Elisabetta (2010): Le tracce di Leonardo, in: Simonetta Monechi and Lorenzo Rook (eds.): Il Museo di Storia Naturale dell'Università degli Studi di Firenze, Vol. 3: Le collezioni geologiche et paleontologiche, Florence, pp. 77–83.

Clauser, Christoph (1531): Das die Betrachtung des Menschenn Harns on anderen Bericht unnütz, und wie doch der Harn zuo empfachen und zuo urteylen am gschicktisten syge ..., Zürich.

Clot, André (2005): Suleiman the Magnificent, London.

Cohen, Floris (2010): Die zweite Erschaffung der Welt. Wie die moderne Naturwissenschaft entstand, Frankfurt/M. and New York.

Collin, Rudolf (1859): Schilderung seines Lebens, verdeutscht von Salomon Vögelin, in: Zürcher Taschenbuch auf das Jahr, pp. 179–220.

Colombat, Bernard and Peters, Manfred (2009): Mithridate – Mithridates (1555). Introduction, texte latin, traduction française, annotation et index, Travaux d'Humanisme et Renaissance 452, Geneva.

Cordus, Valerius (1561): Annotationes in Pedacii Dioscoridis Anazarbei de Medica materia libros v., ed. by Conrad Gessner, Strasbourg.

Cordus, Valerius (1563): Stirpium descriptionis liber quintus, qua in Italia sibi visas describit ..., Strasbourg.

Cordus, Valerius (1565): De halosantho seu spermate ceti ..., in: Gessner (1565a).

Corrodi-Sulzer, Adrian (1923): Zwinglis Vermögensverhältnisse, in: Zwa 4, pp. 174–188.

Crato von Krafftheim, Johannes (1563): In Cl. Galeni divinos libros methodi therapeutices ..., Basel.

Crousaz, Karine (2012): L'Académie de Lausanne entre Humanisme et Réforme (ca. 1537–1560), Education and Society in the Middle Ages and Renaissance 41, Leiden.

Curnis, Michele (2008): L'antologia di Giovanni Stobeo: una biblioteca antica dai manoscritti alle stampe, Minima philologica, Serie greca, 4, Alessandria.

Cuvier, Georges (1841): Histoire des sciences naturelles, depuis leur origine juaqu'à nos jours ..., vol. 2, Paris.

Danzi, Massimo (2004): Il "De balneis" di Conrad Gessner, l'Italia e l'"Internazionale" medico-umanistica delle terme, in: Gli Umanisti e le terme, Atti del Convegno internazionale di studio Lecce – Santa Cesarea Terme, 23–25 maggio 2002, Lecce, pp. 313–348.

Danzi, Massimo (2012): La tradizione scientifica delle acque nel *De Germaniae et Helvetiae thermis* di Conrad Gessner (1553), in: Didier Boisseuil and Hartmut Wulfram (eds.): Die Renaissance der Heilquellen in Italien und Europa von 1200 bis 1600 – Il Rinascimento delle fonti termali in Italia e in Europa dal 1200 al 1600, Frankfurt/M. etc., pp. 177–192.

Dauser, Regina and Ferber, Magnus Ulrich (2010): Die Fugger und Welser. Vom Mittelalter bis zur Gegenwart, Augsburg.

De Toni, Giovanni Battista (1911/1912): Illustrazione del Quarto Volume dell'Erbario di Ulisse Aldrovandi, in: Atti del Reale Istituto Veneto di Scienze, Lettere ed arti 71, pp. 39–131.

Delaunay, Paul (1962): La zoologie au seizième siècle, Histoire de la pensée 7, Paris.

Delisle, Candice (2004): The Letter: Private Text or Public Place? The Mattioli-Gesner Controversy about the aconitum, in: Gesnerus 61, pp. 161–176.

Delisle, Candice (2007): De la lettre au traité: l'*Historia animalium* de Conrad Gesner, in: Philip Ford (ed.): L'animal sauvage à la Renaissance. Colloque international organisé par la société française d'étude du XVIe siècle et Cambridge French Colloquia, Cambridge, 3–6 septembre 2004, Cambridge, pp. 289–301.

Deubelbeiss, Beat (1991): Beschreibung des Mons Fractus durch Conrad Gessner, in: Mitteilungen der Naturforschenden Gesellschaft Luzern 32, pp. 34–51.

Dietl, Cora (2005): Die Dramen Jacob Lochers und die frühe Humanistenbühne im süddeutschen Raum, Quellen und Forschungen zur Literatur- und Kulturgeschichte 37, Berlin and New York.

Dill, Ueli (2010): „Nautile, symbolum atque larva nostra". Theodor Zwinger unterwegs im Meer des Lebens, in: Basler Zeitschrift für Geschichte und Altertumskunde 110, pp. 177–207.

Diner, Dan (2016): Versiegelte Zeit. Über den Stillstand in der islamischen Welt, Berlin (5th ed.).

Dioscorides, Pedanius (1565): Ἐυπόριστα. De curationibus morborum per medicamenta paratu facilia, libri II. Nunc primum et graece editi, & partim a Ioanne Moibano Medico Augustano, partim vero post huius mortem a Conrad Gesnero ..., Strasbourg.

Dioscurides, Pedanius (2002): Fünf Bücher über die Heilkunde. Aus dem Griechischen übersetzt von Max Aufmesser, Altertumswissenschaftliche Texte und Studien 37, Hildesheim etc.

Dioskurides, Pedanios (1902): Des Pedanios Dioskurides aus Anazarbos Arzneimittellehre in fünf Büchern, übersetzt und mit Erklärungen versehen von Julius Berendes, Stuttgart.

Dobay, Akos et al. (2013): Renaissance Model of an Epidemic with Quarantine, in: Journal of Theoretical Biology 317, pp. 348–358.

Dobler, Friedrich (1955): Conrad Gessner als Pharmazeut, Diss. ETH Zürich, Zürich.

Döpp, Siegmar (2012): Neulateinische Wissenschaftspoesie. Ioannes Fabricius Montanus (1527–1566) über Engadiner Heilquellen, Speyer.

Dominici, Stefano (2010): Storia della Toscana, storia della Terra, in: Simonetta Monechi and Lorenzo Rook (eds.): Il Museo di Storia Naturale dell'Università degli Studi di Firenze, Vol. 3: Le collezioni geologiche et paleontologiche, Florence, pp. 3–17.

Doran, Susan (2015): Elizabeth I. and Her Circle, Oxford.

Dotzauer, Winfried (1971): Deutsche Studenten an der Universität Bourges. Album et liber amicorum, Meisenheim am Glan.

Dubravius, Janus (1559): De piscinis et piscis qui in eis aluntur …, item Xenocratis de alimento ex aquatilibus …, ed. by Conrad Gessner, Zürich.

Dürst, Arthur (1993): Die Karte des Heiligen Landes in der Froschauer-Bibel 1525, in: Alfred Cattani (ed.): Alte und neue Schätze. Zentralbibliothek Zürich, Zürich, pp. 50–53.

Dupéron, Jean et al. (2008): *Ulminium diluviale* Unger: historique de la découverte et nouvelle étude, in: Annales de paléontologie 94, pp. 1–12.

Durling, Richard J. (1961): A Chronological Census of Renaissance Editions and Translations of Galen, in: Journal of the Warburg and Courtauld Institutes 24, pp. 230–305.

Durling, Richard J. (1965): Conrad Gesner's *Liber amicorum* 1555–1565, in: Gesnerus 22, pp. 134–159.

Eddy, John A. (1976): The Maunder Minimum, in: Science 192, no. 4245 (June 18), pp. 1189–1202.

Edson, Evelyn et al. (2005): Der mittelalterliche Kosmos. Karten der christlichen und islamischen Welt, Darmstadt.

Egli, Emil (1879): Aktensammlung zur Geschichte der Zürcher Reformation in den Jahren 1519–1533, Zürich.

Egli, Emil (1901): Analecta Reformatoria, vol. 2: Biographien: Bibliander, Ceporin, Johannes Bullinger, Zürich.

Egli, Emil (1904): Ein griechisches Schauspiel an Zwinglis Schule, in: Zwingliana 1, pp. 11–13.

Egmond, Florike (2013): A Collection within a Collection. Rediscovered Animal Drawings from the Collections of Conrad Gessner and Felix Platter, in: Journal of the History of Collections 25/2, pp. 149–170.

Egmond, Florike (2018): Conrad Gessners "Thierbuch". Die Originalzeichnungen, Darmstadt.

Egmond, Florike and Kusukawa, Sachiko (2016): Circulation of Images and Graphic Practices in Renaissance Natural History: the Example of Conrad Gessner, in: Gesnerus 73 (2016), pp. 29–72.

Egmond, Florike and Kusukawa, Sachiko (2019): Gessner's Fish: Images as Objects, in: Leu/Opitz (2019), pp. 581–605.

Eickhoff, Ekkehard (2007): Venedig – Spätes Feuerwerk: Glanz und Untergang der Republik 1700–1797, Stuttgart.

Ein Christenlich ansehen und ordnung von den Eersamen Burgermeister und Radt und dem grossen Radt der statt Zürich, ouch Propst und Capitel zum grossen münster daselbst, von der priesterschafft und pfruonden wegen ermessen und angenommen ..., Zürich 1523.

Eire, Carlos M. N. (2020): Protestantism, Disenchantment, and the Birth of Modernitiy, in: Matthias Pohlig and Detlef Pollack (eds.): Die Verwandlung des Heiligen: Die Geburt der Moderne aus dem Geist der Religion, Berlin, pp. 41–54.

Enenkel, Karl A. E. (2007): Zur Konstituierung der Zoologie als Wissenschaft in der Frühen Neuzeit: Diskursanalyse zweier Grossprojekte (Wotton, Gesner), in: Karl A. E. Enenkel and Paul J. Smith (eds.): Early Modern Zoology. The Construction of Animals in Science, Literature and the Visual Arts. Intersections 7, Leiden and Boston, pp. 15–74.

Epiphanius of Salamis (1565): De XII gemmis quae erant in veste Aaronis ..., in: Gessner (1565a).

Epiphanius of Salamis (2014): Über die zwölf Steine im hohepriestlichen Brustschild „De duodecim gemmis rationalis, edited by Felix Albrecht and Arthur Manukyan, Gorgias Eastern Christian Studies 31, Piscataway (NJ).

Erasmus of Rotterdam (1963): Ausgewählte pädagogische Schriften, ed. by Anton J. Gail, Paderborn.

Erasmus of Rotterdam (1967): Ratio – Theologische Methodenlehre, in: Erasmus von Rotterdam, Ausgewählte Schriften, ed. by Werner Welzig, vol. 3, Darmstadt, pp. 117–495.

Erasmus of Rotterdam (1978): De copia, in: Craig R. Thomson (ed.): Collected Works of Erasmus, vol. 24, Toronto, pp. 279–659.

Erasmus of Rotterdam (1983): Adagia (lateinisch/deutsch), Auswahl von Anton J. Gail, Stuttgart.

Ernst, Ulrich (1879): Geschichte des zürcherischen Schulwesens bis gegen das Ende des sechzehnten Jahrhunderts, Diss. phil. I, Universität Zürich, Winterthur.

Escher, Hermann (1937): Conrad Gessner über Aufstellung und Katalogisierung von Bibliotheken, in: Mélanges offerts à Marcel Godet, Directeur de la Bibliothèque Nationale Suisse à Berne, Neuchâtel, pp. 119–127.

Estienne, Henri (1564): Dictionarium medicum vel, expositiones vocum medicinalium ..., Geneva.

Etter, Walter and Schmidt, Olivier (2016): Conrad Gessner als Paläontologe, in: Leu and Ruoss (2016), pp. 177–187.

Etter, Walter and Schmidt, Olivier (2020): Gessner's fossil crab. An icon for the early history of palaeontology, in: Journal of the History of Collections 32, pp. 411–430 (https://doi.org/10.1093/jhc/fhz039).

Fabricius Montanus, Johannes (1555): Differentiae animalium quadrupedum secundum locos communes ..., Zürich.

Fatio, Olivier (1987): Les débuts de l'Ecole de Lausanne, in: De l'Académie à l'Université de Lausanne 1537–1987: 450 ans d'histoire, Lausanne, pp. 49–54.

Faust, Ingrid (1998–2010): Zoologische Einblattdrucke und Flugschriften vor 1800, 6 vols., Stuttgart.

Ferri, Sara (2001²): Il „Dioscoride", i „Discorsi", i „Commentarii": gli amici e i nemici, in: Sara Ferri (ed.): Pietro Andrea Mattioli, Siena 1501 – Trento 1578. La vita e le opere con l'identificazione delle piante, Perugia.

Fischel, Angela (2009): Natur im Bild. Zeichnung und Naturerkenntnis bei Conrad Gessner und Ulisse Aldrovandi, humboldt-schriften zur kunst- und bildgeschichte 9, Berlin.

Fischel, Angela (2010): Collections, Images and Form in Sixteenth Century Natural History: The Case of Conrad Gessner, in: Intellectual History Review 20, pp. 147–164.

Fischer, Hans (1940): Johannes Fabricius Montanus (1527–1566) und Conrad Gessner, in: Vierteljahrsschrift der Naturforschenden Gesellschaft in Zürich 85, pp. 322–328.

Fischer, Hans (1940a): Zu einem unveröffentlichten Brief Conrad Gessner's an Fabricius Montanus, in: Vierteljahrsschrift der Naturforschenden Gesellschaft in Zürich 85, pp. 329–336.

Fischer, Hans (1946): Ein unveröffentlichter Brief Conrad Gessners (1516–1565) an Johannes Fabricius Montanus (1527–1566) in Chur, in: Gesnerus 3, pp. 125–130.

Fischer, Hans (1946a): Conrad Gessner und Johann Jakob Scheuchzer, in: Vierteljahrsschrift der Naturforschenden Gesellschaft 91, pp. 163–196.

Fischer, Hans (1966): Conrad Gessner, Neujahrsblatt auf das Jahr 1966 als 168. Stück von der Naturforschenden Gesellschaft in Zürich, Zürich.

Fischer, Hans et al. (1967): Conrad Gessner 1516–1565. Universalgelehrter, Naturforscher, Arzt, Zürich.

Flashar, Helmut (2013): Aristoteles. Lehrer des Abendlandes, Munich.

Flohn, Hermann (1949): Klima und Witterungsverlauf in Zürich im 16. Jahrhundert, in: Vierteljahrsschrift der Naturforschenden Gesellschaft in Zürich 94, pp. 29–41.

Frati, Lodovico (1907): Catalogo dei manoscritti di Ulisse Aldrovandi, Bologna.

Fretz, Diethelm (1948): Conrad Gessner als Gärtner, Zürich.

Frey, Stefan (2017): Fromme feste Junker. Neuer Stadtadel im spätmittelalterlichen Zürich, Mitteilungen der Antiquarischen Gesellschaft in Zürich 84, Zürich.

Freytag, Theodor Friedrich (1831): Virorum doctorum epistolae selectae, Leipzig.

Friedrich, Udo (1995): Naturgeschichte zwischen artes liberales und frühneuzeitlicher Wissenschaft. Conrad Gessners „Historia animalium" und ihre volkssprachliche Rezeption, Frühe Neuzeit 21, Tübingen.

Frohne, Renate (2010): Das Welt- und Menschenbild des St. Galler Humanisten Joachim von Watt / Vadianus (1484–1551). Dargestellt anhand ausgewählter Exkurse in den Scholien zu Pomponius Mela: *De chorographia*, Basel 1522. Ein Lesebuch (Lateinisch/Deutsch) mit Kommentaren und Interpretationen, Die Antike und ihr Weiterleben 8, Remscheid.

Fuchs, Leonhart (1542): De historia stirpium commentarii …, Basel.

Fuchs, Leonhart (1543): New Kreüterbuoch …, Basel.

Fuchs-Eckert, Hans Peter (1973): Die „Icones plantarum et analyses partium" des Casimir Christoph Schmidel, in: Festschrift für Claus Nissen zum siebzigsten Geburtstag 2. September 1971, Wiesbaden, pp. 185–276.

Fueter, Eduard R. and Gülland, Erich (eds.): Der Einblattdruck "Hermaphrodit" von Christoph Froschuer mit Erläuterungen von Urs B. Leu, Faksimile-Ausgabe, Reitnau.

Funk, Holger (2017): John Caius's contributions to Conrad Gessner's *Historia animalium* and "Historia plantarum": a survey and commentaries, in: Archives of Natural History 44, pp. 334–351.

Funk, Holger (2018): Caspar Wolf and Conrad Gessner's Unfinished History of Plants. Essay and Translation, in: Gesnerus 75 (2018), pp. 40–94.

Furlan, Italo (1997): Codici greci illustrati della Biblioteca Marciana VI, Padua.

Furno, Martine (2008): Le catalogue des livres imprimés par Sébastien Gryphe dans la préface du livre XII des *Pandectes* de Conrad Gesner: catalogue d'imprimeur ou catalogue de savant?, in: Raphaële Moren (ed.): Quid novi? Sébastien Gryphe, à l'occasion du 450e anniversaire de sa mort. Actes du colloque 23 au 25 novembre 2006, Lyon-Villeurbanne, Bibliothèque municipale de Lyon, Villeurbanne, pp. 33–56.

Gamper, Rudolf (2006): Paracelsus und Vadian. Ihre Begegnung in St. Gallen, in: Gantenbein/Holenstein Weidmann (siehe unten), pp. 117–129.

Gamper, Rudof (2017): Joachim Vadian (1483/84–1551). Humanist, Arzt, Reformator, Politiker, Zürich.

Gantenbein, Urs Leo (1997): Am Scheideweg zwischen Fiktion und Realität: Das Phantastische in Conrad Gessners Naturbetrachtung, in: Hans-Konrad Schmutz (ed.): Phantastische Lebensräume, Phantome und Phantasmen. Aufsätze des Zürcher Symposions der Schweizerischen Gesellschaft für Geschichte der Medizin und der Naturwissenschaften vom 10. und 11. Oktober 1996, Marburg an der Lahn, pp. 173–196.

Gantenbein, Urs Leo and Holenstein Weidmann, Pia (2006): Paracelsus. Der Komet im Hochgebirge von 1531. Ein Himmelszeichen aus St. Gallen für Zwingli, Zürich.

Garofalo, Ivan (2004): Agostino Gadaldini (1515–1575) et le Galien italien, in: Véronique Boudon-Millot et al. (eds.): Lire les médecins grecs à la Renaissance. Aux origines de l'édition médicale. Actes du colloque international de Paris (19–20 septembre 2003), Paris, pp. 283–321.

Gascon, Richard (1971): Grand commerce et vie urbaine au XVIe siècle. Lyon et ses marchands (environs de 1520 – environs de 1580), Civilisations et Sociétés 21 et 22, 2 vols., Paris.

Gemelli Marciano, M. Laura (2009): Die Vorsokratiker, vol. 2: Parmenides, Zenon, Empedokles, Zürich.

Germann, Martin (1994): Die reformierte Stiftsbibliothek am Grossmünster Zürich im 16. Jahrhundert und die Anfänge der neuzeitlichen Bibliographie. Rekonstruktion des Buchbestandes und seiner Herkunft, der Bücheraufstellung und des Bibliotheksinventars. Mit Edition des Inventars von 1532/1551 von Conrad Pellikan, Beiträge zum Buch- und Bibliothekswesen 34, Wiesbaden.

Gerritsen, Johan (1991): Printing at Froben's: An Eye Witness Account, in: Studies in Bibliography 44, pp. 144–163.

Gerstinger, Hans (1926): Johannes Sambucus als Handschriftensammler, in: Festschrift der Nationalbibliothek in Wien. Herausgegeben zur Feier des 200jährigen Bestehens des Gebäudes, Vienna, pp. 251–406.

Gerstinger, Hans (1968): Die Briefe des Johannes Sambucus (1554–1584), Österreichische Akademie der Wissenschaften, Philosophisch-Historische Klasse, Sitzungsberichte, 255. Band, Vienna.

Gessner, Conrad (1541): Historia plantarum et vires ex Dioscoride, Paulo Aegineta, Theophrasto, Plinio, et recentioribus Graecis, iuxta elementorum ordinem …, Basel.

Gessner, Conrad (1541a): Compendium ex Actuarii Zachariae libris de differentiis urinarum, iudiciis et praevidentiis … Sylvula Galeni experimentorum, et aliorum aliquot …, Zürich.

Gessner, Conrad (1541b): Michaelis Ephesii scholia in Aristotelis libros aliquot …, Basel.

Gessner, Conrad (1541c): De lacte et operibus lactariis …, Zürich.

Gessner, Conrad (1542): Catalogus plantarum latine, graece, germanice et gallice …, Zürich.

Gessner, Conrad (1542a): Apparatus et delectus simplicium medicamentorum …, Lyon.

Gessner, Conrad (1542b): Moralis interpretatio errorum Ulyssis Homerici. Commentatio Porphyrii Philosophi de Nympharum antro in XIII. Libro Odysseae Homericae … Ex Commentariis Procli Lycii, Philosphi Platonici in libros Platonis de Republicae apologiae quaedam pro Homero …, Zürich.

Gessner, Conrad (1542c): Compendium de syllogismis [Autor unbekannt, übersetzt von Gessner], in: Porphyrii institutiones quinque vocum, ad Chrysaorium. … His coniunximus in dialecticae studiosorum gratiam, Michaelis Pselli peripatetici, Blemmidae, Georgii Pachymerii, in eosdem Porphyrii et Aristotelis libros epitomen, atque Paraphrasin …, Basel, pp. 339–357.

Gessner, Conrad (1545): Bibliotheca universalis …, Zürich.

Gessner, Conrad (1545a): Lexicon graecolatinum …, Zürich (4th ed.).

Gessner, Conrad (1548): Pandectarum sive partitionum universalium … libri XXI, Zürich.

Gessner, Conrad (1549): Partitiones theologicae …, Zürich.

Gessner, Conrad (1549a): Onomasticon propriorum nominum …, Basel.

Gessner, Conrad (1551): Historia animalium liber I. de quadrupedibus viviparis, Zürich.
Gessner, Conrad (1552): Thesaurus Euonymi Philiatri, de remediis secretis ..., Zürich.
Gessner, Conrad (1553): Excerptorum et observationum de thermis, in: Tommaso Giunta (ed.): De balneis omnia quae extant ..., Venice, f. 289r–298v.
Gessner, Conrad (1553a): Icones animalium quadrupedum viviparorum et oviparorum ..., Zürich.
Gessner, Conrad (1553b): Tabulae collectionum in genere et particulatim per XII menses, in usum pharmacopolarum conscriptae, in: David Kyber: Lexicon rei herbariae trilingue ..., Strasbourg, pp. 467–548.
Gessner, Conrad (1553c): De Germaniae et Helvetiae thermis, in: Tommaso Giunta (ed.): De balneis omnia quae extant apud Graecos, Latinos, et Arabas, tam medicos quoscunque ceterarum artium probatos scriptores ..., Venice, f. 289r–299r.
Gessner, Conrad (1554): Historiae animalium liber II. De quadrupedibus oviparis ... Zürich.
Gessner, Conrad (1555): De raris et admirandis herbis, quae sive quod noctu luceant, sive alias ob causas, lunariae nominantur ..., Zürich.
Gessner, Conrad (1555a): Historiae animalium liber III., qui est de avium natura ..., Zürich.
Gessner, Conrad (1555b): Icones avium ..., Zürich.
Gessner, Conrad (1555c): Mithridates. De differentiis linguarum ..., Zürich.
Gessner, Conrad (1555d): Chirurgia. De chirurgia scriptores optima quique veteres et recentiores ..., Zürich.
Gessner, Conrad (1555e): Ein kostlicher theürer Schatz Euonymi Philiatri / darinn behalten sind vil heymlicher guotter stuck der artzny ..., Zürich.
Gessner, Conrad (1556): De piscibus et aquatilibus omnibus libelli III ... I. Scholia & emendationes in Halieuticon P. Ovidii Nasonis, II. Aquatilium animantium enumeratio ..., III. Eorundem nomenclator germanicus ..., Zürich.
Gessner, Conrad (1556a): Sanitatis tuendae praecepta cum aliis, tum literarum studiosis hominibus, & iis qui minus exercentur, cognita necessaria, Zürich.
Gessner, Conrad (1557): Vogelbuoch ..., Zürich.
Gessner, Conrad (1558): Historia animalium liber IIII. qui est de piscium et aquatilium animantium natura, Zürich.
Gessner, Conrad (1559/60): Theologorum aliquot graecorum veterum orthodoxorum libri ..., Zürich.
Gessner, Conrad (1560): Icones animalium quadrupedum viviparorum et oviparorum ..., Zürich (2nd ed.).
Gessner, Conrad (1560a): Nomenclator aquatilium animantium. Icones animalium aquatilium in mari et dulcibus aquis degentium ..., Zürich.
Gessner, Conrad (1560b): Icones avium ..., Zürich (2nd ed.).
Gessner, Conrad (1562): De libris a se editis epistola ad Guilelmum Turnerum ..., Zürich.

Gessner, Conrad (1563): De anima liber …, in: Conrad Gessner (ed.): Ioannis Lodovici Vivis Valentini de anima et vita libri tres. Eiusdem argumenti Viti Amerbachii …, Zürich, pp. 719–951.
Gessner, Conrad (1563a): Thierbuoch …, Zürich.
Gessner, Conrad (1563b): Fischbuoch …, Zürich.
Gessner, Conrad (1565): De rerum fossilium, lapidum et gemmarum maxime, figuris et similitudinibus liber …, in: Gessner (1565a).
Gessner, Conrad (1565a): De omni rerum fossilium genere, gemmis, lapidibus, metallis et huiusmodi libri aliquot …, Zürich.
Gessner, Conrad (1577): Epistolarum medicinalium … libri III. His accesserunt eiusdem Aconiti primi Dioscoridis asseveratio, et de Oxymelitis elleborati utriusque descriptione & usu libellus, ed. by Caspar Wolf, Zürich.
Gessner, Conrad (1577a): De aconito primo Dioscoridis, asseveration, et … de oxymelitis elleborati utriusque description et usu libellus, ed. by Caspar Wolf, Zürich.
Gessner, Conrad (1584): Epistolarum medicinalium … liber quartus, Wittenberg.
Gessner, Conrad (1586): Physicarum meditationum annotationum et scholiorum libri V. Nunc recens ex variis Gesnerianae diligentiae relictis schedis et libris collecti … per Casparum Wolphium, Zürich.
Gessner, Conrad (1587): Historiae animalium lib. V. qui est de serpentium natura … adiecta est … scorpionis insecti historia …, Zürich.
Gessner, Conrad (1587a): De stirpium collectione tabulae …, Zürich.
Gessner, Conrad (1589): Schlangenbuoch …, Zürich.
Gessner, Conrad (1604): Historiae animalium liber IV. …, Frankfurt/M.
Gessner, Conrad (1746): Epistola de constantia in fide, adversus apostasiam. D. Jacobo Dalechampio, …, in: Museum Helveticum I, pp. 133–150.
Gessner, Conrad (1974): Deutsche Namen der Fische und Wassertiere. Neudruck der Ausgabe Zürich 1556, herausgegeben und eingeleitet von Manfred Peters, Aalen.
Gessner, Conrad (1974a): Mithridates. De diffenrentiis linguarum tum veterum tum quae hodie apud diversas nationes in toto orbe terrarium in usu sunt. Republication of the edition Zürich 1555, edited and introduced by Manfred Peters, Aalen.
Gessner, Conrad (1996): Büchlein von der Milch und den Milchprodukten, German translation by Siegfried Kratzsch, Mönchengladbach.
Gessner, Conrad (2009): Mithridate/Mithridates (1555). Introduction, texte latin, traduction française, annotation et index par Bernard Colombat et Manfred Peters, Travaux d'Humanisme et Renaissance 452, Geneva.
Gessner, Conrad and Guilandinus, Melchior (1557): De stirpium aliquot nominibus … et alia complura, epistolae II, una Melchioris Guilandini Borussi, altera Conradi Gesneri Tigurini …, Basel.
Gilly, Carlos (2001): Die Manuskripte in der Bibliothek des Johannes Oporinus. Verzeichnis der Manuskripte und Druckvorlagen aus dem Nachlass Oporins anhand

des von Theodor Zwinger und Basilius Amerbach erstellten Inventariums, Schriften der Universitätsbibliothek Basel 3, Basel.

Glesinger, Lavoslav (1950): Conrad Gessner's Beziehungen zu einem kroatischen Gelehrten, in: Gesnerus 7, pp. 27–50.

Gmelig-Nijboer, Caroline A. (1971): Conrad Gessner's „Historia animalium". An Inventory of Renaissance Zoology, Communicationes Biohistoricae Ultrajectinae 72, Meppel.

Göbel, Severin (1565): De succino libri duo, in: Gessner (1565a).

Göing, Anja-Silvia (2003): *Physica* im Lehrplan der Schola Tigurina in Zürich 1541–1597, in: Hans-Ulrich Musolff and Anja-Silvia Göing (eds.): Anfänge und Grundlegungen moderner Pädagogik im 16. und 17. Jahrhundert, Beiträge zur Historischen Bildungsforschung 29, Cologne etc., pp. 73–91.

Göing, Anja-Sivlia (2013a): Storing to Know: Conrad Gessner's De Anima and the Relationship between Textbooks and Citation Collections in Sixteenth-Century Europe, in: Göing, Anja-Silvia et al. (eds.): Collectors' Knowledge: What Is Kept, What Is Discarded. Aufbewahren oder wegwerfen – Wie Sammler entscheiden, Leiden and Boston, pp. 209–242.

Göing, Anja-Sivlia (2017): Storing, Archiving, Organizing. The Changing Dynamics of Scholarly Information Management in Post-Reformation Zurich, Library of the Written Word 56, Leiden and Boston.

Goody, Jack (2010): Renaissances. The One or the Many? Cambridge.

Gouron, Marcel (1957): Matricule de l'Université de Médecine de Montpellier (1503–1599), Travaux d'Humanisme et Renaissance XXV, Geneva.

Grafton, Anthony (2011): The Culture of Correction in Renaissance Europe, London.

Grafton, Anthony and Leu, Urs B. (2014): Henricus Glareanus's (1488–1563) Chronology of the Ancient World. A Facsimile Edition of a Heavily Annotated Copy Held in Princeton University Library, Studies in Medieval and Reformation Traditions 177, Leiden and Boston.

Grafton, Anthony (2019): Conrad Gessner and John Caius: The Meanings of Learned Friendship in Renaissance Europe, in: Leu/Opitz(2019), pp. 355–376.

Graux, Charles (1880): Essai sur les origines du fonds grec de l'Escurial, Épisode de l'histoire de la renaissance des lettres en Espagne, Paris.

Grebel, Hans Rudolf von (1964): Antistes Johann Jakob Breitinger 1575–1645. Neujahrsblatt auf das Jahr 1964 zum Besten des Waisenhauses Zürich, 127. Stück. Als Fortsetzung der Neujahrsblätter auf der Chorherrenstube, no. 186, Zürich.

Greene, Edward Lee (1983): Landmarks of Botanical History, 2 vols., Stanford (CA).

Grendler, Paul F. (1977): The Roman Inquisition and the Venetian Press 1540–1605, Princeton (NJ).

Grimm, Heinrich (1966): Die Buchführer des deutschen Kulturbereichs und ihre Niederlassungsorte in der Zeitspanne 1490 bis um 1550, in: Archiv für die Geschichte des Buchwesens 7, Sp. 1153–1172.

Grupp, Peter (2008): Faszination Berg. Die Geschichte des Alpinismus, Cologne.

Gunnoe, Charles (2019): Gessner's Plague: The Bubonic Plague Epidemic of 1562–1566, in: Leu/Opitz (2019), pp. 295–309.

Gunther, Robert T. (1959): The Greek Herbal of Dioskorides illustrated by a Byzantine A.D. 512, New York.

Gutwald, Thomas (2002): Prodigium hoc cum nostro seculo inusitatum sit … Das Nordlicht vom 28. Dezember 1560 als Gegenstand vernetzter Wahrnehmung durch frühneuzeitliche Informationssysteme, in: Wolfgang Harms et al. (eds.): Wahrnehmungsgeschichte und Wissensdiskurs im illustrierten Flugblatt der Frühen Neuzeit (1450–1700), Basel, pp. 239–261.

Guyer, Paul (1973): Wo wohnte der Naturforscher Conrad Gessner?, in: Zürcher Chronik 41/3, pp. 79–82.

Hackethal, Sabine and Hans (1994): Zoologische Klebebände als erste faunistische Sammlungen, in: Andreas Grote (ed.): Macrocosmos in Microcosmo. Die Welt in der Stube – Zur Geschichte des Sammelns 1450–1800, Berliner Schriften zur Museumskunde 10, Opladen, pp. 283–299.

Hanhart, Johannes (1824): Conrad Gessner. Ein Beytrag zur Geschichte des wissenschaftlichen Strebens und der Glaubensverbesserung im 16ten Jahrhundert. Winterthur.

Hanno Carthaginiensis (1559): Ducis navigatio, qua maximam Libycae orae partem ultra Herculis columnas lustravit, ed. by Conrad Gessner, Zürich.

Hardmeyer, Adrian (1980): Zur Balneologie bei Conrad Gessner, Diss. med. Universität Basel, Basel.

Harms, Wolfgang (1989): Bedeutung als Teil der Sache in zoologischen Standardwerken der frühen Neuzeit (Conrad Gesner, Ulisse Aldrovandi), in: Hartmut Boockmann et al. (eds.): Lebenslehren und Weltentwürfe im Übergang vom Mittelalter zur Neuzeit. Politik – Bildung – Naturkunde – Theologie. Bericht über Kolloquien der Kommission zur Erforschung der Kultur des Spätmittelalters 1983 bis 1987, Abhandlungen der Akademie der Wissenschaften in Göttingen, Philologisch-historische Klasse, 3. Folge, no. 179, Göttingen, pp. 352–369.

Harvey, E. Netwon (1957): A History of Luminiscence. From the Earliest Times Until 1900, Memoirs of the American Philosophical Society 44, Philadelphia (PA).

Hauser, Albert (1962): Vom Essen und Trinken im alten Zürich. Tafelsitten, Kochkunst und Lebenshaltung vom Mittelalter bis in die Neuzeit, Zürich.

Hausinger, Angela (2009): Conrad Gesner: Botanische Werke, in: Die Entdeckung der Pflanzenwelt. Botanische Drucke vom 15. bis 19. Jahrhundert aus der Universitätsbibliothek Johann Christian Senckenberg, Frankfurt/M., pp. 86–87.

Hausinger, Angela (2009a): Vom Naturselbstdruck zum Naturdruck, in: Die Entdeckung der Pflanzenwelt. Botanische Drucke vom 15. bis 19. Jahrhundert aus der Universitätsbibliothek Johann Christian Senckenberg, Frankfurt/M., pp. 119–129.

Heini, Janine (2014): Pilatus. Eine sagenumwobene Natur- und Kulturgeschichte, Lucerne.

Hejnic, Josef and Bok, Václav (1989): Gesners europäische Bibliographie und ihre Beziehung zum Späthumanismus in Böhmen und Mähren, Vienna, Cologne and Graz.

Helm, Johannes (1971): Johannes Kentmann 1518–1574: ein sächsischer Arzt und Naturforscher, Wiesbaden.

Helmich, Egon (1938): Die Briefe Conrad Geısners an Crato von Krafftheim nach der Briefsammlung von 1566, Düsseldorf.

Henglein, Dominik K. M. (2010): Casimir Christoph Schmidel (1718–1792) und seine Korrespondenz mit Jhann Ambrosius Beurer (1716–1754), Diss. med. Universität Erlangen-Nürnberg (https://opus4.kobv.de/opus4-fau/frontdoor/index/index/docId/1512).

Hennig, John (1987): Goethes Kenntnis des naturwissenschaftlichen Schrifttums Italiens, in: ders.: Goethes Europakunde. Goethes Kenntnisse des nichtdeutschen Europas: Ausgewählte Aufsätze, Amsterdam.

Henrich, Rainer (2006): „Wunder selzam Vögel" – die ältesten Schweizer Nachrichten über Masseneinflüge von Seidenschwanz (*Bombycilla garrulus*) und Bergfink (*Fringilla montifringilla*), in: Zürcher Taschenbuch 126, pp. 175–193.

Héraclite (1962): Allégories d'Homère, ed. par Félix Bufière, Paris.

Herakleides Pontikos (1544): Allegoriae in Homeri fabulas de diis, nunc primum e graeco sermone in latinum translatae, Conrado Gesnero medico interprete …, Basel.

Hertel, Gerhard (1969): Die Allegorie von Reichtum und Armut. Ein aristophanisches Motiv und seine Abwandlungen in der abendländischen Literatur, Erlanger Beiträge zur Sprach- und Kunstwissenschaft 33, Nuremberg.

Hess, Daniel (2012): Die Natur als vollkommene Lehrmeisterin der Kunst, in: Daniel Hess and Thomas Eser (eds.): Der frühe Dürer. Ausstellung im Germanischen Nationalmuseum vom 24. Mai bis 2. September 2012, Nuremberg, pp. 117–131.

Hess, Daniel (2019): Wissenschaft oder Kunst – Wahrheit oder Verführung? Conrad Gessner und das neue Pflanzenbild im 16. Jahrhundert, in: Leu/Opitz (2019), pp. 161–194.

Hieronymus, Frank (1992): En Basileia polei tēs Germanias. Griechischer Geist aus Basler Pressen, Publikationen der Universitätsbibliothek Basel 15, Basel.

Hieronymus, Frank (1997): 1488 Petri – Schwabe 1988. Eine traditionsreiche Offizin im Spiegel ihrer frühen Drucke, Zwei Halbbände, Basel.

Hieronymus, Frank (2005): Theophrast und Galen – Celsus und Paracelsus. Medizin, Naturphilosophie und Kirchenreform im Basler Buchdruck bis zum Dreissigjährigen Krieg, Publikationen der Universitätsbibliothek Basel 36, 5 vols., Basel.

Hinkel, Artur (1999): "… in einem einöden Münster in Meyssen". Conrad Gessners Bericht über Fledermäuse, in: Mitteilungen des Landesvereins Sächsischer Heimatschutz e. v., pp. 63–65.

Hobson, Anthony (1999): Renaissance Book Collecting. Jean Grolier and Diego Hurtado de Mendoza. Their Books and Bindings, Cambridge.

Holenstein Weidmann, Pia (2013): Monströse Erscheinungen im 16. Jahrhundert zwischen Naturwissenschaft und Gottesfurcht, in: Paul Michel (ed.): Genese und Symbolik von Kompositwesen, Schriften zur Symbolforschung 16, Zürich, pp. 209–226.

Holthuis, Lipke B. (1996): Original watercolours donated by Cornelius Sittardus to Conrad Gesner, and published by Gesner in his (1558–1670) works on aquatic animals, in: Zoologische Mededelingen Leiden 70, pp. 169–196.

Holzberg, Niklas (2010): Aristophanes. Sex, Spott und Politik, Munich.

Houiller, Jacques (1565): Viaticum novum. De omnium fere particularium morborum curatione liber, authoris innominati … nunc primum in lucem editus per Casparum Wolphium …, Zürich.

Hryniewiecki, Boleslaw (1938): Anton Schneeberger (1530–1581) ein Schüler Conrad Gesners in Polen, Veröffentlichungen des Geobotanischen Instituts Rübel in Zürich, 13. Heft, Zürich.

Hubicki, Włodzimierz (1960): Alexander von Suchten, in: Sudhoffs Archiv für Geschichte der Medizin und der Naturwissenschaften 44, pp. 54–63.

Hug, Arnold (1874): Aufführung einer Griechischen Komödie in Zürich am 1. Januar 1531. Vortrag, gehalten an der Jahresversammlung des schweizerischen Gymnasiallehrervereins, den 4. Oktober 1873, Zürich.

Huth, Manuel and Walter, Tilmann (2019): Der Briefwechsel zwischen Conrad Gessner und Johannes Crato von Krafftheim, in: Leu/Opitz (2019), pp. 377–404.

Im Hof, Ulrich (1978): Die Entstehung der reformierten Hohen Schule – Zürich (1525) – Bern (1528) – Lausanne (1537) – Genf (1559), in: Peter Baumgart and Notker Hammerstein (eds.): Beiträge zu Problemen deutscher Universitätsgründungen der frühen Neuzeit, Wolfenbütteler Forschungen 4, Nendeln/Liechtenstein, pp. 243–262.

Infelise, Mario (2014): I padroni dei libri. Il controllo sulla stampa nella prima età moderna, Bari.

Jarvis, Charlie (2007): Order out of Chaos. Linnean Plant Names and their Types, London.

Jedin, Hubert (1957): Geschichte des Konzils von Trient, vol. 2, Die erste Trienter Tagungsperiode 1545/47, Freiburg.

Jenny, Beat R. (1959): Graf Froben Christoph von Zimmern. Geschichtsschreiber, Erzähler, Landesherr. Ein Beitrag zur Geschichte des Humanismus in Schwaben, Lindau and Constance.

Jenny, Beat R. (1964): Arlenius in Basel, in: Basler Zeitschrift für Geschichte und Altertumskunde 64, pp. 5–45.

Jones, Howard (2004): Printing the Classical Text, Bibliotheca Humanistica & Reformatorica 62, Utrecht.

Jostock, Ingeborg (2007): La censure négociée. Le contrôle du livre à Genève 1560–1625, Travaux d'Humanisme et Renaissance N° 430, Geneva.

Julia, Dominique et Revel, Jacques (1989): Les Universités européennes du XVIe au XVIIIe siècle. Histoire sociale des populations étudiantes, Tome 2, France, Recherches d'histoire et de sciences sociales 18, Paris.

Julia, Dominique (1995): L'Université de médecine de Montpellier à l'époque moderne, in: L'Université de Montpellier: ses maîtres et ses étudiants depuis sept siècles 1289–1989. Actes du 61e Congrès de la Fédération Historique du Languedoc Méditerranéen et du Roussillon, colloque historique tenu à la Faculté de Médecine de Montpellier les 23 et 24 octobre 1989, Montpellier, pp. 77–108.

Junod, Louis et Henri Meylan (1947): L'Académie de Lausanne au XVIe siècle. Leges scholae Lausannensis 1547. Lettres et documents inédits, Études et documents pour servir à l'histoire de l'Université de Lausanne, Cinquième fascicule, Lausanne.

Kapp, Friedrich (1886): Geschichte des deutschen Buchhandels bis in das siebzehnte Jahrhunderts, vol. 1, Leipzig.

Kaufmann, Pius (2009): Gesellschaft im Bad. Die Entwicklung der Badefahrten und der „Naturbäder" im Gebiet der Schweiz und im angrenzenden südwestdeutschen Raum (1300–1600), Zürich.

Keller, Carl C. (1893): Zürcherische Apotheken und Apotheker, Separatdruck aus der Festschrift zur Erinnerung an die fünfzigjährige Stiftungsfeier des Schweizerischen Apotheker-Vereins in Zürich, Zürich.

Keller, Georg et al. (1564): Ein kurtzer bericht wie man sich mit der hilff Gottes vor dem praesten der Pestilentz hueten unnd bewaren / unnd so der mensch damit angriffen / was mittel er darfür bruche solle: mengklichem zuo guotem / durch die S[t]attartzet der S[t]att Zürych gestellt, Zürich.

Keller, Hildegard Elisabeth (2008): Fischsprüche, in: dies. et al. (2008): Jakob Ruf. Leben, Werk und Studien, vol. 4, Zürich, pp. 967–993.

Keller, Hildegard Elisabeth (2008a): Kalender zwischen 1543 und 1558, in: Keller, Hildegard Elisabeth et al. (2008), vol. 4, pp. 899–926.

Keller, Hildegard Elisabeth et al. (2008): Jakob Ruf. Leben, Werk und Studien, 5 vols., Zürich.

Keller, Otto (1979): Apotheken und Apotheker der Stadt Schaffhausen, in: Schaffhauser Beiträge zur Geschichte 56, pp. 29–142.

Kentmann, Johannes (1565): Nomenclaturae rerum fossilium quae in Misnia praecipue, et in aliis quoque regionibus inventur, in: Gessner (1565a).

Kentmann, Johannes (1565a): Calculorum qui in corpore ac membris hominum innascuntur genera XII ..., in: Gessner (1565a).

Kettler, Wilfried (2008): Untersuchungen zur frühneuhochdeutschen Lexikographie in der Schweiz und im Elsass: Strukturen, Typen, Quellen und Wirkungen von Wörterbüchern am Beginn der Neuzeit, Bern.

Koch, Ernst (1968): Die Theologie der Confessio Helvetica Posterior. Beiträge zur Geschichte und Lehre der Reformierten Kirche 27, Neukirchen-Vluyn.

Köhler, Walther (1937): Medizinisches bei Zwingli, in: Viertelsjahrsschrift der Naturforschenden Gesellschaft in Zürich 62, pp. 457–468.

Koelbing, Huldrych M. (1961): Ophtalmologisches bei Conrad Gessner (1516–1565), in: Gesnerus 18, pp. 13–21.

Koelbing, Huldrych M. (1984): Zur Geschichte der Pest in der Schweiz, in: Jahrbuch für solothurnische Geschichte 57, pp. 5–12.

Koelbing, Huldrych M. (1986): Pestsperre am Gotthard. Der mailändisch-schweizerische Vertrag von 1585 im Lichte neuerer Forschungen, in: Praxis – Schweizerische Rundschau für Medizin 75, pp. 228–231.

Kölbl, Bernhard A. (2012): Autorität und Autorschaft. Heinrich Glarean als Vermittler seiner Musiktheorie, Elementa musicae 6, Wiesbaden.

Kohler, Alfred (2003): Ferdinand I. 1503–1564. Fürst, König und Kaiser, Munich.

Krüger, Nilüfer (1978): Suppellex epistolica Uffenbachii et Wolfiorium = Katalog der Uffenbach-Wolfschen Briefsammlung, Katalog der Handschriften der Staats- und Universitätsbibliothek Hamburg, 2 Teilbände, Hamburg.

Krumenacker, Yves (2009): Lyon 1562, capitale protestante. Une histoire religieuse de Lyon à la Renaissance, Lyon.

Kühlmann, Wilhelm and Telle, Joachim (2001/2004): Der Frühparacelsimus, 2 vols., Frühe Neuzeit 59 und 89, Tübingen.

Kusukawa, Sachiko (2009): Image, Text and Observatio: The Codex Kentmanus, in: Early Science and Medicine 14, pp. 445–475.

Kusukawa, Sachiko (2010): The Sources of Gessner's Pictures for the *Historia animalium*, in: Annals of Science 67, pp. 303–328.

Kusukawa, Sachiko (2011): Patron's Review. The Role of Images in the Development of Renaissance Natural History, in: Archives of Natural History 38/2, pp. 189–213.

Kusukawa, Sachiko (2012): Picturing the Books of Nature. Image, Text, and Argument in Sixteenth-Century Human Anatomy and Medical Botany, Chicago.

Kusukawa, Sachiko (2014): Conrad Gessner on an "Ad Vivum" Image, in: Pamela H. Smith et al. (eds.): Ways of Making and Knowing. The Material Culture of Empirical Knowledge, Ann Arbor (MI), pp. 330–356.

Kyber, David (1553): Lexicon rei herbariae trilingue …, Strasbourg.

Lai, Yu-chih (2018): Domesticating the Global and Materializing the Unknown. A Study of the *Album of Beasts* at the Qianlong Court, in: Anna Grasskamp and Monica Juneja (eds.): EurAsian Matters. China, Europe, and the Transcultural Object, 1600–1800, Cham, pp. 125–171.

Landolt, Elias (2013): Flora des Sihltals: von der Stadt Zürich bis zum Höhronen, Zürich (PDF online: chrome-extension://efaidnbmnnnibpcajpcglclefindmkaj/https://www.botlib.uzh.ch/static/pdf/Flora_des_Sihltals_Publ_03_2013.pdf).

Lavater, Ludwig (1559): De ritibus et institutis ecclesiae Tigurinae opusculum, Zürich.

Leemann-van Elck, Paul (1940): Die Offizin Froschauer. Zürichs berühmte Druckerei im 16. Jahrhunderts, Mitteilungen der Antiquarischen Gesellschaft 33/2, 104. Neujahrsblatt, Zürich.

Leemann-van Elck, Paul (1935): Der Buchschmuck in Conrad Gessners naturgeschichtlichen Werken, Zürich.

Leemann-van Elck, Paul (1939): Die Offizin Gessner zu Zürich im 16. Jahrhundert, in: Der Schweizer Sammler 13, pp. 1–7 and pp. 153–160.

Legré, Ludovic (1904): La botanique en Provence au XVIe siècle. Les deux Bauhin Jean-Henri Cherler et Valerand Dourez, Marseille.

Lehmann, Paul (1956): Eine Geschichte der alten Fuggerbibliotheken, vol. 1, Tübingen.

Lendemann, Fritz (1996): Die wirtschaftliche Entwicklung im Stadtstaat Zürich, in: Niklaus Flüeler and Marianne Flüeler-Grauwiler (eds.): Geschichte des Kantons Zürich, vol. 2: Frühe Neuzeit, 16. bis 18. Jahrhundert, Zürich, pp. 126–171.

Leu, Urs B. (1990): Conrad Gesner als Theologe, Zürcher Beiträge zur Reformationsgeschichte 14, Bern etc.

Leu, Urs B. (1992): Conrad Gessner und die Neue Welt, in: Gesnerus 49 (1992), pp. 279–309.

Leu, Urs B. (1993): Marginalien Conrad Gessners als historische Quellen, in: Gesnerus 50, pp. 27–47.

Leu, Urs B. (1994): Originalien aus Conrad Gessners erdwissenschaftlicher Sammlung, in: Ernsting, Bernd (ed.): Georgius Agricola – Bergwelten 1494 1994, Essen, pp. 263f.

Leu, Urs B. (1999): Geschichte der Paläontologie in Zürich, in: Paläontologie in Zürich. Fossilien und ihre Erforschung in Geschichte und Gegenwart, Zoologisches Museum der Universität Zürich, Zürich, pp. 11–76.

Leu, Urs B. (2001): Heinrich Bullingers Widmungsexemplare seiner Schrift „Der Widertoeufferen ursprung …“ von 1560. Ein Beitrag zur europäischen Wirkungsgeschichte des Zürcher Antistes, in: Zwingliana 28, pp. 119–163.

Leu, Urs B. (2003): Die Privatbibliothek von Johannes Fries (1505–1565), in: Martin Graf and Christian Moser (eds.): Strenarum lanx. Beiträge zur Philologie und Geschichte des Mittelalters und der Frühen Neuzeit. Festgabe für Peter Stotz zum 40-jährigen Bestehen des Mittellateinischen Seminars der Universität Zürich, Zug, pp. 311–329.

Leu, Urs B. (2004): Die Zürcher Buch- und Lesekultur 1520–1575, in: Emidio Campi (ed.): Heinrich Bullinger und seine Zeit. Eine Vorlesungsreihe, Zürich, pp. 61–90 (= Zwingliana 31).

Leu, Urs B. (2007): Aneignung und Speicherung enzyklopädischen Wissens. Die Loci-Methode von Erasmus, in: Christ-von Wedel, Christine and Leu, Urs B. (eds.): Erasmus in Zürich. Eine verschwiegene Autorität, Zürich, pp. 327–342.

Leu, Urs B. et al. (2008): Conrad Gessner's Private Library, History of Science and Medicine Library 5, Leiden and Boston.

Leu, Urs B. (2008a): Textbooks and Their Uses – an Insight into the Teaching of Geography in 16th-Century Zurich, in: Emidio Campi et al. (eds.): Scholarly Knowledge. Textbooks in Early Modern Europe, Travaux d'Humanisme et Renaissance 447, Geneva, pp. 229–248.

Leu, Urs B. (2012): Natura sacra. Der Frühaufklärer Johann Jakob Scheuchzer (1672–1733), Zug.

Leu, Urs B. (2013): Der Basler Buchdruck von 1500 bis 1700: Versuch eines Überblicks, in: Corina Lanfranchi (ed.): Gut zum Druck! Streifzüge durch 525 Jahre Druck- und Verlagsgeschichte in Basel, Basel, pp. 136–145.

Leu, Urs B. (2014): Moral Treatment of Immoral Texts from Classical Antiquity: Conrad Gessner's Martial-Edition of 1544, in: Luca Baschera et al. (eds.): Following Zwingli. Applying the Past in Reformation Zurich, Surrey and Burlington, pp. 197–208.

Leu, Urs B. (2014a): Die Bedeutung Basels als Druckort im 16. Jahrhundert, in: Christine Christ-von Wedel et al. (eds.): Basel als Zentrum des geistigen Austauschs in der frühen Reformationszeit, SMHR 81, Tübingen, pp. 53–78.

Leu, Urs B. (2016): The Rediscovered Third Volume of Conrad Gessner's *Historia plantarum*, in: Ann Blair and Anja Goeing (eds.): For the Sake of Learning: Essays in Honor of Anthony Grafton, vol. 1, Leiden and Boston, pp. 415–422.

Leu, Urs B. (2016a): Conrad Gessners Netzwerk, in: Urs B. Leu and Mylène Ruoss (eds.): Conrad Gessner 1516–2016, Zürich, pp. 62–75.

Leu, Urs B. (2016b): Fünfzig unbekannte Briefe an Conrad Gessner, in: Librarium 59, S. 15–23.

Leu, Urs B. (2019): Gegen Quacksalberei und Aberglauben: Ein Gutachten Conrad Gessners und seiner Kollegen zuhanden des Zürcher Rats, in: Leu/Opitz (2019), pp. 311–322.

Leu, Urs B. (2022): Johann Jakob Scheuchzer (1672–1733) – Pionier der Alpen- und Klimaforschung, Zürich.

Leu, Urs B. (2023): Conrad Gessner als Naturforscher, in: Conrad Gessner: Fossilienbuch, hsg. und übersetzt von Petra Schierl, Basel (in print).

Leu, Urs B. and Opitz, Peter (2019): Conrad Gessner (1516–1565). Die Renaissance der Wissenschaften / The Renaissance of Learning, Berlin.

Leu, Urs B. and Ruoss, Mylène (2016): Conrad Gessner 1516–2016, Zürich.

Leu, Urs B. and Weidmann, Sandra (2004): Heinrich Bullingers Privatbibliothek, Heinrich Bullinger Werke, Abteilung 1: Bibliographie, vol. 3, Zürich.

Leu, Urs B. and Weidmann, Sandra (2019): Huldrych Zwingli's Private Library, Studies in Medieval and Reformation Traditions 215, Leiden and Boston.

Leu, Urs B. and Weidmann, Sandra (2020): Der bibliophile Reformator. Die Privatbibliothek Rudolf Gwalthers, Bibliotheca Bibliographica Aureliana 255, Baden-Baden.

Ley, Willy (1929): Conrad Gesner. Leben und Werk, Münchener Beiträge zur Geschichte und Literatur der Naturwissenschaften und Medizin, Heft 15/16, Munich.

Lienhard, Luc (2014): Ein 400-jähriger Schatz. Pflanzen und Illustrationen aus dem Herbarium von Felix Platter (1536–1614), Ausstellungskatalog, Bern.

Locher, Gottfried W. (1979): Die Zwinglische Reformation im Rahmen der europäischen Kirchengeschichte, Göttingen.

Longeon, Claude (1979): La description du Mont Pilat par Jean Du Choul (1555), in: Etudes foréziennes 10, pp. 31–61.

Louthan, Howard (1994): Johannis Crato and the Austrian Habsburgs. Reforming A Counter-Reform Court, Studies in Reformed Theology and History Volume 2, Number 3, Princeton (NJ).

Lutz, Hans (1937): Conrad Gesners Beziehungen zu den Verlegern seiner Zeit nach seinen Pandekten von 1548, in: Mélanges offerts à Marcel Godet, Directeur de la Bibliothèque Nationale Suisse à Berne, Neuchâtel, pp. 109–117.

Mägdefrau, Karl (1992): Geschichte der Botanik. Leben und Leistung grosser Forscher, Stuttgart etc.

Magno, Alessandro Marzo (2012): L'alba dei libri. Quando Venezia ha fatto leggere il mondo, Milan.

Magnus, Olaus (1555): Historia de gentibus septentrionalibus …, Rome.

Magnus, Olaus (2006): Die Wunder des Nordens, erschlossen von Elena Balzamo und Reinhard Kaiser, Frankfurt/M.

Mahlmann-Bauer, Barbara (2002): Die Krise der Reformation. Johann Jakob Wicks Chronik aussergewöhnlicher Natur- und Himmelserscheinungen, in: Wolfgang Harms et al. (eds.): Wahrnehmungsgeschichte und Wissensdiskurs im illustrierten Flugblatt der Frühen Neuzeit (1450–1700), Basel, pp. 193–236.

Maler, Josua (1561): Die Teütsch spraach. Alle Woerter, namen, und arten zuo reden in hochteütscher spraach, dem ABC nach ordenlich gestellt, unnd mit guotem Latein gantz fleissig unnd eigentlich vertolmetscht …, Zürich.

Marcus Aurelius (1559): De seipso seu vita sua libri XII., ed. by Conrad Gessner, Zürich.

Mariotti, Mauro Giorgio (2004): Pier Andrea Mattioli Botanico, in: Daniela Fausti (ed.): La complessa scienza die semplici. Atti delle celebrazione per il V centennario della nascita di Pietro Andrea Mattioli, Siena, 12 Marzo–19 Novembre 2001, Siena [(= Gli atti dell' accademia delle scienze di Siena detta de' Fisiocritici, Serie XV, Tomo XX (Suppl.) 2001)], pp. 149–158.

Martial, Marcus Valerius (1544): Epigrammata …, ed. by Conrad Gessner, Zürich.

Martin, Henri-Jean (1983): Classements et conjonctures, in: Roger Chartier et Henri-Jean Martin (eds.): Histoire de l'édition française, Le livre conquérant, Tome I, Paris, pp. 429–457.

Mattioli, Pietro Andrea (1554): Commentarii, in libros sex Pedacii Dioscoridis Anazarbei, de medica material, Venice.

Mattioli, Pietro Andrea (1558): Commentarii secundo aucti, in libros sex Pedacii Dioscoridis Anazarbei de medica materia, Venice.

Mauelshagen, Franz (2003): Entzauberung vor der Entzifferung. Conrad Gessner und die zoologische Rezeption der 'Hieroglyphika' Horapollons im 16. Jahrhundert, in: Aleida and Jan Assmann (eds.): Hieroglyphen. Stationen einer abendländischen Grammatologie. Archäologie der literarischen Kommunikation VIII, Munich, pp. 221–243.

Mauelshagen, Franz (2010): Klimageschichte der Neuzeit 1500–1900, Darmstadt.

Mauelshagen, Franz (2011): Wunderkammer auf Papier. Die Wickiana zwischen Reformation und Volksglaube, Frühneuzeit-Forschungen 15, Tübingen.

Mela, Pomponius (1522): De orbis situ libri tres, Basel.

Melissa, Antonius and Maximus the Confessor (1546): Sententiarum sive capitum, theologicorum praecipue, ex sacris et profanis libris, tomi tres ..., Theophili Sexti Antiochensis episcopi de Deo et fide Christianorum contra gentes institutionum libri tres ..., Tatiani Assyri ... oratio contra Graecos ..., Zürich.

Menzel, Ulrich (2015): Die Ordnung der Welt. Imperium und Hegemonie in der Hierarchie der Staatenwelt, Berlin.

Meyer, Andreas (1986): Zürich und Rom. Ordentliche Kollatur und päpstliche Provisionen am Frau- und Grossmünster 1316–1523, Bibliothek des Deutschen Historischen Instituts in Rom, vol. 64, Tübingen.

Meyer, Helmut (1976): Der Zweite Kappeler Krieg. Die Krise der Schweizerischen Reformation, Zürich.

Meyer, Helmut (1981): Der Zweite Kappeler Krieg. Gedenkschrift zur 450. Wiederkehr des Todestages von Huldrych Zwingli, 11. Oktober 1531–1. Oktober 1981, Zürich.

Meyer, Hermann (1883): Die schweizerische Sitte der Fenster- und Wappenschenkung vom XV. bis XVII. Jahrhundert. Nebst Verzeichniss der Zürcher Glasmaler von 1540 an und Nachweis noch vorhandener Arbeiten derselben, Frauenfeld.

Meyer-Ahrens, Conrad (1860): Geschichte des medizinischen Unterrichtes in Zürich, in: Denkschrift der medizinisch-chirurgischen Gesellschaft des Kantons Zürich zur Feier des fünfzigsten Stiftungstages den 7. Mai 1860, Zürich, pp. 1–40.

Miani, Laura (2004): Ulisse Aldrovandi. Pandechion epistemonicon, in: Biancastella Antonino (ed.): Tesori della Biblioteca Universitaria di Bologna: Codici, libri rari e altre meraviglie, Bologna, pp. 140f.

Michel, Paul (2015): Monströses in der frühneuzeitlichen Zoologie, in: Peggy Grosse et al.: Monster. Fantastische Bilderwelten zwischen Grauen und Komik, Nuremberg, pp. 47–59.

Midelfort, H. C. Erik (1982): Witchcraft, Magic, and the Occult, in: Steven Ozment (ed.): Reformation Europe. A Guide to Research, St. Louis (MO), pp. 183–209.

Milt, Bernhard (1928): Epilepsie. Medizinhistorische Dokumente aus dem 16. Jarhundert, in: Schweizerische Medizinische Wochenschrift 58, no. 36, pp. 890–906.

Milt, Bernhard (1929): Conrad Gesner und Paracelsus, in: Schweizerische Medizinische Wochenschrift 59, no. 18/19, pp. 486–488 und 506–509.

Milt, Bernhard (1931): Beitrag zum Problem der Kropfkrankheit im 16. Jahrhundert, in: Schweizerische Medizinische Wochenschrift 61, no. 12, pp. 280–281.

Milt, Bernhard (1945): Conrad Gessner als Balneologe, in: Gesnerus 2, pp. 1–16.

Milt, Bernhard (1947): Zürichs Vergangenheit in Naturwissenschaft und Medizin (Mittelalter und 16. Jahrhundert), in: Gesnerus 4, pp. 19–43.

Milt, Bernhard (1950): Gilg Tschudi als Patient Conrad Gessners (1516–1565), in: Vierteljahresschrift der Naturforschenden Gesellschaft in Zürich 95, pp. 58–64.

Milt, Bernhard (1953): Chemisch-alchemistische Heilkunde und ihre Auswirkungen in Zürich, in: Vierteljahrsschrift der Naturforschenden Gesellschaft in Zürich 98, pp. 178–215.

Modestino, Riccardo and Nascimbeni, Andrea (2014): La cultura ebraica e la società civile a Ferrara, in: Medici ebrei e la cultura ebraica a Ferrara, Ferrara, pp. 62–78.

Mörikofer, Johann Caspar (1867/1869): Ulrich Zwingli nach den urkundlichen Quellen, 2 vols., Leipzig.

Moledina, Sheza (2012): Sur les pas des imprimeurs lyonnais, Lyon.

Montaigne, Michel de (2007): Tagebuch einer Reise nach Italien über die Schweiz und Deutschland. Aus dem Französischen von Ulrich Bossier. Mit einem Vorwort von Wilhelm Weigand, Zürich.

Montgomery, Scott L. and Kumar, Alok (2016): A History of Science in World Cultures. Voices of Knowledge, London and New York.

Moreni, Alberto (1989): La Bibliotheca Universalis di Conrad Gesner e gli Indici dei libri proibiti, in: La Bibliofilìa 91, pp. 131–150.

Morgenstern, Karl (1821): Vom Sprachstudium, besonders dem Studium der Griechischen und Lateinischen Sprache, und des classischen Alterthums ueberhaupt, als einem wesentlichen Haupttheile der akademischen Studien, in: Karl Morgenstern (ed.): Doerptische Beytraege fuer Freunde der Philosophie, Litteratur und Kunst, vol. 3, Dorpat and Leipzig, pp. 333–395.

Moss, Ann (1996): Printed Commonplace-Books and the Structuring of Renaissance Thought, Oxford.

Müllener, Eduard-Rudolf (1965): Conrad Gessners Illustrationen zu "De Anima", in: Gesnerus 22, pp. 160–175.

Müller, Clemens (2008): Goldpillen und gestählter Wein. Pharmakotherapie und Pharmazeutik im medizinischen Werk Jakob Rufs, in: Keller, Hildegard Elisabeth et al. (2008): Jakob Ruf. Leben, Werk und Studien, vol. 5, Zürich, pp. 268–340.

Müller, Clemens (2011): „... so wil ich üch ain liepplich Pülferli schiken ...". Doktor Joachim von Watts arabistische Arztpraxis in St. Gallen, in: Karl Schmuki et al. (eds.): Zeit für Medizin! Einblick in die St. Galler Medizingeschichte, Neujahrsblatt des Historischen Vereins des Kantons St. Gallen, St. Gallen, pp. 37–44.

Müller, Clemens (2017): Josua Kesslers Trauergedichte auf Vadian, in: Gamper (2017), pp. 355–360.
Muller, Frank (1990): Heinrich Vogtherr der Ältere (1490–1556). Aspekte seines Lebens und Werkes, in: Jahrbuch des historischen Vereins Dillingen an der Donau, 92 (1990), pp. 173–274.
Muller, Frank (1997): Heinrich Vogtherr l'Ancien. Un artiste entre Renaissance et Réforme, Wolfenbütteler Forschungen 72, Wolfenbüttel.
Näf, Beat (2004): Traum und Traumdeutung im Altertum, Darmstadt.
Näf, Werner (1944): Vadian und seine Stadt St. Gallen, 2 vols., St Gallen.
Naegeli, Marianne et al. (1981): Zürcher Kunst nach der Reformation. Hans Asper und seine Zeit, Katalog zur Ausstellung im Helmhaus, Zürich, 9. Mai bis 28. Juni 1981, Zürich.
Nebinger, Gerhart and Rieber, Albrecht (1978): Genealogie des Hauses Fugger von der Lilie: Stammtafeln, Schwäbische Forschungsgemeinschaft bei der Kommission für bayerische Landesgeschichte, Reihe 4, Band 17, Studien zur Fuggergeschichte 26, Tübingen.
Nuovo, Angela (2013): The Book Trade in the Italian Renaissance, Library of the Written Word 26, Leiden and Boston.
Nutton, Vivian (1985): Humanist Surgery, in: Andrew Wear et al. (eds.): The Medical Renaissance of the Sixteenth Century, Cambridge, pp. 75–99.
Nutton, Vivian (2008): The Fortunes of Galen, in: Robert J. Hankinson (ed.): The Cambridge Companion to Galen, Cambridge, pp. 355–390.
Nutton, Vivian (2018): An Autobibliography by John Caius, London.
Ohl, Michael (2015): Die Kunst der Benennung, Berlin.
Olariu, Dominic (2022): Johannes Kentmann und der „Codex Kentmanus". Bilder im Netzwerk der Naturkunde im 16. Jahrhundert, in: Jürgen Herzog (ed.): Johann Kentmann und die Torgauer Gärten, Schriften des Torgauer Geschichtsvereins 16, Königsbrück, S. 61–82.
Oliver, Eugène (1951): Les années Lausannoises (1537–1540) de Conrad Gesner, in: Schweizerische Zeitschrift für Geschichte 1, pp. 369–428.
Olmi, Giuseppe (1992): L'inventario del mondo. Catalogazione della natura e luoghi del sapere nella prima età moderna, Annali dell'Istituto storico italo-germanico, Monografia 17, Bologna.
Olson, Roberta J. M. and Mazzitelli, Alexandra (2007): The Discovery of a Cache of over 200 Sixteenth Century Avian Watercolors. A Missing Chapter in the History of Ornithological Illustration, in: Master Drawings 45, pp. 435–521.
Omont, H. (1891): Une lettre de Conrad Gesner à David Chytraeus (1543), in: Centralblatt für Bibliothekswesen 8, pp. 122f.
Orelli, Johann Caspar von (1837): Der Briefwechsel zwischen Conrad Gessner und Henricus Stephanus, Neujahrsblatt der Stadtbibliothek Zürich auf das Jahr 1837, Zürich.

Palmer, Richard (1985): Medical Botany in Northern Italy in the Renaissance, in: Journal of the Royal Society of Medicine 78, pp. 149–157.

Palmer, Richard (1985a): Pharmacy in the Republic of Venice in the Sixteenth Century, in: Andrew Wear et al. (eds.): The Medical Renaissance of the Sixteenth Century, Cambridge, pp. 101–117 and 303–312.

Partridge, Loren (2015): Art of Renaissance Venice 1400–1600, Oakland (CA).

Peine, Josef (1941): Das dritte Buch der medizinischen Briefe des Züricher Philosophen und Arztes Conrad Gesner, Düsseldorf.

Pellikan, Conrad (1892): Die Hauschronik, Deutsch von Theodor Vulpinus, Strasbourg.

Perfetti, Stefano (2000): Aristoteles Zoology and its Renaissance Commentators (1521–1601), Ancient and Medieval Philosophy, De Wulf-Mansion Centre, Series 1, vol. 27, Leuven.

Peters, Manfred (1984): Theodor Bibliander, De ratione communi omnium linguarum et literarum commentarius, Zürich 1548, in: Archiv für das Studium der neueren Sprachen und Literaturen 221, pp. 1–18.

Peters, Manfred (2002): Das 16. Jahrhundert als sprachtheoretische Epochenschwelle. Theodor Bibliander und Conrad Gessner, in: Peter Wiesinger and Hans Derkits (eds.): Akten des X. Internationalen Germanistenkongresses Wien 2000. „Zeitenwende – Die Germanistik auf dem Weg vom 20. ins 21. Jahrhundert", vol. 3 (= Jahrbuch für Internationale Germanistik, Reihe A, Kongressberichte, vol. 55), Bern etc., pp. 113–118.

Peters, Manfred (2019): Conrad Gessner, Bahnbrecher der Allgemeinen Linguistik, in: Leu/Opitz (2019), pp. 499–516.

Pettegree, Andrew (2010): The Book in the Renaissance, New Haven and London.

Pfister, Rudolf (1952): Die Seligkeit erwählter Heiden bei Zwingli. Eine Untersuchung zu seiner Theologie, Zürich.

Pinon, Laurent (1995): Livres de zoologie de la Renaissance: une anthologie (1450–1700), Paris.

Pinon, Laurent (2002): Clématite bleue contre poissons séchés. Sept lettres inédites d'Ippolit Salviani à Ulisse Aldrovandi, in: Mélanges de l'école francaise de Rome. Italie et Méditerranée 114, pp. 477–492.

Pinon, Laurent (2005): Conrad Gessner and the Historical Depth of Renaissance Natural History, in: Gianna Pomata and Nancy G. Siraisi (eds.): Historia. Empiricism and Erudition in Early Modern Europe, Cambridge (MA), pp. 241–267.

Platter, Felix (1976): Tagebuch (Lebensbeschreibung) 1536–1567, ed. by Valentin Lötscher, Basler Chroniken 10, Basel.

Platter, Thomas (1999): Lebensbeschreibung, hsg. v. Alfred Hartmann, 2. Aufl. durchgesehen und ergänzt von Ueli Dill, mit einem Nachwort von Holger Jacob-Friesen, Basel.

Pollack, Detlef and Pohlig, Matthias (eds., 2020): Die Verwandlung des Heiligen. Die Geburt der Moderne aus dem Geist der Religion, Berlin.

Pollet, Jacques Vincent (1977): Julius Pflug Correspondance ..., vol. 3, Leiden.

Poskett, James (2022): Neue Horizonte. Eine globale Geschichte der Wissenschaft, Munich.

Prescher, Hans; Helm, Johannes Helm and Fraustadt, Georg (1980): Johannes Kentmanns Mineralienkatalog aus dem Jahre 1565, Abhandlungen des Staatlichen Museums für Mineralogie und Geologie zu Dresden, Band 30: Beiträge zur Geschichte der Mineraliensammlungen und des Bergbaus im 16. und frühen 17. Jahrhundert, Leipzig.

Pyle, Cynthia M. (1984): Das Tierbuch des Petrus Candidus, geschrieben 1460, ausgemalt im 16. Jahrhundert, Codices e Vaticanis selecti quam simillime espressi iussu Ioannis Pauli PP II consilio et opera curatorum bibliothecae Vaticanae 60, Zürich.

Pyle, Cynthia M. (2000): Conrad Gessner on the Spelling of his Name, in: Archives of Natural History 27, pp. 175–186.

Raimondi, Claudio (1906): Lettere di P. A. Mattioli ad Ulisse Aldrovandi, in: Bullettino Senese di Storia Patria 13, pp. 121–185.

Rand, Mare (2014): „Ramulus cum flore melius pingitur in meis ...". Verschollene Pflanzenzeichnungen aus dem Gessner-Nachlass in der Universitätsbibliothek Tartu/Dorpat, in: Hanspeter Marti (ed.): Kulturaustausch. Baltisches Echo auf Gelehrte in der Schweiz und in Deutschland. Freundesgabe für Arvo Tering, Cologne, pp. 13–48.

Rath, Gernot (1950): Die Briefe Conrad Gessners aus der Trewschen Sammlung, 1. Teil, in: Gesnerus 7, pp. 140–170.

Rath, Gernot (1951): Die Briefe Conrad Gessners aus der Trewschen Sammlung, 2. Teil, in: Gesnerus 8, pp. 195–215.

Reeds, Karen M. (2006): Leonardo da Vinci and Botanical Illustration: Nature Prints, Drawings and Woodcuts ca. 1500, in: Jean A. Givens et al. (eds.): Visualizing Medieval Medecine and Natural History, 1200–1500, AVISTA Studies in the History of Medieval Technology, Science and Art 5, Aldershot, pp. 205–237.

Régnier-Roux, Daniel (2014): Les éditions lyonnaises entre 1554 et 1559 du Trésor des remèdes secrets de Conrad Gessner, in: Revue de l'enssib 2014/2, pp. 1–43 (nur elektronisch einsehbar unter: http://revue.enssib.fr/les-%C3%A9ditions-lyonnaises-entre-1554-et-1559-du-tr%C3%A9sor-des-rem%C3%A8des-secrets-de-conrad-gessner-r%C3%A9gnier).

Reichholf, Josef H. (2015): Einhorn, Phönix, Drache. Woher unsere Fabeltiere kommen, Frankfurt/M.

Reske, Christoph (2007): Die Buchdrucker des 16. und 17. Jahrhunderts im deutschen Sprachgebiet, Auf der Grundlage des gleichnamigen Werkes von Josef Benzing, Beiträge zum Buch- und Bibliothekswesen 51, Wiesbaden.

Reyer, Heinz-Ulrich (2021): Rendezvous der Fabelwesen. Drache, Einhorn & Co zwischen Mythos und Wirklichkeit, Darmstadt.

Richter, Günter (1965): Verlegerplakate des 16. und 17. Jahrhunderts bis zum Beginn des Dreissigjährigen Krieges, Wiesbaden.

Richter, Günter (1986): Konzessionspraxis und Zahl der Druckereien in Frankfurt a. M. um 1600, in: Archiv für Geschichte des Buchwesens 27, pp. 131–156.

Riedl-Dorn, Christa (1989): Wissenschaft und Fabelwesen. Ein kritischer Versuch über Conrad Gessner und Ulisse Aldrovandi, Perspektiven der Wissenschaftsgeschichte 6, Vienna and Cologne.

Riedweg, Christoph (2000): Ein Philologe an Zwinglis Seite. Zum 500. Geburtstag des Zürcher Humanisten Jacob Wiesendanger, gen. Ceporinus (1500–1525), in: Museum Helveticum 57, pp. 201–219.

Robinson, Hastings (1842): The Zurich Letters … During the Early Part of the Reign of Queen Elizabeth [1558–1579], Cambridge.

Roeck, Bernd (2005): Geschichte Augsburgs, Munich.

Rohls, Jan (2012/2013): Der Aristotelismus an den reformierten Ausbildungsstätten, in: Wolfenbütteler Renaissance-Mitteilungen 34 (2012/2013), pp. 65–83.

Rossiaud, Jacques (2012): Lyon 1250–1550. Réalités et imaginaires d'une métropole, Seyssel.

Roth, Friedrich (1907): Augsburgs Reformationsgeschichte, vol. 3, 1539–1547, bezw. 1548, Munich.

Roth, Friedrich (1928): Zur Lebensgeschichte des Augsburger Stadtadvokaten Dr. Claudius Pius Peutinger (1509–1552), in: Archiv für Reformationsgeschichte 25, pp. 99–127 und 162–255.

Rozzo, Ugo (2008): Sébastien Gryphe editore di umanisti ed "eretici" italiani (1524–1542), in: Raphaële Moren (ed.): Quid novi? Sébastien Gryphe, à l'occasion du 450e anniversaire de sa mort. Actes du colloque 23 au 25 novembre 2006, Lyon-Villeurbanne, Bibliothèque municipale de Lyon, Villeurbanne, pp. 111–152.

Rudwick, Martin (1972): The Meaning of Fossils. Episodes in the History of Palaeontology, Chicago and London.

Rüegg, Michael (2014): An den Grenzlinien der Wissenschaft. Zur Kritik fataler Zukunftsversprechen von Ganzheitsmedizin und Hirnforschung, Zürcher Arbeiten zur Philosophie 3, Basel.

Rütiner, Johannes (1996): Diarium 1529–1539, Lateinischer Text und Übersetzung, 3 vols., ed. by Ernst Gerhard Rüsch, St Gallen.

Ruf, Jakob (1556): Libellus de tumoribus quibusdam phlegmaticis non naturalibus …, Zürich.

Ruoss, Mylène (2016): „Fenestras XV. in quibus … piscium species … in vitro depictae sunt." Zu Conrad Gessners Fensterschmuck in seinem Studierzimmer, in: Leu and Ruoss (2016), pp. 197–211.

Ruoss, Mylène (2019): Conrad Gessners Einfluss auf die nachreformatorische Kunst, in: Leu/Opitz (2019), pp. 195–233.

Sabba, Fiammetta (2012): La ‚Bibliotheca universalis' di Conrad Gesner. Monumento della cultura europea, Il Bibliotecario 25, Rome.

Salzmann, Charles (1940): Thaddeo Duno von Locarno als Stadtarzt in Zürich. Eine biographische Studie, in: Vierteljahrsschrift der Naturforschenden Gesellschaft in Zürich 85, pp. 337–344.
Salzmann, Charles (1956): Ein Brief von Gerolamo Cardano an Conrad Gessner 1555, in: Gesnerus 13, pp. 53–60.
Salzmann, Charles (1959): Francesco Calzolari, der Erforscher der Flora des Monte Baldo, und seine Pflanzensendungen an Conrad Gessner in Zürich, in: Gesnerus 16, pp. 81–103.
Salzmann, Charles (1960): Die Editio princeps des Marc Aurel durch Conrad Gessner 1559, in: Verhandlungen der Schweizerischen Naturforschenden Gesellschaft. Wissenschaftlicher und administrativer Teil, vol. 140, pp. 153f.
Salzmann, Charles (1965): Conrad Gessners Persönlichkeit, in: Gesnerus 22, pp. 115–133.
Samson, Alexander (2006): Diego Hurtado de Mendoza and the Jewess of Venice: Tolerance, Interfaith Sexuality and *Converso* Culture, in: Richrad J. Pym (ed.): Rhetoric and Reality in Early Modern Spain, London, pp. 57–71.
Schierl, Petra (2019): Gessner und Agricola über „Fossilien“: Naturforschung zwischen Autopsie und Tradition, in: Leu/Opitz (2019), pp. 145–158.
Schiess, Traugott (1905): Bullingers Korrespondenz mit den Graubündnern, 2. Teil, Basel.
Schiess, Traugott (1910): Briefwechsel der Brüder Ambrosius und Thomas Blaurer 1509–1548, Band 2, August 1538–Ende 1548, Freiburg i. Br.
Schindler, Alfred (1996): Huldrych Zwingli e Giovanni Pico della Mirandola, in: Dall'Accademia neoplatonica fiorentina alla riforma. Celebrazioni del V centenario della morte di Lorenzo il Magnifico, Convegno di studio, Firenze, Palazzo Strozzi, 30 ottobre 1992, Florence, pp. 51–65.
Schindling, Anton (1977): Humanistische Hochschule und frei Reichsstadt. Gymnasium und Akademie in Strassburg 1538–1621, Veröffentlichungen des Instituts für europäische Geschichte Mainz 77, Wiesbaden.
Schlegel, Birgit and Schlegel, Kristian (2011): Polarlichter zwischen Wunder und Wirklichkeit. Kulturgeschichte und Physik einer Himmelserscheinung, Heidelberg.
Schmid-Forrer, Jacqueline (1983): Die Familie Forrer von Winterthur, 3 vols., verfielfältigtes Typoskript, Winterthur.
Schmiedel, Casimir Christoph (1754): Conradi Gesneri … Opera botanica …, vol. 1, Nuremberg.
Schottenloher, Karl (1953): Die Widmungsvorrede im Buch des 16. Jahrhunderts, Reformationsgeschichtliche Studien und Texte 76/77, Münster.
Schouppé, Anton von (1991): Episoden aus der Korallenforschung vor 1800. VI. International Symposium on Fossil Cnidaria including Archaeocyatha and Porifera, Inaugural Lecture, Münster.
Schwarz-Zanetti, Gabriela and Fäh, Donat (2011): Grundlagen des makroseismischen Erdbebenkatalogs der Schweiz, vol. 1, 1000–1680, Zürich.

Sergeev, Michail (2019): Der *Mithridates* (1555) zwischen Sprachmuseum und neulateinischem Onomastikon: einige Überlegungen zur Konzeption und zum Genre des Gessnerschen Handbuchs, in: Leu/Opitz (2019), pp. 517–532.

Serrai, Alfredo (1990): Conrad Gesner, Il Bibliotecario 5, Rome.

Simmler, Josias (1555): Epitome bibliothecae Conradi Gesneri ..., Zürich.

Simmler, Josias (1555a): Appendix bibliothecae Conradi Gesneri ..., Zürich.

Simmler, Josias (1566): Vita clarissimi philosophi et medici excellentissimi Conradi Gesneri ..., Zürich.

Siraisi, Nancy G. (1987): Avicenna in Renaissance Italy. The *Canon* and Medical Teaching in Italian Universities after 1500, Princeton (NJ).

Somogyi, Joseph de (1950): Ad-Damīrī's Hayāt al-hayawān. An Arabic Zoological Lexicon, in: Osiris 9, pp. 33–43.

Sotomaior, Antonio (1667): Index librorum prohibitorum et expurgandorum novissimus ..., Madrid.

Spillmann, Kurt (1962): Zwingli und die Zürcher Schulverhältnisse, in: Zwingliana 11, pp. 427–448.

Spinks, Jennifer (2008): Jakob Ruf's 1554 Trochbüchle: A Zurich Physician Explains and Interprets Monstrous Births, in: Intellectual History Review 18, pp. 41–59.

Sprecher, Johann Andreas von (1976): Kulturgeschichte der Drei Bünde im 18. Jahrhundert, bearbeitet von ... Rudolf Jenny, Chur.

Springer, Katharina B. and Kinzelbach, Ragnar K. (2009): Das Vogelbuch von Conrad Gessner (1516–1565). Ein Archiv für avifaunistische Daten, Berlin and Heidelberg.

Steiger, Rudolf (1978): Conrad Gessners Itinerar seiner Bündner Reise von 1561, in: Gesnerus 35, pp. 214–223.

Steinke, Hubert (2008): Medizinische Karriere im städtischen Dienst, in: Keller, Hildegard Elisabeth et al. (2008), vol. 1, pp. 91–102.

Steinke, Hubert (2008a): Ein gelehrter Chirurg, in: Keller, Hildegard Elisabeth et al. (2008), vol. 1, pp. 105–117.

Steinke, Hubert (2008b): Flugblatt von der Schaffhauser Missgeburt, in: Keller, Hildegard Elisabeth et al. (2008), vol. 2, pp. 415–429.

Steinke, Hubert and Müller, Clemens (2008): Tumorbüchlein, in: Keller, Hildegard Elisabeth et al. (2008), vol. 4, pp. 786–882.

Steinmann, Judith (2000): Züricher Einbände aus dem 16. Jahrhundert, in: Einbandforschung 6, pp. 10ff.

Steinmann, Martin (1967): Johannes Oporinus. Ein Basler Buchdrucker um die Mitte des 16. Jahrhunderts, Basler Beiträge zur Geschichtswissenschaft 105, Basel.

Steinmann, Martin (1969): Aus dem Briefwechsel des Basler Druckers Johannes Oporinus, in: Basler Zeitschrift für Geschichte und Altertumskunde 69, pp. 103–203.

Stettler, Bernhard (2014): Überleben in schwieriger Zeit. Die 1530er und 1540er Jahre im Spiegel von Vadians Korrespondenz, Zürich.

Stevenson, Joseph (1866): Calendar of State Papers of the Reign of Elizabeth, 1561–1562 …, vol. 4, London.

Stobaeus, Johannes (1543): Sententiae ex thesauris Graecorum delectae …, ed. by Conrad Gessner, Zürich.

Stobaeus, Johannes (1549): Sententiae ex thesauris Graecorum delectae …, ed. by Conrad Gessner, Basel.

Stolberg, Michael (2009): Die Harnschau. Eine Kultur- und Alltagsgeschichte, Köln etc.

Stotz, Peter (1967): Gessners Autobiographie in der „Bibliotheca universalis" 1545, in: Hans Fischer et al.: Conrad Gessner 1516–1565. Universalgelehrter, Naturforscher, Arzt, Zürich, pp. 212–214.

Strohl, Jean (1917): Conrad Gessner's Waldrapp, in: Vierteljahrsschrift der Naturforschenden Gesellschaft in Zürich 62, pp. 501–538.

Stucki, Heinzpeter (1996): Das 16. Jahrhundert, in: Niklaus Flüeler and Marianne Flüeler-Grauwiler (eds.): Geschichte des Kantons Zürich, vol. 2: Frühe Neuzeit, 16. bis 18. Jahrhundert, Zürich, pp. 172–281.

Suter-Meyer, Katharina (2020): Wissensvermittler – Kritiker – Autor: Joachim Vadians Kommentare zu Pomponius Mela (Basel 1522), Zürich.

Taffetani, Fabio (2012): Herbaria. Il grande libro degli erbari italiani, Florence.

Thylesius, Antonius (1545): Opuscula aliquot …, ed. by Conrad Gessner, Basel.

Tilemann, Hero (1939): Das zweite Buch der medizinischen Briefe des Züricher Philosophen und Arztes Conrad Gessner, Diss. med. Düsseldorf, Düsseldorf.

Treviranus, Ludolph Christian (1830): Caroli Clusii … et Conradi Gesneri … epistolae ineditae, Leipzig.

Trevisan, Renata (1995): Melchiorre Guilandino, in: Alessandro Minelli (ed.): L'Orto botanico di Padova 1545–1995, Venice, pp. 59–62.

Tuilier, André (1994): Histoire de l'université de Paris et de la Sorbonne, Tome 1, Paris.

Van den Abeele, Badouin (2002): Les albums ornithologiques de Jacques Daléchamps, médecin et naturaliste à Lyon (1513–1588), in: Archives internationales d'histoire des sciences 52, pp. 3–45.

Vandewiele, L. J. (1993): Wat groeide er in de tuin van Pieter van Coudenberghe, in: F. de Nave (ed.): De Botanica in de zuidelyke Nederlanden (einde 15de eeuw–ca. 1650), Publikaties van het Museum Plantin-Moretus en het Stedelyk Prentenkabinet 27, Antwerp, pp. 23–31.

Vischer, Daniel-Lucas (2015): Schiffe, Flösse und Schwemmholz. Unterwegs auf Schweizer Gewässern: eine Technikgeschichte, Baden.

Vischer, Manfred (1991): Bibliographie der Zürcher Druckschriften des 15. und 16. Jahrhunderts, Bibliotheca Bibliographica Aureliana 124, Baden-Baden.

Vischer, Manfred (2001): Zürcher Einblattdrucke des 16. Jahrhunderts, Bibliotheca Bibliographica Aureliana 185, Baden-Baden.

Vögelin, Salomon (1878): Das alte Zürich, vol. 1, Zürich (thoroughly revised 2nd ed.).

Vogel, Katja (2019): *Thrinodiae herois Huldrychi Zwinglii* – Conrad Gessners dichterischer Nachruf auf Huldrych Zwingli, in: Leu/Opitz (2019), pp. 465–484.

Von Moos, Peter (2014): Heiden im Himmel? Geschichte einer Aporie zwischen Mittelalter und Früher Neuzeit, Schriften der philosophisch-historischen Klasse der Heidelberger Akademie der Wissenschaften 54, Heidelberg.

Wackernagel, Rudolf (1881): Rechnungsbuch der Froben und Episcopius, Buchdrucker und Buchhändler zu Basel 1557–1564, Basel.

Wackernagel, Hans Georg (1956): Die Matrikel der Universität Basel, vol. 2, 1532/33–1600/01, Basel.

Waldis, Vera (1979): Obrigkeitliche Massnahmen gegen die Pest in Stadt und Herrschaft Rheinfelden im 16. und 17. Jahrhundert, in: Gesnerus 36, pp. 206–227.

Waldis, Vera (1983): Der „Stich" von 1564 – eine primäre Lungenpest, in: Gesnerus 40, pp. 223–228.

Walravens, Hartmut (1973): Conrad Gessner in chinesischem Gewand, in: Gesnerus 30, pp. 87–98.

Wartmann, Beat (2019): Conrad Gessners Orchideen in der *Historia plantarum*, in: Leu/Opitz (2019), pp. 83–118.

Wartmann, Wilhelm (1908): Les vitraux suisses au Musée du Louvre. Catalogue critique et raisonné, Paris.

Weber, Berchtold (2016): Die Wappen des Conrad Gessner und seiner Verwandten, in: Leu and Ruoss (2016), S. 212–220.

Weber, Max (2019): Wissenschaft als Beruf, Stuttgart.

Weber, Peter Xaver (1913): Der Pilatus und seine Geschichte, Lucerne.

Webster, Charles (1990): Conrad Gessner and the Infidelity of Paracelsus, in: Henry, John and Hutton, Sarah (eds.): New Perspectives on Renaissance Thought. Essays in the History of Science, Education and Philosophy in Memory of Charles B. Schmitt, London, pp. 13–23.

Wehrli, Gustav Adolf (1924): Der Zürcher Stadtarzt Dr. Christoph Clauser und seine Stellung zur Reformation der Heilkunde im XVI. Jahrhundert, Veröffentlichungen der Schweizerischen Gesellschaft für Geschichte der Medizin und der Naturwissenschaften 2, Zürich.

Wehrli, Gustav Adolf (1934): Die Krankenanstalten und die öffentlich angestellten Ärzte und Wundärzte im alten Zürich, Mitteilungen der Antiquarischen Gesellschaft in Zürich, Bd. 31, Heft 3, 98. Neujahrsblatt, Zürich.

Weitschat, Wolfgang and Wichard, Wilfried (2002): Atlas of Plants and Animals in Baltic Amber, Munich.

Wellisch, Hans H. (1978): Early Multilingual and Multiscript Indexes in Herbals, in: The Indexer 11/2, pp. 81–102.

Wellisch, Hans H. (1981): How to Make an index. 16th Century Style: Conrad Gessner on Indexes and catalogs, in: International Classification 8, pp. 10–15.

Wellisch, Hans H. (1984): Conrad Gessner – A Bio-Bibliography, Zug.
Wellisch, Hans H. (1986): The First Arab Bibliography: Fihrist al-ʿUlum, University of Illinois Occasional Papers 175, December 1986, Champaign (IL).
Widmann, Hans (1966): Nachwort zu: Bibliotheca universalis und Appendix [Faksimile], Milliaria 5, Osnabrück.
Wieland, Melchior and Gessner, Conrad (1558): De stirpibus aliquot epistolae, Padua.
Willich, Jodocus (1563): Ars magirica, hoc est, coquinaria, de cibariis, ferculis, opsoniis, alimentis et potibus diversis parandis … accedit, Iacobi Bifrontis Rhaeti de operibus lactariis epistola …, Zürich.
Winteler, Jakob (1956): Landschreiber Jakob Vogel von Glarus, in: Zwingliana 10, pp. 267–297.
Wirtz, Carolin (2006): Köln und Venedig. Wirtschaftliche und kulturelle Beziehungen im 15. und 16. Jahrhundert, Beihefte zum Archiv für Kulturgeschichte 57, Köln.
Wolf, Caspar (1566): ΥΠΟΣΧΕΣΙΣ sive de Conradi Gesneri stirpium historia ad Ioannem Cratonem … pollicitatio, in: Simmler (1566), f. 45r–56r.
Wolf, Caspar (1581): Alphabetum empiricum, sive, Dioscoridis et Stephani Atheniensis philosophorum et medicorum de remediis expertis liber, iuxta alphabeti ordinem digestus, Zürich.
Wolf, Norbert (2002): Trecento und Altniederländische Malerei, Kunst-Epochen 5, Stuttgart.
Wolfschmidt, Gudrun (2010): Astronomie in Nürnberg – Zentrum des Instrumentenbaus, in: Wolfschmidt, Gudrun (ed.): Astronomie in Nürnberg, Nuncius Hamburgensis: Beiträge zur Geschichte der Naturwissenschaften 3, Hamburg, pp. 19–143.
Wolkenhauer, Anja (2002): Zu schwer für Apoll. Die Antike in humanistischen Druckerzeichen des 16. Jahrhunderts, Wolfenbütteler Schriften zur Geschichte des Buchwesens 35, Wiesbaden.
Wotschke, Theodor (1908): Der Briefwechsel der Schweizer mit den Polen, Leipzig.
Wotschke, Theodor (1926): Polnische Studenten in Heidelberg, in: Jahrbücher für Kultur und Geschichte der Slaven, Neue Folge, vol. 2 (1926), pp. 46–67.
Zäh, Helmut (1997): Wolfgang Musculus und der Ankauf griechischer Handschriften für die Augsburger Stadtbibliothek 1543/44, in: Rudolf Dellsperger et al. (eds.): Wolfgang Musculus (1497–1563) und die oberdeutsche Reformation, Colloquia Augustana 6, Berlin, pp. 226–245.
Zanchi, Girolamo (1613): Opera theologica, vol. 8, Geneva.
Zaunick, Rudolph (1915): Fragmente der ältesten sächsischen Fischfauna des Dr. Johannes Kentmann (1518–1574), in: Abhandlungen der naturwissenschaftlichen Gesellschaft Isis in Dresden, pp. 15–36.
Zemon Davis, Natalie (1966): Publisher Guillaume Rouille, Businessman and Humanist, in: R. J. Schoeck (ed.): Editing Sixteenth Century Texts. Papers given at the Editorial Conference, University of Toronto, October, 1965, Toronto, pp. 72–112.

Zedelmaier, Helmut (1992): Bibliotheca universalis und Bibliotheca selecta. Das Problem der Ordnung des gelehrten Wissens in der frühen Neuzeit, Beihefte zum Archiv für Kulturgeschichte 33, Cologne.

Zoller, Heinrich et al. (1972–1980): Conradi Gesneri Historia plantarum. Faksimileausgabe, 8 vols., Dietikon.

Zoller, Heinrich and Steinmann, Martin (1987–1991): Conradi Gesneri Historia plantarum. Gesamtausgabe, 2 vols., Dietikon.

Zoller, Heinrich (1989): Zum Wandel der Pflanzendarstellung während der Renaissance. Vom Beginn des 15. Jahrhunderts zu Leonardo da Vinci, Albrecht Dürer und Conrad Gessner, in: Bauhinia 9/2, pp. 109–123.

Zuliani, Federico (2017): Gessner, Mainardo, un anonimo e la *Tora venenata*. Notizie di storia valdese da due carteggi a brandelli, in: Riforma e movimenti religiosi. Rivista della società die studi valdesi, 2017/2, pp. 207–225.

Zwingli, Huldrych (1929): Sämtliche Werke, ed. by Emil Egli et al., vol. 10, Zürich.

Zwingli, Huldrych (1983): De providentia Dei, in: Huldreich Zwinglis sämtliche Werke, vol. 6/3, Zürich.

Zwingli, Huldrych (1995): Erklärung des christlichen Glaubens 1531, in: Thomas Brunschweiler et al. (eds.): Huldrych Zwingli Schriften, vol. 4, Zürich, pp. 281–361.

Index of Names

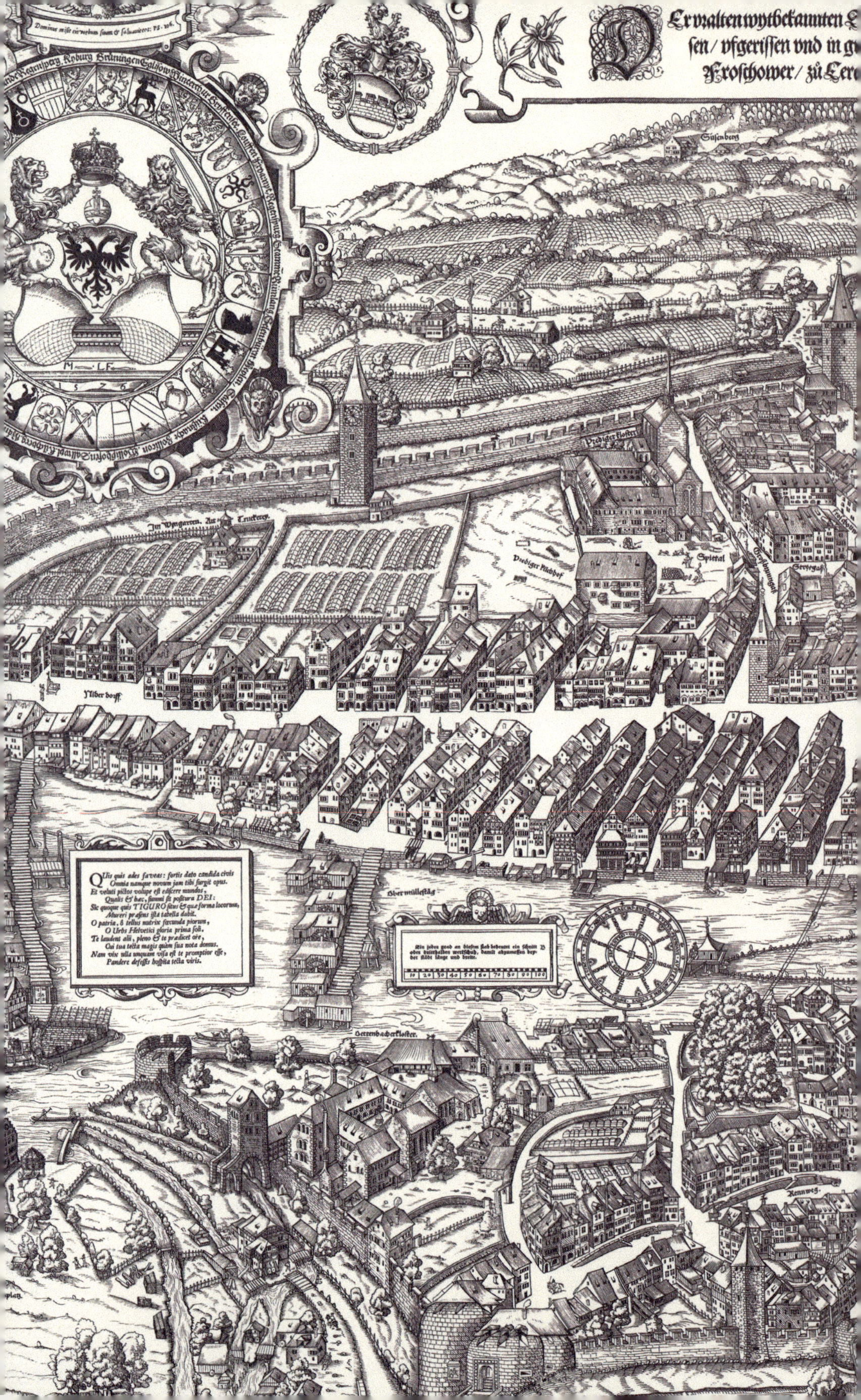
DEr valten wytbekannten
sen / vfgerissen vnd in g
Froschower / zů Ler
Süsenberg
Prediger Kilchhof
Spital
Nider dorff
Ober müllestäg
QUis quis ades faveas: fortis dato candida civis
Omnia namque novum jam tibi surgit opus.
Et veluti pictos volupe est ediscere mundos,
Qualis & hæc, summi sit positura DEI:
Sic quoque quis TIGURO situs & quæ forma locorum,
Mureri præsens ista tabella dabit.
O patria, ô tellus nutrix fœcunda piorum,
O Urbs Helvetici gloria prima soli.
Te laudent alii, pleno & te prædicet ore,
Cui tua tecta magis quàm sua nota domus.
Nam vix ulla unquam visa est te promptior esse,
Pandere defessis hospita tecta viris.
Ein jeden grad an diesem stab bedeutet ein schritt B oder drittehalben werckschuh, damit abzumessen beyder städt länge und breite.
10 20 30 40 50 60 70 80 90 100
Oetenbacherkloster.
Rennwig.